GUITAR AND BASS SPEAKERS & CABS

How they work, how to improve them or build your own to get the best tone

IGOR S. POPOVICH

B.Sc. (El. Eng.)

DISCLAIMER & COPYRIGHT NOTICE

The information contained in this book is to be taken in the context of a general overview, not specific advice. You should not act on the information contained herein without seeking professional advice. Neither the author nor the publisher (or any other person involved in the publication, distribution, or sale of this book) accepts any responsibility for the consequences that may arise from readers acting in accordance with the material given in the book. Professional advice about each particular case/instance should be sought.

Our choice of designs, parts, models, and brands was based on their availability, cost and educational value. We were not influenced or induced by anybody in our selection, and their use does not mean we endorse or recommend them. You should satisfy yourself that a particular device, component or method is suitable for your intended purpose.

Commercial speakers and designs were used here for review and discussion purposes as "fair dealing," permitted by international copyright laws.

Designs marked with a copyright symbol are the intellectual property of their copyright holders and should not be used without their permission. They are discussed here from an educational perspective only.

SAFETY WARNING & INDEMNITY NOTICE

The consequences of any modifications to your speakers and cabinets are at your own risk, and no responsibility will be accepted by us.

Woodwork involves the use of power tools and other safety hazards. By purchasing and reading this book, you agree to indemnify its author, publisher, and retailer against any claims of any nature and for any reason. If unsure about any aspect of testing, repairing, or constructing a tube amplifier, engage the services of a suitably qualified and authorized expert.

Published by Career Professionals Australia

First edition

Bulk purchases

This book may be purchased in larger quantities for educational, business or promotional use. Please e-mail us at sales@careerprofessionals.com.au

National Library of Australia Cataloguing-in-Publication Data:

Popovich, Igor S.

GUITAR AND BASS SPEAKERS & CABS

ISBN: 978-0-6482982-2-9

1. Electrical engineering 2. Electronics 3. Electroacoustics

I Igor S. Popovich II Title III Index

621.38

BOOK CONTENTS

BOOK CONTENTS, continued

THE PRELIMINARIES (a.k.a. "THE INTRODUCTION")

Why this book?

When you start your Ph.D. or doctorate studies, the first thing they tell you is to do a literature search or review in your chosen topic. A "review" is an euphemism for serious study and analysis - who studied what, why and how, which researcher carried such studies on, and which went a different way. Then, hopefully, you'd identify the holes in the cheese, so to speak, the areas not properly investigated or not studied at all.

After reading dozens of books on hi-fi amps and audiophile speakers, and many more on guitar amps, it dawned on me that books on guitar a speaker cabinets ("cabs" in further jargon) simply don't exist.

A rational person would at that juncture ask himself why - either guitar speakers are so rare, obscure things that nobody would be interested in them or in a book about them, or, they are relatively common, but being such simple, mundane objects means there's nothing about them that can or should be studied.

Both possibilities can be immediately eliminated; there are literally millions of guitar amps and speaker cabs in the world, and no, they are not simple, boring boxes whose only job is to enclose a loudspeaker of some sort and serve as shelves on which rowdy guitar players would keep their drinks and hash-smoking paraphernalia.

To understand their operation and to be able to design and build your own guitar or bass speaker enclosures (or hi-fi speakers for that matter), you must understand certain physical and engineering principles - there are no ifs or buts about it! And, unfortunately for you (and for me), neither physics nor engineering can be understood or mastered without mathematics.

For that reason, out of necessity, some of the chapters in this book will be maths-heavy, while others will be almost entirely practical. I have tried to strike the right balance between the two whenever and wherever I could.

If you find a certain topic so maths-heavy that your head starts spinning and even doing your tax return seems more fun than reading my endless mathematical masturbation (pardon, calculations), stop and move onto the more practical parts of the book. Life is too short to suffer for no good reason, and, I must admit, I cannot for the life of me see why a guitar player (assuming you are one) would need to know how to analyze an acoustical equivalent circuit of his guitar cab and its impulse response.

But, just in case other types of dudes read this book (such as those who like solving partial differential equations but cannot play three simple chords on an electric guitar), a few sausage-length formulas and perplexing electro-mechanical-acoustical circuits are there to whet their insatiable appetites for deeper understanding.

How to read this book

One way to read a technical book like this one is to go to a section or topic that interests you immediately and then keep jumping back-and-forth to the related issues and chapters. This will, I suspect, be the way the more experienced designers and constructors will approach this book.

A more systematic reading approach is sequential, starting from the beginning. This is what I would recommend. Although it seems more time-consuming (since you will read about issues you may already know a lot about), paradoxically, this approach is often faster. You will not miss anything, and you will not waste time flipping forward and backward trying to clarify an issue that you've overlooked and perhaps not fully understood.

Who am I to tell you what to do and how to do it?

I graduated in 1986 in Zagreb, the capital city of Croatia, then one the republics in the federation called Yugoslavia. The Faculty of Electrical Engineering at that time had ten or so majors (the first two years were common, the second two years one specialized in one of these majors. They ranged from high voltage and power engineering, though to industrial electronics (my major), computing, automation & control, telecommunications, radio communications, ending with the smallest and least popular section of all, which, I remember, had six (6!) students. It was called "Department of Electroacoustic", and it was the only department that had more academic staff than students.

These dudes studies microphones, amplifiers, speakers, room acoustics, studio techniques and other fascinating subjects. I must admit the thought of studying that field had never even crossed my mind.

Engineers majoring in electroacoustic could only get a job at radio or TV stations, or, perhaps, in a theater or music hall. As you can imagine, there weren't many such places around; in my home town of Osijek (around 120,000 people then) there were only two - a Croatian National Theater (HNK) and RTV Osijek, meaning a combined radio & TV station. Both were government-run, didn't need young electroacoustic experts, and didn't pay well.

My audio journey started on two seemingly separate paths, electronics and music. My Dad, Dr. Slobodan Popovich, better known as Dr. Bob, was a physics professor by day and an electronics guru by night, so I grew up surrounded by tube amplifiers, DIY speakers, radios and all sorts of electronic equipment and test instruments, some "modern" (at the time, we are talking 1960s and 70s), some older (from 1940s), most of which he bought from military surplus stores or got for free from his students and neighbors.

I vividly remember WWII American, Russian and German (Wermacht) radio stations, vintage tape recorders, turntables and PA amplifiers, all with vacuum tubes, of course, and loudspeaker boxes that came with them, some with permanent Alnico magnets, others still with electrodynamic (field-coil) speakers.

Since I loved music, as I became a teenager, I pestered my dad to buy me an electronic organ, Italian, of course, since they were physically the closest to us. Elka, Farfisa, Crumar and other lovely keyboards could be smuggled in a car over the border between Italy and Yugoslavia, as my old man did a few times over the next few years.

My first keyboard amplifier was a well-worn Farfisa version of a Leslie cabinet. It was better designed and more modern than (even then) the archaic Leslies, looked like fridge and weighted a tonne, but it sounded awesome. Due to a perplexing grounding issue that my Dad couldn't quite fix, it also received local radio stations quite well, so it doubled as a radio receiver!

The journey continued in earnest once my Dad joined me in Australia in 1995. We started a company designing and building hi-fi amplifiers and preamps, only tube, of course, we would not touch solid state stuff with a wooden pole! Don't you love such English expressions, although you never hear the opposite, like "Hmmm, nice, I would touch her with a wooden pole!"

Manufacturing anything in Australia is a losing proposition; labor rates are astronomically high, the technical expertise is almost nonexistent and the audiophile community was, and still is, extremely small.

For instance, there were a few transformer winding places, but only making small power (industrial) transformers. Nobody knew how to design audio transformers in Perth, a city of 1.2 million people then, 2 mil. now, so Dad and I had no other choice but to learn how to do it ourselves.

After we'd closed that audio business, we were (so to speak) all dressed up with nowhere to go; we had all that knowledge, we'd designed and built hundreds of tube amps, of every possible topology, power level and with every conceivable tube type available, yet nothing to show for it.

Then, one sunny day in 2005, Dr. Bob had a life changing idea. I am smart (and modest, obviously), but in some ways he still outranks me. Why don't we build a tube amp, he said, and record the whole process with my AU$1,499 digital video camera (my heart still breaks when I remember how much I paid for it 20 years ago - add stingy to my list of attributes!), burn it onto DVDs and sell such programs worldwide on eBay?

And that's exactly what we did. We recorded and produced not one, but 15 different DVD programs, on how to wind your own power and audio transformers, how to make your own tube guitar amp, how to troubleshoot and fix guitar amps, a few programs on tube testers, preamps, electronic measurements, and similar audio topics.

For some reason, it took me another ten years to have an idea of my own (perhaps I am not as smart as I'd like to think I am), and that was to convert the material from these DVD programs into books, a written medium. From 2015 to 2021 (6 years for the mathematically challenged) I wrote nine technical books, among others, two on guitar amps and three on hi-fi amps.

In a shameless act of blatant self-promotion, they are all listed at the end of this book. All are available for purchase from Amazon and other online bookshops, on eBay, and directly from me, with a gratis signature and dedication, of course. You never know, when I kick the bucket, my signature could be worth a fortune!

Getting in touch with us

If you've liked the book and benefited from it, the best way to repay a favor is to recommend it to your friends and to write a glowing online review. Also, if you spot an error or an omission or have any constructive criticism of the book, I'd like to hear from you to fix it together.

If you want to contribute ideas or projects for the next edition, please let me know if you have ideas on how to make the next edition better.

My e-mail is **igorpop@careerprofessionals.com.au**

I sincerely hope that this book will answer, if not all, then at least some of your questions about the design, building, fixing, and modifying of speaker cabs. I wish you every success on your journey through the fascinating world of electroacoustic!

Dr. Iggy

ABBREVIATIONS, FRAMES, AND SYMBOLS USED

AC	Alternating current	MAX	Maximum	GND	Ground terminal
DC	Direct current	MIN	Minimum	COM	Common terminal
THD	Total harmonic distortion	NFB	Negative feedback	AWG	American Wire Gauge
EM	Electromagnetic	PCB	Printed Circuit Board	SPL	Sound Pressure Level
FET	Field effect transistor	PP	Peak-to-peak (AC signal)	SW	Switch
VCA	Voice coil assembly	VC	Voice coil	DF	Damping factor
CW	Clockwise	CCW	Counterclockwise	SS	Solid state (semiconductor)
WWII	World War Two	FC	Field coil	DCR	Direct Current Resistance
LOG	Logarithmic scale	LIN	Linear scale	EMF	Electromotive force (voltage)
NO	Normally open (contact)	NC	Normally closed (contact)	TPI	Threads per inch
TR, VR	Turns or voltage ratio (of a transformer)	NP	Non-Polarized (electrolytic capacitor)	XLR	Type of electrical connector used in audio equipment
RC	Resistive-Capacitive circuit or filter	LC	choke-capacitor or inductive-capacitive filter	IEC	International Electrotechnical Commission
IC	Integrated circuit	RPM	Rotations per minute	AUX	Auxiliary (contact)
IN	Input	OUT	Output	VOL	Volume
PA	Public address	TS	Thiele-Small parameters	MDF	Medium density fiberboard
RCA	Radio Corporation of America or unbalanced audio connector	RMS	"Root-Mean-Square", effective value of an AC signal	LCR	Inductance-capacitance-resistance meter or circuit
AF	Audio frequency	CPS	Cycles per second	MDF	Medium density fiberboard
dB	Decibel	BP	Bipolar (capacitor)	SAC	Sound absorption coefficient
PCB	Printed circuit board	XO	Crossover	DIY	Do it yourself
HF	High frequencies	LF	Low frequencies	OPT	Output transformer
SE	Single-ended (amplifier)	PM	Permanent magnet	PP	Push-pull (amplifier)

Symbols

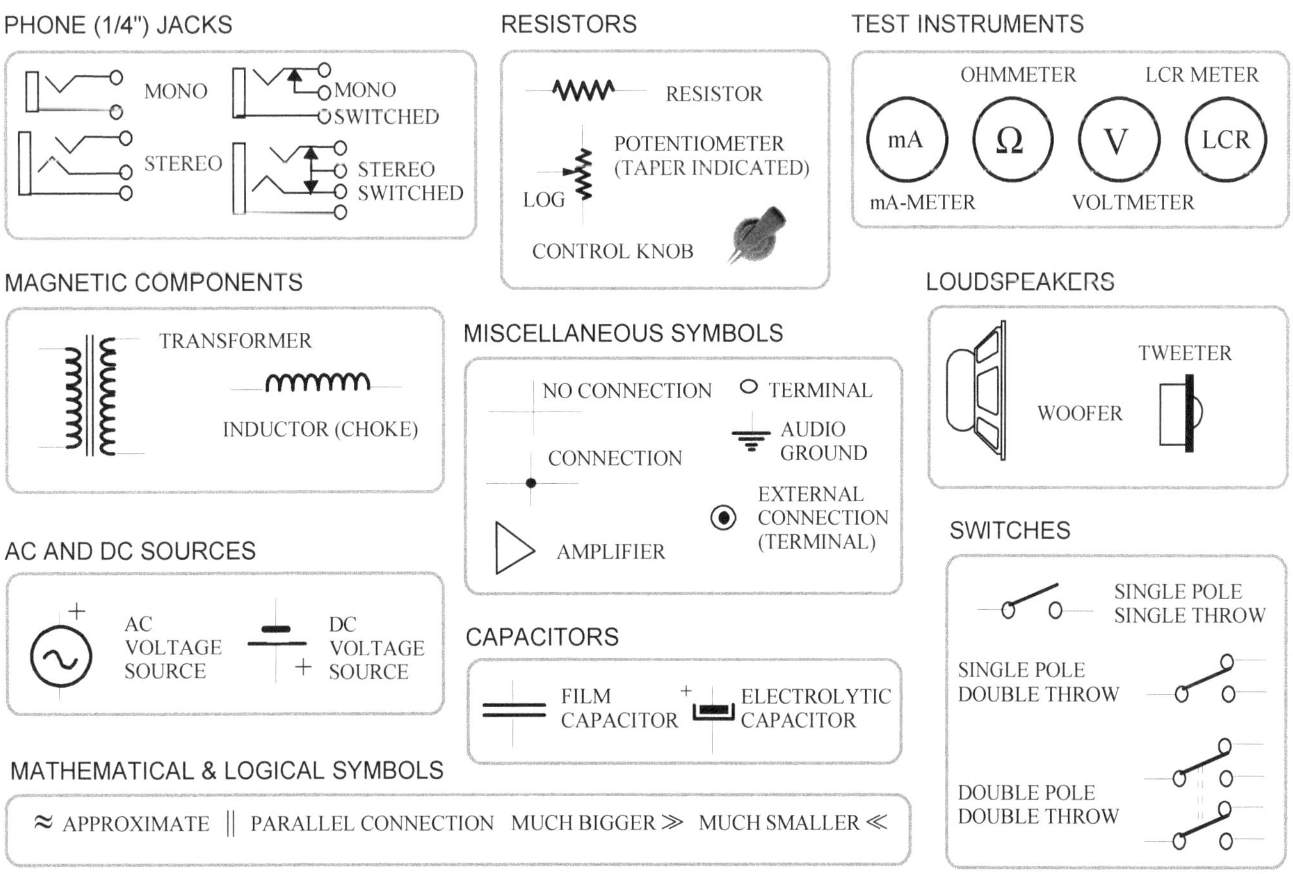

Frames

 IMPORTANT FORMULA

 TONE TALK

 BAD, WRONG

 GOOD, CORRECT

 IMPORTANT POINT or WARNING

 DIY PROJECT

 MEASURED RESULTS

 FURTHER READING

 SPECIFICATIONS or SUMMARY

 RULE OF THUMB

Test instruments

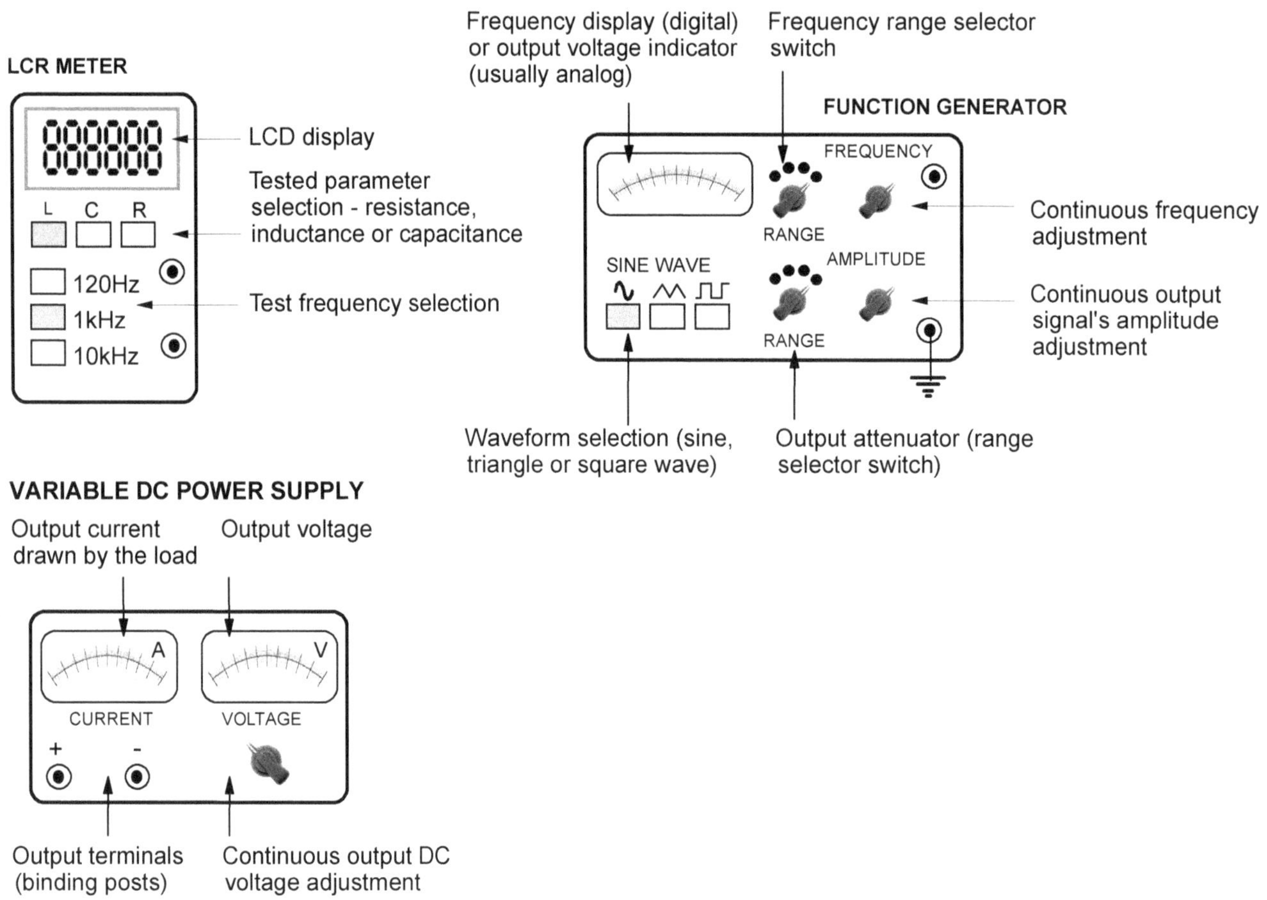

ACOUSTICS 101 - STUDIO & VENUE ACOUSTICS, AND TIPS FOR RECORDING GUITAR SPEAKERS

1

To understand speakers and speaker cabs, and also the acoustic behavior of listening rooms and concert halls), we need a speed course on the fundamentals of acoustics.

While this is by no means an extensive coverage, it should serve as a convenient starting point for your further study, should you wish to deepen your understanding of why things are the way they are and why they sound the way they do.

Also included are a few acoustic tips & tricks for getting the best sound while recording guitar speaker cabs in a studio.

- SOUND WAVES - WHAT THEY ARE AND HOW THEY PROPAGATE & BEHAVE
- HUMAN HEARING 101
- DIMENSIONS, PROPORTIONS AND RESONANT MODES OF ROOMS AND SPEAKER ENCLOSURES
- GUITAR SPEAKERS IN A STUDIO - RECORDING TIPS & TRICKS
- DECIBELS AND LOGARITHMIC SCALES
- HOW TO CALCULATE & CONVERT SOUND PRESSURE LEVELS
- GUITAR SPEAKERS IN A STUDIO - RECORDING TIPS & TRICKS

SOUND WAVES - WHAT THEY ARE AND HOW THEY PROPAGATE & BEHAVE

What are sound waves?

A wave is a disturbance that propagates in time (1) and space (2). The nature of the disturbance is described in the name of a particular wave. Electromagnetic waves are traveling changes in the electric and magnetic fields (which always coexist tougher). E/M waves are heat, infrared light, visible light, UV-light, X-rays, gamma rays, microwaves, and radio waves.

An eye is an electromagnetic sensor or detector of a relatively narrow bandwidth, sensitive only to the visible light; it cannot detect electromagnetic waves of the frequencies below or above that of the visible light spectrum.

With transverse waves, such as electromagnetic waves, the changes in the amplitude of the disturbance happen at right angles, in a perpendicular plane(s) to the direction of the wave's propagation. For instance, a disturbance may be oscillating (vibrating) up and down while its traveling (spreading) left-to-right.

Interestingly, electromagnetic waves do not need a medium to exist and propagate. For instance, they can propagate even in a complete vacuum (no air or any other substance present).

Some well known examples of mechanical waves are water waves and sound waves. Sound waves are longitudinal waves, where, by definition, all the particles of the medium through which the disturbance (sound) travels, such as gas, liquid or solid, vibrate in the same direction as the wave (propagates).

Sound and all other mechanical waves need a medium to propagate; sound cannot exist in or travel through vacuum!

An ear (and associated internal organs) is a mechanical sensor or transducer of a wide frequency bandwidth, from 20 Hz to 20,000 Hz. A Hertz is the international unit for frequency (or the number of oscillations per second); in the USA the archaic albeit self-explanatory usage of CPS (cycles per second) still persists. A "period" of a wave is the inverse of frequency: T=1/f, symbol "T" is used for period, as in "Time", so the unit for a period is a second.

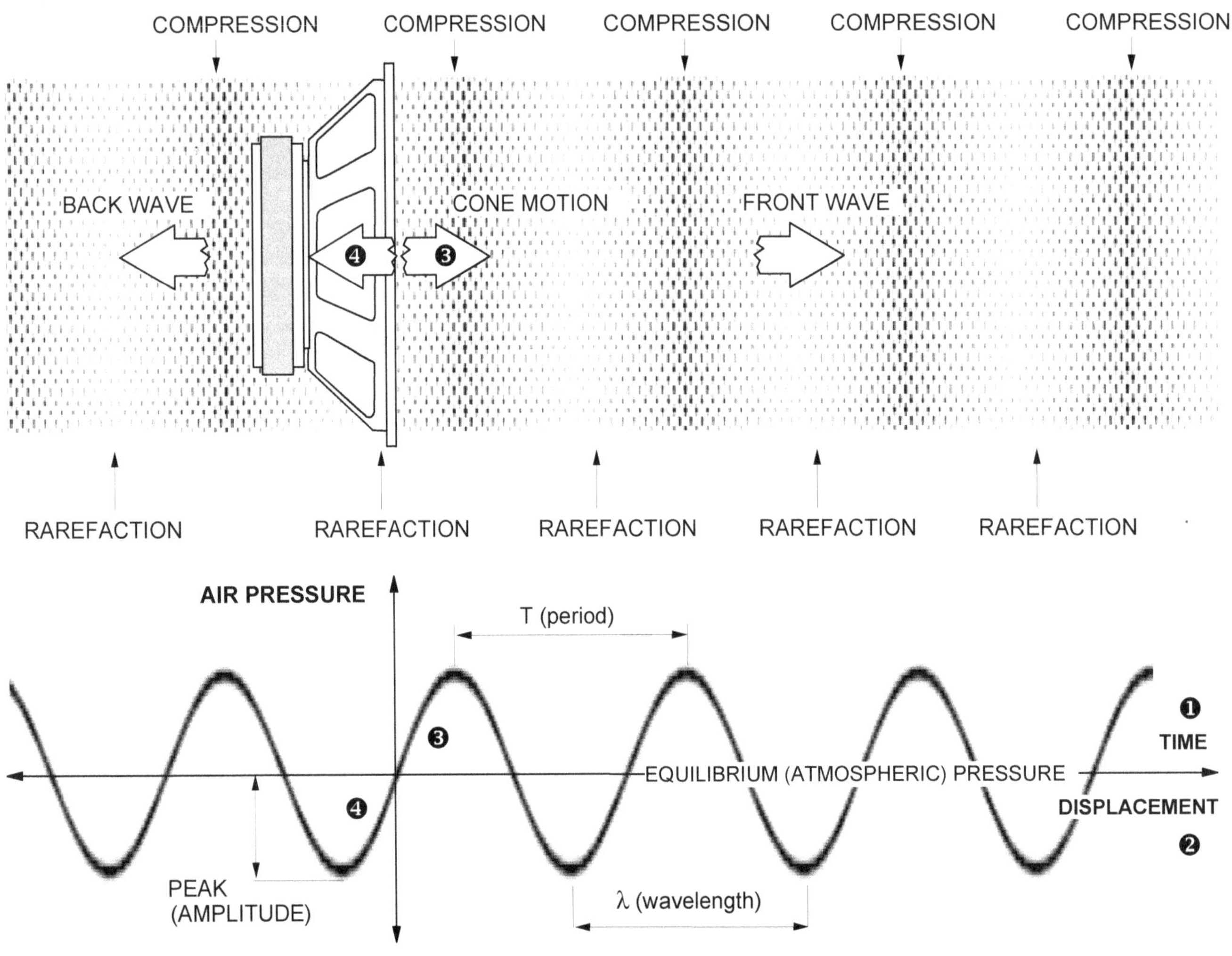

ABOVE: How a moving loudspeaker cone produces a pressure wave (disturbance) that our ears detect as "sound". The horizontal axis can be either time (1) or distance (2) the wave has traveled (displacement)

The relatively simple diagram on the previous page illustrates quite a few important points and concepts. We are assuming that the signal from a sine wave generator (or function generator or oscillator) is amplified by an amplifier (not shown), which in turn drives a moving coil speaker. The speaker's cone vibrates forward and backward X times per second, that X being called the frequency of the signal.

When the cone moves forward, it compresses the air in front of it (3). Compression simply means that air particles are moving closer together, creating regions of high pressure. Likewise, rarefactions occur in low-pressure areas when air particles are spread apart from each other.

This pressure disturbance moves forward, but the speaker's cone also creates identical pressure changes from its back (4), but of the opposite "phase". While the front pressure is rising, the back pressure is falling, and vice versa. A dynamic speaker is a dipole radiator, "di" meaning two, and "pole" referring to two directions of radiation.

The pressure wave can be depicted the same way we depict current or voltage waveforms, as a sine wave whose amplitude at any point in time is proportional to the signal driving the loudspeaker. The compressions are "positive" humps (3), above the equilibrium or atmospheric air pressure level (no disturbance or sound), while the rarefactions are "negative" humps (4), the areas of lower pressure than the equilibrium or atmospheric air pressure level.

Grand piano as an acoustical dipole

Piano's soundboard is a large panel resonator, but not made of paper or similar material (as in loudspeakers), but usually made of spruce. The board is curved, so its concave shaped "crown" behaves as a diaphragm, arching upward toward the strings.

The vibration of the strings is transferred to the soundboard through two bridges, over which the strings are stretched, the long bridge, crossed by treble strings, and the short bridge for bass strings. If you are a guitar player, you will be familiar with this concept, bridges are also used in electric and acoustic guitars.

The soundboard isn't one solid piece of timber, but is made by edge-gluing together strips of spruce 2-5" wide, which were previously air- and then kiln-dried to a specific moisture content.

The protective lid (2) covering the piano's internals can be held open by the lid prop. A piano with its lid open projects its sound further and wider. The volume (loudness) is also increased, as the tonal balance is changed.

An open lid makes the sound brighter and more resonant through better dispersion of high frequencies. Closing the lid is like dialing down a treble or presence control on a guitar amp, and such high frequency tone control results in a warmer, more mellow sound.

The lid can even be completely removed (detached from the piano), something often done in recording studios to make microphone placements easier and to achieve a certain type of sound.

The soundboard vibrates in the vertical plane (up & down), radiating the sound towards the floor (A) and upwards, towards the lid (B), in a typical acoustic dipole fashion (the waves are equal in amplitude but of the opposite phase). Opening the lid helps the reflection of the upward wavefront and its dispersion and even phase inversion.

We will use the same principle in one of our projects, where we will have two drivers "back-to-back", firing in the opposite directions, with the flexibility of connecting them in- and out-out-of-phase by a flick of the switch, thus achieving two very different sonic characters. See the "Double Trouble" project on page 161.

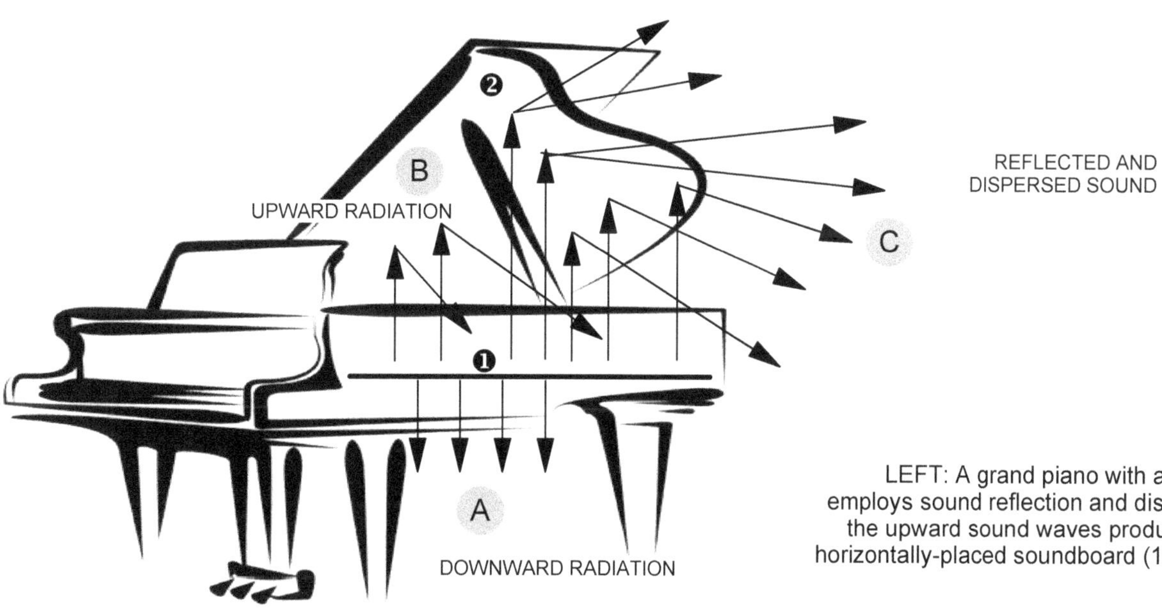

LEFT: A grand piano with an open lid employs sound reflection and dispersion of the upward sound waves produced by its horizontally-placed soundboard (1) to enrich its sonics

The fundamental acoustic law

Looking at the time domain (the wave's travel as a function of time), the frequency and thus the period T of our simple test signal do not change. A period is simply the time that takes for the sound wave to move from one peak to another, or one minimum to another, or even one zero crossing to another.

In terms of speed of propagation and displacement, the wavelength is a distance corresponding to one period, or how far one full wave travels in the unit of time.

The three physical quantities are related by this simple formula: $\lambda = v/f$, where λ (Greek letter lambda) is the wavelength (the distance between two identical peaks or zeroes of a periodic wave), v is the speed of sound in free air (or any other medium the speaker is in, for instance water for underwater speakers), and f is the sound frequency in Hz.

THE FUNDAMENTAL ACOUSTIC FORMULA

$$\lambda = v/f$$

λ = the wavelength (the distance between two identical peaks or zeroes of a periodic wave) in meters or feet

v = the speed of sound in free air, in meters per second or feet per second

f = sound frequency in Hz.

How sound travels through air and other media

We can calculate the speed of the sound in a solid material from its density ρ (Greek letter "rho") and its Young's modulus (elasticity) E: $v=E/\rho$. So, in essence, the denser the material (medium of propagation), the slower the speed of sound. The higher the material's modulus of elasticity, the faster the sound travels through it.

That simple formula is not valid for gases, including air. For instance, the speed of sound varies from around 5,189 m/s in steel and 3,536 m/s in concrete to 1,517 m/s (meters per second) in water and 344 m/s (or 1130 ft/s) in air.

In gases (such as air), the speed of sound depends also on the temperature and the molecular weight of the gas.

An interesting case is solid timber, whose elasticity module isn't constant or consistent; it is much higher along the grain than across it.

For instance, Young's modulus of beech wood is 14×10^9 N/m^2 along and only 0.9×10^9 N/m^2 across the grain. With a density of 680 kg/m^3, the speed of sound along the grain will be about four times faster than across it!

This fact is important in the construction of musical instruments and the design and construction of loudspeaker cabinets made of solid wood. This difference in speed propagation will also impact speaker cabinet vibrations in various directions and the damping of its resonances. So, it does matter how you orientate those pine boards you are making a speaker cab out of!

In contrast to solid timber, the elasticity modulus of man-made timber products such as medium density fiberboard (MDF) and plywood does not vary with direction.

Sound diffraction, reflection and absorption

When a sound wave of specific frequency "f" hits a solid object such as a cabinet's wall or a wall in the room the cabinet is in, depending on the density and other properties of that material, part of it reflects back, and part of it is absorbed by the material (3).

That portion that enters the material also reflects back at the other (rear) boundary, with only part of it exiting back into the air on the other side (4).

RIGHT: Reflected, refracted, absorbed, and transmitted sound, plus the re-radiated "sound" (vibrations of unrelated frequency)

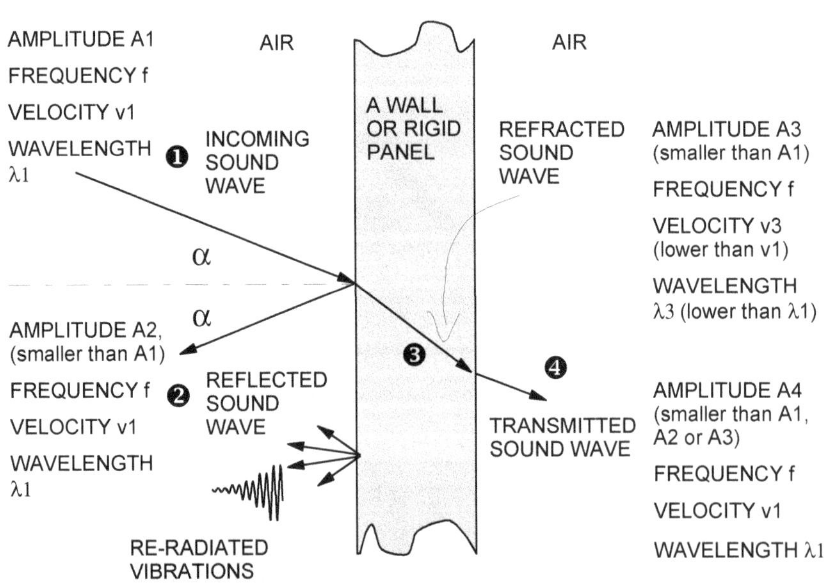

First of all, the incoming or incident angle with reference to the vertical (dotted line in the illustration) is always equal to the reflected angle. Both are thus marked α (Greek letter alpha).

Secondly, each material has its sound absorption coefficient (SAC). SAC is not a single figure but is frequency-dependent; there are SAC-versus-frequency curves for various materials. You can find them online for building materials (concrete, timber flooring, etc.), home furnishing materials (carpets, drapes, sofas, etc.), and various room acoustic treatments such as panel absorbers.

Recording studio owners and managers will be familiar with this important concept.

Notice that the frequency of the incoming, refracted, reflected, and the transmitted sound wave is the same. The frequency does not change, but the wave's wavelength and speed do change in inverse proportions. Sonically, the most damaging is the re-radiated sound or vibrations of unrelated frequency (5).

Each object or structure (a wall, a panel, a piece of furniture, a door or a window, etc.) absorbs the sonic energy from the speakers. The sonic waves "excite" the structure and cause it to vibrate at its own natural frequency. That structure can be a speaker enclosure, your listening room at home, or a concert venue (hall).

Such vibrations are then re-radiated back into the room as a discordant, unpleasant sound since its frequency is unrelated to the frequency of the original sound.

Why the size of speaker enclosures and other objects matters and how it affects the sonics

Using our fundamental formula $f=v/\lambda$, we can calculate the wavelength corresponding to a few selected frequencies across the audio range of human hearing.

A 20Hz signal has a wavelength of 17.2 meters. Most objects in a typical practice or listening room (even a small concert venue) are much smaller, so they don't represent an obstacle to such long wavelengths - the sound waves will envelope all smaller objects.

Even at 200 Hz, an upper bass/lower midrange frequency, the wavelength is 1.72m. A guitarist standing directly in front of a speaker cab still won't attenuate or reflect such waves to any noticeable degree (he or she may be 1.8m tall, so perhaps the vertical dispersion may be impacted, but in terms of horizontal dispersion his legs are only 0.2-0.4m wide, much smaller than the signal's wavelength.

At a typical treble frequency of 2kHz (2,000 Hz), the 17.2cm wavelength is small enough to be impacted by the objects around the sound source - the musicians' bodies, guitars, other speakers, etc. If a guitar player stands right in front of a speaker, other band members will notice a pronounced muffling of the sound, the loss of treble if you will. At higher frequencies such an absorption effect will be even more pronounced.

Even without any obstacles in front of a speaker or any other sound source (a voice, acoustic instrument, etc.), bass and treble frequencies ravel differently. You must have noticed that as you approach a pub with live music or a concert venue, the first thing you'd hear outside would be the boom-boom of the bass instruments; no high frequencies would be present at such distant locations.

First, high frequencies are attenuated faster and to a greater extent than low frequencies because they are subject to more scattering and absorption effects of the medium (air) they travel through than low frequency (long wavelength) sounds. This is due to the medium's viscosity and the sound's interaction with air molecules.

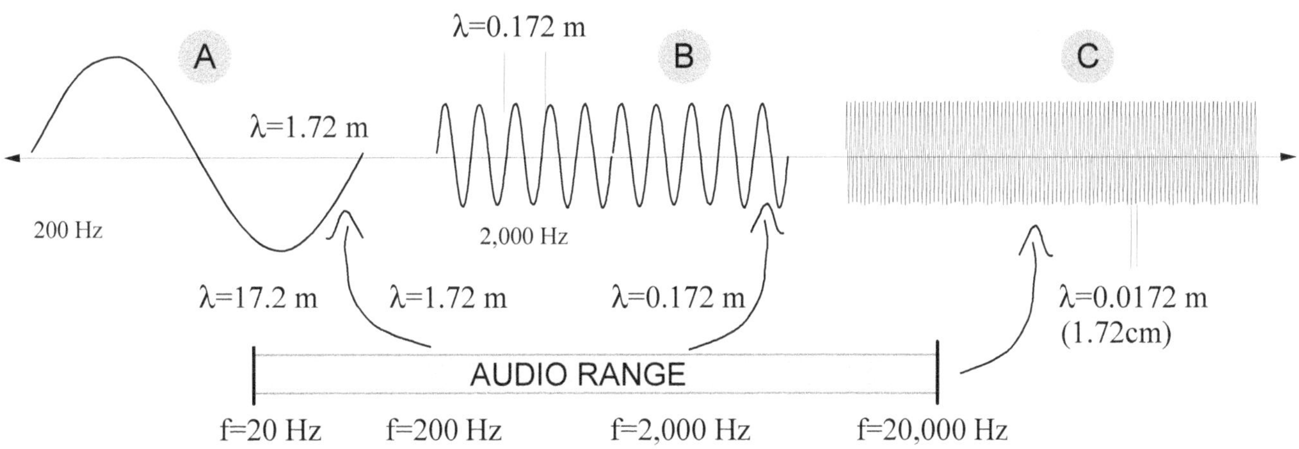

ABOVE: With a speed of sound of v=344 m/s (at 20 deg C), its wavelength varies from 17.2m at 20Hz to 1.72 cm at 20 kHz!

The bass production problem

Producing high frequencies at high power or sound pressure levels isn't difficult, although beaming is an issue, so spreading or dispersing such high frequencies uniformly is a challenge. Producing deep bass is another story.

90% of all efforts, including all the models and designs of various types of speaker enclosures are all essentially trying to solve the bass production problem. As we will see later on, the results of all such efforts are a compromise, resulting in half-baked solutions to the (so far) seemingly unsolvable problem.

The crux of the matter is the size relationship between the diameter of a dynamic speaker's cone and the wavelength of the sound it's trying to reproduce.

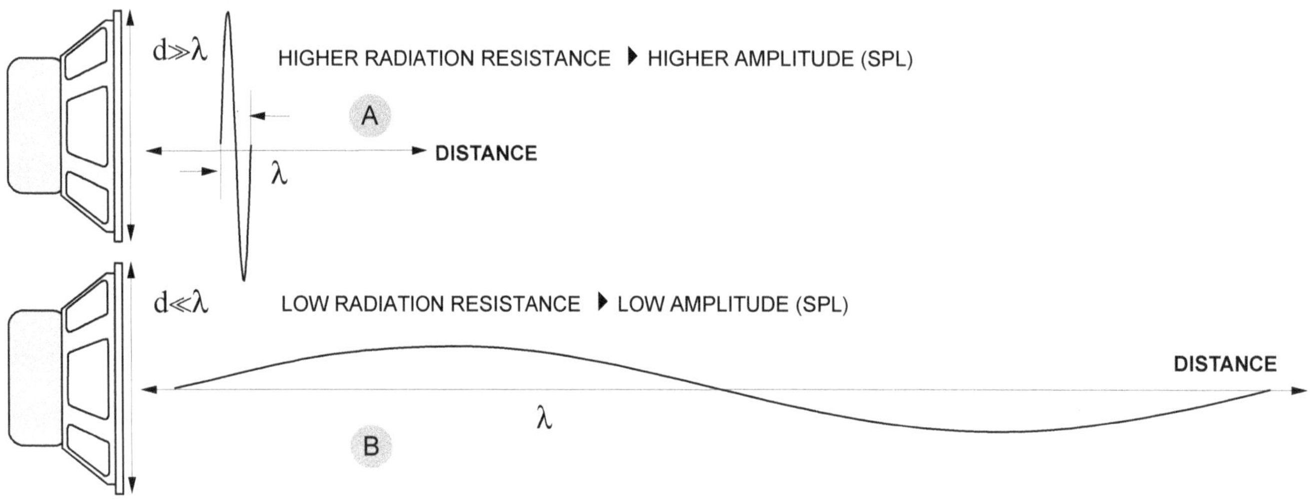

When the cone diameter is much larger than the wavelength of the sound wave (A), the cone faces a higher radiation resistance of the surrounding air, so higher efficiencies and larger amplitudes (SPL levels) can be achieved for such high frequencies.

However, at the opposite end, where the extremely long wavelength of low bass frequencies is much larger than the cone's diameter, the comparatively tiny cone cannot move much air. In other words, it faces a very low radiation resistance of the surrounding air, so the poor "coupling" results in low efficiency and significantly smaller amplitudes (SPL levels), as in (B).

To maintain the same sound pressure level, for each octave (2:1 ratio of frequencies) lowering of the lowest reproduced frequency by a speaker enclosure, a four times larger volume displacement (and cone excursion) is required.

How hard that is to achieve is best illustrated by a simple log-log graph, with the X_{PP} (peak-to-peak cone excursion) on the Y-axis and the frequency f on the X-axis.

15" drivers are not very common in either guitar or bass amps, but even they struggle in that regard. To radiate 1 acoustic watt of acoustic power at 50Hz, a 15" driver needs a PP excursion of 10mm (point A). That's quite achievable.

However, at 20Hz, the same driver needs to flap around at approx. 8cm peak-to-peak. That is 4cm each way, something even a heavyweight such as a 15" driver cannot do.

Spare a thought for a poor 8" speaker. To produce 1 acoustic watt at 20Hz (C), its cone must move 50 cm (almost 20 inches) in total, something clearly impossible.

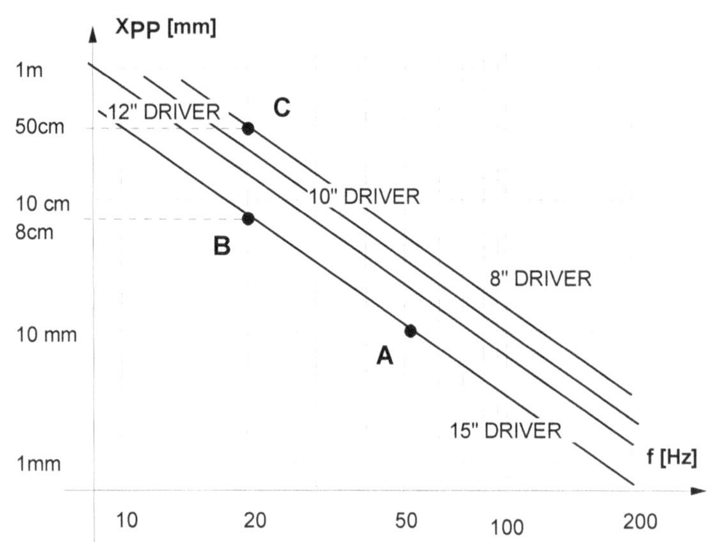

ABOVE: Peak-to-peak cone excursion (displacement) versus frequency required of the most common guitar speaker sized cones in order to radiate 1 watt of acoustic power (log-log axis)

Speaker directivity - how the sound changes as we move around the speaker's front

An ideal speaker would spread all frequencies equally in all direction; real speakers don't. We can see their shortcomings in that regard from both SPL-vs-frequency characteristics and from polar diagrams. Speaker manufacturers don't usually publish such curves, but guitar players notice differences between different drivers and cabinets while they move around them.

Some sound awesome when you stand directly in front of them ("on axis"), illustrated as "A", but muffled and dull when you move sideways ("off axis"), point "B" on the polar diagram below. Others sound harsh, aggressive and irritating when you stand right in front, but pleasant and tonally-rich when you move 30-45 degrees to the side.

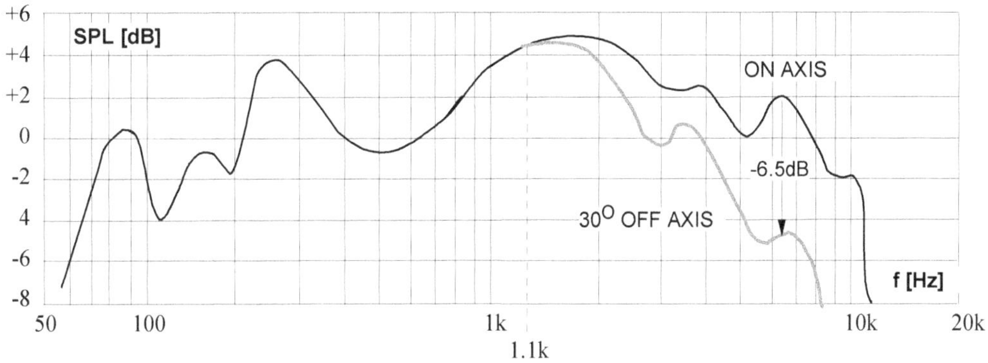

LEFT: Typical frequency response (SPL vs. frequency characteristics) of a 10" speaker cab on axis and 30 degrees off axis (thick gray line)

The SPL curve of this particular ten incher is fairly typical of all guitar speakers with a single cone (speakers with dual cones, acoustic lenses and similar improvements measure better in this regard). Bass- and lower midrange frequencies (in this case up to about 1,100 Hz) are not affected. After that point, the higher the frequency, the bigger the SPL drop as we move off axis (away from the center line).

At 6kHz the difference is a whopping -6.5dB, and this is at only 30 degrees of the center line, you can imagine the attenuation of high frequencies at larger angles, as we move more sideways towards "C".

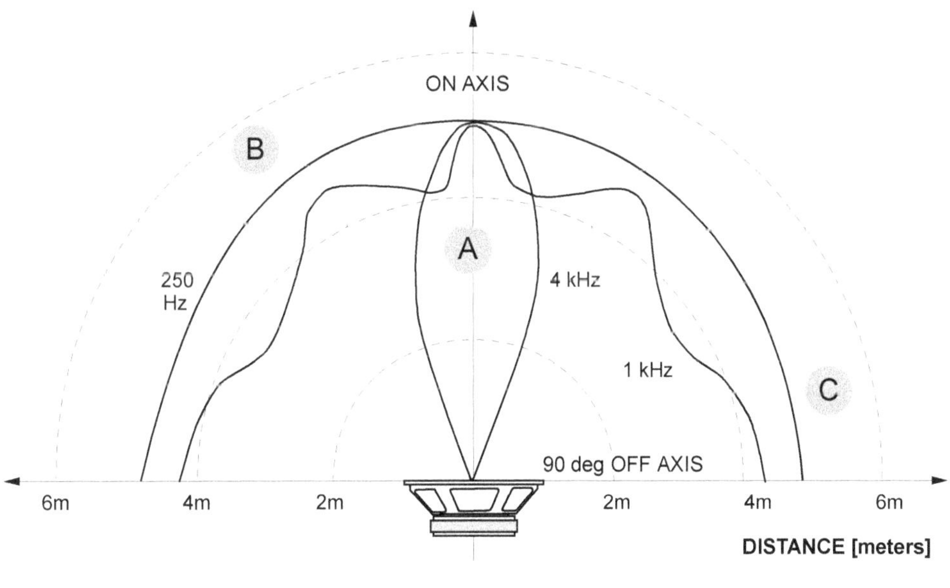

LEFT: Typical polar diagrams of a 10" driver in an open-back enclosure, showing dispersion patterns for three representative frequencies, 250 Hz, 1kHz and 4 kHz

Harmonics and spectral "signatures" of instruments

We have already mentioned simple harmonic motion and its representation, namely the sine wave. French mathematician Joseph Fourier (1768-1830), explained mathematically how a signal of any shape can be broken down into a sum of sine waves of various frequencies and amplitudes; such sine waves are called "harmonics".

A square wave, for instance, is comprised of an infinite number of sine waves, but only odd harmonics (1st, 3rd, 5th, 7th, etc.) are present; there are no even harmonics.

The 1st harmonic (called "the fundamental") is a sine wave of the same frequency as the square wave, the frequency of the 3rd harmonic is three times the frequency of the first, and so on. Notice how the amplitude of the harmonics drops off in an exponential fashion.

Although the illustration only shows spectral components up to the 11th harmonic, the harmonics continue indefinitely. The amplitudes of the harmonics are $A_N = A/N$, so the amplitude of the third harmonic is 1/3 of the fundamental's amplitude, the fifth harmonic has a 1/5 of the 1st harmonic amplitude, etc.

This other illustration shows the contribution of only the first three harmonics, the 1st, the 3rd, and the 5th. Notice how their sum (thicker line) already resembles a square wave. As the successive higher harmonics are added, the waveform will more and more approach the shape of a square wave.

Although amplifiers and speakers are modeled, analyzed, and tested using sine waves of a fixed frequency, there are no such pure tones in nature. Sounds have complex waveforms, comprised of sine waves of various amplitudes and frequencies.

Music signals (both live and reproduced) are complex waveforms whose amplitudes vary enormously, primarily due to the presence of short transients or abrupt changes in the signal's amplitude.

A spectrum of a signal is depicted in an X-Y manner with frequency on the horizontal or X-scale and the amplitude of various harmonics on the vertical scale, usually in dB. Two examples of musical instrument spectra are depicted here, violin's and flute's.

Notice that the amplitudes of some of the harmonics produced by musical instruments are higher than the fundamental tone; that is perfectly normal. Also, not all harmonics are depicted in the violin's case, only those up to 1,500 Hz.

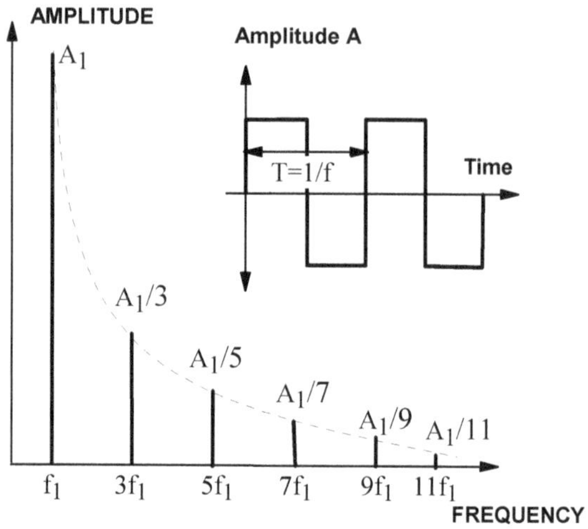

ABOVE: The spectrum of a square wave

BELOW: The fundamental A_1 and first two harmonics (A_3 and A_5) of a square wave added together.

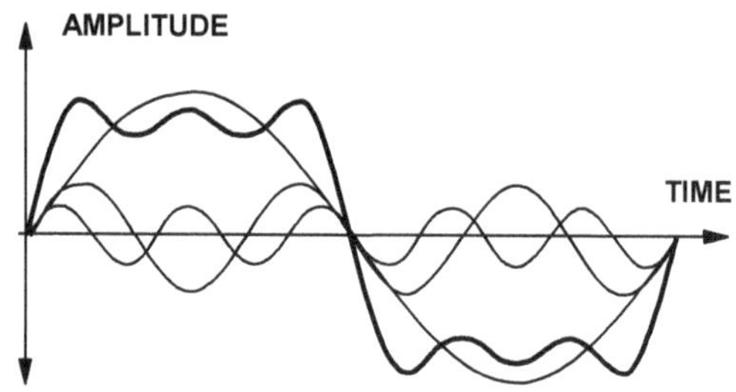

BELOW: The spectrum of a violin compared to that of a flute

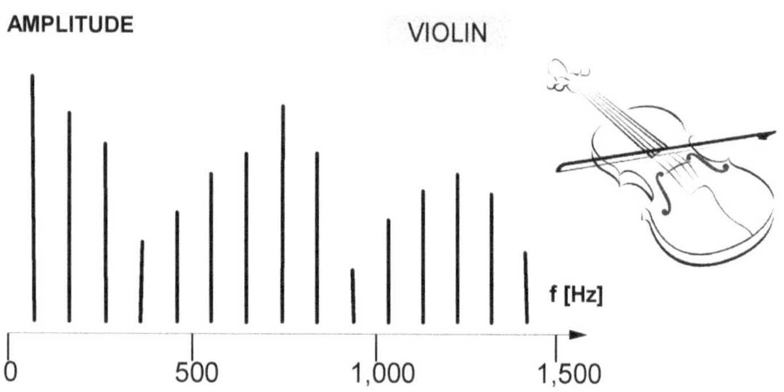

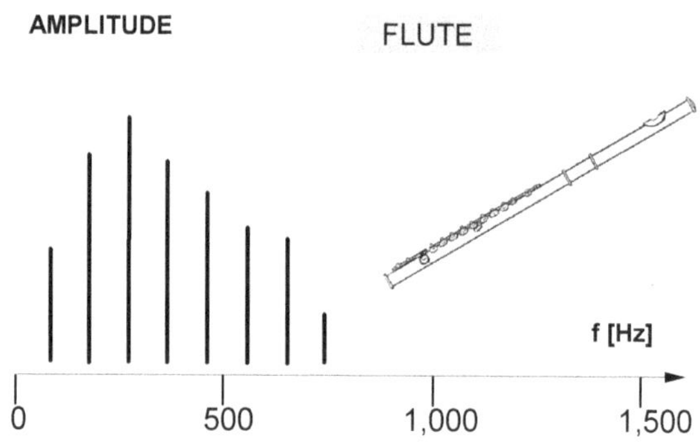

The difference between small listening rooms and studios, and large venues (concert halls)

The diagram (next page) illustrates a typical echogram of a large room or a concert hall. Direct sound and its amplitude D undergoes no reflections. Still, in comparison to the SPL at the sound source (amplitude S), it is attenuated, so the sound at the listening position has a reduced amplitude (lower SPL). This attenuation follows the so-called "inverse square law." Doubling the distance between the source and listener results in a square drop in amplitude - twice the distance, four times lower the SPL.

Sometime after time t_T (transit or travel time needed for the sound to travel from its source to the listener), the first reflection R1 arrives. The time delay between the direct sound and R1 is a critical parameter of a room, called the Initial Time Delay Gap or ITDG.

In great-sounding concert halls, ITDG is in the order of 20 or so milliseconds (0.02 seconds), but in small listening rooms that we audiophiles use, it is pretty short, typically only 1-3 ms (milli seconds or 1/1,000 of a second).

The exception is the LEDE approach (Live End - Dead End), used in recording studios and some listening rooms, where higher ITDG figures can be achieved, approaching those of live concert halls.

The front part of the room (where the speakers are positioned) is made as "dead" or absorbent as possible to minimize reflections. The back of the room is made as reflective and diffusive as possible ("live").

The first reflection is followed by a series of discrete reflections of various surfaces, the walls, floor, and ceiling. This is followed by a third component, the late reflections that compose the so-called reverberant field. The amplitudes of such echoes reduce or decay exponentially.

In the diagram, the reduction seems linear, and it is because the vertical axis is amplitude in dB (decibels) which is a logarithmic unit. More on decibels very soon, at the end of this chapter.

The direct to reverberant sound ratio will depend on the distance between the listener and the sound source. The distance at which the direct sound and the reverberant sound are of equal level (loudness) is called the critical distance d_C.

This parameter is of interest in large rooms only, such as auditoria and concert halls; it is usually larger than the longest dimension of a typical home listening room.

In the rooms of interest to us in this book, the reverberant field (echoes) doesn't develop; the rooms are too small for such a relatively uniform field to exist.

Thus, we are left with the discrete early reflections, which are behind many acoustic problems of listening rooms.

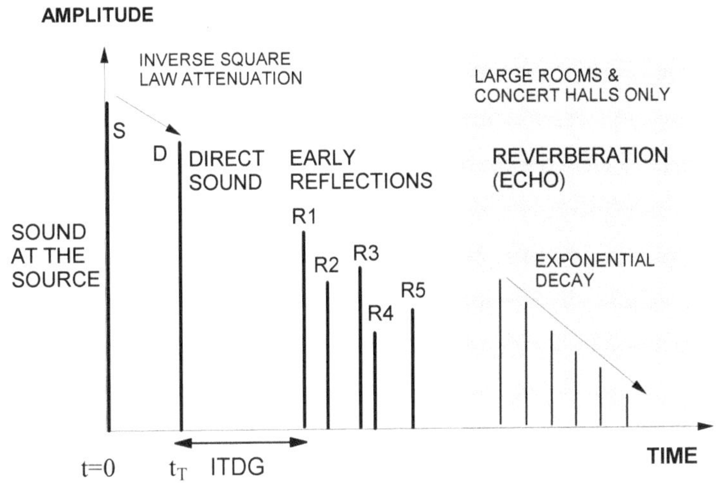

ABOVE: The idealized (in principle) acoustic response (echogram) of a large room

HUMAN HEARING

The ear as an amplifier and A/D converter

The human ear and associated audio "circuitry" is a complex and sensitive instrument. Its frequency range spans four decades (20 to 20,000 Hz) or 10 octaves (20 - 40 - 80 - 160 - 320 - 640 - 1,280 - 2,560 - 5,120 - 10,240 - 20,480 Hz).

Its dynamic range is even wider; it can detect and process sounds from 0 dB to 120 dB, a full six decades!

Our auditory system is comprised of three main parts or subsystems. The external or outer ear starts with the earlobe or pinna, the visible part of the ear. It collects sounds and passes them through the ear canal to the tympanic membrane, a.k.a. eardrum.

The ear canal is akin to a resonant tube. About 25 mm (1") long, it resonates at a frequency of around 3-3.5 kHz, making the ear most sensitive in the midrange.

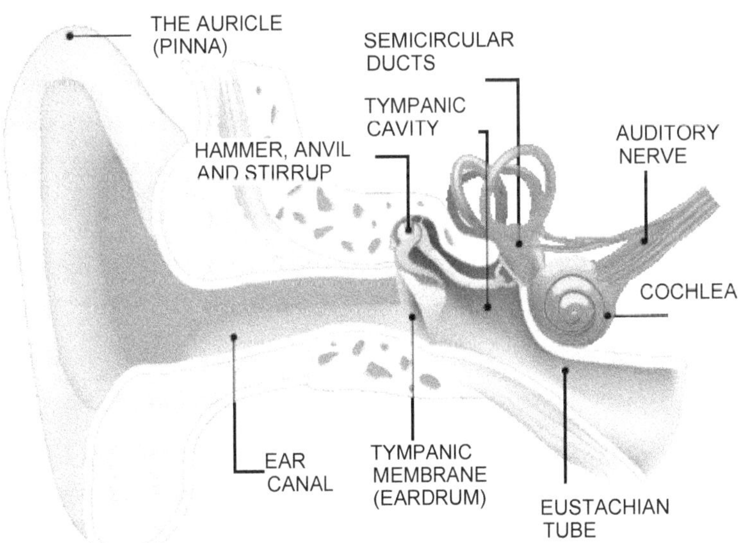

ABOVE: The simplified anatomy of the human ear, with the main components marked and named

The eardrum is a vibrating membrane whose task is to convert sound pressure variations into mechanical motion (1-2 mm excursions).

The middle ear consists of the bones called the ossicles (the three are popularly known as the hammer, the anvil, and the stirrup) located in an air-filled tympanic cavity. In contact with the eardrum, these bones act as a mechanical lever system that further amplifies the force from the eardrum about 30 times.

The inner ear is made up of the cochlea and the auditory nerve. The cochlea is a complex coil-shaped organ (2 and 3/4 turns) filled with fluid.

When subjected to pressure changes from the middle ear, it produces an electrical signal, which is then transmitted to the brain through the auditory nerve. That nerve is actually a bunch of 4,000-odd nerve fibers, and the signal the cochlea produces is not a continuous or analog signal but a series of short and sharp pulses. The greater the SPL (sound pressure level), the higher the number of pulses sent to the brain.

The cochlea could be thus considered a sophisticated analog-to-digital (A/D) converter, converting analog pressure changes into digital electric pulses sent to the brain.

Why do audio systems and guitar amps sound different at lower volume levels?

The "equal loudness" or Fletcher-Munson curves, illustrated here, show how human hearing varies with signal frequency and sound pressure levels (SPL).

In audiometry, the unit of loudness is a phon, but in acoustics, a decibel (dB) is used. The curves for various phon levels have a similar shape. Two things are apparent.

Our hearing is frequency-dependent. Human ears are most sensitive in the frequency range between 500 Hz and 5,000 Hz, where most of the human speech content falls. This is the critical "midrange."

The highest sensitivity is around 3.5 kHz (1). Interestingly, most of the electric guitar tones and harmonics also fall into that frequency band.

Secondly, as the frequency of sound drops towards the bass region, higher and higher SPL levels are needed to achieve the same perception of "loudness" by human test subjects (2).

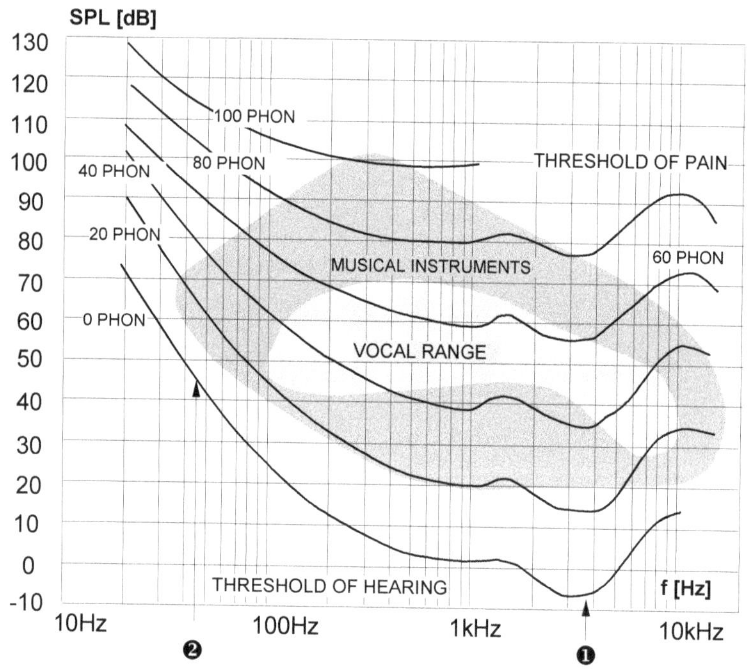

ABOVE: The equal-loudness curves, ISO226:2003 revision of Fletcher- Munson curves

These curves were derived by asking people to adjust a volume to achieve the same subjective level of "loudness." Bass usually needs to be boosted at lower volumes to restore the original balance across the frequency range.

The same conclusion applies to higher frequencies, although to a lesser extent. As a result, bass frequencies need to be amplified more than the midrange signals to achieve the same perception of loudness, especially at lower loudness levels.

This becomes a factor of great annoyance to audiophiles, whose hi-end amplifiers and preamplifiers usually have no tone controls, so it is impossible to compensate for this nonlinearity. Thus, even the best audio system will sound "thin" and lacking in the bass at low levels.

The importance of auditioning guitar amps and speakers at your usual playing levels

I've just returned from a visit to a local hi-end audio shop. Two setups were proudly displayed. Very expensive Sonus Faber speakers were driven by Audio Research tube amps (monoblocks) in one, while a McIntosh hybrid amp (tubes in the preamp stages, transistors in the output stage) was the pièce de résistance in the other.

Both pairings sounded "clean," but ordinary, even anemic. Both amps were of push-pull persuasion with plenty of power, so the lack of dynamics or emotional engagement was surprising. Perhaps it was the music played or my mental state at the time, which reminds me of another issue.

Many speakers, guitar and hi-fi alike, need a certain amount of power to "come alive"; they don't sound their best when played at low volumes. So, if you play your guitar or listen to your hi-fi system listen primarily at low volumes, you must find a speaker that will sound its best at such low levels.

The drawback of some of those speakers is that they suffer from distortion and compression at higher volumes. You have to decide early on what is more important to you and search for speakers of that type.

The same applies to some amplifiers, which at low volumes sound "thin" and "anemic. I call them anorexic teenagers and avoid them as much as possible - they aren't much fun to listen to or be around.

DIMENSIONS, PROPORTIONS AND RESONANT MODES OF ROOMS AND SPEAKER ENCLOSURES

Standing waves

The sound wave produced by the rear of the speaker's cone propagates through the enclosure until it hits one of its four sides or its back cover. Most of its reflects back at the same angle, so, at certain frequencies, where half the wavelength equals the distance between the two parallel sides, the incoming and reflecting waves coincide, always falling in the same spot, so it seems that the wave isn't traveling at all.

Named standing waves for that reason, they are most pronounced between the cab's parallel surfaces, its two sides, top and bottom, and front and back cover (baffle).

The points of minimum vibration, where the incoming and reflected wave (theoretically, assuming a complete reflection off the cabinet's wall) cancel each other, are called "nodes", while those where the vibration is at maximum (where the incoming and reflected wave double up), are the "antinodes.

The same problem happens in listening rooms, recording studios and concert halls - they are all "boxes' of sorts.

Axial, tangential and oblique reflections in a box (speaker enclosure, listening room, recording studio)

The axial modes are reflections and standing waves between opposite walls and between the floor and ceiling (or the sides of the speaker cab). They are dominant (of highest magnitudes) and, as such, are of primary concern.

Tangential modes involve four surfaces in one plane (two pairs of opposite surfaces), such as all four walls or sides of the enclosure, often called the frame (1-3-2-4), two walls (front & back), ceiling and floor, or two side walls, ceiling and floor.

The oblique modes are seldom an issue and can be disregarded for most purposes.

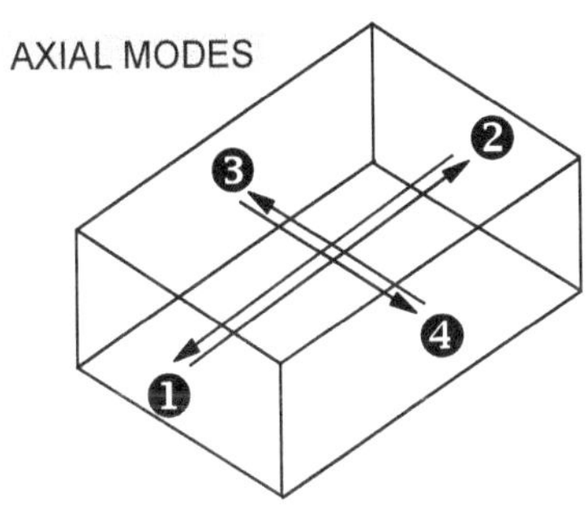

Three axial modes (standing waves between two opposite surfaces): Between the front (1) and the back wall (2), side walls (3) and (4), and the ceiling and floor (not drawn for clarity)

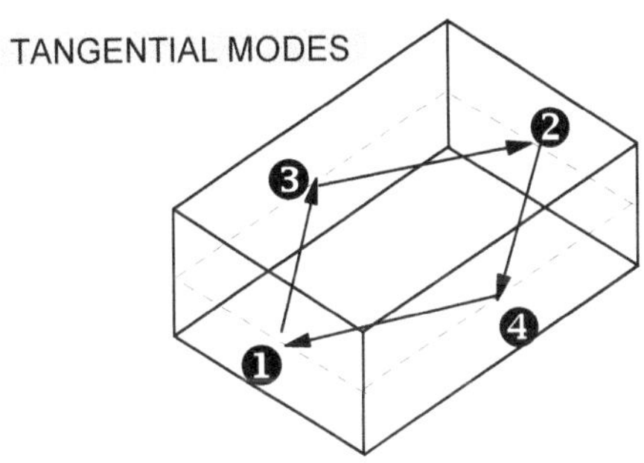

Tangential modes involve two pairs of opposite surfaces: all four walls (1-3-2-4), two walls (front & back) + ceiling + floor, and two side walls + ceiling + floor. All reflections are in one plane (dotted line).

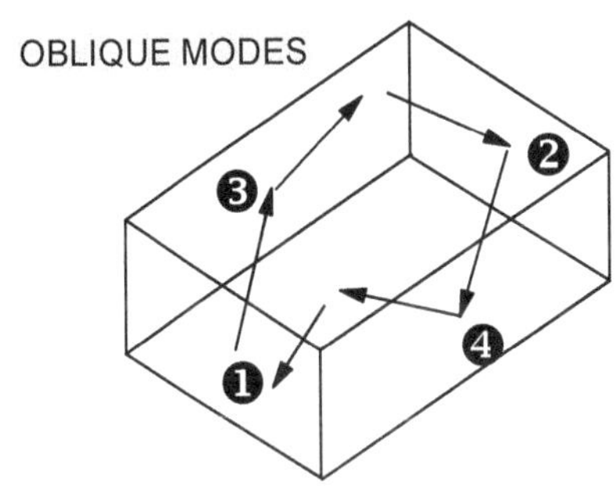

The oblique modes involve all six surfaces, the four walls, floor (5), and ceiling (6). Only one of the many possible paths is illustrated.

Calculate and analyze your room's axial modes

Providing the room or speaker internal dimensions are known, calculating axial modes and resonances isn't difficult. It can be done manually or in a spreadsheet. There are even online calculators, but only for regularly shaped rooms or boxes.

Again, the fundamental formula is $\lambda = v/f$, where λ (lambda) is the wavelength (the distance between two identical peaks or zeroes of a periodic wave), v is the speed of sound in free air, and f is the sound frequency in Hz.

Since every room has three orthogonal dimensions (length, height, and width), standing waves appear in all of those dimensions. So, there are vertical standing waves, horizontal standing waves, and those propagating sideways (lateral or transversal standing waves).

We will study them through a few examples of real speaker cabinets later on in the book. For now, the only principle we need to understand is that equal dimensions of a "box" will "equalize" the wavelengths of standing waves and make the resonant frequencies across its height, depth and width equal. That is highly undesirable, because it would make the frequency response of such speakers and rooms highly irregular, with highly pronounced peaks at some and sever dips at other frequencies.

In other words, some bass frequencies (the standing waves are an "issue" mostly at lower frequencies) will be so strong and "boomy", while others would be so attenuated that the sonic balance would be, to use a highly technical expression, totally out-of-whack!

Since standing waves can never be completely eliminated, the name of the game is not, as we have just mentioned, just to spread them around, but also to make those frequencies as equally spaced as possible. Obviously, that would require that the three dimensions of the "box" be different.

The standing wave frequencies repeat at equal intervals (multiples), so the three dimensions should also not be a multiple of each other either. For instance if a speaker cab is 700mm wide, it should not be 350mm tall (1/2 of its width) and 175mm deep (1/2 of its height).

What makes a well-proportioned listening room, recording studio or a loudspeaker enclosure?

Humanity has been grappling with the question of the optimal room dimensions and building proportions for millennia, not just in terms of their acoustics but also in terms of pleasing proportions, and even regarding the health benefits they bring their occupants. Various "optimal" designs have been suggested.

Similar efforts were made in the design of musical instruments, indicating that we could benefit from using such ratios in the design of loudspeaker enclosures. Let's have a brief look at a few.

One set of "magic" numbers is 2:1:1.118, used for the shape of "King's Chamber" in the Great Pyramid of Giza. For a 6 m long listening room, the width would be 3 m, and the ceiling height would need to be 3.354 m.

"Golden Solid" or "Golden Cuboid" proportions were used for Egyptian tombs and medieval churches. Φ (Greek letter "phi") is the "golden ratio." The dimensions progress in a 1 - Φ - Φ^2 order, also called the Fibonacci sequence.

For a 6 m long listening room, the width would be 3.71 m, while the ceiling height needs to be 2.29 m.

These exact ratios have been successfully applied to the dimensioning of musical instruments, loudspeaker boxes, and even the placement of loudspeakers in a room.

Other acoustics experts suggest ratios of 0.6 : 1 : 1.6 (the same as 2.618:1.618:1 ratio), which results in quite a pleasing speaker cabinet's appearance.

The 0.8 : 1 : 1.25 ratio results in a much deeper, squat box, not to my liking.

ABOVE: This recommended speaker enclosure dimension ratio is practical and of visually pleasing proportions. For a 50cm tall speaker cab, the depth should then be 50/1.68 = 29.8 or 30cm, and the width should be 30*2.68 = 80cm.

Speakers, speaker cabs and guitars sound different because they are made of different materials

Speaker drivers with their cones made of different materials (paper, aluminium, polypropylene, flax fibers, fiberglass, or Kevlar®) sound different, just as speaker cabinets made of MDF (medium-density fiberboard) sound different from those made of solid pine, or like my hi-fi audiophile speakers, of solid mahogany, for instance.

Place your amp and speaker cab on a timber floor of an older timber house and then on the concrete floor of a recording studio - the same equipment will sound quite different. The answer is in the different materials used to construct the walls and floors in these two buildings.

Your house is a timber house with hollow stud walls covered by gyprock sheets. Your floor is also timber, raised off the ground on studs. The studio is of double brick construction (external walls) with single brick internal walls. The floor is a concrete slab over which a carpet was laid.

While the walls and the floor of the studio are solid and rigid, your walls and the ceiling at home are effectively large resonant panels that will vibrate when you play music. They will absorb some of the sound energy, resonate at specific frequencies, and re-radiate such energy back into the room.

In essence, your practice room will behave almost like the resonant body of an acoustic guitar, while the studio room would be akin to the solid body of an electric guitar. The two will sound very different.

ALL MATERIALS WERE NOT CREATED EQUAL a.k.a. THE WOODEN LAW

You cannot make a violin out of a tree that grew to be used as rake handle. The same applies to people.

DECIBELS AND LOGARITHMIC SCALES

We have already mentioned the amazing wideband instrument called the human ear, whose dynamic frequency range is four decades wide (20 Hz to 20,000 Hz). To be able to cope with such a wide frequency range, instruments of this type (both "natural," such as the ear, and "man-made", such a dB-meter or SPL meter) respond not in a linear but a logarithmic manner to such stimuli.

Perhaps even more importantly, the air pressure levels and sound intensity changes vary in such wide a range, that the use of a linear scale would be impossible or at least impractical. At 1 kHz, the threshold of hearing corresponds to sound intensity of 10^{-12} watts/m^2. On the other end of the intensity scale, the threshold of pain is around 10 watts/m^2. That is a ratio of 10^{13}, or 1 followed by 13 zeros!

Decibel defined

Decibel or "dB" is not a unit of measurement, but a dimensionless number, a unit for a power ratio or gain: $P = 10*\log(P/P_0)$ [dB], or, in acoustics, $P = 10*\log(I/I_0)$ [dB] for sound intensity.

P is the power produced by a system or a device (an amplifier, for instance) or a pressure level and P_0 is the referent power level chosen for a particular purpose, comparison or measurement. In terms of sound intensity, $I_0 = 1*10^{-12}$ watts/m^2.

However, a dB does not have to be used as a ratio of a measured value and the referent value (such as the threshold of hearing), any referent value can be used. For instance, we can compare an increase in power levels before and after a certain change to an amplifier or a speaker cab: $dB = 10\log(P_{NOW}/P_{BEFORE})$.

The beauty of decibels is that we can add and subtract them, which are the two easiest mathematical operations.

Say we had a cab and modified it in some way. We measured the SPL levels the speaker cab produced (with the same input signal and at the same test signal frequency) and got 86 dB before and 89 dB after the mod. The increase in SPL is simply 89 - 86 = +3dB!

A 3 dB change is easily detectable. Interestingly, 1 dB is the smallest audible change in sound level, but not all people are able to detect such a small change, only some musicians, audio technicians, and seasoned audiophiles.

It can be proven mathematically that for voltage, current or sound pressure ratios, a factor 20 is used instead of 10 (since the power is proportional to voltage and current squared, that's where the factor 2 comes in), so VOLTAGE $V = 20\log(V_2/V_1)$ [dB], CURRENT $I = 20\log(I_2/I_1)$ [dB], and SOUND PRESSURE $SPL=20\log(p_2/p_0)$ [dB]. Notice that the symbols for current and intensity are the same (I), which is mildly confusing to say the least! Also notice that we will use capital "P" for power and lower case "p" for sound pressure!

In acoustics, the referent level P_0 is usually 20 microPascals (20μPa).

The logarithmic power curve

To understand the logarithmic nature of the decibel, the logarithmic power or intensity curve is the visual representation of the equation

$P = 10\log(P_2/P_1)$ [dB]

The ratio P_2/P_1 is on the horizontal axis and the decibels are on the vertical axis (the ordinate).

When $P_1=P_2$ their ratio is 1, and log1=0 (because $10^0 = 1$). Any number to the power of zero equals one (point A).

As P_2 becomes larger than P_1, the log curve is in the positive territory, and as P_2/P_1 drops under one, the curve goes into the negative values.

In point E, P_2 is twice as large as P_1, which is +3 dB. In point D, P_2 is half as large as P_1, which is -3 dB.

These are also called half-power points, because the ratio of the two power levels is 2 or 1/2.

In point C, $P_2/P_1=10$, and log10=1, so $P_2/P_1=10$ dB. Similarly in point B, $P_2/P_1 = 0.1$ so $P_2/P_1 =$ -10 dB.

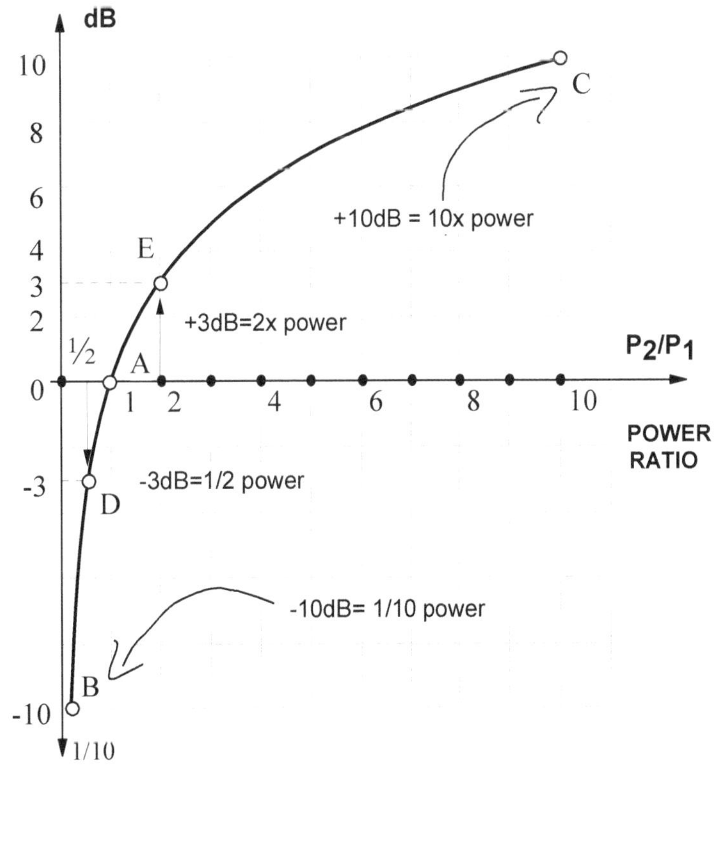

Linear (LIN) and logarithmic (LOG) scales

Many multimeters and other instruments have linear and logarithmic scales, so it's of paramount importance to understand logarithms and logarithmic scales.

Log scales are in widespread use in audio. A VU-meter, typically used in mixing consoles, hi-fi amplifiers and tape-decks, has a dB scale and a % scale. Does it show power or voltage (signal amplitude) levels?

We can tell by comparing the dB and % scales. 0dB (A) is 100%, meaning a full input signal is fed to the output.

The -3dB (point B) is just above the dot representing the 70% level, which we know is the 71% mark (0.71 ratio). Since log0.71= -0.1487, to get to -3 dB we must multiply -0.1487 (approx. -0.15) by 20.

This means this VU-meter scale is in voltage levels: $V = 20\log(V_2/V_1)$ [dB] !

Point C or -10dB at 31.6% of the bottom linear scale (0-100). Point D marks the +3dB point, what number on the bottom scale should we write under it? Notice the designer or manufacturer of this scale and this meter did not mark the percentages above 100.

Just as -3dB means a ratio of 0.71 (71%), +3dB is its inverse or 1/0.71=1.41 or a 41% increase, so the number on the linear scale should be 141!

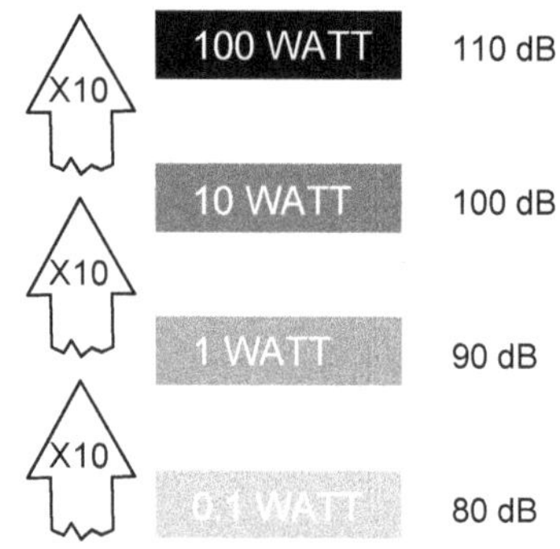

ABOVE: Each 10 dB increase in SPL requires a tenfold (10 times) increase in the amplifier power!

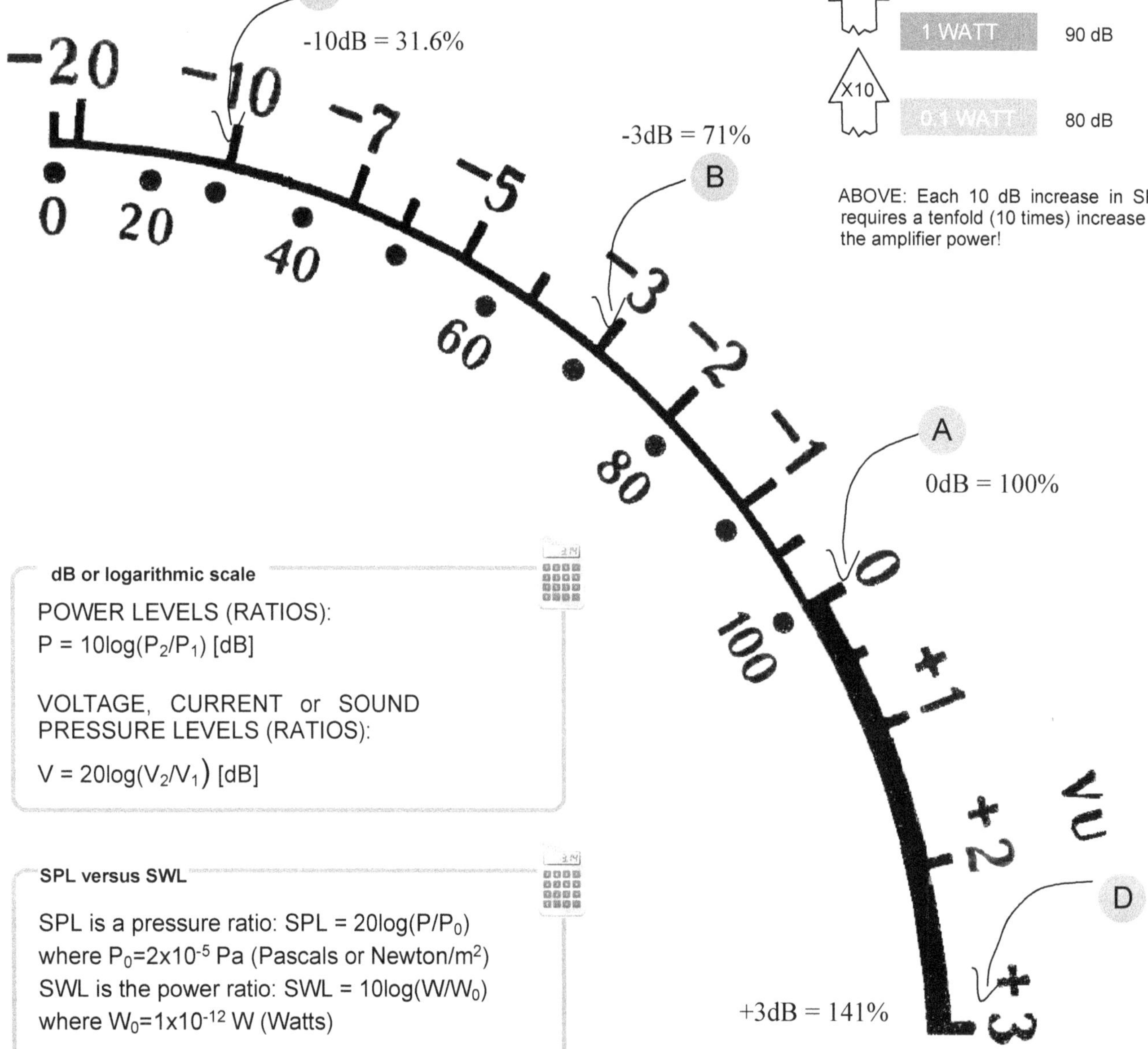

dB or logarithmic scale

POWER LEVELS (RATIOS):

$P = 10\log(P_2/P_1)$ [dB]

VOLTAGE, CURRENT or SOUND PRESSURE LEVELS (RATIOS):

$V = 20\log(V_2/V_1)$ [dB]

SPL versus SWL

SPL is a pressure ratio: $SPL = 20\log(P/P_0)$
where $P_0=2\times10^{-5}$ Pa (Pascals or Newton/m^2)
SWL is the power ratio: $SWL = 10\log(W/W_0)$
where $W_0=1\times10^{-12}$ W (Watts)

HOW TO CALCULATE & CONVERT SOUND PRESSURE LEVELS

The Inverse Square Law: How to convert the SPL (loudness) rating at one distance to another distance

The sensitivity of a particular guitar speaker was declared as 100 dB/watt @ 0.5m. Usually, the specs are given as dB/watt@1m. Is it possible to convert the two figures to the same distance? It certainly is!

The pressure created by a sound wave propagating through air drops with distance according to the Inverse Square Law: $SPL_2/SPL_1 = (d_1/d_2)^2$

If we take a logarithm of both sides and rearrange the formula we get: SPL @ d_2 = SPL @d_1 - 20·log10 (d_2/d_1)

In our case SPL @d_1 is 100 dB, d_1 is 0.5m, d_2 is 1m, so we get

SPL @ 1m = SPL @ 0.5m - 20·log10 (1/0.5) = 100 - 20log2 = 100- 20*0.3 = 100-6 = 94 dB! Thus, the SPL rating of this speaker is 94 dB/watt@1m!

Sound attenuation calculation using Inverse Square Law

SPL @ d_2 = SPL @d_1 - 20·log10 (d_2/d_1)
SPL @ d_1 = known sound pressure level at the first location (distance d_1 from the source)
SPL @ d_2 = unknown sound pressure level at distance d_2 from the source
d_1 = distance from the sound source to the location of known sound pressure level
d_2 = distance from the sound source to the second location

How to calculate speaker's SPL at higher power inputs if their sensitivity is known

Using the same speaker as in the last example, we know its sensitivity rating is 94 dB/watt@1m! That is, again, with a small signal (1 watt of electrical power) driving the speaker. What if we are playing a VOX AC30 amp in large pub, and we crank it up to its nominal maximum. So now we have 30 watts of power driving that speaker; what SPL level can we expect?

Again, our good buddy the decibel comes to the rescue. $SPL_{30} = SPL_1 + 10\log 30/1 = 94 + 14.77 = 108.77$ dB

94 dB is quite a loud sound, especially if practicing at home. Does the equation work in the other direction, for power levels below the nominal 1 watt? It certainly does.

What if we use, say, only 10% (1/10) of that power (0.1 watt) to drive this speaker? The SPL in that case will be

$SPL_{0.1} = SPL_1 + 10\log 0.1/1 = 94 + 10^* (-1) = 94 - 10 = 84$ dB

SPL at higher power levels if the sensitivity at a lower level is known

If sensitivity at any power level "Y" is known: $SPL_X = SPL_Y + 10\log X/Y$
If Y=1 (watt), or nominal SPL @ 1 watt, $SPL_X = SPL_1 + 10\log X$
where X = higher input power level in watts

How to calculate SPL levels produced by two or more speaker cabs

We have two guitar players in rock band, each with his own amp, driving its own speaker cab.

During a recording session in a large studio, their manager gets his trusted SPL meter out and measures the SPL of the first guitarist as 71dB (playing on his own), while the other played more loudly at 73 dB.

What SPL level will the manager's SPL meter indicate with both guitar players playing together?

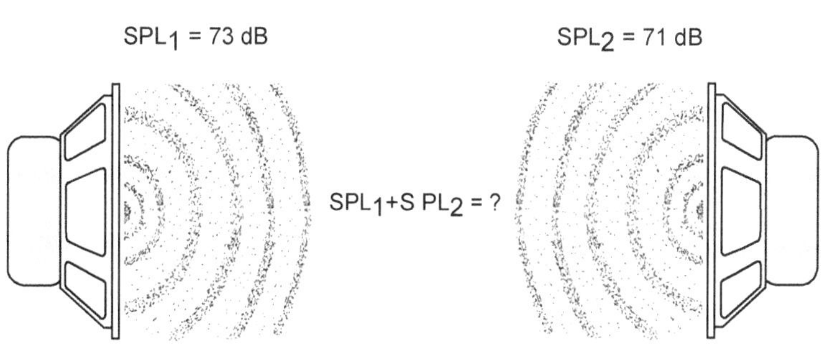

ABOVE: How do we add SPLs of two or more sources (such as two guitarists playing together)? With great difficulty, as you will soon see ...

Not all SPL level and similar calculations using decibels are simple. This problem seems simple, but the calculation isn't easy because we cannot simply add decibel figures together; we must convert them back to their original ratios before adding them together, and then convert the result back into decibels.

For a single sound source SPL=20log(P/P_0), where p_0 is the agreed or referent pressure level and $P_0^2 = 4x10^{-10}$ N^2m^{-4}. For multiple uncorrelated sources (not for two or more speakers in the same cabinet, for instance, these are acoustically coupled, and thus "correlated"):

$$SPL=20\log\left(\frac{\sqrt{(P_1^2+P_2^2+ \dots +P_N^2)}}{P_0}\right) = 10\log\left(\frac{(P_1^2+P_2^2+ \dots +P_N^2)}{P_0^2}\right)$$

By squaring the whole single sound source equation SPL=20log(P/P_0) and expressing P^2, we get $P^2=P_0^2 x10^{(P/10)}$ or, for each of the two sources

$P_{71dB}^2=P_0^2x10^{(71/10)} = 4x10^{-10}x10^{7.1} = 5.04\ x10^{-3}\ N^2m^{-4}$ and $P_{73dB}^2=P_0^2x10^{(73/10)} = 4x10^{-10}x10^{7.3} = 7.98\ x10^{-3}$ N^2m^{-4}

$$SPL= 10\log\frac{P_1^2+P_2^2}{P_0^2} = 10\log\frac{5.04\ x10^{-3} + 7.98\ x10^{-3}}{4x10^{-10}} = 10\log 32{,}550{,}000 = 10x7.513 = 75.13\ dB$$

You can look at this way: adding another source 2dB "louder" than the original 71dB source only increased the total SPL by just over 2dB (2.13dB), and not by 73dB as some would claim without deeper thought. This makes intuitive sense. If adding 71dB and 73dB resulted in an SPL of 144dB, life on earth would be unbearably loud.

Also, we know that doubling the acoustic power (two speakers producing identical SP levels) or doubling the amplifier power into the same speaker would only add +3dB to the overall SPL, so adding 71dB to 73dB has to result in an increase of less than 3dB!

GUITAR SPEAKERS IN A STUDIO - RECORDING TIPS & TRICKS

When you record a guitar or bass cab in a studio (or any other instrument), you add two more variables to the cab's tone, that of the microphone itself, and the acoustics of the studio. Just as different speakers sound different, different microphone types and even brands sound different.

As with all other links in the music production chain, you can use the sonic signature of various mikes to your benefit, to achieve a certain synergy, an effect or tone you desire, or to your detriment, where an otherwise awesome sounding speaker cab sounds thin, shrill or unpleasant on the actual recording.

Microphone types and technologies

Dynamic microphones work as miniature loudspeakers, only in reverse, which means they also have a diaphragm attached to a voice coil, both moving in a magnetic field.

The cheapest to produce and the most rugged of all microphone types, dynamic microphones are rugged and able to handle the highest possible sound pressure levels without distortion. However, their high frequency extension is inferior to other types, namely condenser and ribbon microphones.

Condenser (an archaic name for a capacitor) microphones have electrically-biased or polarized diaphragm, which, when struck by sound waves, changes its distance from the fixed back plate. Since he capacitance between two conductive plates is directly proportional to the distance between them, such pressure variations are transduced into capacitance and then voltage variations.

The AC signal voltage must be amplified inside the mike by a very high impedance preamplifier, either vacuum tube or FET (Field Effect Transistor).

Condenser mikes have a higher sensitivity than dynamic ones, and have a smoother and more extended high frequency response that adds detail, transparency and microdynamics to voices and instruments. A smaller membrane has a resonant frequency towards the hi end of the usable frequency range. It also means these mikes capture transients better.

If one had to generalize, compared to dynamic mikes' warmer and softer tone, condenser mikes have a harder, more crisp, airy and detailed sonic signature.

Ribbon mikes have been around for a century. A thin corrugated (folded) aluminium or plastic ribbon (coated with a deposited metallic layer) is suspended in the strong magnetic field.

The ribbon is the membrane and the voice-coil in one, acting as 1/2 a turn. When sound waves hit the ribbon, a minuscule voltage is induced across its length. that voltage and the ribbons very low impedances stepped up by an inbuilt transformer, typically from 0.15 ohms to 200-300 ohms.

The corrugations increase the length and thus the impedance of the ribbon, allow for optimal tensioning of the ribbon, which acts as a spring, and finally, they scatter the incoming sound waves, making the ribbon movement more uniform across its whole length.

If both sides of its ribbon are open to the incoming sound, the mike will work as pressure-gradient transducer with figure 8 characteristics.

Due to an almost negligible moving mass, such mikes have an excellent dynamics and transient response. However, they are fragile and very sensitive to both mechanical impact and wind exposure.

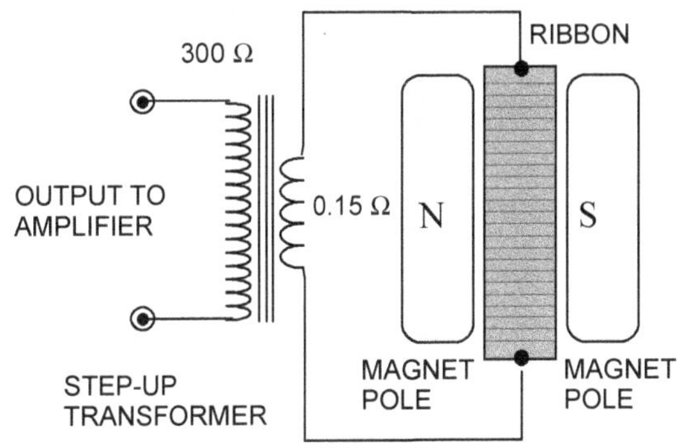

ABOVE: The operating principle of ribbon microphones

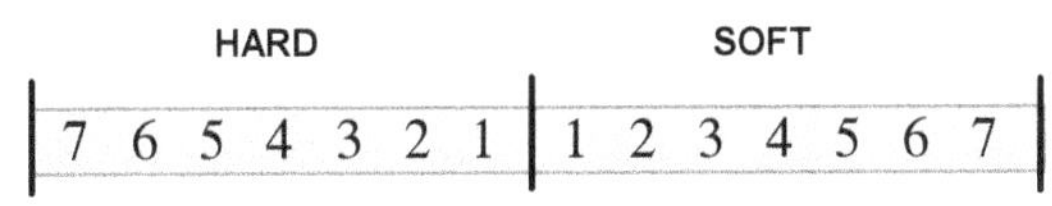

ABOVE RIGHT: grading microphones by their tone in terms of "hardness" (hard, treble-heavy tonal balance) and "softness" (soft, mellow, midrange dominant tone)

Choose the microphone that will complement the speaker's tone

We all know that Marshall and Hiwatt amps sound "harder" than the more mellow Fender lovelies, while the shimmery or "brilliant" VOX sonics sit somewhere in-between. That is a very simplified statement of course, but it does apply to speakers and speaker cabs as well.

We also know that too much of anything generally isn't good, sonically or in any other sense. This soft-hard balancing act is based on the "opposites attract" principle in matchmaking speaker cabs and microphones.

Studio owners, recording engineers and many guitar players have realized that not all microphones in their studio sound the same (result in the recorded sound of the same "hardness), and neither do speaker cabs.

If you have a sonically hard cab, record it with a softer mike, and vice versa. Unless, of course, you are deliberately aiming for either an extremely soft, mellow recording, lacking treble and crispiness, or an overly aggressive, hard metallic sound.

Phantom power

Phantom power is a method of sending DC voltage to a balanced microphone through the same two lines that feed the audio AC signal back to the microphone preamp or a mixing console. 12V, 24V and +48 V_{DC} voltages are used, with +48 V_{DC} being the most common

Balanced XLR connectors are always used, pin 1 is always GND or "ground", also known as COM or "common", pins 2 and 3 are "hot" or signal pins.

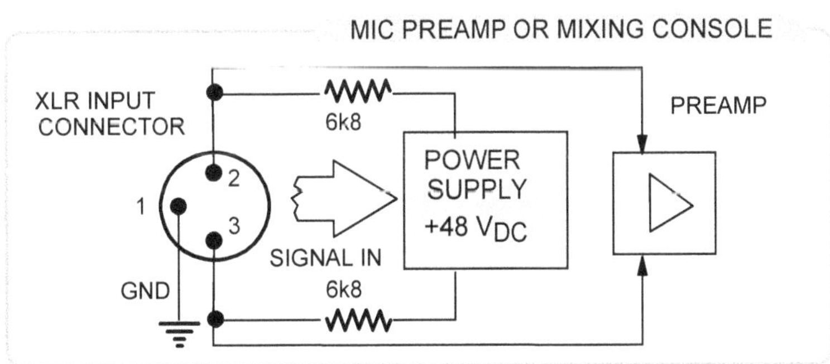

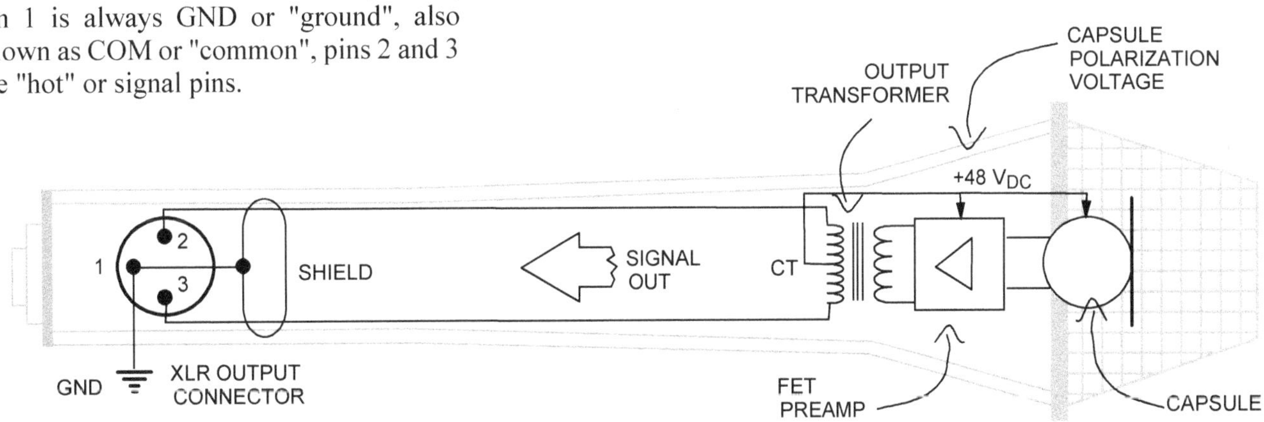

The positive voltage is applied to both pins 2 & 3 through precision resistors or through an output transformer inside a console (not shown on the schematics). Inside the microphone, that voltage is taken from the center tap (CT) of the output transformer and used to polarize (bias) the capsule and also to power up the mike's internal FET preamp.

Coverage patterns

Dynamic microphones whose diaphragm's back is sealed have an omnidirectional pickup characteristic (illustration on the following page). A directional coverage such as the cardioid response can be achieved by allowing the incident sound to reach the back of the diaphragm through some sort of an acoustic labyrinth that will delay that component of incident sound so it appears with an opposite phase and thus cancel the front wave.

The same applies to condenser mikes. The ones with a sealed back (except for a small pressure-equalizing path) are omnidirectional, while those whose back plate is a porous sintered disk operate on a pressure gradient (velocity) principle and are directional. The porous disk provides the required acoustic delay for rear sound waves.

Inherently, ribbon mikes generally have a bi-directional pickup characteristic. Sound waves arriving from its sides act equally on both sides of the ribbon, canceling each other.

By placing two gold-plated Mylar diaphragms back-to-back with a center plate between them, it is possible to have a condenser mike with switchable characteristics. The front diaphragm is always positively biased (+DC voltage).

A positively biased back diaphragm will produce an omnidirectional pickup pattern, with a zero bias the pattern will change to cardioid, while negatively biasing the rear diaphragm results in a figure-8 characteristic.

Some microphones (such as Audio-Technica AE2500) have two elements (condenser and dynamic) in a single-housing, and, since the elements are so close together, provide a coincidental phase relationship between the two outputs, something impossible to achieve with two separate mikes.

The condenser element does require 11- 52V_{DC} phantom power, so a 5-pin XLR-type connector is used. The balanced condenser mike signals (and phantom power input) are across pins 2 and 3, pins 4 & 5 are the dynamic mike's output, and pin 1 is ground (shield).

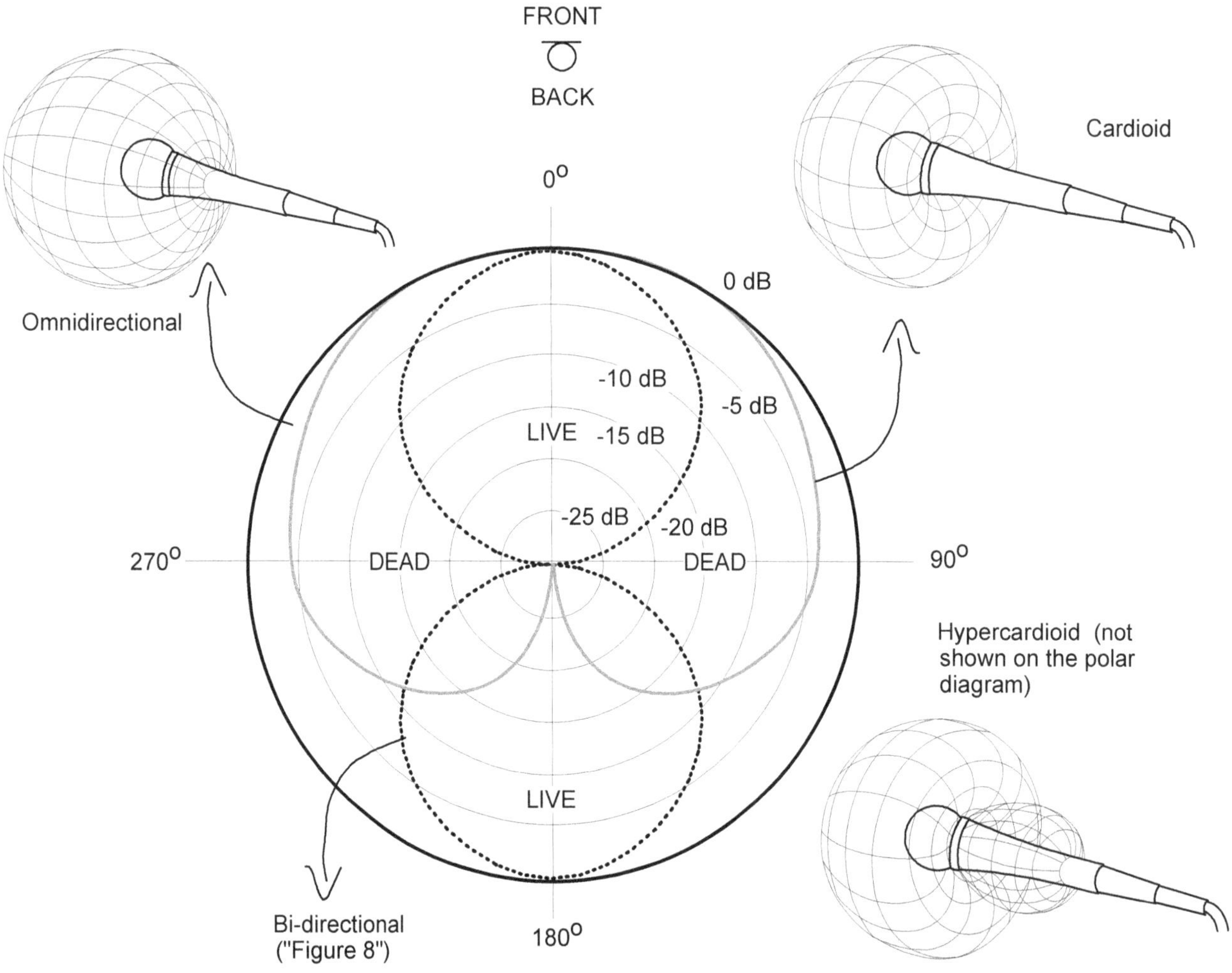

End-fire and side-fire mikes

Most dynamic (moving coil) and small-diaphragm handheld condenser mikes are top-fire or top-address type. They have elongated casings, with the capsule at one end (the "top") and the output cable connection at the other end (the "bottom").

END-FIRE MIKE

Side-fire or side-address microphones are usually large diaphragm condenser or ribbon mics. The primary axis is perpendicular to the center of the side-facing diaphragm, so these mikes are placed in an upright position.

A common mistake is to place a side-fire mike backwards, with its back facing the speaker cab instead of its front. For that reason, most such mikes have a symbol marked on their case, indicating the front,

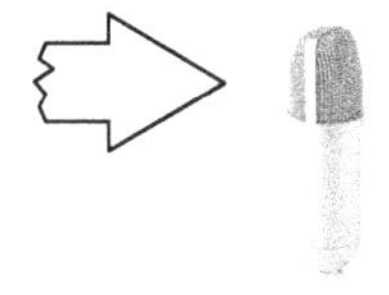

SIDE-FIRE MIKE

Capturing "the flame"

We have already discussed speaker directionality and the difference between their on-axis (center line through the middle of the driver) and off-axis performance. Since the simplified shape of the projected sound field reminds people of a candle flame (just horizontal instead of vertical), this type of recording tonal control is often referred to as chasing the flame or capturing different parts of the flame.

Since higher frequencies are attenuated faster than the low (bass) frequencies, the further away from the cone we position the microphone, the less HF we will capture and the smoother and more midrange-heavy the sound will be. Of course, as always, the further the mike is from the cab, the more of the surrounding (ambient) sounds it will pick up, and the mote attenuated the SPL will be, so don't go crazy!

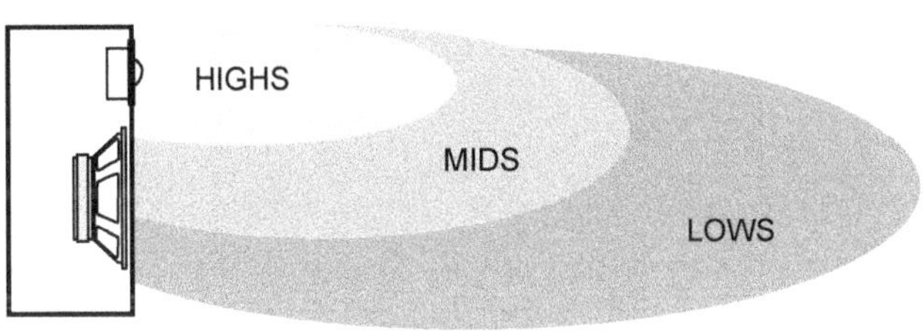

ABOVE: A typical hi-fi loudspeaker shows significant beaming and difference in projection patterns between various audio bands.

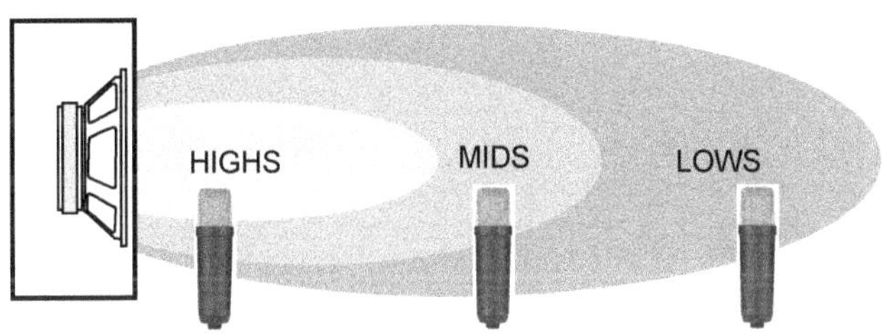

ABOVE. Although guitar speaker cabs suffer from the same problem, this presents an opportunity for getting different tonal balances from the same speaker by placing the mike at different distances along "the flame"

Miking multiple speaker cabinets

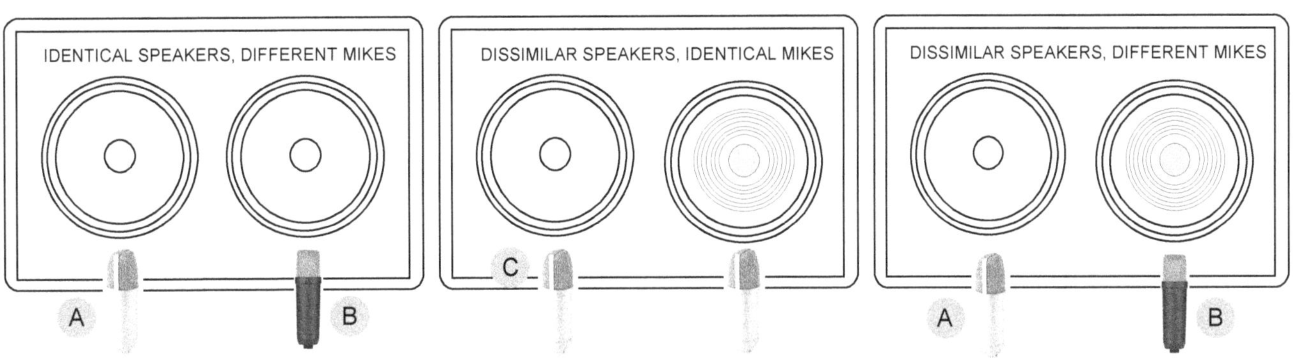

Multiple speaker cabs (two or more drivers in one enclosure), especially if they use dissimilar drivers, present a multitude of possible sounds and their combinations that can be captured. Combine that with a possibility of using two different mikes, you get a wide variety of awesome tones from only a single cab and a pair of mikes.

The mikes could be of the same type, but different "hardness", such as one S4, the other H3, for instance, and/or they can be different types as well, one dynamic, the other condenser.

Mike position as tone control

Moving the mike across the speaker's radius is akin to adjusting tone control on an amp. just like dialing back the "treble" or "brilliance" control, it changes from the bright and treble-heavy sound when directly across the speaker's center, to a warmer, fatter and more mellow tone as the mike is moved towards the speaker's rime.

RIGHT and BELOW: Due to the beaming effect where high frequencies are primarily produced by the inner part of the cone (around the dust cap), placing a mike directly along the center line of a speaker results in a bright sound with a lot of "bite". As the mike is moved sideways towards the outer suspension (the speaker's edge), the recorded tone mellows and becomes warmer, more rounded and full-bodied.

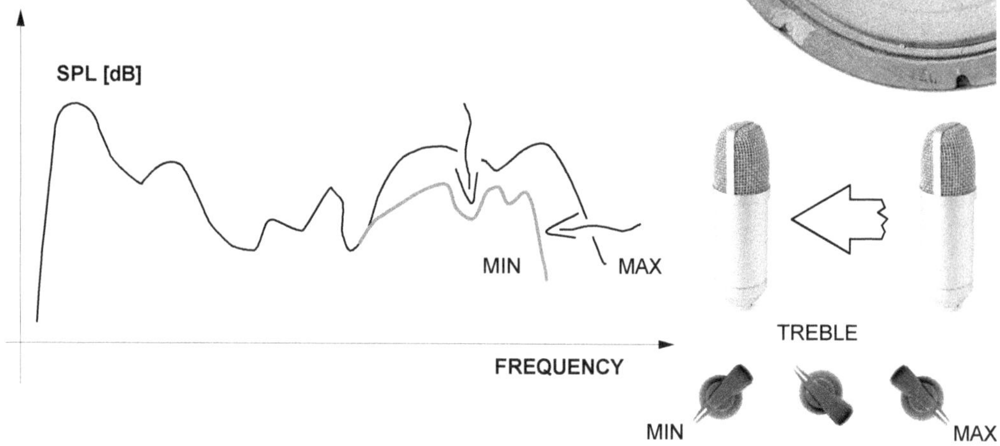

Back-of-the cab miking

Closed back cabinets are more controlled and controllable in a studio, with a more directional sound stage. Open back speakers radiate in a more unpredictable way, with the rear sound waves bouncing off the walls. Such reflections may or may not be desirable in the final mix.

On the other hand, open back cabs present unusual opportunities, such as miking the back of the cab in addition to the front, and even placing the rear mike inside the cab.

RIGHT: Back-of-the cab-miking
1) Main mike (cardioid)
2) Rear mike (cardioid or omni) - usually requires phase inversion (polarity reversal) at the mixing desk or preamplifier

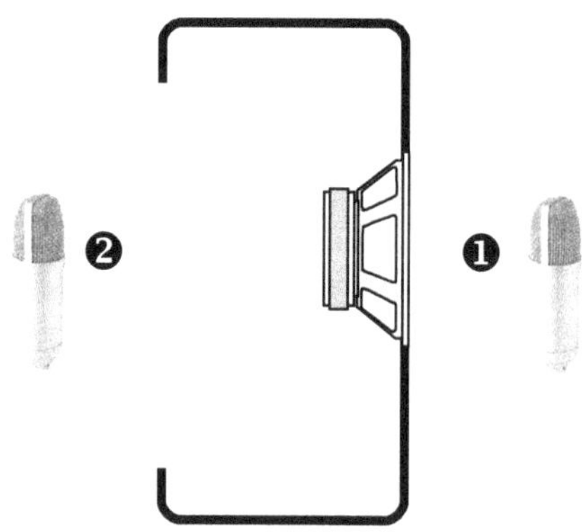

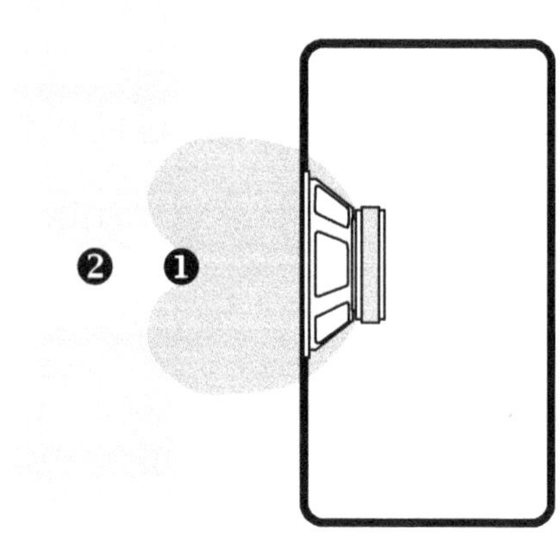

ABOVE: Miking the ambiance
1) Main mike (cardioid)
2) Ambient or side mike (Figure 8)

Miking the ambient sound

Figure-8 microphones, placed some distance away from the cab, are ideal for capturing the ambient sound. Obviously, they live sides should be placed to face tot eh left and right of the speaker, so the dead side will be facing the speaker (2).

The main mike should be a directional one such a cardioid or supercardioid.

HOW LOUDSPEAKERS WORK

2

To comprehend speaker specs and parameters, and, more importantly, to figure out their capabilities and limitations, we need to understand the physical and engineering principles behind their construction and operation.

Although this chapter is (out of necessity) 80% theory, keep reading for those twenty or so practical percent - I will show you the only way to reduce the power of your amp-cab combination while preserving its awesome tone even at very low volumes ...

Stay tuned ...

- THE ANATOMY OF PERMANENT MAGNET DYNAMIC LOUDSPEAKERS
- TWEETERS AND COMPRESSION DRIVERS
- THE PHYSICS BEHIND THE OPERATION OF DYNAMIC LOUDSPEAKERS
- ELECTRODYNAMIC (FIELD COIL) SPEAKERS
- THE WAY IT USED TO BE: THE WONDERFUL WORLD OF VINTAGE DRIVERS

THE ANATOMY OF PERMANENT MAGNET DYNAMIC LOUDSPEAKERS

Speakers 101 - the naming of the parts

A dynamic loudspeaker is an electromechanical transducer, transforming the electrical energy of the signal current flowing through its voice coil ("spool") into a mechanical sound wave (air pressure) produced by its cone. The voice coil is wound on a cylindrical voice coil "former," attached to the cone and placed in a uniform magnetic field in the "gap" created by either a permanent magnet or an electromagnet (more on those soon). It moves linearly in that narrow air gap between the poles of a magnet.

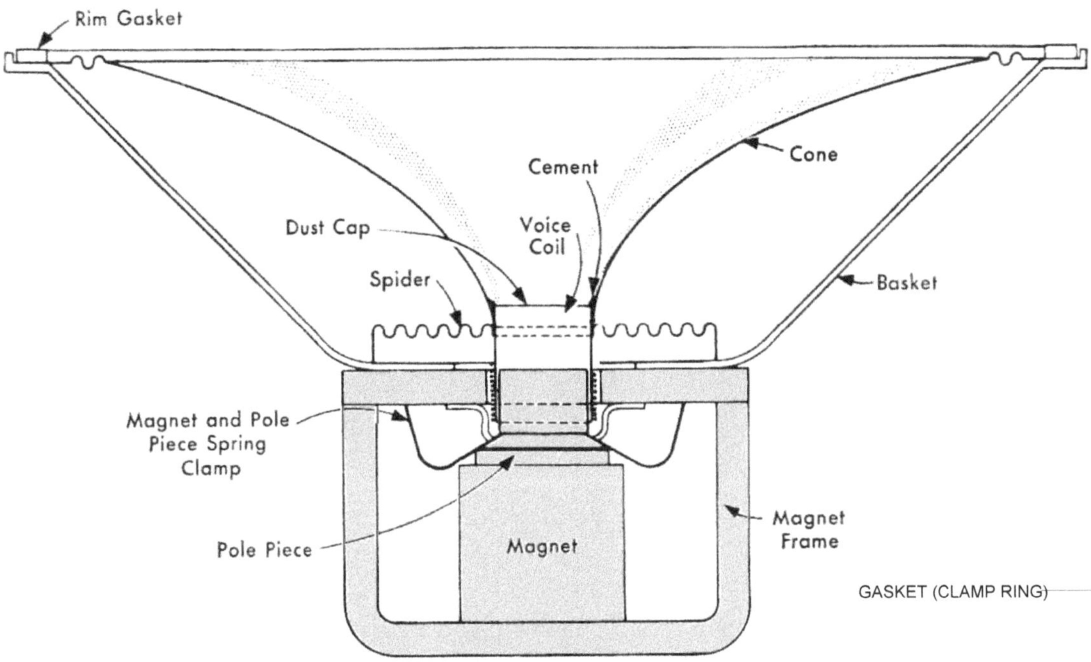

ABOVE: A cross-sectional view of a typical Alnico loudspeaker.

RIGHT: A cross-sectional view of a typical ceramic magnet loudspeaker. Ring magnet (1), spider or inner centering suspension (2), paper cone (3), terminal strip (4) riveted to the metal frame (7), incoming wires (5), voice coil connection wires (6) and the "surround" or outer centering suspension (7). The "yoke" or magnet frame (8) closes the magnetic circuit and concentrates its magnetic flux through the narrow gap in which the voice coil sits. Finally, the metal frame (9) holds everything together.

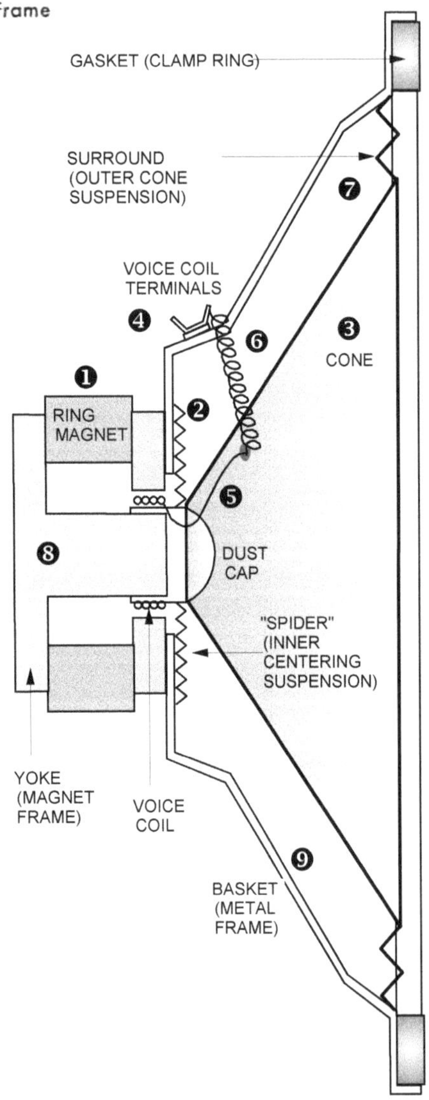

The VC former is a tube on which the voice coil was wound. It is usually made of paper or modern materials such as Kapton or titanium. The former is cemented or glued to the cone, and this whole assembly is held in place and centered in the air gap by the inner and outer suspension.

The inner suspension is often referred to as a spider, due to the shape of the early ones, which were stamped out of thin sheets of pressed bakelite and resembled a spider (next page).

The modern spider or rear suspension is a fabric disc with multiple concentric corrugations, impregnated with a phenolic resin. Its dimensions, profile, the number of corrugations, the fabric's weave type and density, and the quantity of the resin used are the main factors that determine its compliance, and its inverse, stiffness.

Such elasticity or "compliance" of the inner and the outer suspension allow the cone to move in a linear motion ("in and out"), while preventing or at least minimizing its lateral (sideways) movements.

The dust cap's purpose is to prevent foreign matter such as dust and small particles from entering the air gap and thus affecting the movement of the voice coil and the cone.

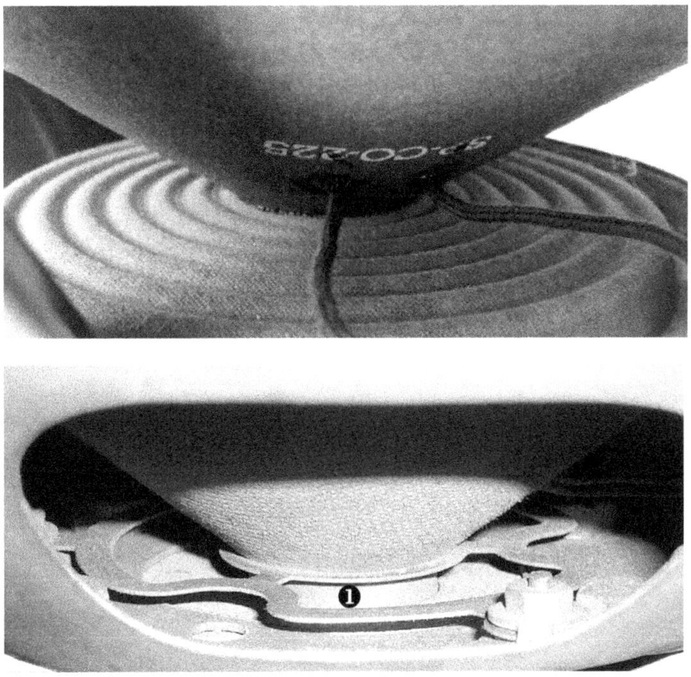

ABOVE: Two most common types of "spiders" or inner cone supports, the stamped Bakelite spider from 1920-40s and the modern "corrugated" support. The voice coil former is visible (1).

BELOW: One of the many similar shapes of a vintage stamped spider

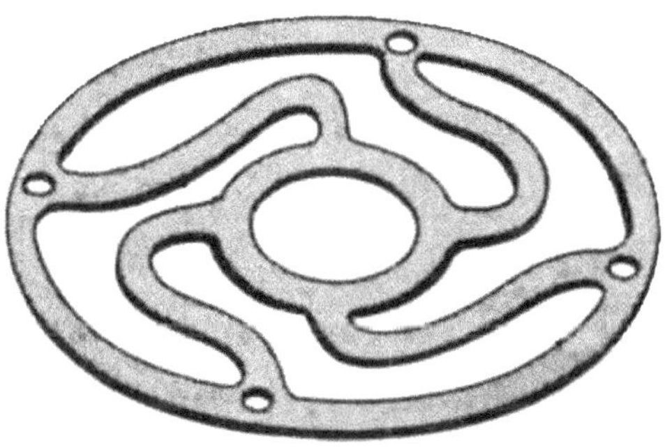

BELOW: After the removal of the dust cap, the voice coil former (4), the central slug (2) and the glued voice coil connection wires (3) are clearly visible.

"Coil rub" happens when the voice coil touches and rubs against the central slug or the magnet frame during movement. It sounds unpleasant and can be detected by gently pushing the cone in by hand, which, apart from the visual checks (for a damaged or torn cone) is the first and most basic test of a speaker's condition.

Alnico and "ceramic" or "ferrite" magnets

Most modern loudspeakers use ferrite magnets, made by pressing ferrite powder into a ring-shaped structure.

The first permanent magnet speakers used Alnico (an acronym for Aluminium-Nickel- Cobalt mix), an alloy whose chemical composition is 8–12% aluminium, 15–26% nickel, 5–24% cobalt, up to 6% copper, up to 1% titanium, and the rest is iron.

In the 1970s, "ceramic" or "ferrite" magnets were developed, which replaced Alnico almost entirely because they were cheaper and easier to manufacture.

However, Alnico speaker drivers are back in fashion. Although more expensive than ceramic magnet drivers, they are considered better sounding (more "musical" and engaging").

Typically, assuming the same volume of a magnet, an Alnico V (Roman number 5) magnet is 75% stronger than the more modern ferrite material, but the strength of neodymium magnets is more than five times higher than the magnetic energy of Alnico V!

Neodymium magnets

In late 1970s, research was underway for another rare earth magnet that would be cheaper than samarium-cobalt (SmCo) magnets.

General Motors (GM) in the USA and Sumitomo Special Metals in Japan independently discovered the neodymium Nd2Fe14B compound almost simultaneously in 1984. GM used the melt- spining technique, while Sumitomo opted for sintered magnet making.

Also known as NdFeB, NIB, or Neo magnets, these permanent magnets are made from an alloy of neodymium, iron, and boron. Neodymium magnets are graded N28 up to N55, according to their maximum energy product, or the magnetic flux per unit volume. The higher the value, the stronger the magnet.

Compared to the same volume ferrite magnets, neodymium ones are 18 times stronger, or 12 times stronger by weight. Sintered neodymium magnets are vulnerable to corrosion, so a protective coating is required, usually nickel or zinc plating.

Speakers with Alnico magnets are instantly recognizable by their UI-shaped magnet frame or "yoke", made of mild steel.

The U-part (1) is spot-welded to the "I-part" or plate (2). Only the central "slug" (3) is the actual permanent magnet made of Alnico, bolted on top to the U-piece. (4) is the voice coil.

One possible reason why Alnico, ceramic, neodymium and electrodynamic speakers sound different

While guitar amplifiers and speaker cabinets with electrodynamic speakers are still rare, most guitar players would agree that speaker boxes with drivers using Alnico magnets sound better than those with ceramic magnets. Likewise, most of those who have heard expensive audiophile electrodynamic speakers swear that they sound even better than the Alnico ones.

All three types of dynamic speakers work on the same principle. A voice coil placed in a stationary and uniform magnetic field, which, when the audio signal (current) flows through it, moves due to the interaction between the variable magnetic field created by the flow of the signal current and the stationary magnetic fields created by the permanent or electrodynamic magnet. More on that soon.

The skeptics or "absolutists" claim that there should be and that there isn't any difference between the three types, that, assuming the magnetic fields are of the same strength, how the field was created should make no difference.

However, all other things being equal (the driver's construction, size, power rating, resonant frequency, etc.), the sonic difference does indeed come from the source of the permanent magnetic field, or, more precisely, how such field was created!

The supposedly "permanent" or steady (DC) magnetic field is not permanent at all. It varies (gets modulated) by its interaction with the AC magnetic field created by the audio signal current flowing in the voice coil. This changes the response and the dynamics of the speaker driver.

As the voice coil moves through a stationary magnetic field, its variable magnetic field (created by the audio signal flowing through it) distorts the stationary magnetic field and initiates a complex dynamic interaction.

A dynamic signal compression results, with consequences such as reduced dynamics, increased driver distortion, and aberrations in its frequency response.

This is the case with ceramic magnets, slightly less so with Alnico drivers (which is most likely why they sound differet) and much less so with electrodynamic drivers.

The magnetic field created by a permanent magnet can be weakened by the varying magnetic field of the voice coil. Since the strength of its magnetic field drops with the increase in the voice coil's field (louder music passages), this results in a lower SPL produced by the driver and a significant dynamic compression. Alnico drivers produce a stiffer magnetic field with minimal compression and improved dynamics.

The electromagnet's flux field density also remains constant when a voice coil moves in the gap of the motor, and such compression and distortion effects are not present!

On this vintage Australian-made Magnaflux driver, the whole magnetic structure (1) can be detached from the frame (2) - VC (3) - spider (4) assembly and removed. You simply twist it a few degrees until the four notches on the magnet structure line up with the lips on the base (5) and lift it up.

The heart of any speaker - its voice coil

The voice coil is a simple inductor (a choke) wound on a cylindrically-shaped tube or "former", since it gives the VC its shape or form. The diameter, the material and the number of turns of the magnet wire used to wind the coil will all impact its tone, as will the size (diameter) and the material used for the former.

Vintage speakers used paper formers, but these were limited in terms of temperature they could withstand and thus were the factor limiting the speakers' maximum power, the main reason vintage speakers were mostly low power transducers (10-30 watts).

The VC diameter and size are directly proportional to their power rating and heat dissipating capability. Simply put, the larger the VC size (diameter), the higher the speaker's power rating. Of course, the size of the magnet and the supporting magnetic structure also goes up, as increasing the costs and the final retail price of larger speakers.

Guitar speakers have VC diameters in the 1” (25mm) - 2.5” (64mm) range. Manufacturers often have a preference for one or two standard sizes and design all of their drivers around them.

For instance, most of Celestion guitar speakers use copper wire on 1¾” (44 mm) voice coil formers, so they all have a similar sonic signature' Their bite and crunch in the midrange is immediately recognizable, and over the early years became the epitome of the "British rock tone".

Jensen speakers, both the vintage American beauties and the currently-produced Italian ones, use smaller diameter voice coils. A smaller VC will have a brighter tone with a higher frequency extension (since it is lighter and easier to move), but with a thinner bass.

Aluminium or copper wire for the voice coil?

The three most common materials for voice coil wire are copper, aluminium, and, wait for it, copper-clad aluminium.

Copper has a higher electrical conductivity and thermal conductivity than aluminum. However, aluminum is significantly lighter and less expensive than copper. While aluminum’s conductivity is lower, it is still a viable option for many applications, especially where weight and cost are concerns.

The electrical conductivity of copper is 101% IACS (International Annealed Copper Standard) at 20°C, while the conductivity of aluminium is only 61%, so the ratio is 101/61=1.656 in favor of copper.

However, aluminium is much lighter than copper, their density being 2.7 grams/cm3 and 8.96 grams/cm3 respectively, so the ratio is 8.96/2.7=3.318 in favor of aluminium. Overall, combining the two ratios we get the figure of merit for aluminium 3.318/1.656 = 2.00 times higher than that of copper.

Aluminium is much cheaper and lighter than copper, but has lower thermal conductivity than copper, meaning aluminium voice coils cannot dissipate the heat as well as those wound with copper wire can.

Also, an aluminum coil with the same resistance and overall length as the copper coil will have a slightly larger volume. So, assuming the same efficiency, an aluminium wound coil would require a larger magnetic structure, which increases the cost.

To have the same resistance, an Al wire need to have a 57% larger cross-section than a copper wire. Such bulkier coil would require a larger magnetic gap and a larger and stronger magnet, increasing cost.

A partial solution is the use of a rectangular-formed Al wire, which will increase the space factor of the voice coil since more conductor can be fit into the gap area. Such ribbon wire can be wound very tightly edgewise in a single layer, and, when turns are glued together, is self-supporting, so there is no need for a VC former.

While this improves the cooling and saves some space and cost, a 30% larger gap and magnet is still required (hence the "partial" solution).

Aluminium oxidizes easily, and is harder to solder and work with. The solution was found in coating an aluminium wire with a thin layer of copper. The use of CCA (copper clad aluminium) wires only slightly increases the overall mass of the voice coil.

The voice coil geometry

Designing a guitar speaker (or any speaker) is full of choices, which, unfortunately, due to the strict laws of physics, also means that it is full of contradictory requirements and inevitable compromises. One of the most important choices that affect the size, weight and cost of a speaker, in addition to its dynamic behavior, sensitivity, distortion levels and sonic character is the size of its voice coil in relation to the length of its magnet's gap "G".

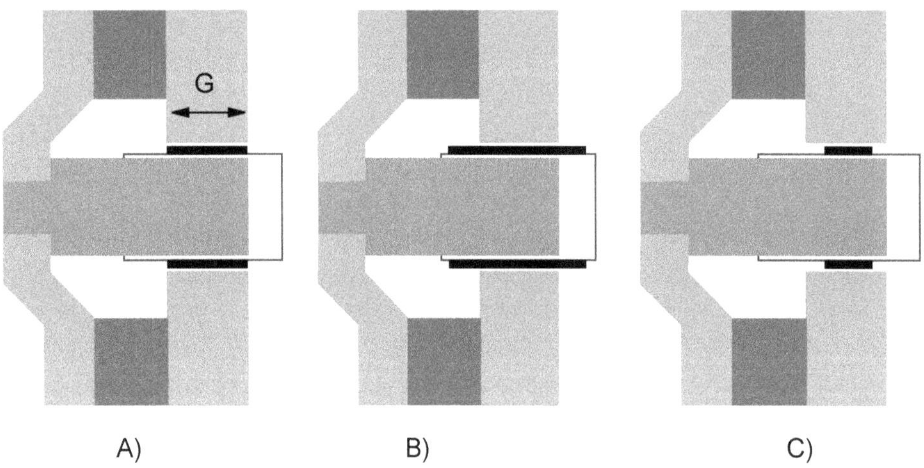

LEFT: Three types of loudspeaker voice coils

A) Conventional design - voice coil length equal to the magnet gap length "G"

B) Overhanging voice coil designs - the voice coil length is larger than the magnet gap length - low efficiency, low distortion

C) Voice coil length smaller than the magnet gap length - high efficiency, low distortion

The length of the conventional voice coils, used in low-compliance drivers, is the same as the gap length "G", so as soon as the conventional VC moves forward or backwards, some turns leave the gap. Turns outside the magnetic field do not contribute to the mechanical force exerted on the cone. In other words, a natural compression occurs and signal peaks are rounded or even clipped, causing significant distortion. The transducer's efficiency is relatively high while the costs can be kept low by using smaller VC inside smaller and cheaper magnetic assemblies.

In contrast, with "overhanging" voice coil designs the VC length is larger than the magnet gap length, so the whole length "G" is always filled with VC turns, and the VC always has energized turns within the magnetic gap. However, so many inactive turns (always outside "G") add weight and reduce the driver's efficiency and cost.

Since they allow bigger excursions, the overhanging VCs are also called "long-throw" coils. On the negative side, while only some of the turns are "active" (in the gap), all its turns must be energized by the driving amp, thus wasting power and lowering the overall efficiency.

Long-throw voice coils are thus only used for low compliance woofers, which must be capable of long excursions at bass frequencies in order to move as much air as possible, but are incapable of producing mid- and high- frequencies due to their big & heavy voice coils and cones.

Finally, you may ask, why don't they make "G" much longer than the VC length? That would make the VC as small and light as practically possible, and even during long excursions all the VC turns would remain inside the magnetic structure?

The answer can be found in any economic textbook, and that is cost. Large magnets and supporting magnetic structures (the yoke, the top and bottom plate, etc.) are costly and heavy, so such designs are only used for hi-end hi-fi drivers. Minimizing speaker distortion isn't on guitar speaker makers' priority list, anyway.

Speaker terminals

The cheapest and thus most common type of instrument speaker terminals is a dual lug, either stamped at a right angle (1) or in-line (4). The braided wire or "tinsel lead" (2) from the voice coil is usually brought in through a hole in the rivet that attaches the lug to a piece of cardboard (2), which is in turn also riveted to the speaker's basket or metal frame (3).

This makes it possible to plug in or solder two wires (each through its own hole), useful when paralleling drivers inside a cab. Etone chose to solder the incoming wire to one of the lugs (4), thus losing such flexibility.

The "+" terminal is marked either by a stamped sign on the cardboard strip or by coloring (painting) one of the rivets, usually in red (5).

A more expensive system, seldom used these days in guitar speakers, involved using spade-terminated wires, tightened inside screw-in type machined binding posts (7) or bare wires fed through a hole in push-release binding posts (9). Such binding posts still had terminal lugs (6), to which the incoming wires were soldered.

The "+" terminal was marked either by a red dot glued tot he top magnet cover (8), or by using pushbuttons (part of the binding posts) of different colors (9).

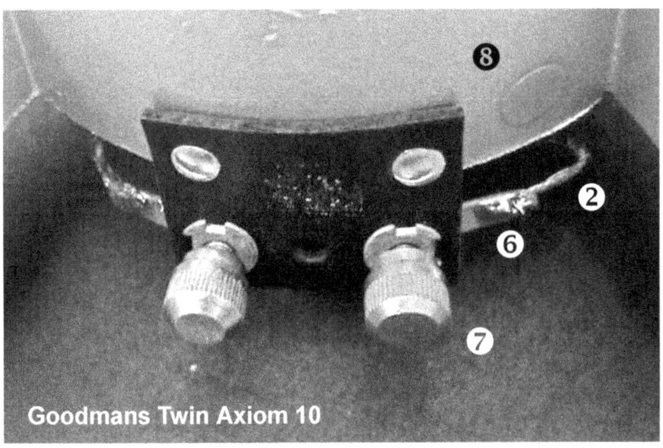

Goodmans Twin Axiom 10

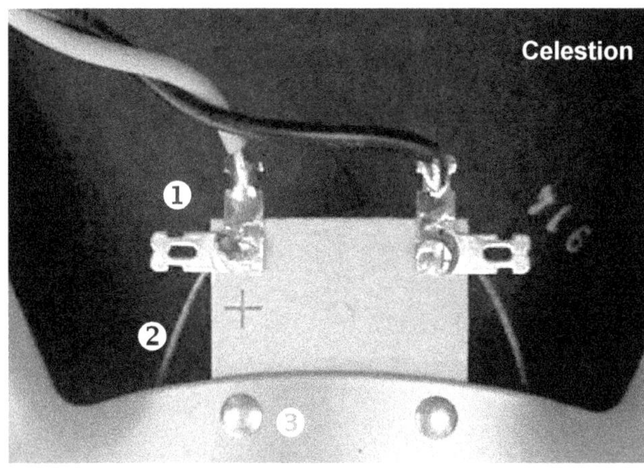

Celestion

JBL K120

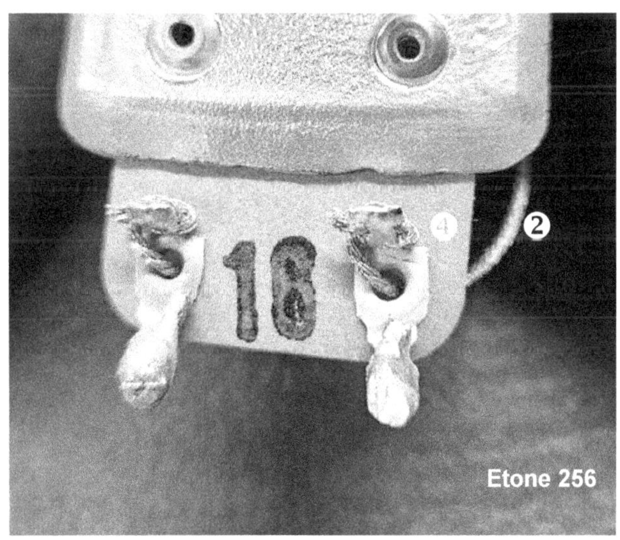

Etone 256

Braided tinsel leads

The solid wire used to wind the voice coil is not soldered directly top the drivers terminals. The constant flexing would cause fractures and an eventual breaks. The VC wire is brought out and glued along the surface of the cone to the termination point where it is soldered to a flexible "tinsel" wire that connects it to the driver terminals.

Made from superfine copper or silver strands, the flexibility is achieved by braiding the multiple strands together and adding interwoven cotton to prevent chafing and fractures.

Failures are still possible, of course. The soldered joints fatigue, crack and fail, resulting in either a total loss of sound or an intermittent crackling. resoldering the connections is an easy fix.

However, the solid VC wire may still fracture in the vicinity of its attachment point to the voice coil, in which case repair may or may not be possible.

TWEETERS AND COMPRESSION DRIVERS

The anatomy of typical conical tweeter

The ultra-compact (66 x 66 x 45 mm) Eminence APT:30 horn loaded tweeter weighs only 0.4 lbs (180 grams) and offers 80° conical dispersion angle and an average sensitivity of 93.4 dB. Construction details include a 2 oz. ferrite magnet, aluminum voice coil former, and Mylar diaphragm.

Electrically, parameters of interest are its 10-watt power rating, 8Ω nominal impedance (7.5Ω minimum impedance) and a resonant frequency of 2.2 kHz. Crossover frequencies above 3,500 Hz are recommended, using a 12 dB/octave slope crossover designs.

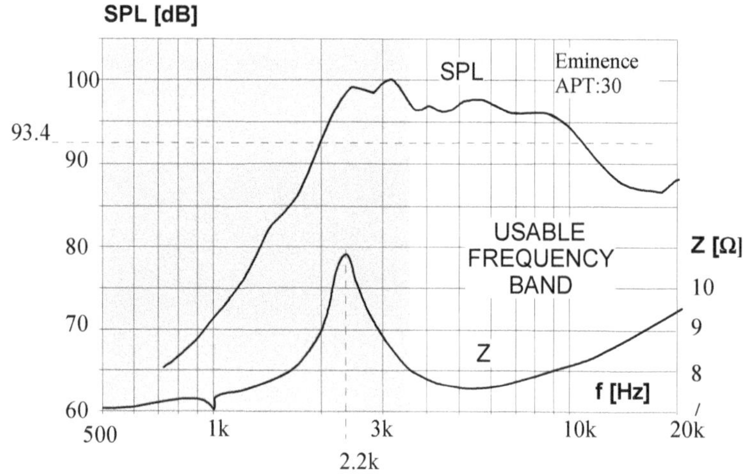

ABOVE: The SPL response and impedance curves of Eminence APT:30 horn loaded tweeter

Dome tweeters

Dome tweeters operate on the same electromagnetic principle as woofers and midrange drivers, but instead of a cone, a much smaller (typically 1" diameter) dome diaphragm (4) is attached to the voice coil former (5).

Soft domes are made by thermoforming polyester film or impregnating silk or polyester fabric with a polymer resin. The hard dome varieties are made of lightweight metals such as titanium, aluminium, or aluminium-magnesium alloys. Due to the high frequencies (short wavelengths) of the sound waves they have to produce, tweeters and midrange drivers don't need large or any enclosures at all.

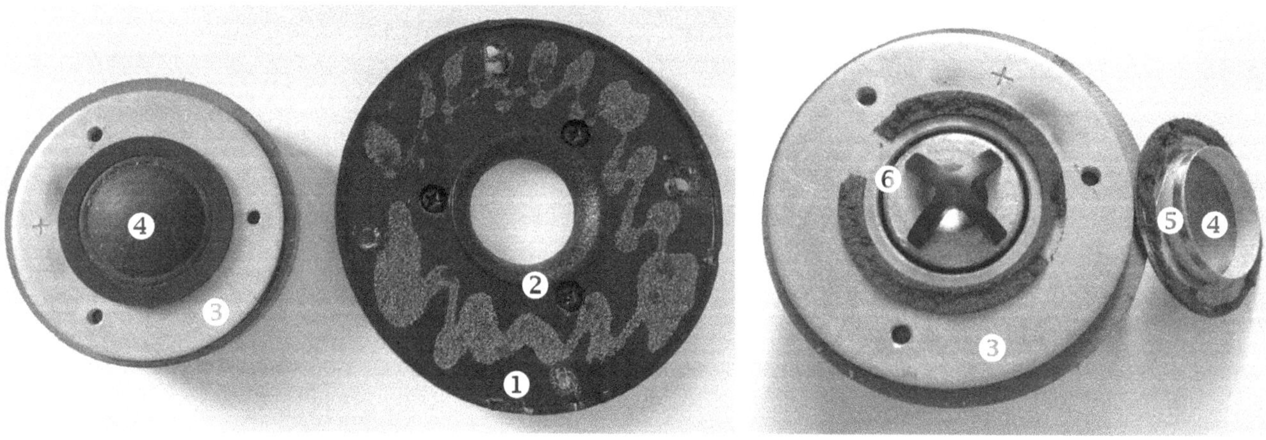

ABOVE: The anatomy of a soft dome tweeter: Plastic faceplate/mounting flange, after the removal of the glued foam layer (1) with waveguide (2), Magnet assembly (3), fabric dome (4), Kapton former with voice coil (5), air gap (6)

Compression drivers

Dome tweeters are in almost all hi-fi speakers, but the PA and instrument speaker world is ruled by their cousin, the compression driver. Essentially a horn-loaded tweeter, a compression driver creates higher sound pressure levels and thus louder volumes. In other words, they are even more sensitive and efficient than the dome tweeters.

The compression in its name indicates that diaphragm's area is significantly larger than the throat opening of the horn, and this compression of an acoustic transformer produces the high sound pressures required, resulting in a much higher acoustic power output from such tiny diaphragm excursions.

The membrane or diaphragm (6) radiates into a compression chamber, through a phasing plug (7) to the driver's exit, attached to the throat of the horn (4). The phasing plug equalizes path lengths from the diaphragm to the exit.

In music venues where speakers need to fill large spaces, compression drivers and horns are essential for wide dispersion and broad coverage.

CDs suffer less dynamic compression and are able to produce dynamics far better than dome tweeters.

Of course, they aren't the most subtle or nuanced of transducers, very few audiophiles would tolerate them in their listening room speakers, although there are some CDs that can compete with dome tweeters in the sound quality.

Typically, 1" VC drivers are rated at 50 watts, 1¾" and 2" units; ratings go up to 100-120 watts, while 3" beasts can handle 80-200 watts of power.

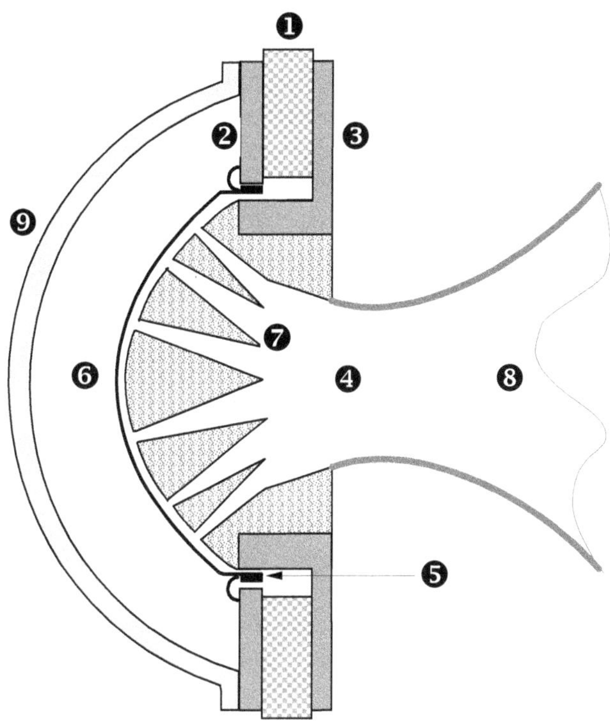

ABOVE: The anatomy of a compression driver

1) Ring magnet 2) Top plate 3) Magnetic yoke 4) Driver's exit and horn's throat 5) Voice coil 6) Membrane (diaphragm) 7) Phasing plug 8) Horn 9) Rear cover

THE PHYSICS BEHIND THE OPERATION OF DYNAMIC LOUDSPEAKERS

A dynamic speaker as an electromechanical transducer

Loudspeakers, just like microphones and phono (turntable) cartridges, are electromechanical transducers. That means they convert one type of energy (electrical) into another (mechanical).

Phono cartridges convert minute vibrations of a stylus (due to its movement in the record's groove) into weak (very low voltage) electrical signals.

Microphones also convert mechanical energy (the pressure changes or "sound waves") into low amplitude electrical signals, which are then amplified.

A third well-known example of a transducer is a guitar pickup. Here, the movement (vibration) of a metal string above a tiny magnet induces a small voltage in the coil wrapped around the four or six magnets (depending on the number of strings the guitar has), and thus converts the mechanical motion of a string into and electrical "motion" or movement of electrons in the copper wire used to make the winding of the transducer. The name "pickup" was obviously concocted due to the fact that it "picks up" or "detects" the minute movements of the string.

Now you can see that, in comparison to those three types of electromechanical transducers (which should really be called mechano-electrical transducers), loudspeakers are truly electromechanical, because their conversion is of electrical energy of the music signal (generated by an amplifier that "drives" the speaker) into the mechanical energy of the sound waves (the movement of air) that the speaker's cone produces.

Interestingly, a loudspeaker can also be used as a microphone ("in reverse"), as in intercoms. Connect it to an input of a combo amplifier and an oscilloscope, and blow, talk or sing into its cone. You will hear a sound produced by the combo amp's speaker and see its waveform on the oscilloscope's screen!

What really happens in the speaker's "motor"?

You may have heard a speaker's "engine" referred to as a "motor", which may seem strange, but is an accurate depiction of its physical behavior. A motor is any device that converts electrical energy into mechanical motion, and speakers certainly do that.

With no audio signal, the voice coil is at rest. An AC signal current flowing in the voice coil (from the output of an amp) generates a pulsating magnetic field of the same waveform as the original signal from the amp.

That varying magnetic field interacts with the magnet's magnetic field, just like two magnets attract or repel each other, depending on their orientation. This attraction-repulsion produces a mechanical force that is transferred onto the speaker's cone. The cone's movement "follows" the audio signal and produces sound waves.

When signal current I_S flows through the voice coil placed in the uniform magnetic field with magnetic flux Φ, the mechanical force acting on the voice coil is $F=B*I_S*\ell_C$.

"B "is the magnetic flux density, I_S audio signal current, and ℓ_C is the total length of the coil wire. Since ℓ_C is constant for each speaker's voice coil, the mechanical force is directly proportional to B.

The relationship between the magnetic flux density B and magnetic flux is $B = \Phi/A$, where A is the cross-section ("A" for area) of the magnetic pole piece (yoke).

The SI unit for magnetic flux is Wb (Weber), while the unit for magnetic flux density is Tesla: Tesla (T) = Weber/m^2. The older, now obsolete unit was G (Gauss). 1T = 10,000 G = 10^4 G! For instance, if a speaker has the magnetic field density of 14,000 Gauss, that is 1.4 Teslas. In 1950s and 60s, it was customary to use the term "lines" as in "Our latest speaker has magnet strength of 14,000 lines!", so it's obvious that "lines" = Gauss!

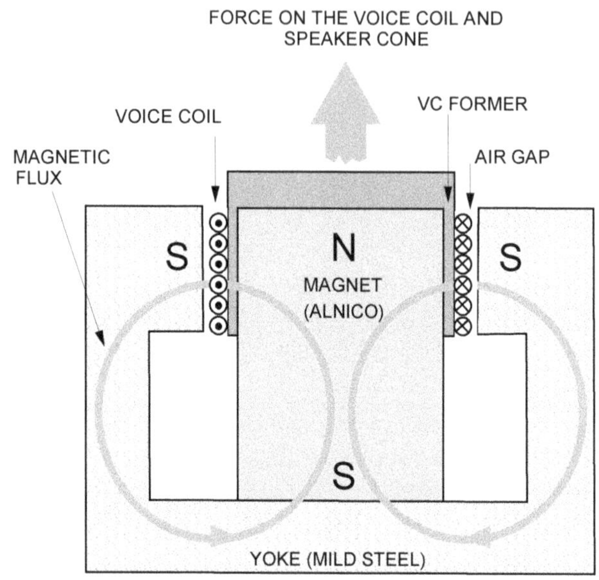

ABOVE: The cylindrical permanent magnet ("slug") has a north N and south S pole. Since the yoke (magnet frame) is attached to it, the slug magnetizes the mild iron of the yoke, whose two pole pieces concentrate the invisible lines of the magnetic flux onto a narrow gap in which the voice coil is situated, and thus close the magnetic circuit.

If B is reduced, the force on the voice coil is also reduced; this results in smaller movement (excursions) of the cone, reduced air pressure created by the cone and lower SPL (loudness).

"B" depends on the type and size of the magnet used and is also fixed, thus determining the speaker's efficiency, sensitivity and "loudness". The force acting on the cone is $F=k*I_S$, directly proportional to the signal current.

THE ARROW CONVENTION

When looking at the cross-section of a wire or a coil (the wire direction is perpendicular to the paper), we can imagine the current flow as a flying arrow. The current flowing away "into the paper" is marked with a cross (the tail of the arrow), and the current flowing "from the paper" towards you is marked with a point (the tip of the arrow).

Counter-EMF, motional impedance, and damping - what do they mean and why are they important

So far we have only considered one phenomenon, which is the mechanical force that results from the interaction of the two magnetic fields, the constant or DC field of the magnet, and the pulsating or AC magnetic field created by the flow of the signal current through the voice coil.

Another physical process happens as soon as the VC starts moving in the magnetic field of the speaker's magnet. A voice coil is an inductor or "choke", trying to oppose or choke any change of current through it by pushing its own, self-induced current in the opposite direction.

That process is called self-induction, whereby a voltage is induced in the VC that is proportional to the speed of its movement and the $B\ell$ factor of the coil: $CEMF = B\ell_C v$.

Another name for voltage is Electromagnetic Force; in one electromechanical analogy force is analogous to voltage and speed (velocity) to current. This voltage is called "back-EMF" or "counter-EMF", indicating that such a force acts "backwards", meaning of the opposite polarity to the signal voltage (from a guitar amp).

Thus, the amp will "see" or "feel" a significant increase in speaker's impedance around the resonant frequency. Such an impedance that only exists when the speaker VC and cone move is thus called motional impedance.

Without motional impedance the speaker's cone would move faster and further, meaning that CEMF and motional impedance act as an electric brake on the speaker's motor, providing electromagnetic damping of the cone.

Dynamic loudspeaker as a complex impedance

A typical dynamic loudspeaker is a complex RCL (resistive-inductive-capacitive) network whose impedance modulus and phase angle vary with frequency. Impedance is a vector, described mathematically by a complex number $Z=R+jX$, where R is its "real" or resistive component, and X is its reactance, or "imaginary" component.

The impedance modulus, a hypotenuse in a right-angle triangle, indicated by symbol Z inside two vertical lines ($|Z|$), can be calculated using Pythagora's Theorem from its resistance and reactance as $|Z| = \sqrt{(R^2 + X^2)}$.

Since the reactances of a capacitor and inductor (both present in our simplified model of a loudspeaker) oppose each other, we can write $X=X_L-X_C = \omega L-1/\omega C$. Thus, the imaginary part of the impedance, the reactance X, can be positive or negative.

ω (Greek letter omega) is angular frequency, which is the "real" frequency f multiplied by 2π (roughly 6.28): $\omega=2\pi f$. Finally we get $|Z| = \sqrt{[R^2 + (X_L-X_C)^2]}$ and its phase angle is $\phi = \arctan(X/R) = \arctan[(X_L-X_C)/R]$. In this case X_L is larger than X_C, so the arrow faces upward. When X_C is larger than X_L, the speaker is predominately capacitive, X is negative and the arrow phases downward!

Again, we see that the reactance of both the inductive and the capacitive component of a loudspeaker vary with frequency, as do the phase angle of its impedance and its modulus (amplitude).

At resonance (f_R), the capacitive and inductive motional reactances cancel out, leaving only the resistance, the sum of the DC resistance of the VC and R_{ES}, the resistive part of the motional impedance.

This typical impedance (modulus) vs. frequency curve of a dynamic loudspeaker (driver itself, without an enclosure) shows a resonant peak around 27Ω at the resonant frequency f_R = 90Hz. If, say, the DC resistance of the speaker's VC is 6Ω, and we have a resonant peak of 27Ω, that means R_{ES} = 27-6 = 21Ω!

The steady rise of impedance at higher frequencies is due to the inductance of the voice coil (the motional impedance is zero), whose reactance (reactive impedance) rises linearly with frequency: $X_L=2\pi L$.

The impedance curve of a speaker makes one realize that the "nominal" speaker impedance, usually 4, 8 or 16Ω, proclaimed by manufacturers, is almost meaningless, but, paradoxically, we still assume such a fixed load when buying amplifiers and trying to match them with speaker cabs (or, sometimes, the other way around).

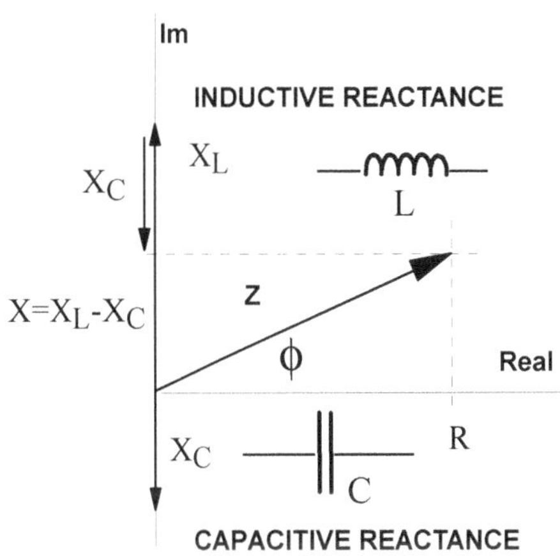

ABOVE: A graphical representation of impedance as a vector in the complex impedance plane.

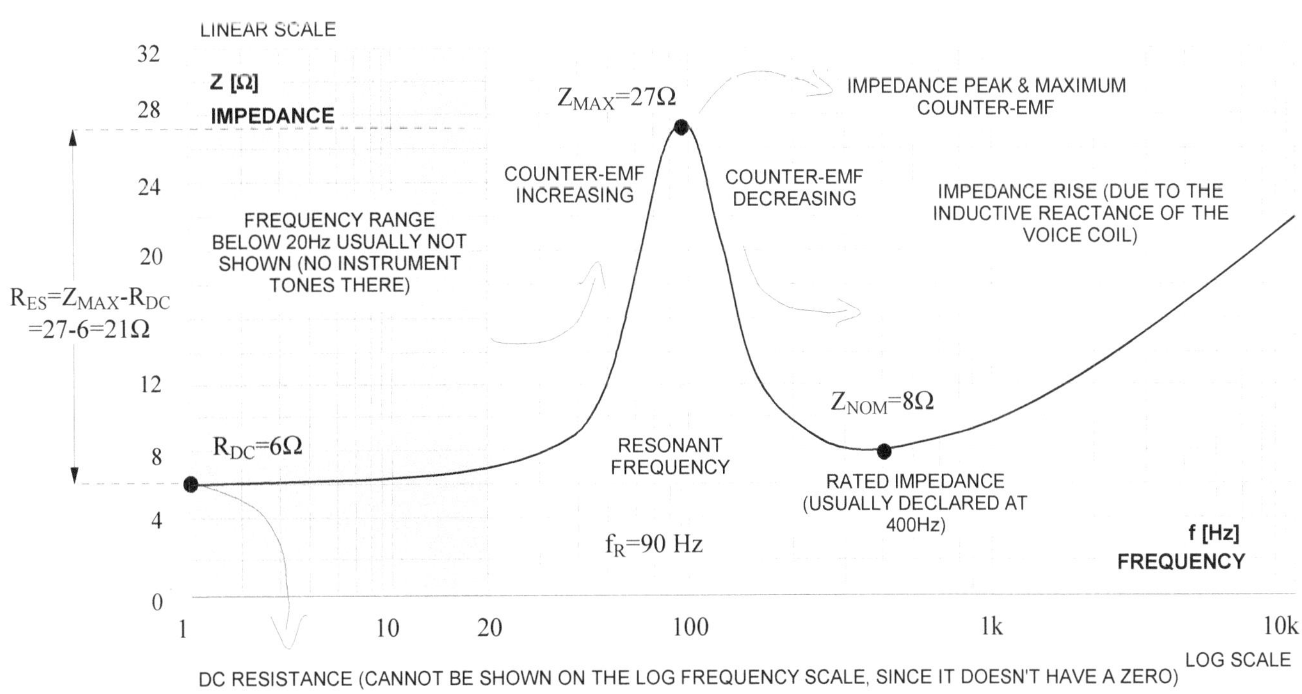

ABOVE: A typical speaker (driver only) impedance (modulus) versus frequency curve bears no resemblance to its nominal (constant) impedance, a figure declared by its manufacturer

Dynamic speaker's impedance phase versus frequency curve, or "phase response" shows that the phase angle is zero at only three frequencies. The frequency f=0 (zero) cannot be shown on a logarithmic horizontal scale. Log scales don't have a zero, but you can see the curve approaching zero (1). A frequency of zero means the signal is DC (direct current), not AC (alternating current), so the phase angle is also zero.

Of more interest are the resonant frequency (2) where the inductive and capacitive aspects of a speaker cancel each other, and some frequency (3), usually around 300-500 Hz, where the same happens again. Between these frequencies (the shaded area), a speaker behaves as a variable capacitor, the phase angle of its impedance is negative.

Outside that relatively narrow band of frequencies (typically between 50-100Hz on the LF side and the 300-500 Hz upper limit), a speaker behaves as variable inductor; after all, it is a linear motor (converts electrical energy into linear motion of its cone).

Notice a very abrupt change of the phase angle around the resonant frequency. The more narrow the impedance peak is, the faster and more abruptly the phase angle will swing from its positive to its negative peak. The sharper the resonance the more rapid the phase change.

RIGHT: The speaker's impedance phase versus frequency curve shows the dual faced (Dr. Jekill and Mr. Hyde) character of dynamic speakers, behaving as an inductor and as a capacitor, depending on the frequency in question!

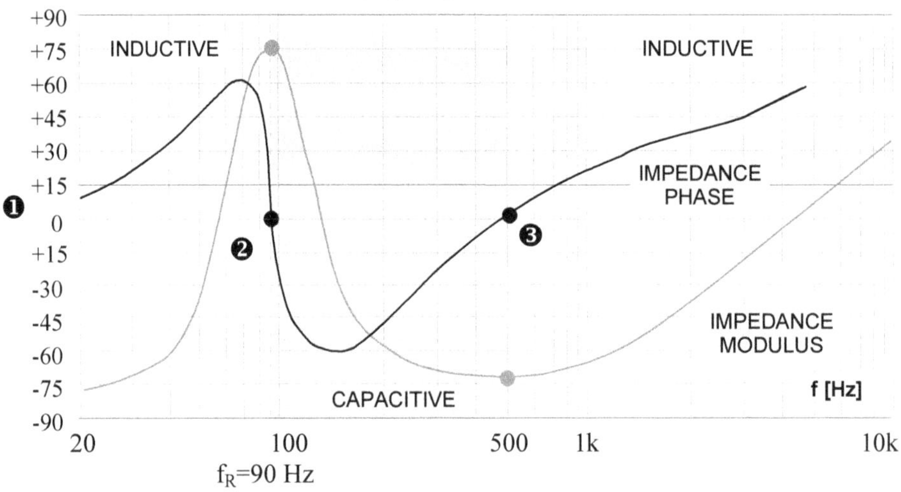

ELECTRODYNAMIC (FIELD COIL) SPEAKERS

How they work

Alnico and ceramic magnet permanent speakers are relatively recent invention. In the first half of the 20th-century, speakers did not have permanent magnets at all. Instead, a magnetic yoke with a center insert was used, and the steady magnetic field was provided by the flow of DC current through the "field coil."

The field coil would have a few thousand turns of a suitably sized lacquer-insulated copper wire, wrapped around a spool or cylindrical bobbin, which would then snugly fit over the center plug (core).

When signal current I_S flows through the voice coil placed in the uniform magnetic field with magnetic flux Φ, the two magnetic fields interact just as with permanent magnet speakers.

This mechanical force acting on the voice coil is $F=B*I_S*\ell_C$. "B "is the magnetic flux density, I_S audio signal current, and ℓ_C is the total length of the coil wire. Since ℓ_C is constant for each speaker's voice coil, the mechanical force is directly proportional to B.

If B is reduced, the force on the voice coil is also reduced; this results in smaller movement (excursions) of the cone, reduced air pressure created by the cone and lower SPL (loudness).

With permanent magnet speakers B depends on the type and size of the magnet ring and is thus fixed, and so is their efficiency and "loudness".

The force is $F=k_1*I_S$ (k_1 is a different constant).

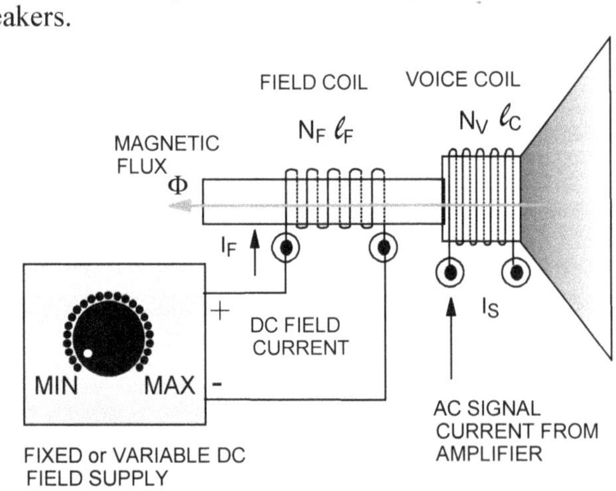

ABOVE: The operational principle of electrodynamic speakers

The magnetic flux density of electrodynamic speakers is proportional to the number of turns in the field coil N_F and its current I_F so the force on the voice coil and the cone is $F = k_2 * I_S * I_F$!

This means that with electrodynamic speakers, we can vary the loudness not just by changing the amplitude of the signal current I_S through the voice coil (by changing the volume control on an amplifier) but also by varying the field current I_F of the electromagnet.

The higher the current through the field coil (I_F), the higher the speaker's SPL (sound pressure level).

BELOW: A cross-sectional view of a typical electrodynamic loudspeaker.

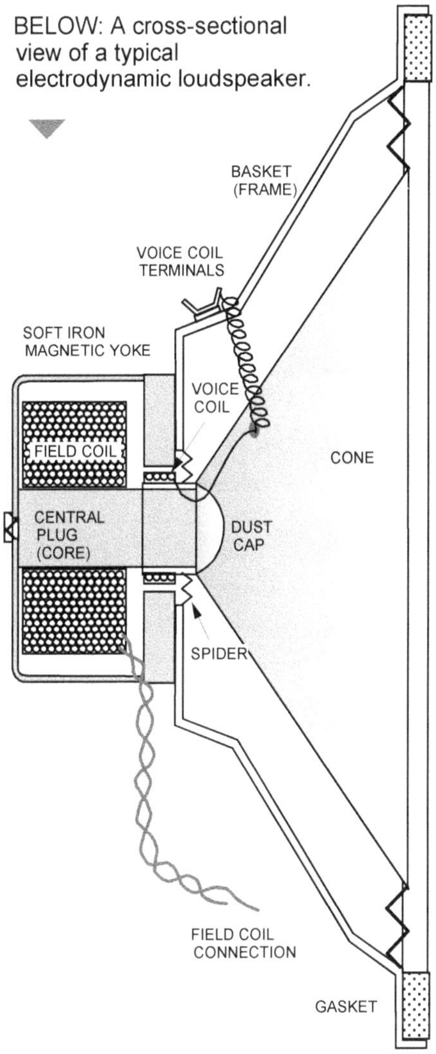

The best way to control the loudness of your amp-speaker combination

With electrodynamic speakers, we can vary the loudness not just by changing the amplitude of the signal current I_S through the voice coil (by changing the volume control on an amplifier) but also by varying the field current I_F of the electromagnet.

The higher the current through the field coil (I_F), the higher the speaker's SPL (sound pressure level).

If your combo amp has no master volume, the only way to control its loudness is by varying the gain, the amplification of its preamp or voltage amplifying stages.

This means that by increasing the gain and the distortion of its preamp and driver stages, you also increase the grid signal on the output tubes and thus the output power level and loudness.

A master volume enables you to push the preamp stages in overdrive and breakup (distortion), while keeping the amp's power output constant or even lowering it, so you get preamp distortion without head-banging loudness levels.

However, to get the output stage to distort, you still need to push it into overdrive (clipping), which again results in unacceptably high SPL (sound pressure levels), especially while practicing or playing large, powerful amps in small venues.

Then came attenuators, which attenuate the electrical signal before it is fed into a speaker. The cheapest units use purely resistive attenuators, using fixed power resistors for a few steps in loudness, or, for continuous control, high power potentiometers, better known as rheostats.

However, no matter how well designed or intricate such an attenuator may be, they always change the sound of the amp-speaker combination since they change the amp-speaker interaction, frequency response, and tonal balance, mostly due to the absence of the speaker-amp interaction.

With resistive loads, there is no induced back EMF (Electromotive Force) or voltage that a dynamic speaker feeds back into the output transformer's secondary.

Then came the idea of incorporating a dummy loudspeaker motor or "voice coil assembly" (VCA) to mimic an actual speaker. Weber is one of the rare attenuator makers that includes such VCAs in their attenuator designs.

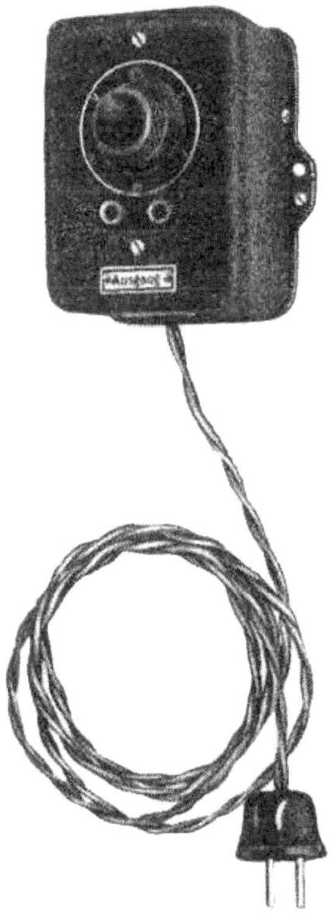

Abb. 553. Lautstärkeregler für dynamische Lautsprecher.

RIGHT: An illustration from a vintage German radio technology book, showing a commercial 1930s rheostat used to vary the field current and thus loudness of an electrodynamic speaker ("dynamische loutsperecher"), "lautstarke-regler" means loudness regulator or, volume control.

VCAs have no metal basket or paper cone, but do behave like a "real" speakers ... up to a point, since there is not acoustic (air) loading on their nonexistent cones. So, it comes close, but it still ain't the real McCoy.

Finally, we come to a conclusion that to preserve the tone of the real distorting power stage and the real distorting speaker, we must use a real speaker, with its cone and all other bits & pieces. But how do we reduce its loudness while keeping the desired tone (distortion)?

The only way is to change the force that acts on the voice coil, which would reduce the excursion of the speaker's cone, thus reducing the loudness. That force is $F=B*I_S*\ell_S$. We cannot reduce "ℓ_S", the length of the winding of speaker's voice coil, and lowering the signal current I_S (as we have just discussed) doesn't sound right, so the only way is to reduce the speaker's sensitivity and thus its loudness by reducing its DC field current I_F - all we need is an electrodynamic speaker and a suitable variable DC power supply.

Most American-made electrodynamic speakers don't specify anything about their field coil on the actual speaker, making it hard to determine the required maximum voltage of the DC power supply. Even the usually pedantic Germans used to specify only one or two parameters.

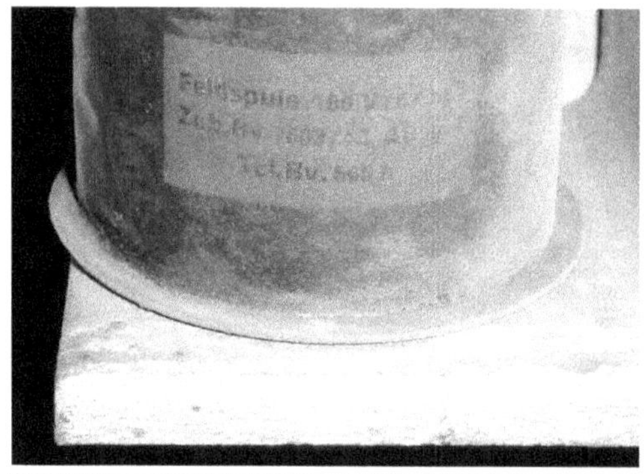

Let's look at three examples. The field coil of a pre-WWII Telefunken speaker is marked "Tel. Bv. 665 a" and "Feldspule 100V/67 mA". *Feld* is German for field and *spule* means coil.

We need a 100V DC power supply, from which the field coil will pull 67 mA of current. Using Ohm's Law we can calculate the field coil's DC resistance, $R=V/I=100/0.067= 1{,}493\ \Omega$ or $1.5\ k\Omega$.

A postwar Radio-Graetz speaker did not have its voltage or current specified, only 1,600 Ω resistance. However, we do gain a further insight into its coil, whose number of windings and wire diameter are printed on the sticker, 14,700 turns of 0.16 mm (diameter) lacquered (L) copper (Cu) wire.

As for the maximum current allowed, various copper wire tables often give you slightly different results for 0.16 mm wire, but they generally fall around 330mA mark.

1,670 ohms (1.67 kΩ) is fairly close to Telefunken's 1.5 kΩ, so we can assume that a similar power supply would be needed here as well, up to $100V_{DC}$.

Finally, the field coil of another German (Sachsenwerk) speaker is marked 75mA 6 Watt (75 mA DC current and 6 Watts power rating). That is not the power rating of the speaker - it is the power consumption of its field coil.

Since the power dissipated in the coil is $P=I^2*R$ we can calculate its resistance as $R = P/I^2 = 6/(0.075^2) = 6/0.005625 = 1{,}067\ \Omega$ or just under 1.1 kΩ. Since power $P=V*I$, we can calculate the required DC voltage as $V= P/I= 6/0.075 = 80$ Volts.

Sadly, most teachers aren't very good at exciting young minds and illustrating the boring theory with real life examples. So, if you ever wondered as a child why in the world you needed to learn Ohm's and Kirchoff's Laws in physics class, now you know: so that twenty (or fifty!) years later, you can confidently calculate other, non-specified parameters of loudspeakers and amps!

Commercial benchmark: FluxTone electrodynamic speakers and "Mechanically adjustable variable flux speaker" by Eminence

USA company FluxTone manufactures a range of electrodynamic speakers and electronic controllers named *Variable Magnet Technology*. A 25 dB range of adjustment is claimed. The controller can be positioned on top of an amp or even inside it, a much user-friendlier solution.

FluxTone claims that they have patented (in the USA) their way of changing SPL of a speaker by adjusting the field coil excitation using a variable power supply with simple one-knob control, but an extensive Google search found no mention of such a patent. Furthermore, how could they patent something that was widely known and practiced before WWII? Whatever happened to the concept of "prior art"?

As always, there is the minor issue of price. For instance, their Model 2 (modeled after the "Original Jensen F-12-N") in 2016 sold for a cool US$1,125! Eight years later, in 2024, it was US$1,500! Yes, that's just for one bare speaker driver and the VMT controller box.

Is there a cheaper, DIY option? There certainly is, and that is to buy a vintage electrodynamic speaker and build your own field current controller.

In contrast to the FluxTone mysterious patent, the USA patent no. US8848968B1, filed in 2012 and assigned to Eminence Speaker, LLC in 2014, titled "Mechanically adjustable variable flux speaker" is real, and Eminence speakers utilizing this principle can be bought for not much more than ordinary (fixed flux density) speakers.

However, the maximum attenuation is only -9dB, hardly worth the bother.

Buying vintage electromagnetic speakers online

There are three problems with buying vintage field coil speakers online. For some reason, they are expensive, much more than similar vintage speakers with permanent magnets. On eBay, they are usually listed at $150-300 each.

Secondly, most are now more than 70 years old, which is a very long time for paper cones to survive. Thus, their old cones are often weak and brittle, even if they are intact with no rips or holes.

Finally, transporting bare speaker drivers by themselves is inherently risky since there is no speaker cabinet to protect them. If the seller does not pack them well, the damage is almost assured, which is what happened to the first 12" speaker we bought online. Most sellers on eBay either have no idea of how to pack fragile items or simply don't care. Anyway, we got reimbursed by eBay and kept the output transformer.

We then bought a smaller 8" American speaker made in 1937 by Motorola, which was well packed by the seller and arrived in perfect condition.

EXPERIMENT: Varying the SPL of a an electromagnetic speaker

Using a variable DC power supply, start with zero volts and gradually ramp the voltage up while monitoring the current. Many benchtop power supplies have dual meters for voltage and current.

You must ensure that the output signal voltage of the amplifier remains constant. Since you are not changing the frequency of the test signal, that should be the case. The voltage will depend on the impedance of the speaker.

The sound pressure meter should be placed in line with the speaker's center (on its "axis").

We used a Legacy combo amp (a re-badged Epiphone Valve Junior) for this test since it also had an 8" PM speaker. First, we measured the SPL of the existing speaker.

Since we weren't interested in the absolute sensitivity values and our test bench had less than 1 meter of space, we set the SPL meter 0.6m away. We chose a much lower input voltage, 0.6 V_{RMS}, because 1 watt of input power produced a deafening and annoying sound level. The relative comparisons still apply, of course. With such a setup, the original speaker for the Legacy combo amp produced 98 dB of sound pressure, which was very loud indeed.

The final question is about the required voltage and current capabilities of the DC voltage source. Most of these vintage speakers were used in pre-WWII radio consoles and didn't have a separate DC power supply.

BELOW: The test setup for measuring the sensitivity of an electromagnetic speaker and its SPL as a function of the excitation of its field coil FC

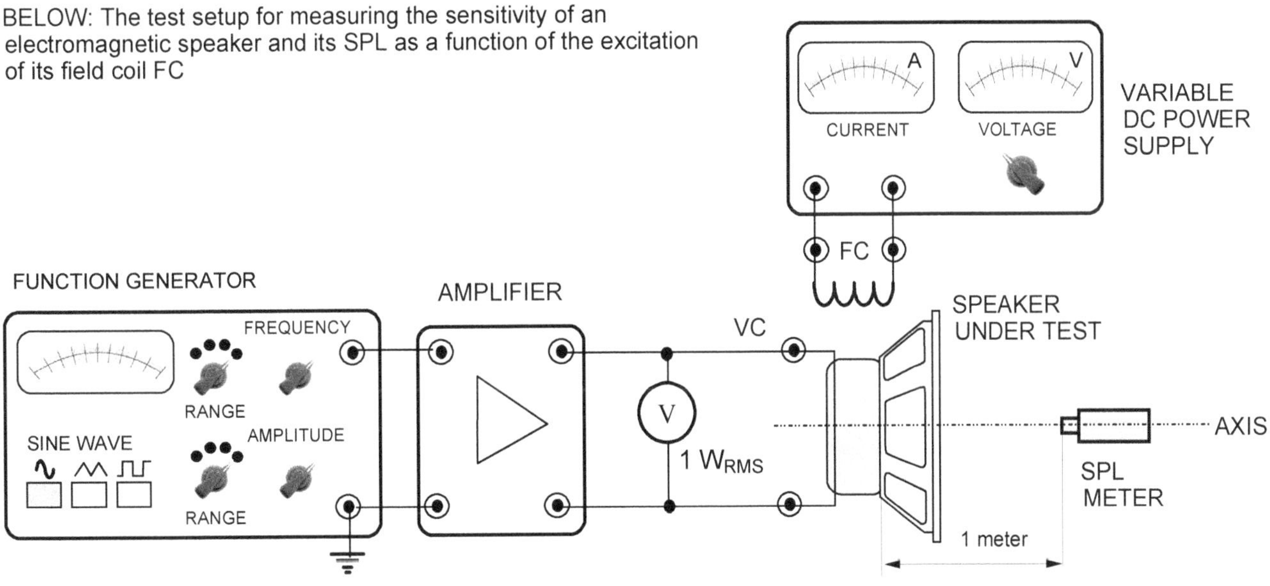

The field coil would be used as a filtering choke in the amp's power supply, resulting in substantial savings. Chokes and power supplies were and still are relatively expensive.

The Motorola driver had a DCR (DC resistance of the field coil) of 700Ω, and assuming 70mA through the coil (that's the order of magnitude those old radios were pulling from their power supplies), that would require an excitation voltage of $V_{DC} = I_{DC} * R_{DC} = 0.07*700 = 49$ V.

So, a DC power supply of the standard 1A current capacity would be fine, but most such low voltage bench power supplies only go up to 20, 25 or 30V_{DC}, you'll need one that can supply up to at least 50V_{DC}!

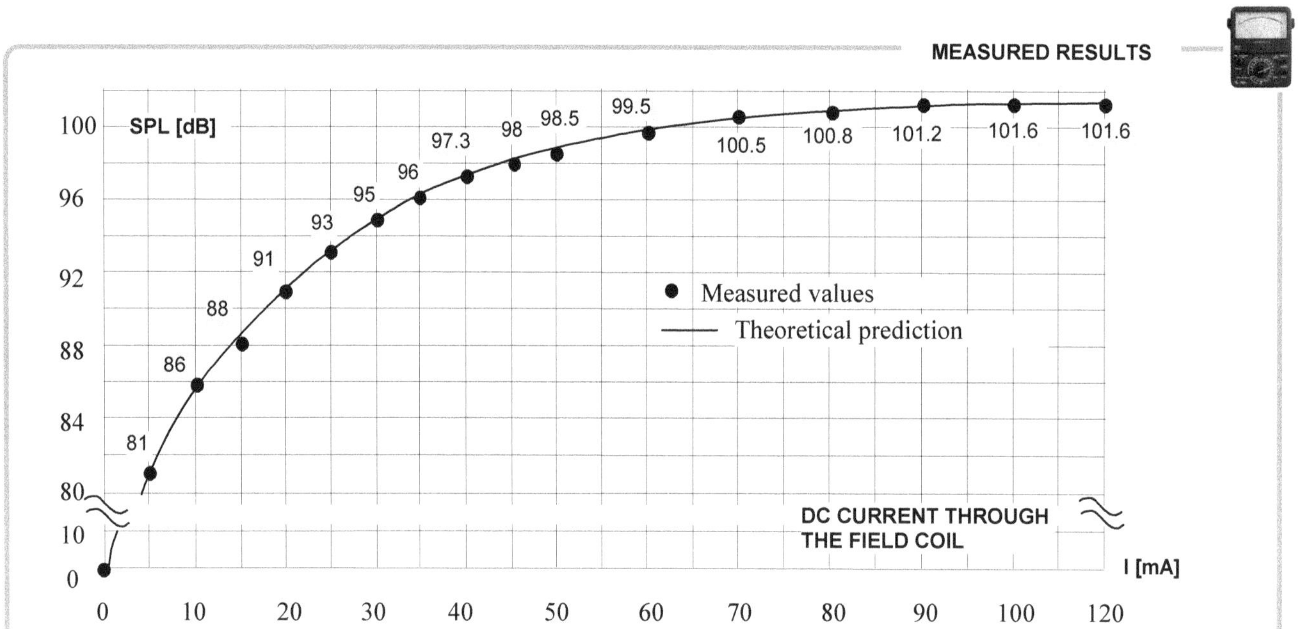

ABOVE: SPL of the 8" Motorola electrodynamic speaker as a function of the DC current through its field coil, measured at 0.6m distance

Some dual-channel DC power supplies have a master-slave or tracking option, and their two outputs can be connected in series, so a 2x30V power supply could provide a 0-60 V single output.

The graph shows the results of our experiment. The horizontal scale (field coil current) is linear, and the vertical scale (SPL in decibels) also seems linear, but that is not the case since dB is a logarithmic unit.

The magnetic field strength and SPL (loudness) increases linearly with the field current, but such a linear curve on a graph with a vertical dB scale has a logarithmic shape.

Even at a very low excitation level of, say, 5mA (the first point measured), the SPL had already reached 81 dB! Doubling the current to 10mA increased the SPL by 5dB to 86 dB. The maximum SPL was around 101.6 dB, reached with 95mA of field current. Further increase in current did not increase the output loudness.

Of course, you need to be careful not to exceed the nominal excitation current too much and for too long since such a high current may burn the field coil out. However, it only took a few seconds to measure each test point, so even at 120 mA, there was no overheating or smoke from the field coil.

The first conclusion is that by replacing a permanent magnet speaker in your amp with an electrodynamic speaker of higher sensitivity, you can raise the loudness of your amp. When we replaced its original, PM speaker with this Motorola electrodynamic driver, the SPL of the Legacy combo amp increased 3.6 dB (from 98 dB to 101.6 dB).

Since power in dB is $P = 10\log(P_1/P_2)$, we get $P_1/P_2 = 10(P/10) = 10(3.6/10) = 100.36 = 2.3$ times! To get the same loudness from the original Legacy PM speaker, you'd need to increase the power of that amp 2.3 times (230%).

Since that was a small, 5 watt single-ended EL84 amp, with the replacement Motorola speaker at its maximum field coil excitation, that little amp would be as loud as a 5*2.3 = 11.5 watt amp!

The first benefit of replacing the original PM speaker in your amp or cab with a field coil speaker is higher output SPL or "loudness," which is what ultimately matters.

The second benefit is that, using variable excitation to such a speaker, an SPL range of around -20 dB is achievable, a significant reduction in loudness.

Since power in dB is $P=10\log(P_1/P_2)$, we get $P_1/P_2=10(P/10) = 10(20/10) = 10^2 = 100$. This means you could effectively attenuate your 5 watt amp down to a 0.05W level or a 100W Marshall amp down to 1 watt!

THE WAY IT USED TO BE: THE WONDERFUL WORLD OF VINTAGE DRIVERS

Why even consider vintage speakers?

The prices of currently-produced "boutique" guitar speakers, especially those of Alnico persuasion, seem to be rising steadily. List prices of US$300-500 are the norm. For instance, Celestion Blue 12" 15 watt Alnico speaker sells for around US$350.-

No wonder musicians see vintage speakers a viable alternative, especially the high-end models made by reputable USA, UK and Japanese speaker makers such as Goodmans, Jensen (the original USA-based firm, not the current Italian one), Utah, CTS, Pioneer, Oxford, Altec Lansing, JBL, Magnavox, Philips, Peerless, and the like.

Lots of audiophiles and guitar players claim that vintage vacuum tubes, produced in the USA, UK, Germany, France, Holland and other mostly European countries, are superior sounding and longer lasting (more reliable) than the currently produced ones from China, Russia and Slovak republic. By the same token, many also claim that vintage speaker drivers sound better than the currently produced ones.

However, time is not loudspeakers' friend. Providing their glass isn't porous and doesn't allow any air inside, vacuum tubes don't age sitting on shelves. The same cannot be said about speakers.

Paper cones are hygroscopic - they attract moisture from the air, lose their stiffness, not to mention mold that can develop. If they sit for years in one position, both cones and their suspensions (spiders) get deformed and the voice coils may start rubbing against the magnet structure.

The glue or cement that attaches the voice coil to the cone is probably the weakest point of vintage speakers. The quality of such adhesives has markedly improved in the last three to four decades, the vintage formulations simply weren't made to last sixty or seventy years. Drive your vintage speaker too hard and the voice coil will separate from the paper cone sooner rather than later.

Speakers with rubber and foam surroundings are especially vulnerable. Likewise, their outside suspensions dry out and crumble away. Of course, most speakers can be re-coned by experts, but that is usually cost-prohibitive - you may as well buy a brand new, currently-produced speaker with warranty.

Bolting two drivers together

Although it seems crude and primitive in comparison with the more complex driver designs that follow on the next few pages, simply placing a tweeter or small speaker (as pictured) in front of the woofer ensures both drivers have the same acoustic center. This avoids the common ailment of 2- and 3-way speaker designs where the drivers are vertically displaced along the height of the enclosure, namely incoherence.

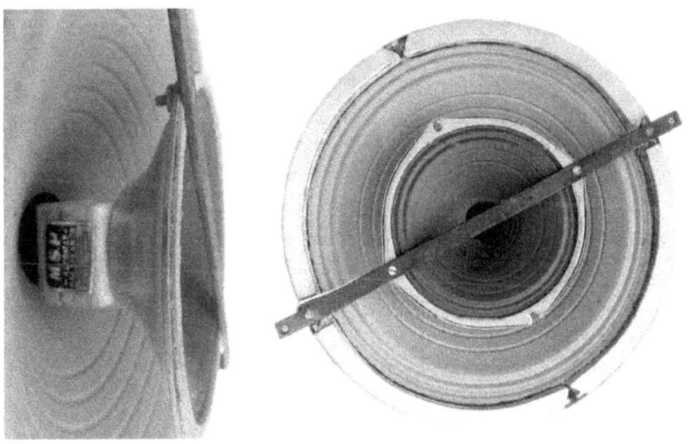

However, no matter how small the tweeter is, it is still an obstacle in front of the woofer's cone, and electronic crossovers are needed, thus increasing complexity.

Dual cone single motor drivers ("whizzer cone")

A "whizzer", auxiliary or high frequency cone is attached to the voice coil former around the same perimeter as the main, much larger cone, but continues to vibrate and produce sound long after the larger and heavier main cone has stopped producing any discernible high frequency output.

Since no modifications to the magnet system or the voice coil are required, and there is no need for electrical crossovers, the whizzer cone approach to boosting high frequencies is the simplest and cheapest of all.

15" Goodmans Audiom 15P twin-cone driver

We salvaged a British-made Goodmans 15" full range Audiom 15-P driver from a Kory cabinet, an Australian take on Leslie rotating speakers of the mid- to late-1970s. Due to its stellar reputation, the driver itself can fetch almost the same price as the whole Kory Leslie amplifier/cab combo with that speaker in it.

So, we substituted a modern 15" driver and made the Goodmans driver available for sale on its own, thus almost doubling the profits from the repair & refurbishment of the Kory Leslie amp.

Although originally recommended for PA and organ amps/cabs, the speaker sounds insane in both guitar and bass cabinets.

Notice a relatively low resonant frequency of 56 Hz (much lower than most currently produced models in the 85-105 Hz range), which means that much lower bass notes can be achieved.

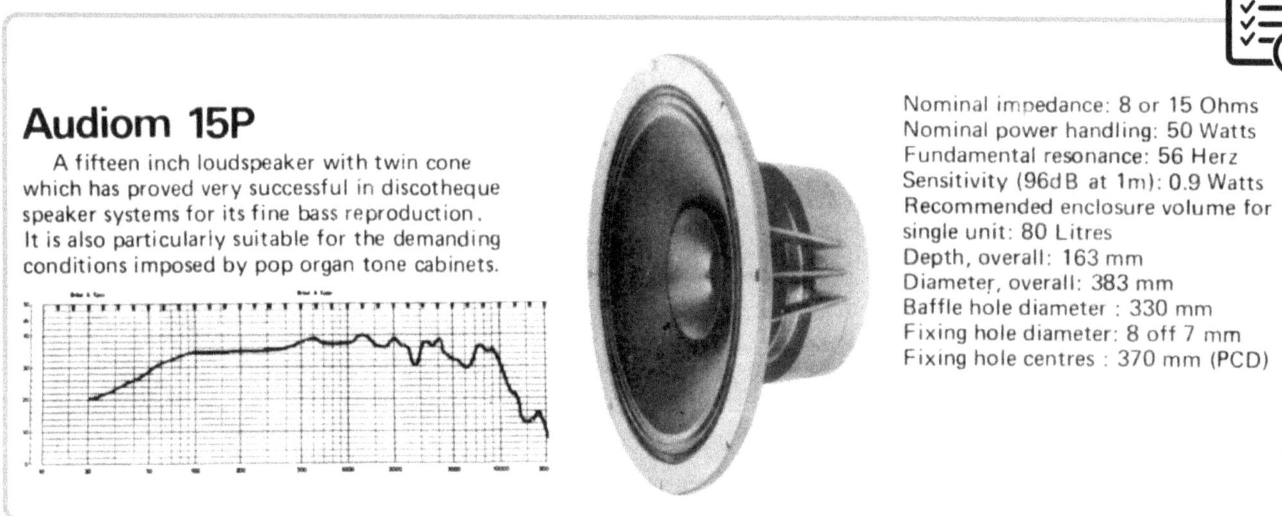

Audiom 15P

A fifteen inch loudspeaker with twin cone which has proved very successful in discotheque speaker systems for its fine bass reproduction. It is also particularly suitable for the demanding conditions imposed by pop organ tone cabinets.

Nominal impedance: 8 or 15 Ohms
Nominal power handling: 50 Watts
Fundamental resonance: 56 Herz
Sensitivity (96dB at 1m): 0.9 Watts
Recommended enclosure volume for single unit: 80 Litres
Depth, overall: 163 mm
Diameter, overall: 383 mm
Baffle hole diameter : 330 mm
Fixing hole diameter: 8 off 7 mm
Fixing hole centres : 370 mm (PCD)

The redrawn SPL-vs.-frequency characteristic (published by Goodmans) shows a flat response from 90 to 300 Hz (1), followed by a remarkably uniform response all the way up to about 10 kHz (2), with a +/-5 dB variations (peaks and dips).

The -3dB frequency bandwidth is from 70 to 10,000 Hz.

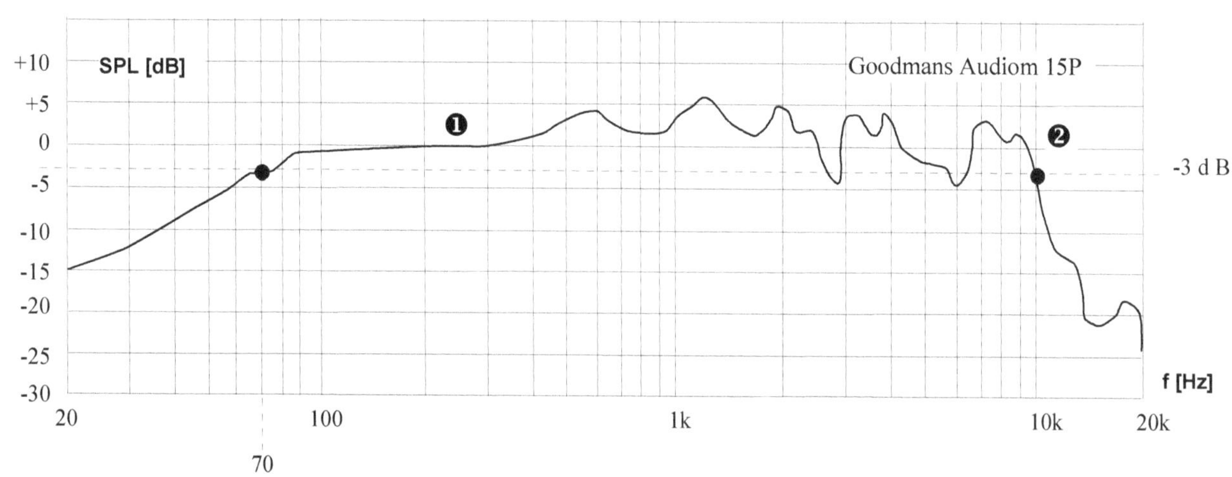

ABOVE: The SPL response of Goodmans Audiom 15P driver

Biflex designs

Altec Lansing Corporation patented their Biflex cone design in 1958 (U.S. patent 2834424).

The idea was to achieve the performance of a double-cone loudspeaker with a single cone by effectively splitting the cone into two sections. At bass and lower midrange frequencies, the two sections work in unison (as one). At a crossover frequency, the larger outer section decouples (mechanically and acoustically) from the smaller inner section, which continues to produce higher frequencies.

This mechanical "crossover" is achieved by introducing a compliance in the cone's middle section. Altec used a double corrugation, whose distance from the center of the cone determined the decoupling or crossover frequency. However, that wasn't enough.

A significant dip in the SPL curve (caused by the compliance acting as an acoustical capacitance) was objectionable. The solution was found in applying a layer of viscous damping liquid to both sides of the cone at the corrugation region.

When dried, it became a viscous-plastic layer which introduced a resistive element and substantially reduced the magnitude of the dip in the frequency characteristic, without substantially increasing the mass of the cone.

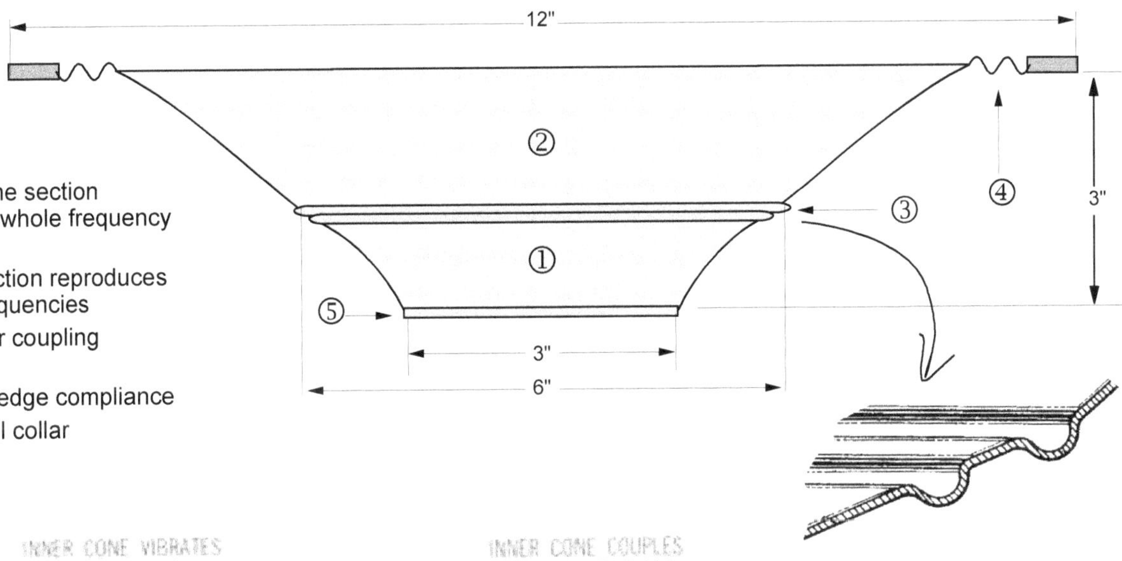

1. The inner cone section reproduces the whole frequency range
2. The outer section reproduces only the low frequencies
3. The middle or coupling compliance
4. The outer or edge compliance
5. The voice coil collar

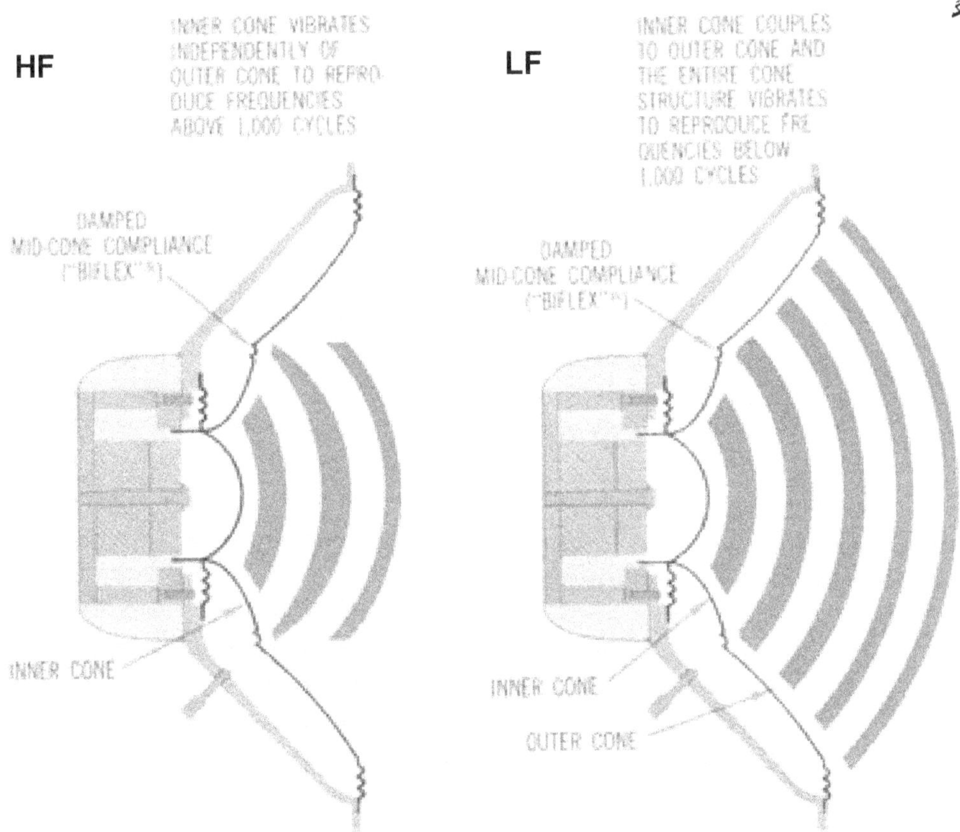

Some designs discussed in this section were abandoned for the complexity and cost reasons - they were simply too expensive compared to the alternatives (two or three standard drivers with cheap crossovers). Biflex designs didn't involve complex magnet profiles or expensive machining of close tolerance parts, but the middle or coupling compliance proved troublesome.

The main issue was repeatability and consistency. Just as manually-assembled vacuum tubes (electronic valves) suffer from a wide dispersion of their parameters, so did those pesky mechanical interfaces. As always, the money monkeys won over engineers, and the concepts was abandoned.

12" vintage Altec Biflex against modern Celestion speaker

Comparing a frequency response of a vintage "full range" high-fidelity speaker with a modern guitar amp speaker may seem futile, but such vintage speakers often come up for sale, so they are available for your DIY project if you feel inclined to try them out. Thousands were sold in the 1950s and 60s, either in commercial hi-fi speaker cabinets or as bare drivers to amateur speaker builders of the era.

As soon as I superimposed the two SPL curves onto the graph above, I noticed a striking similarity in the response of the two speakers in the 120 Hz (1) to 1 kHz (7) range. However, this is where any similarity ends.

If, for the sake of simplicity, we consider the bandwidth using -5dB points (instead of the customary -3dB drop in the SPL), the nominal frequency range of Altec 412A ranges from about 62 Hz (2) to 12 kHz (3). Celestion G12T100 only manages 100 Hz (4) to 6 kHz (5).

The Celestion driver has a pronounced dip at 1,300 Hz (8), followed by a peak at 2,100 Hz (9), while the response of the Altec 412A is incredibly smooth in that critical midrange (1-4 kHz), where the human ear is most sensitive.

Altec's bass between 40 and 100 Hz will be 3-5 dB louder, something of little consequence if used in a guitar cab, but very valuable if used as a hi-fi or bass guitar speaker. Also, Altec's top end (above 6 kHz) will be sparkly and detailed, while Celestion will reproduce only a whisper around 12 kHz, the best it could manage is to come within 10dB of the Altec's response (3).

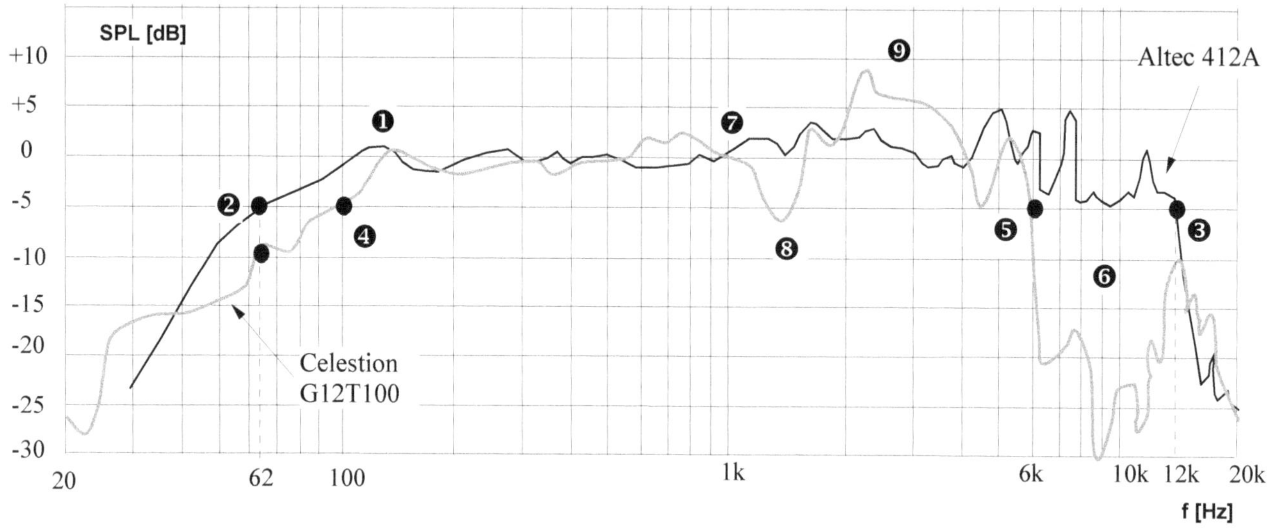

ABOVE: Frequency response SPL curves of Altec 412A, a vintage "full range" high-fidelity speaker, compared to the modern Celestion G12T100 guitar speaker

Mechanical 2-way drivers (dual cone, single voice coil)

At the first glance, this interesting approach, used only by the Fukuin Electric Works Ltd., a.k.a. "Pioneer", seems to be a hybrid between a simple whizzer cone and the Altec's Biflex design. It is a much more elaborate and thus interesting design.

Two separate cones are attached to the same voice coil, but the outer edge of the smaller, tweeter cone is attached to the midsection of the larger woofer cone with a corrugated edge. The apex of the woofer cone, behind the tweeter cone, has a number of oval openings or apertures, which work as a mechanical crossover. Two of them are visible on the photo (next page).

Below the crossover frequency both cones move in unison, but above it, the larger cone does not move due to the pressure equalizing action of those apertures.

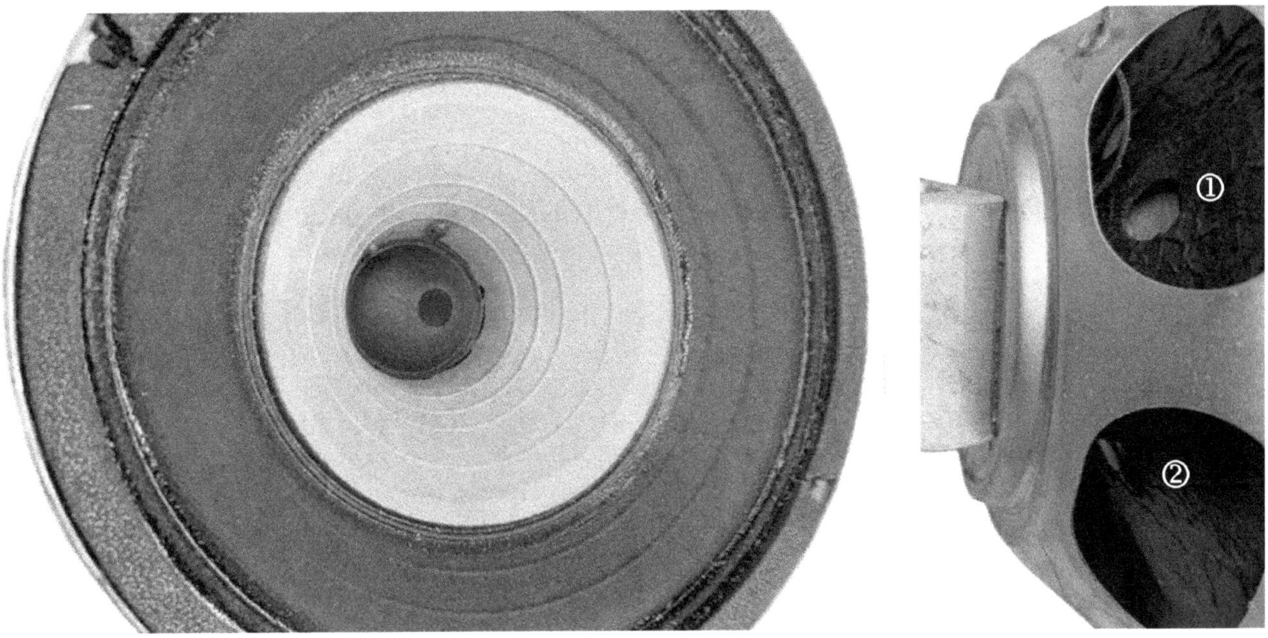

ABOVE: Pioneer PIM8L mechanical 2-way Alnico speakers (8", 6W, 16Ω). It may look like a Biflex design, but the darker LF cone does extend all the way to the voice coil underneath the lighter-colored HF cone, which can be seen from the side through the oval pressure-equalizing holes (1) and (2).

BELOW: The SPL curve of Pioneer PIM-8 8" speaker, with f_R between 55 and 70 Hz and the incredibly high declared sensitivity of 99.5 dB/W! The SPL curve hovers above 100 dB between 100Hz and 10kHz, so this figure is very conservative.

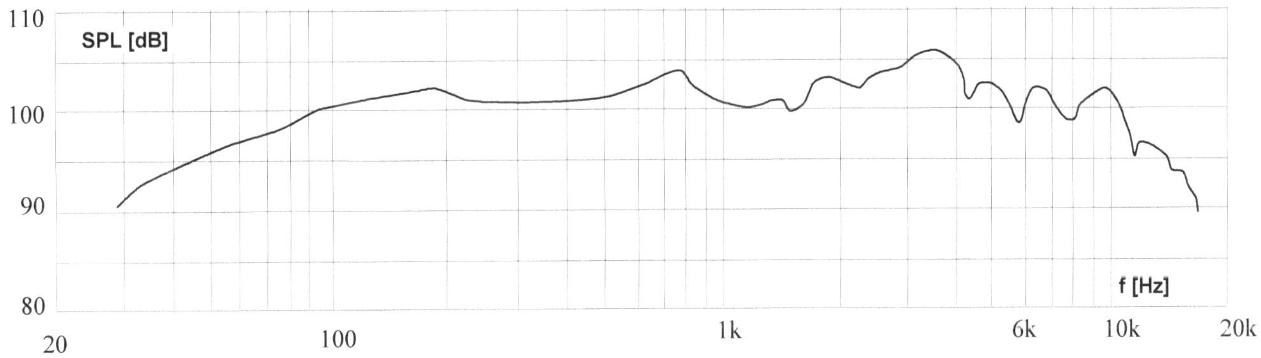

The mechanical crossover and coaxially mounted cones improve the distribution pattern of the speaker at all frequencies, while the fiberglass used in the cone paper mixture helped achieve a smooth frequency response, free from the significant peaks or dips that plague most other speakers.

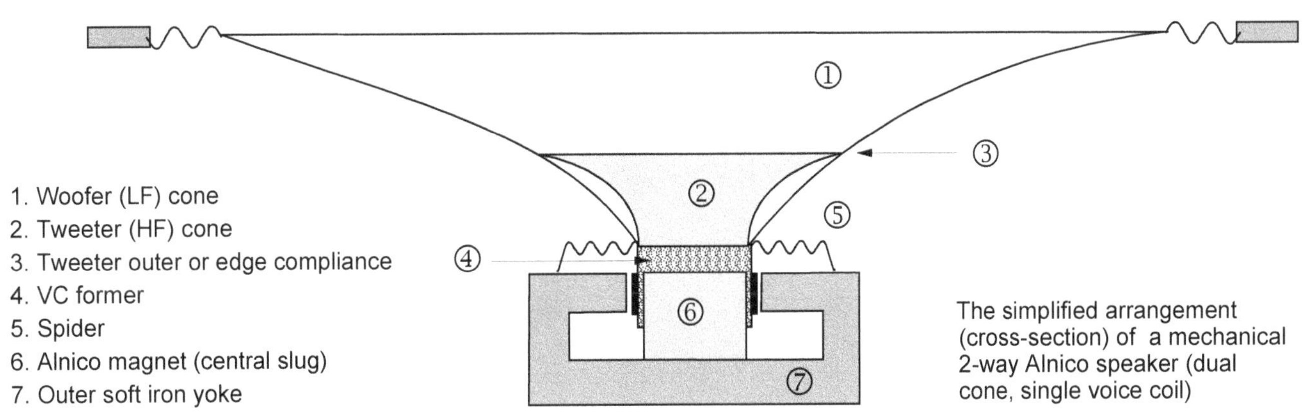

ABOVE: The simplified arrangement (cross-section) of a mechanical 2-way Alnico speaker (dual cone, single voice coil)

Dual cone dual voice-coil drivers

Dr. Harry Ferdinand Olson, (1901 – 1982), was a true luminary of the acoustical engineering field. Born in the USA of Swedish immigrant parents, Dr. Olson was the author of ten books and over a hundred patents. As the Acoustic Research Director of RCA Laboratories, Olson designed a range of full range drivers, whose performance was truly impressive for the period, late 1940s and early 1950s, and would be considered first class even today.

His duo-cone designs, made and sold by RCA, incorporated two coaxial cones, each driven by its own voice coil. Cheaper models, such as RCA 515S1 used a single magnet to energize the air gaps for both VCs, with a magnetic bridge arrangement to divert the appropriate amount of flux to each of the air gaps. The more elaborate and expensive drivers, such as RCA LC-1A, had two separate magnets for the LF and HF motors.

In 515S1, the outside suspension of the small cone is attached to the large cone, with a mechanical network that prevents interaction between the two cones. The small cone does not vibrate at the low frequencies because its inner suspension is much stiffer than the outer suspension. Also, the space between the two cones is vented in the large cone to prevent acoustic coupling due to the compliance of this volume of air.

LS-1A had a 2.5"voice coil and a magnet, weighing 4 lbs or 1.8 kg! The whole 15" speaker weighed 21 lbs (10 kg)! Its frequency response was specified as 25 Hz - 16 kHz with a very conservative 40 watt power rating. The resonant frequency of the LF cone was an incredibly low 22 Hz, meaning LS-1A was a high compliance design.

1. Woofer (LF) cone
2. Woofer spider
3. Woofer voice coil
4. Woofer VC former
5. Tweeter (HF) cone
6. Tweeter outer or edge compliance
7. Tweeter voice coil
8. Tweeter VC former
9. Alnico magnet (central slug)

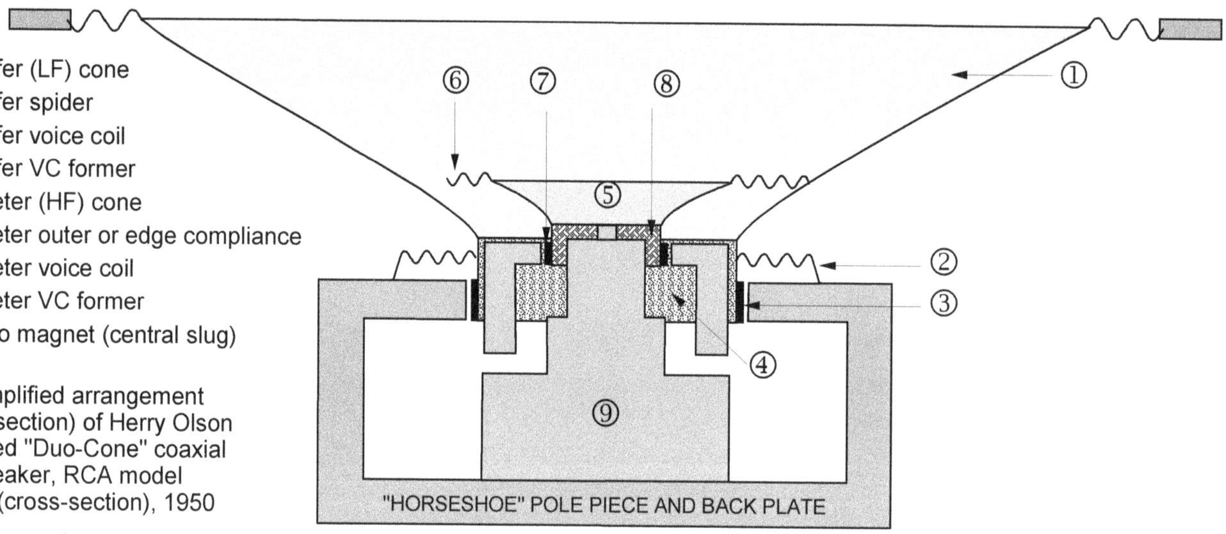

The simplified arrangement (cross-section) of Herry Olson designed "Duo-Cone" coaxial loudspeaker, RCA model 515S1 (cross-section), 1950

515S1 had a narrower, 40 Hz - 15 kHz frequency range .

The 1,600 Hz inbuilt electronic crossover was a single 4μF capacitor in series with the HF coil to reduce its current at low frequencies. The LF voice coil had such a high inductance, that its reactance was more than sufficient to adequately limit its current at the high frequencies without the need for an addition series inductance, as is often the case in such first order crossovers.

LS-1A achieved a wide dispersion angle (120°), among other aspects of its design, by gluing oval domes to its larger cone. From the RCA promotional blurb:

"The low-frequency cone is stiffened with seven acoustical domes which, because of their shape and relative location, contribute to the wide dispersion of the high frequencies from the tweeter, at the center, by spoiling the symmetry. This eliminates the interference normally characteristic of such shapes without loss of either high or low frequencies."

ABOVE: RCA LS-1A had seven oval-shaped acoustical domes glued to its low frequency cone

The best coaxial drivers adopted a complex and thus expensive "thru-bore" construction. The pole piece (6) of the low frequency magnetic structure was shaped to act as a high frequency horn, in addition to the LF cone (9) which continued the same contour, thus maximizing the length of the tweeter "horn".

The longer horn length enabled the lowering of the crossover frequency down to around 2,000 Hz.

Some models also included an acoustic lens in front of the tweeter horn, which helped widen the HF dispersion and reduce beaming.

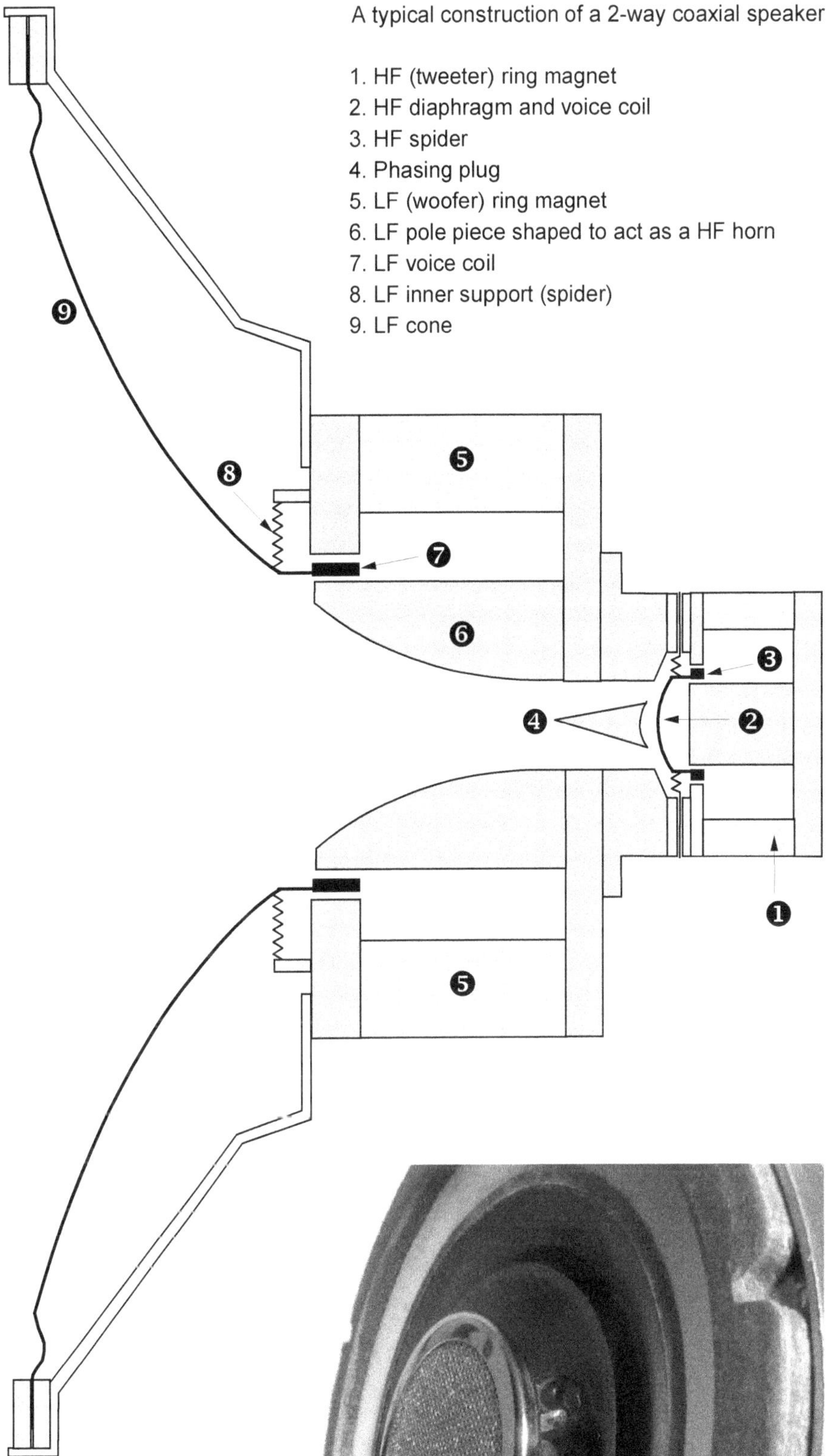

A typical construction of a 2-way coaxial speaker

1. HF (tweeter) ring magnet
2. HF diaphragm and voice coil
3. HF spider
4. Phasing plug
5. LF (woofer) ring magnet
6. LF pole piece shaped to act as a HF horn
7. LF voice coil
8. LF inner support (spider)
9. LF cone

BELOW: An aluminium acoustic diffuser by Jensen

RIGHT: A close-up of a vintage full range speaker by Utah Electronics, 8" 2-way coaxial with an acoustic lens, frequency range 35-19,000 Hz, 55 Hz resonant frequency, 10 oz. magnet with 1" voice coil, rated at 30 watts

Elliptical (oval) speakers

While round cones dominate the guitar and hi-fi speaker cabs, oval speakers rule the car audio world, even to this very day. They are still being produced in smaller (car audio) sizes, 3x5", 4x6" and 5x7". But, just as hi-fi speakers, all have rubber surrounds and are mostly of 2-way or 3-way construction, unsuitable for guitar cab duty.

The vintage oval speaker of interest to us here are bigger ones, 6x9", especially those with Alnico magnets. Since most predate the dawn of the ceramic magnet age, they are invariably Alnico beauties and have cloth or paper surrounds.

Musical instrument brands such as Lowrey, Rodgers, Conn and Hammond used them as midrange drivers in their organs, and since those heavy and bulky beasts are superfluous these days, most such organs can be had for free, so keep an eye on your local online marketplace ads.

For instance, 1970s Rodgers church organs used ten oval drivers in each of their "flat" M13 panels, with four such panels, a whopping 40 drivers in one installation! The ideas was to create a "wall of sound" and wide dispersion throughout the hall or church. Obviously, ten closely placed speakers also benefited from a significantly increased sensitivity due to the mutual acoustical coupling effect (see page 128), and achieved higher SPL levels.

We have already explained why a cube (three equal sides) is the worst shape of a speaker cabinet. Speakers with a round cone profile suffer from the same problem, but in two dimensions, just like a speaker mounted in the center of a square baffle.

Both distances are equal (center-to-top & bottom and center-to-left & right), and the standing waves occur at the same frequency, thus doubling their effect.

With oval speakers that si not the case, their length and width are very different and so are the standing waves frequencies, which are now better distributed along the frequency spectrum, and thus less pronounced or objectionable .

Unless you remember your high school algebra, calculating the cone area of an ellipse isn't easy. Just keep in mind that the effective (radiating) area of four 6"x 9" drivers is way bigger than that of a 12 incher, it almost equals the radiating area of a 15" driver!

Furthermore, in terms of their high frequency extension, the performance of oval speakers is way superior, as is their transient response.

Even without any whizzer cones or added tweeters, they easily reach 14 kHz, something even 8" round drivers can only dream about , not to mention the 12" or 15" fatsos!

BTW, do speakers dream, and if they do, what do they dream about? Nice, sexy amps to drive them, perhaps?

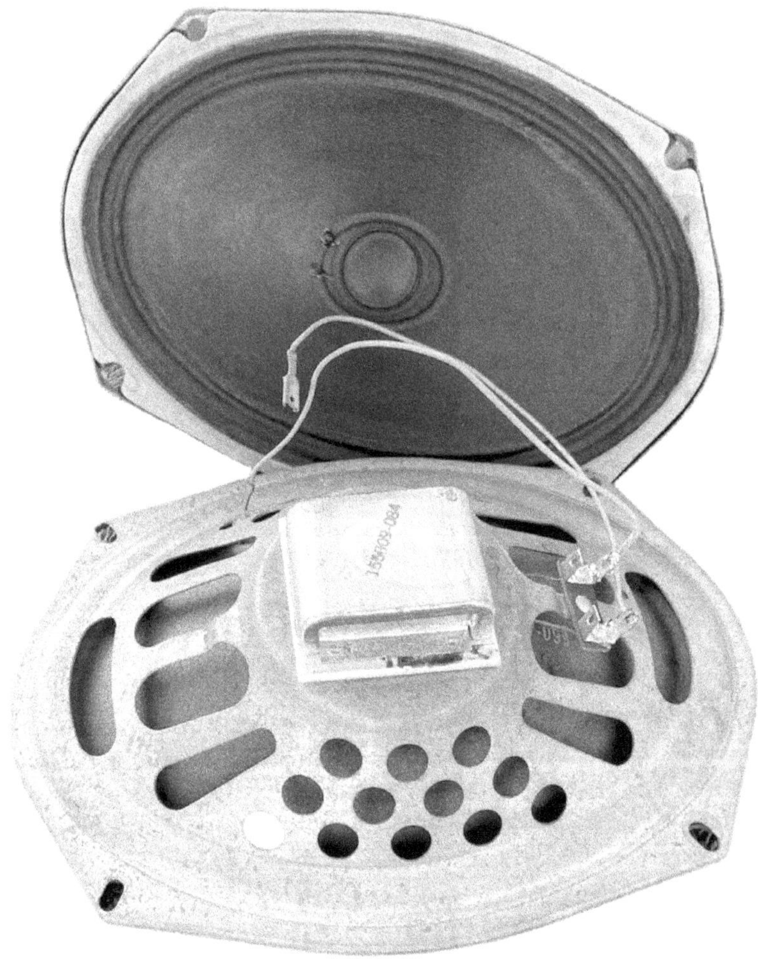

ABOVE: A pair of vintage USA-made Stromberg-Carlson 6"x9" drivers (circa 1962, my vintage) in incredible condition. See how we used in a DIY cab modification project on page 199.

DECIPHERING SPEAKERS' SPECIFICATIONS AND PARAMETERS

3

A speaker (or "driver") is a complex electromechanical and acoustical transducer, and so are their parameters.

However, don't despair, you don't need a 4-year degree in physics or electronics to understand their specifications! Dr. Iggy will try to explain it all here, in as few pages as humanly possible.

Beware, though, this chapter is maths-laden, and will get pretty heavy going, so don't blame me afterwards. Remember, things can be simplified only up to a certain point, but no further!

- THIELE-SMALL AND OTHER DRIVER PARAMETERS
- SENSITIVITY AND EFFICIENCY
- HOW TO MEASURE, CALCULATE OR ESTIMATE SPEAKER'S BASIC PARAMETERS
- ELECTRICAL EQUIVALENT MODELS OF DYNAMIC SPEAKERS (DRIVERS)
- THE SAME MODEL SPEAKERS BUT OF DIFFERENT NOMINAL IMPEDANCE - HOW IS IT DONE, AND ARE THERE ANY SONIC DIFFERENCES BETWEEN THEM?

THIELE-SMALL AND OTHER DRIVER PARAMETERS

In the early 1970s several research papers by Neville Thiele and Richard Small were published in the Journal of the AES (Audio Engineering Society). Thiele and Small were the first to systematize the speaker parameters in order for designers to be able to calculate the end results relatively easily, saving them hours of trial-and-error attempts to optimize the interaction between the driver and the enclosure.

These ten or so parameters since then became known as Thiele-Small (TS) parameters, and are measured and calculated by speaker manufacturers, and published in speakers' data sheets. As a practical example, we will use the current (2024) production data for Jensen P12R, a 12" Alnico guitar speaker, made in Italy.

CASE STUDY: Jensen P12R driver

Both the 8Ω and 16Ω versions share the basic characteristics and parameters: nominal VC diameter 25 mm (0.98"), surface area of the cone S_D = 491 cm^2, magnet weight of 200 g (7 oz), overall weight (1.5 kg or 3.31 lbs), magnetic flux density of B=0.96 T, VC winding depth (8 mm, 0.31"), and magnetic gap depth (6mm or 0.24").

A copper voice coil is wound on a Kapton former, attached to a paper cone with an untreated integral solid paper surround. The rated power is 25 watts, with "musical" power rating of 50 watts.

PARAMETER	SYMBOL	8 Ω	16 Ω	UNIT
Voice coil DC resistance	R_E	6.05	12.0	ohms
Resonance frequency	f_S	80	77	Hz
Mechanical Q factor	Q_{MS}	13.07	12.53	
Electrical Q factor	Q_{ES}	2.67	2.67	
Total Q factor	Q_{TS}	2.1	2.2	
Moving mass	M_{MS}	25.2	19.9	g
Mechanical compliance	C_{MS}	0.157	0.214	mm/N
Force factor	$B\ell$	5.53	6.68	Wb/m
Equivalent acoustic volume	V_{AS}	54	72.6	liters
Maximum linear displacement	X_{MAX}	+/-1	+/-0.75	mm
Reference efficiency	η_0	1.56	1.19	%
Losses electrical resistance	R_{ES}	39	58	ohms
Voice coil inductance @ 1kHz	L_E	0.5	0.84	mH
Sensitivity @1W, 1m		93.7	94.0	dB

f_S - the resonant frequency of the driver

For an electrical oscillator the resonant frequency depends on the inductance L and capacitance C:

$$f_R = \frac{1}{2\pi\sqrt{LC}}$$

A mass M_{MS} suspended on an ideal (frictionless and mass-less) spring with stiffness S (or it reciprocal, compliance or elasticity C_{MS}=1/S) is a mechanical resonant circuit, and its resonant frequency is:

$$f_S = \frac{1}{2\pi\sqrt{M_{MS}/S}} = \frac{1}{2\pi\sqrt{M_{MS}C_{MS}}}$$

It is the same formula, from which it is obvious that the mechanical equivalent of electrical capacitance C is compliance C_{MS}, and that the mass M_{MS} is the mechanical equivalent of inductance L.

Instead of f_R, f_S is used in TS parameters, with "S" for speaker and not stiffness, so don't get confused.

The mass M_{MS} is in kg (kilograms), the unit for compliance C_{MS} is meter/Newtons, and frequency is in Hz (hertz).

The moving part of the loudspeaker motor (the cone+dust cap+voice coil) have a total mass M_{MS}, while the inner and outer suspension act as a spring with stiffness S or compliance C, so the same formula applies to this resonant system.

At the resonant frequency the weight of the moving parts becomes balanced with the restraining motional force of the suspension. At that frequency the counter-EMF (voltage) induced in the driver is at its maximum, and so is the driver's impedance, resulting in the characteristic peak in its impedance curve.

Quick check: for the 16Ω Jensen P12R speaker we enter 0.0199 kg for M_{MS} and 0.214*10^{-3} m/N for C_{MS} and get f_S=77.12 Hz, as declared.

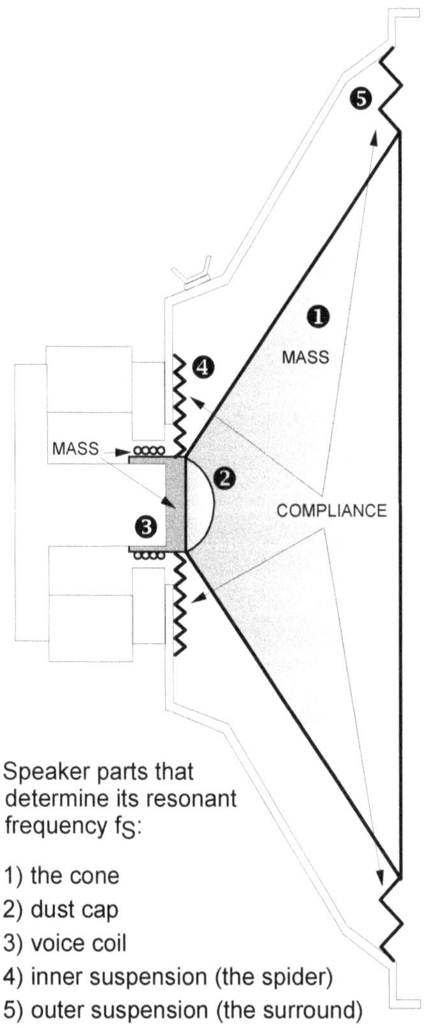

Speaker parts that determine its resonant frequency f_S:

1) the cone
2) dust cap
3) voice coil
4) inner suspension (the spider)
5) outer suspension (the surround)

M_{MS} - mechanical mass of the VCA and air load

This mechanical parameter is the sum of the voice coil assembly's (VCA) mass, which is the VC mass plus the cone mass (all moving parts), and the air load. The "air load" or "driver radiation mass load" is the air mass the cone has to displace (push). M_{AS} is its acoustic equivalent: $M_{MS}=M_{AS}S_D^2$ (S_D is the surface area of the cone).

C_{MS} - mechanical compliance of driver's suspension

This mechanical parameter is a measurement of the stiffness (or its inverse, compliance) of the driver's suspension. Measured in meters per Newton (the unit for force), you can understand C_{MS} as the force required to push the speaker's cone a certain distance, while overcoming the stiffness of the suspension, which, just like a spring you are trying to compress, is resisting such a push.

The compliance of the suspension is inversely related to the speaker's resonant frequency: a higher compliance (C_{MS}) means lower f_S, which is usually desirable. However, there is a downside and thus a limit to how far compliance should be reduced - a more compliant suspension also has less ability to prevent undesirable lateral movements.

C_{AS} is its acoustic equivalent: $C_{MS}=C_{AS}/S_D^2$ (S_D is the surface area of the cone).

V_{AS} - driver's acoustic compliance expressed as "an equivalent volume of air"

Instead of dealing with and measuring C_{MS} or C_{AS} directly, speaker dudes prefer to use "Compliance Equivalent Volume", the volume of air that when compressed to one cubic meter exerts the same force as the driver's compliance C_{MS}. Since air (a combination of gasses) pressure changes with both humidity and temperature, V_{AS} is difficult to measure accurately.

V_{AS} can also be defined as the volume of air in a sealed enclosure that will raise the resonant frequency of the boxed driver to a value of 41% higher than its free air value, $f_E=1.41f_S$.

$V_{AS}=\rho*v^2*S_D^2C_{MS}$ where $\rho=1.18$ kg/m^3 (the density of air), v=344 m/s (the speed of sound), and S_D is the effective radiating area of the cone. We can write $V_{AS}=1.4*10^5*S_D^2C_{MS}$ [m^3] or $V_{AS}=1.4*10^8*S_D^2C_{MS}$ [liters]. For the 16Ω Jensen P12R speaker we enter $S_D=491\text{cm}^2=491*10^{-4}$ m^2 and $C_{MS}=0.214*10^{-3}$ m/N and get $V_{AS}=0.07223$ m^3 or 72.23 liters.

R_E - the DC resistance of the driver

When you take your trusty multimeter out and measure the DC resistance of a speaker, this is the parameter you are measuring. Since writing subscripts is awkward, R_E ("E" for electrical, since there is also a mechanical and acoustical resistance) is also referred to as '"DCR", an acronym for Direct Current Resistance.

Keep in mind that both the nominal speaker's impedance (proclaimed by the manufacturer) and the actual impedance modulus at any frequency is always higher than DCR. As we have seen when discussing the impedance - vs.- frequency speaker curves (page 39), the load on your guitar amp's output will never drop to such a low value. Even in the area of almost pure resistive load (frequencies above the resonant peak, after the impedance drops down to the its minimum), that resistance is at least 20% higher than DCR. That is why, for example, 8Ω speakers will usually have DCR in the 6.0-6.3 Ω range (6.05Ω for Jensen P12R).

L_E - the voice coil inductance

This is the figure you will get when you use your LCR-meter to measure the inductance of a speaker's voice coil. Since inductance changes linearly with frequency, the industry standard is to measure it at 1kHz, which is one of the standard test frequencies on small digital LCR meters (usually 120Hz, 1kHz and 10kHz).

Remember, a voice coil is noting but a moving inductor or "choke". It has a DC resistance, and AC reactance which is almost purely inductive, but there is also a small capacitance formed by so many turns wound in a "sandwich".

For Jensen P12R L_E is 0.5 mH (milli henry) for the 8Ω version, and 0.84 mH for its 16Ω version. Why is the inductance 0.84/05 = 1.68 or 68% higher? Think about it for a while, we'll get back to that issue soon ...

$B\ell$ - force factor (motor strength)

Since $F=B*I_S*\ell_C$, the product of the flux density (in Teslas) and the total length of the voice coil wire in the magnetic gap (in meters) is an indication of the speaker's motor strength: $B\ell=F/I$. The higher the mechanical force F the speaker produces for a given audio signal current "I", the more powerful the speaker. The unit is Wb/m (Weber per meter).

R_{MS} - mechanical losses of the speaker

R_{MS} is mechanical resistance that models friction losses of speaker's suspension. Measured in kilograms per second (kg/s), R_{MS} is the main factor in determining driver's mechanical Q-factor (Q_{MS}).

R_{ES} is its equivalent in the electrical model, while R_{AS} is its equivalent in the acoustic model. R_{MS} was not specified by Jensen, but R_{ES} was. As we will see soon in the analysis of various speaker models, the relationship is $R_{MS}=(B\ell)^2/R_{ES}$, so for the 16Ω Jensen P12R we have $R_{MS}=(B\ell)^2/R_{ES} = 6.68^2/58 = 0.769$ kg/s.

R_{ES} can be read from the impedance curve, it is the difference of the peak impedance R_{MAX} and the speakers DC resistance R_E: $R_{ES}=R_{MAX}-R_E$.

X_{MAX} - Maximum Linear Excursion (MLE)

When the VC moves beyond X_{MAX} compression sets in and distortion increases. The speaker's output becomes nonlinear because some turns of the VC are outside the magnetic gap. X_{MAX} is thus the maximum linear distance the cone should travel in one direction, a usable excursion limit usually defined by the manufacturer. The usual criterion is the point at which THD (Total harmonic Distortion) of the driver reaches 10%, but other speaker makers use other criteria.

X_{LIM} is a related parameter, the longest distance the cone can travel without the driver suffering a mechanical or structural failure. It is the lowest (shortest) of the four physical travel limits: VC hitting the back plate, spider hitting the top plate, VC completely jumping out of the gap, and the physical (structural) limitation of the cone.

V_D - Peak volume displacement

The peak volume displacement is the product of X_{MAX} and S_D, the effective piston area of the cone or diaphragm: $V_D = X_{MAX}S_D$.

EBP - the Efficiency Bandwidth Product

Calculated as a ratio of the resonant frequency f_S and the electrical Q-factor, (EBP = f_S/Q_{ES}), and measured in Hz (Hertz), EBP is used as a rough ("rule of thumb") indicator of speaker's suitability for use in vented (bass reflex) or sealed enclosures.

Speakers with EBP closer to 50Hz (or less) are more suitable for a closed box use, while those with EBP towards a figure of 100Hz (or above) are better suited for vented enclosures. Obviously, with a careful designs, drivers with EBP in the 50-100 Hz range can perform well in both cab types.

Q-factors - Q_{MS}, Q_{ES}, and Q_{TS}

The Q-factors ("Q" for quality) are indicators of speakers damping or how well the cone is controlled around the resonant frequency. Both mechanical and electrical damping (breaking) are present in a dynamic speaker. The mechanical damping is like trying top slow down by slamming the breaks on your moving car, while the electrical breaking would be analogous to shifting the gear down so the engine and the transmission (gear box) provide an additional load on the driving wheels.

The speakers suspension (the spider and the surround) provide mechanical damping or breaking, while its "motor" (the moving part, VC and the cone) acts as an electrical brake.

Q_{MS} is the mechanical Q-factor due to the damping made by the speaker's inner and outer suspension- the spider and the surround. A higher Q_{MS} indicates lower mechanical losses, resulting in a higher resonant impedance peak.

The electrical Q or Q_{ES} , describes the electrical damping or breaking made by the VC – magnet assembly (the moving part of speaker's "motor") at the resonant frequency. A voltage or CEMF is induced in the voice coil moving through the magnetic field, resulting in a current which opposes such voltage and acts as a brake, opposing such motion.

Q_{TS} is the total Q. It is calculated the same way we calculate the resistance of two resistors in parallel, we add their reciprocal values (1/R): $1/Q_{TS} = 1/Q_{MS} + 1/Q_{ES}$.

Expressing Q_{TS} from there we get $Q_{TS} = (Q_{ES} \times Q_{MS})/(Q_{ES} + Q_{MS})$.

Q_{MS} is always higher than Q_{ES}, which means that the total Q or Q_{TS} is always lower than Q_{ES}! It also means that Q_{ES} or electrical breaking is the dominant factor in voice coil damping.

for the 16Ω Jensen P12R we have Q_{MS}=12.53 and Q_{ES}=2.67, so and $Q_{TS} = (Q_{ES} \times Q_{MS})/(Q_{ES} + Q_{MS})$ = 12.53*2.67/(12.53+2.67) = 33.46/15.2 = 2.2

SENSITIVITY AND EFFICIENCY

Speaker's efficiency versus sensitivity - what is the difference?

Language is a tool to describe concepts, feelings and other aspects of our lives and the world we live in. Just like any other tool, there are proper and improper uses of language; in other words, instead of use with respect, precision and efficiency (no pun intended), language can be misused, misapplied and, the ultimate problem, misunderstood.

The efficiency of a typical speaker driver by itself is extremely low. the order of magnitude is 0.5-1%. That means that between 99% and 99.5% of the electrical power the amplifier feeds into a speaker is wasted into heat in the speaker's voice coil. Only 0.5-1.0% is converted by the speaker into acoustic power (sound pressure changes).

Efficiency: η = POWER OUT/POWER IN

η or "eta" is a Greek letter used for the coefficient of efficiency in physics and engineering

This means that the efficiency of anything (your car's engine, air conditioner, refrigerator, heater, etc.) is always less than 100% - energy out is always lower than energy in!

Sensitivity, on the other hand, is more specific, and, in this case, more useful to us. If a sales dude in a guitar shop told you, pointing to the two cabs you narrowed your choice down to, "This cab is 1% efficient, but this one is 1.5% efficient!", you'd be none wiser.

However, if Dr. Iggy was by your side and said, "The first speaker has 97dB/watt @1m average sensitivity, while the other one's sensitivity figure is 99 dB/watt @1m!", you would immediately know that a 3 dB higher sensitivity means the second speaker cab would sound noticeably louder (assuming the same amp power) than the first!

Did I just write the longest and most convoluted sentence in a nonfiction book?!? Would I get a Nobel Prize for verbosity?

While the two concepts are related - a more efficient speaker will always have a higher sensitivity figure - they are not interchangeable.

Efficiency is about energy in versus energy out, a ratio or percentage (no dimension), while sensitivity is the SPL level in dB (decibels) produced by a speaker (driver itself) or a whole cab with a specific input (either in Volts RMS or Watts) at a specific frequency of the test signal, measured by an SPL meter at a specific distance from the front baffle (usually 1 meter).

How is SPL related to efficiency?

The sound pressure level is logarithmically-related to efficiency: $SPL=112 + 10\log\eta_0$

Expressed the opposite way, the free air reference efficiency at any frequency can be calculated from the speaker's SPL at that frequency: $\eta_0 = 10^{(SPL-112)/10}$

For instance, for a driver producing an SPL of 97dB, the efficiency is $\eta_0 = 10^{(97-112)/10} = 10^{-1.5} = 0.0316$ or 3.16%.

Remember, the free air reference efficiency is a ratio of its output or radiated acoustic power and the input (electrical) power required to produce that output: $\eta_0 = P_{AR}/P_E$.

HOW IS SPL RELATED TO EFFICIENCY?

$SPL=112 + 10\log\eta_0$

$\eta_0 = 10^{(SPL-112)/10}$

η_0 = free air reference efficiency

Sensitivity declared in dB/watt @ 1m versus dB/2.83V @ 1m? Are those speaker dudes fu CENSORED ng around with unsuspecting guitar players?!?

Now that we've clarified that confusion, we are faced with another. Some manufacturers use dB/watt @ 1m, while others use a voltage figure as the input into the speaker or cab - dB/2.83V @ 1m.

First of all, both are equally problematic because neither tells us at which frequency that measurement was taken.

Secondly, where does that awkward number, 2.83 Volts comes from? Why don't those dudes at Celestion, Jensen and Eminence use 1 Volt, a nice and simple figure that even math-challenged (notice I didn't say meth-challenged) musicians would understand?

The clue lies in Ohm's law, V=R*I or I=V/R, and since power is P=V*I, we can get rid of current by substitution and get $P=V^2/R$ (power equals voltage squared divided by resistance)! If, and only if, R=8Ω, for 1 watt of input power into the speaker, the voltage would be $V=\sqrt{(PR)} = \sqrt{(1*8)} = 2.83$ V!

So, under that assumption (the manufacturer measured the SPL level at a frequency where the speaker's impedance was exactly 8 ohms), the two figures are identical. However, that is usually not the case.

Driven by their marketing departments and upper management's incessant thirst for increased profits and return to shareholders, some (most) speaker makers use whichever figure is higher. Very few publish the frequency at which the sensitivity figure was measured and calculated.

Some even declare "average" sensitivity, which, just like proclaiming a speaker to be an 8 ohms speaker (while its impedance jumps up and down between 6 and 66 ohms, like a yo-yo) , is meaningless!

CASE STUDY: Converting speaker sensitivity ratings at one distance to another

The two no-brand Chinese-made speakers used in Marshall MG100FX combo amp (see page 189, where we ditch the solid state amp and modify its cab) had a declared sensitivity of 100 dB/watt @ 0.5m. A standard practice is to declare specs as dB/watt@1m. How do we convert specs using different distances?

The formula for converting the SPL (loudness) rating at one distance to another distance is

$SPL\ @\ d_2 = SPL\ @d_1 - 20 \cdot \log(d_2/d_1)$

In our case SPL $@d_1$ is 100 dB at 0.5m, d_2 is 1m, so we get SPL @ 1m = SPL @ 0.5m - $20 \cdot \log(1/0.5)$ = 100 - 20log2 = 100- 20*0.3 = 100-6 = 94 dB

The sensitivity rating of these speaker drivers was thus a lowish 94 dB/watt@1m. The figure at 0.5m was used to make it look better (higher), an old marketing trick.

Converting speaker sensitivity ratings at one distance to another

$SPL\ @\ d_2 = SPL\ @d_1 - 20 \cdot \log(d_2/d_1)$

99% heater, 1% speaker

Since a typical dynamic loudspeaker converts more than 95% of the input electrical power as heat, how is that heat distributed or carried away, and how do different parts of the speaker get?

All that heat must be dissipated by the coiled wire in the speaker's VC, so the voice coil and its former are the hot spot - the temperature there is of the 170-180°C magnitude.

The winding copper wire is usually insulated with one or more coats of polyimide ML insulation, which belongs to the 240°C thermal class, which is the highest temperature the insulation can withstand for prolonged periods of time.

Having a wire that can withstand high temperature is great, but the same applies also to the VC former material and the adhesives (glues) used to hold the whole coil together - they get equally hot.

If the former buckles due to heat, the VC will deform and start rubbing against the magnetic structure, and if the adhesive melts in the heat, the whole VC assembly will become loose and the same rubbing problem will develop.

Soon after, the VC will actually fall apart, detach itself from the spider and the cone, to which it is supposed to be glued, and arrivederci Roma ...

Vintage speakers used paper formers, which can only withstand up to around 100°C; again, modern materials such as "Nomex" (an aromatic polyamide) and polyimide go up to 150°C and 240°C respectively.

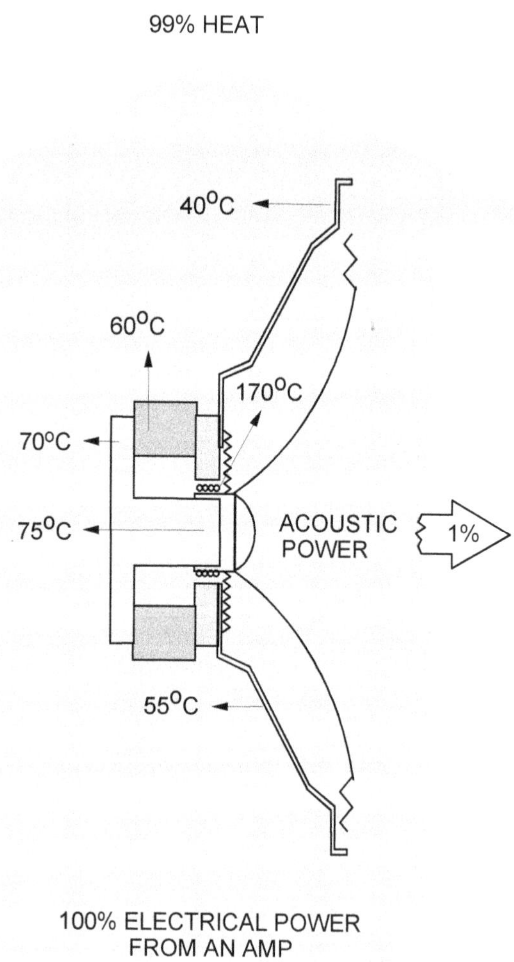

ABOVE: Typical (order-of-magnitude) temperatures of various speaker parts during continuous operation

The efficiency and impedance (mis)matches along the signal chain

The main problem of direct radiators is the interface between the (relatively) heavy and rigid speaker cone and the low density air that it tries to transfer the acoustic power to. The situation is akin to having an amplifier with an output impedance of 800Ω driving an 8Ω speaker load; only about 1% of power would be transferred to the load, while 99% would be wasted on the amp's internal resistance as heat.

Looking at the signal chain (below), a voltage signal from a high impedance voltage source (guitar pickup) enters a tube amp with the same order-of-magnitude input impedance, which isn't ideal (the ideal is a low impedance source and high impedance load), but isn't a problem either, since the transfer of power isn't required here. The amp amplifies that voltage but also supplies high current to the load (a speaker) - power amplification takes place.

Solid state amps have very low output impedances (0.01-0.1 Ω), so they are more efficient power sources compared to tube amps whose output impedance is in the 1-10 Ω ballpark. However, even one or two ohms coupled to an 8- or 16-Ω load is still OK from the impedance matching point-of-view.

The third interface (3) is the radiation resistance. A heavy and dense speaker cone is trying to move light, low density gas (air), with very little success - the efficiency of such transfer is only in the ½% to 2% ballpark.

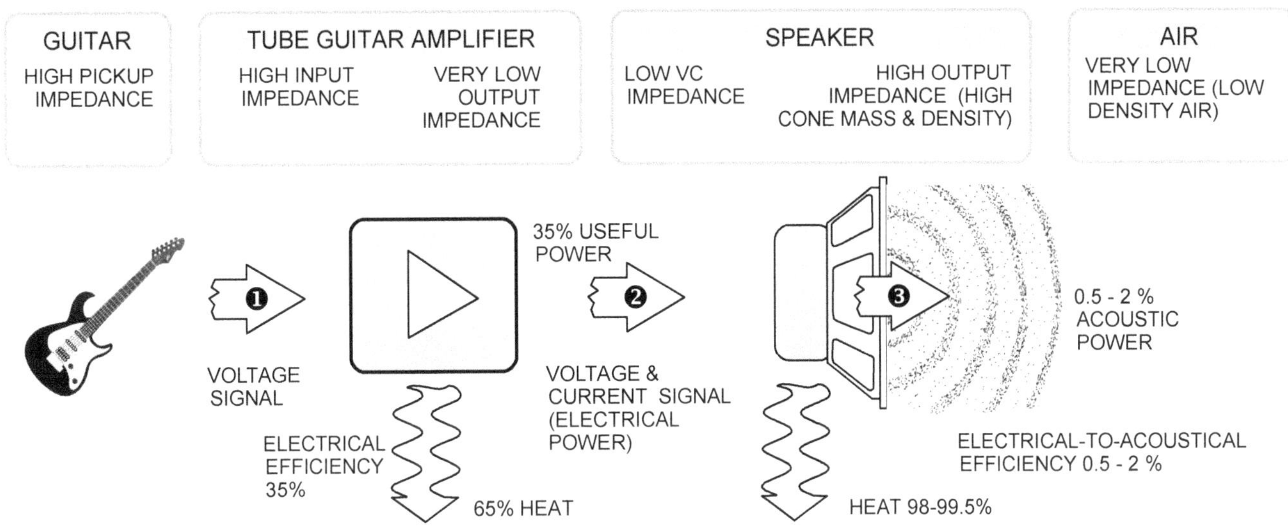

ABOVE: The three interfaces in the guitar-amp-speaker-air signal chain, with the corresponding impedances and efficiencies

Declared and real sensitivity figures - truth telling in speaker business

We have already mentioned inconsistent sensitivity declaring practices across speaker manufacturers. Let's look at two typical examples.

Both Celestion G12T100 and Dai-ichi IS12000 are 12" speakers selling around the US$129 mark (but often discounted down to $70-90 range), so our comparison will be meaningful.

Celestion's sensitivity specification omits the frequency at which it applies and is thus incomplete; so is Dai-ichi's! The quoted SPL of 97dB/W is achieved at 400-500Hz, and, if an average figure across a certain frequency band, it seems correct in the 100Hz-1kHz range (2). Below that it drops almost uniformly down to 70dB/W at 20Hz (1).

There is a wide fluctuation in the critical 1-2 kHz midrange region (3) and a few peaks and dips in the 2-5 kHz range (4). After that, the efficiency drops abruptly, followed by a not-so-pretty curve in the 5-20 kHz treble range (5).

Dai-ichi's curve is smoother, but its sensitivity is 2-5dB lower. The smoother curve could just be a marketing trick - different speaker makers use different smoothing techniques when presenting their measured results.

The lower bass (6) and upper bass-lower midrange behavior (7) is similar to Celestion's, but there is a significant drop in sensitivity from 1 to 3 kHz (8), followed by a mild peak around 5 kHz. This will result in a scooped-out midrange.

Celestion G12T100 12" Guitar Speaker

- Power rating 100 W_{RMS}
- Nominal impedance: 4, 8 or 16Ω
- DCR: 3.4Ω, 7.4Ω or 14.8Ω
- Sensitivity: 97 dB/W
- Magnet: 1kg ceramic
- Frequency range 80Hz-5kHz
- Resonant frequency: 86Hz
- Weight: 3.2kg

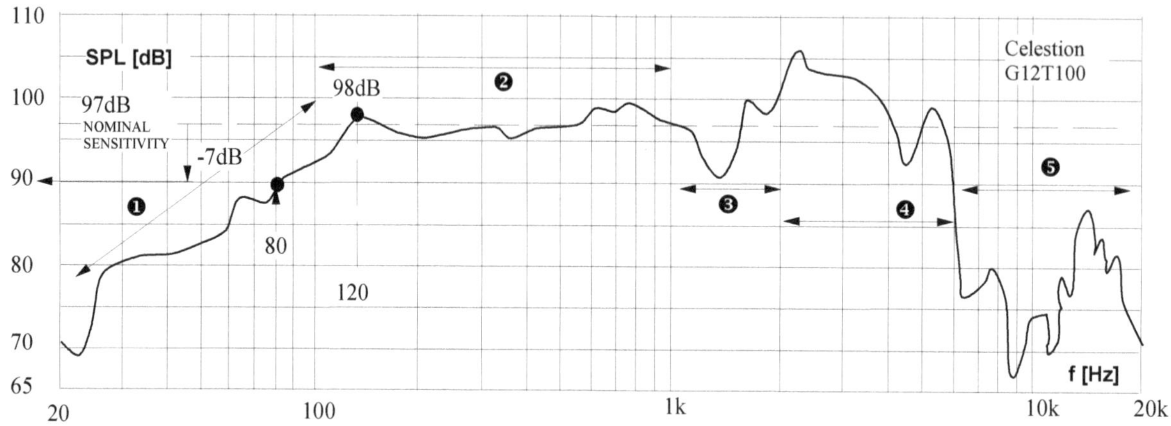

ABOVE: SPL vs. frequency characteristics of Celestion G12T100 speaker

Notice a much lower resonant frequency of IS12000, which can be explained by its almost 50% heavier magnet and a much bigger and heavier voice coil.

So, in contrast to the bright Celestion (notice the pronounced peak between 2 and 4 kHz, a jump from 97 to 106dB!) this speaker is likely to sound bass-heavy.

Dai-ichi specify the frequency range extending all the way down to 32 Hz, which seems way superior to Celestion's 80Hz. Let's see if that claim holds water so to speak. At 32 Hz, the Dai-ichi's SPL curve is at around 78 dB, so the SPL dropped from 95 to 78 dB, a 17 dB drop!

Dai-ichi IS12000 12"

- Power rating 250 W_{RMS}
- Nominal impedance: 8Ω
- DC resistance: 5.7Ω
- Sensitivity: 95dB/W
- Magnet: 1.42 kg (50 oz) ceramic
- Voice coil diameter: 32mm (1.25")
- Frequency range 32Hz-5kHz
- Resonant frequency: 45Hz

BELOW: SPL vs. frequency characteristics of Dai-ichi IS12000 speaker

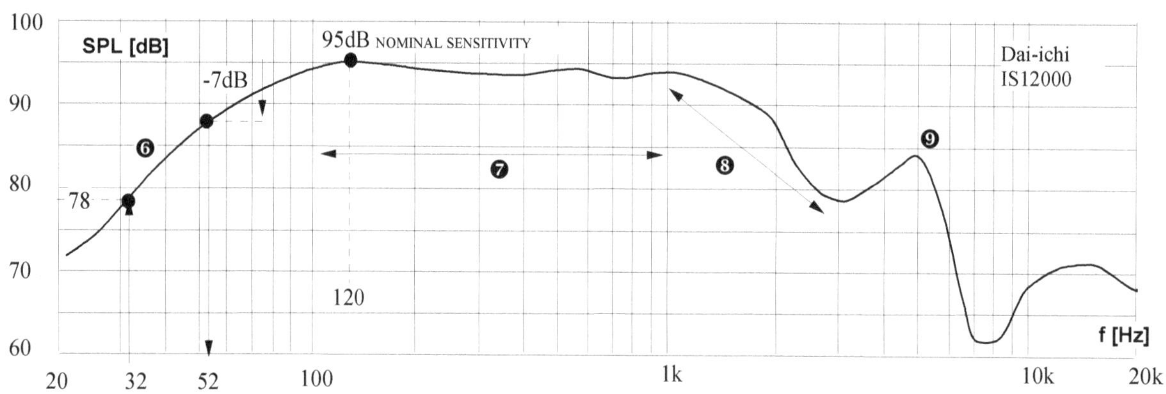

In Celestion's case, at the specified 80Hz the SPL curve is down to just under 90Hz, so a drop of 7Hz from its nominal (or average?) SPL level. Obviously, comparing a 7dB and 17dB drops is meaningless. Applying the identical -7dB drop to IS12000 speaker, 95-7 = 88dB, we find that it happens at the frequency of around 52 Hz.

Notice that Dai-ichi's nominal SPL level is only achieved at the peak of its curve, at 120Hz. At all other frequencies its speaker's sensitivity is (much) lower, so Celestion seems to be much more transparent in their specs.

In other words, the Dai-ichi's nominal SPL level is its absolute maximum, while the Celestion's nominal SPL sensitivity seems to be some sort of average. As we have seen, the Celestion's peak SPL level is 106dB - that is the figure that should be compared to the Dai-ichi's 95dB!

MISLEADING MANUFACTURERS' SPECIFICATIONS RESULT IN ERRONEOUS COMPARISONS

Simply comparing speaker parameters as specified by their manufacturers is not just useless, it can be misleading. For a meaningful comparison, always consult and compare speakers' SPL and impedance curves.

HOW TO MEASURE, CALCULATE OR ESTIMATE SPEAKER'S BASIC PARAMETERS

The starting point - speaker's impedance curve

Buying a brand new, currently produced driver for your DIY enclosure or as a replacement driver for your existing cab usually means you'd be able to find its parameters online or directly from its manufacturer. But what do we do about vintage or obscure drivers (no name or model number)?

The first step is to measure speaker's impedance at 20 or so frequencies in the bass region and draw the speaker's impedance-vs.-frequency curve around its resonant peak. If you want to have the whole curve, measure another ten or twenty points in the midrange-treble region, but that part of the curve is not needed for parameter determination.

How to determine the resonant frequency and Q-factors of a speaker driver from its impedance curve

The height and shape of the speaker's impedance curve (how steep the sides of the resonant peek are) plus a speaker's DC resistance is all we need to calculate its mechanical, electrical and overall Q-factor.

1. Measure DCR of R_{DC} or R_E (same thing) using a multimeter (on "Ohms"): In our case that is 6.0 ohms
2. Read the resonant frequency from the Z-curve: $f_R=f_S=90$ Hz
3. Read the impedance peak from the Z-curve: $R_{MAX}=27\Omega$
4. Calculate coefficient "k": $k=R_{MAX}/R_E= 27/6 = 4.5$
5. Calculate the "reference" impedance R_X: $R_X= \sqrt{k}*R_E = \sqrt{4.5}*6 = 2.12*6= 12.73\Omega$
6. On the impedance curve draw a horizontal line through $Z=R_X = 12.73\Omega$ and read the two frequencies at which the impedance equals $R_X=12.7\Omega$: $f_1=52$ Hz and $f_2=152$ Hz.

   In our case the horizontal scale isn't very precise, but let's assume for a moment that it is, and. conveniently that upper frequency is 152Hz, so the frequency difference f_2-f_1 is a nice and round number 100.

   If your reading of f_1 and f_2 is accurate, calculate $\sqrt{(f_1*f_2)}$, which should give you $f_S \pm 1$dB!

   In this example we have $\sqrt{(f_1*f_2)} = \sqrt{(52*152)} = 88.9$ Hz, close enough for illustration purposes using only a sketch of an impedance curve of no particular speaker!
7. Calculate the mechanical Q-factor: $Q_{MS}= \sqrt{k}*f_S/(f_2-f_1) = \sqrt{4.5}*90/(152-52) = 191/100 = 1.91$
8. Calculate the electrical Q-factor is $Q_{ES}= Q_{MS}/(k-1) = 1.91/(4.5-1) = 0.55$
9. Finally, the total Q is $Q_{TS}= (Q_{ES}*Q_{MS})/(Q_{ES}+Q_{MS}) = (0.55*1.91)/(0.55+1.91) = 0.42$

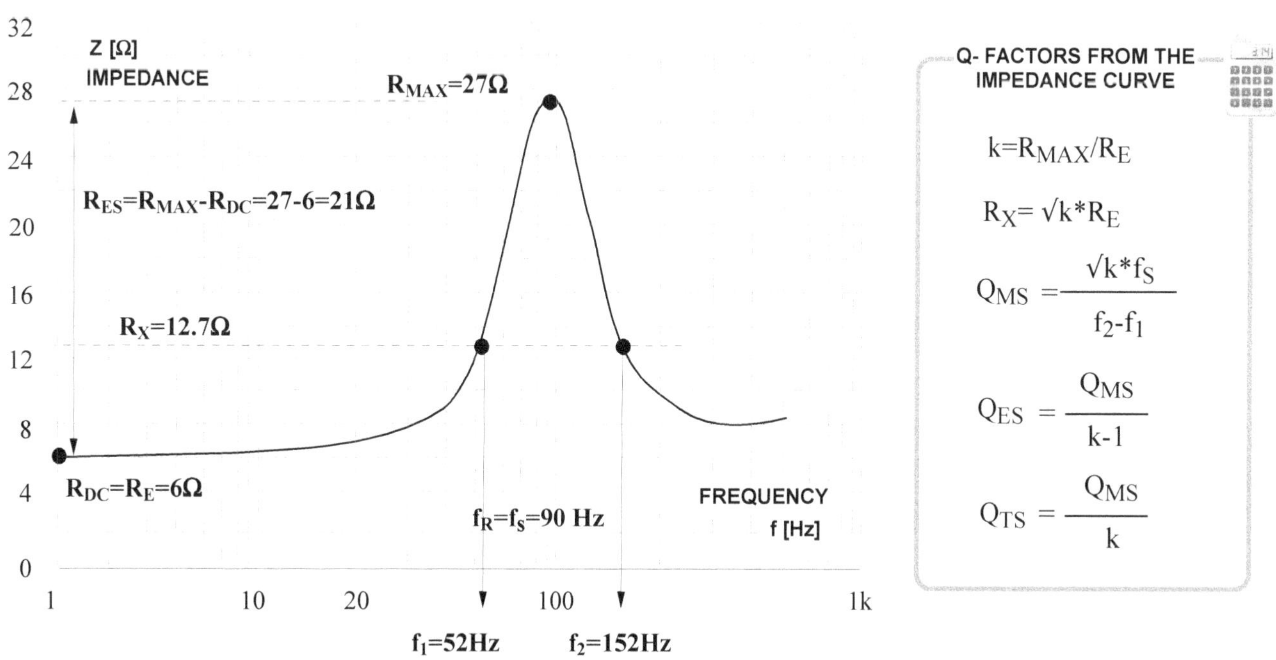

ABOVE. A precisely drawn impedance curve around the speaker's resonant frequency f_S can tell us many of its parameters!

How to determine driver's C_{MS} and V_{AS} using the added mass method

There are a few methods of determining V_{AS} of a speaker but most require the use of test boxes (enclosures), so the beauty of this approach is that it's simpler and faster.

By adding precisely weighed coins or other similar small objects around the perimeter of the dust cap we will lower the resonant frequency of the driver from f_{S1} to f_{S2}. After repeating the impedance curve measurements (but only around the resonant frequency) we will determine the new resonant frequency f_{S2}, and then determine the compliance of the driver:

$$C_{MS}=1/(4\pi^2M)*(f_{S2}+f_{S1})*(f_{S1}-f_{S2})/(f_{S2}*f_{S1})^2$$

In our case, our kitchen scales had only a 1 gram resolution (ideally we'd need scales with 0.1 g resolution), so we weighted six Australian 20 cent coins and got M=66 grams or 0.066 kg, meaning each coin was 11 grams.

With six coins added, the resonant frequency dropped from 90 to 42 Hz. The reduction in the resonant frequency must be at least 25%; if it isn't, add more weight and repeat the test. However, if the reduction is too large, as in this case (42/90 = 46/7%), reduce the weight and test again.

Too heavy a weight would push the suspension into the nonlinear area and deform them so much that the results wouldn't be valid at all.

We ended up using 4 coins (33 grams) and got f_{S2}=65 Hz, which was a reduction of more than 25% but it wasn't too large.

$C_{MS}= 0.066/6.28^2*(90+42)*(90-42)/(90*42)^2 = 0.3838*132*48/14288400 = 0.17*10^{-3}$ m/N

V_{AS} is related to C_{MS} as $V_{AS}=1.4*10^5*S_D{}^2*C_{MS}$, where S_D is the effective piston or cone area in square meters (m^2), so the calculated V_{AS} will be in cubic meters (m^3).

Since $S_D=\pi*d^2/4$, we need to precisely measure the effective cone diameter "d". That is impossible (where does the "active" cone area stop and the compliance or surround start?), which is the drawback of this method. In the photo below the cone ends at "D", and the double-corrugated suspension ends at (A), so the usual practice is to measure the radius as in "B" (before the start of the outer corrugation), or "C" (after the first corrugation - its outer diameter).

To make matters worse, the formula features a square of the diameter "d", so any inaccuracy is compounded.

However, in most cases the results will be meaningful and "in the ballpark".

C_{MS} and V_{AS} USING THE ADDED MASS METHOD

$C_{MS}=1/(4\pi^2M)*(f_{S2}+f_{S1})*(f_{S1}-f_{S2})/(f_{S2}*f_{S1})^2$

$V_{AS}=1.4*10^5*S_D{}^2*C_{MS}$,

S_D = "effective" piston or cone area in m^2

C_{MS} = compliance of the driver

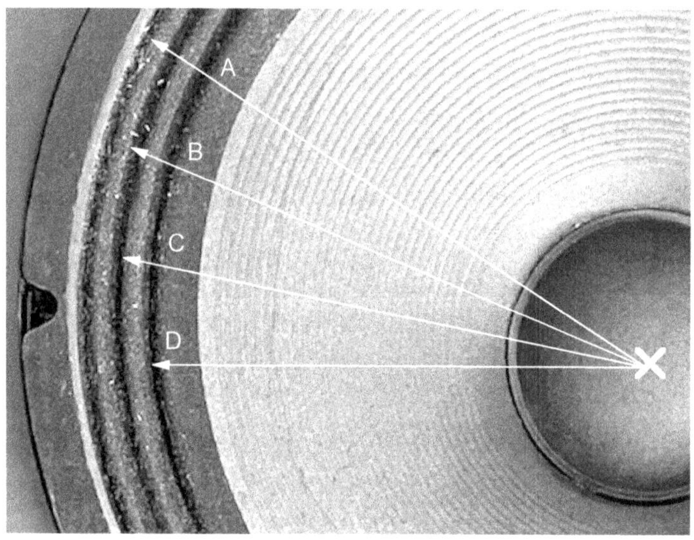

ABOVE: What distance from the center (X) of the cone is the radius we should use to calculate the effective cone area - A, B, C, or D? Or somewhere in between?

LEFT: Precisely determine the weight of the coins or other weights you are going to use

RIGHT: Place the weights symmetrically and equidistantly from the center. The best option is usually around the perimeter of the dust cap.

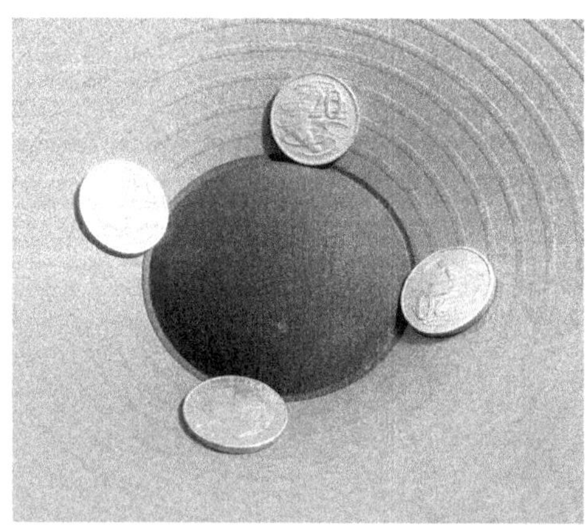

How to measure the Bℓ product (motor strength) of a speaker

This is one of the simplest recipes in this book, but its results are often disappointing, a perfect example of how simple does not mean easy. Place a "naked" speaker (driver) on its magnet, with the cone facing up. Connect a low voltage DC power supply to speaker's VC terminals.

If your power supply has a precise mA meter (3), fine, but most don't, so you will also need a multimeter on "mA" range in the circuit (4).

Place a weight of precisely determined mass on the cone (1); the cone and the VC will move downwards. That force is the weight of the added mass: F_M=mg, a product of the mass and the gravitational acceleration, which is approx. 9.81 m/s^2 . At 45° latitude it is actually 9.80665 m/s2, and varies up to 0.5% - it is about 0.5% greater at the poles than at the Equator, where it is the lowest (due to the Earth's deformity).

Turn the DC power supply on and slowly increase the voltage until the cone starts moving upward and stop when it reaches the previous reference level (with no weight added). The electrical force produced by the magnetic field created by such flowing current is F_E=BℓI. When the forces are identical, we have $F_E=F_M$ or BℓI = mg, so we can calculate the motor strength as Bℓ = mg/I.

For instance, for the 16Ω Jensen P12R speaker (see page 54) with Bℓ = 6.68 Wb/M, assuming a 20g added weight (0.02 kg), the required current supplied by the DC power supply should be around I=mg/Bℓ = 0.02*9.81/6.68 = 0.029371 Amperes or 29.4 mA.

Apart from the error of knowing the actual "g" (up to 0.5%), the most significant source of uncertainty in this measurement is the (in)ability to precisely measure or determine the referent level of the cone.

Various methods, jigs and DIY level meters have been used, but to describe them here would take too much time and space, so I will leave it to you to research it, in case you decide to perform this test.

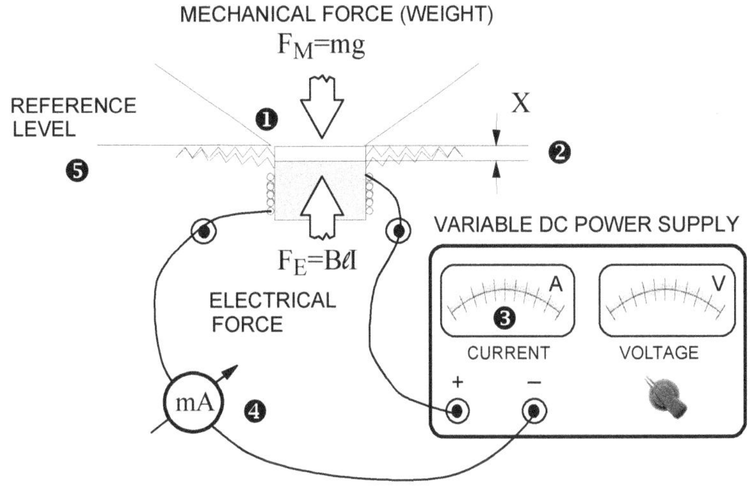

ABOVE: A setup for measuring the Bℓ strength of a speaker is simple, but precise level or distance measurement is key!

ELECTRICAL EQUIVALENT MODELS OF DYNAMIC SPEAKERS AND CABS

The electro-mechanical-acoustical analogies

As an electrical engineering student in the 1980s Yugoslavia, I'd noticed a strange paradox. While my friends who studied mechanical engineering could not understand intangible concepts such as PN-junctions (transistors) or thermionic emission (tubes), complex impedances and circuits, my fellow electrical engineers had trouble with the more obvious (to them!) mechanical stuff such as levers, pulleys and gearboxes, not to mention the more complex mechanical contraptions.

Loudspeakers are electroacoustic transducers (thus the name for this field of study, "electroacoustics"), but the mechanical part is hidden in that hyphen. They are in fact electro-mechanical-acoustical transducers, three-headed scary monsters very few engineers understand, be they of electrical or mechanical persuasion.

The good news for us with electrical or electronics backgrounds is that speaker behavior is studied using electrical diagrams and methods, so for us the hardest part is understanding how mechanical and acoustical properties and parameters should be converted into basic passive electrical components (resistors, capacitors, inductors and transformers) and the so called "equivalent" circuits.

I have tried to do that in this section to the best of my abilities, as concise as I could, in the simplest language possible. Simplification was needed, of course; this is not an academic treatise or a university textbook, so this whole section is actually totally optional. If it all starts to make your head spin or eyes glaze over, skip it without feeling guilty or inadequate.

To complicate matters, there are two types of analogies, and they are totally opposite (can something actually be "partially opposite?) In the Impedance Analogy, the mechanical force is represented by a voltage source, and the velocity (speed) is represented by current.

ELECTRICAL SYMBOL	COMPONENT'S NAME	ELECTRICAL	MECHANICAL	ACOUSTIC
	RESISTOR	Resistance	Resistance	Resistance
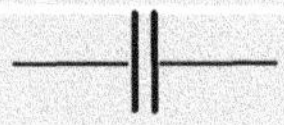	CAPACITOR or CONDENSER	Capacitance	Compliance	Compliance
	COIL, INDUCTOR or CHOKE	Inductance	Mass	Inertance

The Mobility Analogy is the electrical "dual" of the Impedance Analogy: the mechanical force is represented by current, and the velocity (speed) is represented by the voltage.

Isn't using two opposite "conventions" (and that's what they are) insane? Only at a fist glance. In fact, it all depends on the circuit ("equivalent diagram") and on what we are trying to illustrate, deduce, or calculate. Sometimes using one analogy makes the maths simpler, at other times it's the other way around. So, as with most human concoctions, it is a matter of convenience, saving time and effort while (or by) simplifying things along the way.

To further complicate matters, some models are for the speaker (driver) only (in free air), others are for the driver mounted on an infinite baffle (not very practical or useful, is it?), while many are for drivers in different type of enclosures, usually in sealed and vented boxes.

Last, but not least, there are complex, universal models (those that are applicable to the whole audio frequency range), but some authors, books and articles deal with simplified models, most of which are simplified for lower (bass) frequencies only and are thus not valid for mid- and high-frequency ranges.

The equivalent series model of a dynamic loudspeaker

The series model below shows that an audio amplifier can be modeled as an AC (sine wave symbol inside a circle) voltage source with its internal or output resistance in series. Google "Thevenin's Theorem" to learn more about how this is done.

The current pushed by the amp then passes through the series RL circuit, the DC resistance of the voice coil (VC) and its inductance. The "E" in the subscript stands for "Electrical" parameters of the VC. Then we enter the mechanical domain and model the speaker's suspension as a series RCL or RLC or CLR (the order does not matter) circuit.

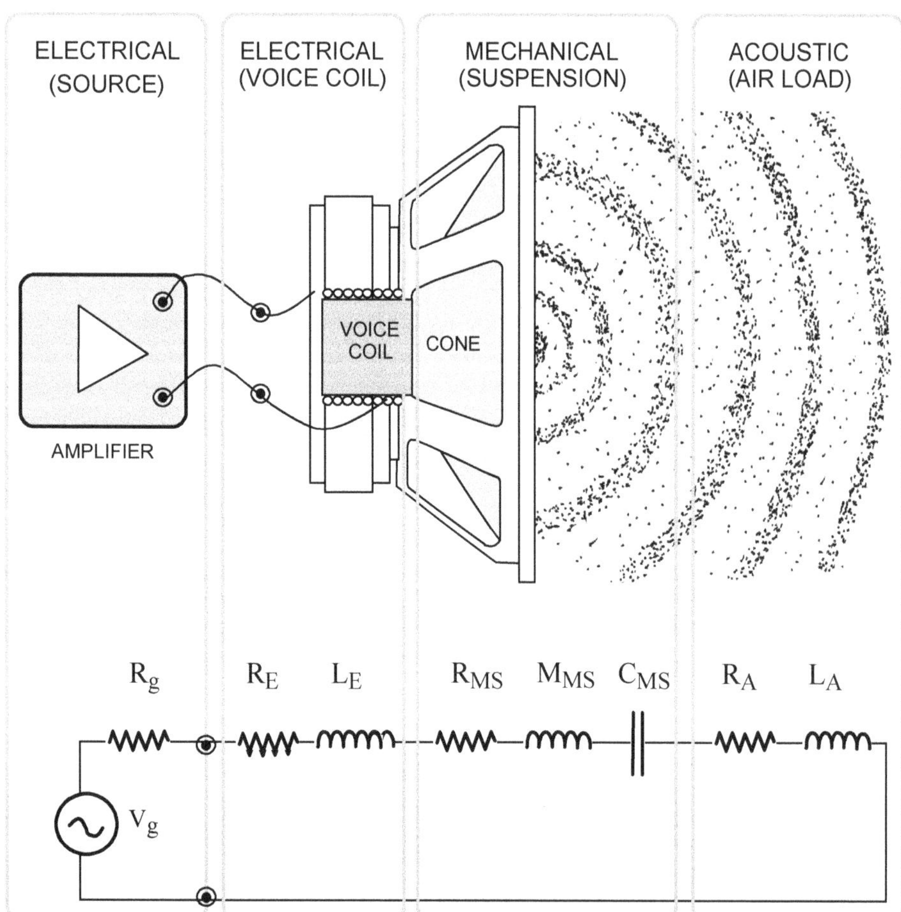

LEFT: The simplified series model of a loudspeaker is a 2nd order electrical system, a RLC resonant circuit!

R_g output resistance of the amplifier

$V_g(t)$ audio signal (time variable AC voltage source)

R_E DC resistance of the voice coil

L_E inductance of the voice coil

R_{MS} mechanical resistance of the suspension

M_{MS} the mass of the moving armature (voice coil+cone)

C_{MS} compliance of the suspension

R_A frictional resistance of the air load

L_A mass of air load

R_{MS} is the mechanical resistance of the cone and suspension (friction), L_{MS} is the equivalent of the mass of the moving voice coil+cone, and C_{MS} is the capacitance equivalent to the compliance (elasticity) of the suspension. Remember, by "suspension" we mean both the spider and the outer suspension together, their sum total!

So far we have only covered the "source", we still haven't modeled the actual load, which is the air that must be moved by the speaker's cone. The acoustic load has a resistive component R_A, the radiation resistance of the air load, and L_A, which the actual mass of air that must be displaced. There is also a small capacitive component, but it is usually disregarded to make the analysis easier.

However, this is a gross simplification, valid for the driver only, without any baffle or enclosure. Soon we will present a more detailed model (equivalent electrical circuit) of the acoustic load, which will apply to sealed boxes and also bass-reflex enclosures.

This "in principle" model isn't dimensionally correct and cannot be used as such, since it mixes three types of parameters, electrical, mechanical and acoustic, without converting them from their respective domains.

Depending on our purpose (what parameters or performance aspects of the speaker we want to analyze), we need to convert it into one of three possible domains, the electrical, the mechanical or the acoustic. That means that we will then have three different models.

As their very name indicates, both electrical-mechanical (EM) and mechanical-acoustic (MA) "transformations" can be modeled as an electrical transformer with a voltage or turns ratio VR=TR, and its square, the impedance ratio $IR=VR^2$!

For the EM transformation the VR ratio is $VR = B\ell : 1$, and for the MA transformation it is $VR = 1: S_D$. The impedance ratios are then $B^2\ell^2 : 1$ and $1: S_D^2$. The model is using the Impedance Analogy, since force F_C on the cone is modeled by the voltage and the velocity of the cone is analogous to the current through the voice coil v_C.

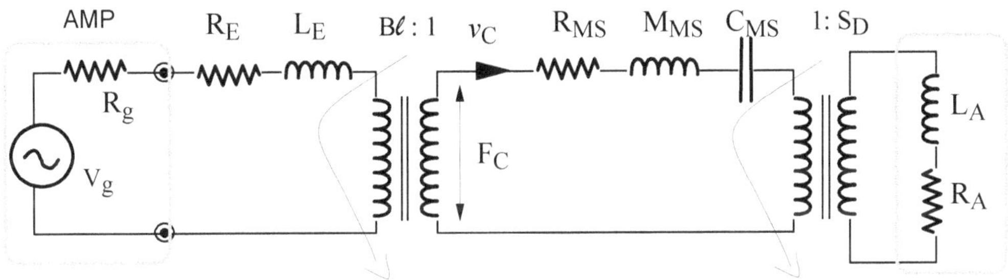

LEFT: The simplified series model of a loudspeaker is a 2nd order electrical system, a RLC resonant circuit!

ACOUSTIC (AIR) LOAD

The force on the moving VC is $F_{MOTIONAL} = B*\ell*I$

The induced CEMF is $E_{MOTIONAL} = B\ell v$

Motional impedance is $Z_{MOTIONAL} = E_{MOTIONAL}/I = (B\ell)^2*(v/F) = (B\ell)^2*Z_M$, where Z_M is mechanical impedance.

Thus, to convert mechanical impedance into a motional one (used in the electrical domain), we need to multiply it by the impedance ratio $(B\ell)^2$, which implies a voltage ratio of $B\ell^2:1$!

The pressure on the surface of the cone is force divided by the area of the cone: $p=F/S_D$.

The volume velocity is velocity multiplied by the surface area of the cone: $v_V=S_D*v$.

Mechanical impedance is the ratio of velocity and force $Z_M = v/F = (v_V/S_D)/(p*S_D) = (v_V/p)/S_D^2 = Z_A/S_D^2$, where acoustical impedance is $Z_A = v_V/p$. Thus, to convert acoustic impedance to mechanical impedance, we need to multiply it by the impedance ratio $1/S_D^2$, which implies a voltage ratio of $1/S_D:1$ or $1:S_D$!

Electrical » mechanical » acoustic transformation

Reflecting the electrical parameters to the mechanical side multiplies the voltage source by $(B\ell)/(R_g+R_E)$ while the resistance R_g+R_E becomes $(B\ell)^2/(R_g+R_E)$. The coil inductance L_E in series with R_E becomes a capacitor in parallel with it, with the value $L_E/(B\ell)^2$.

Reflecting the parameters once more, this time to the acoustical side is relatively simple, the resistances and inductances are divided by S_D^2, the impedance ratio of the transformer, while the capacitances are multiplied by it. The voltage source is only divided by the voltage ratio S_D.

Please understand that the acoustical load R_A-L_A is a gross simplification, used for the time being. We will elaborate more on various equivalent circuits of different types of enclosures in the next chapter, since it is the enclosure that will determine the exact equivalent circuit Z_A.

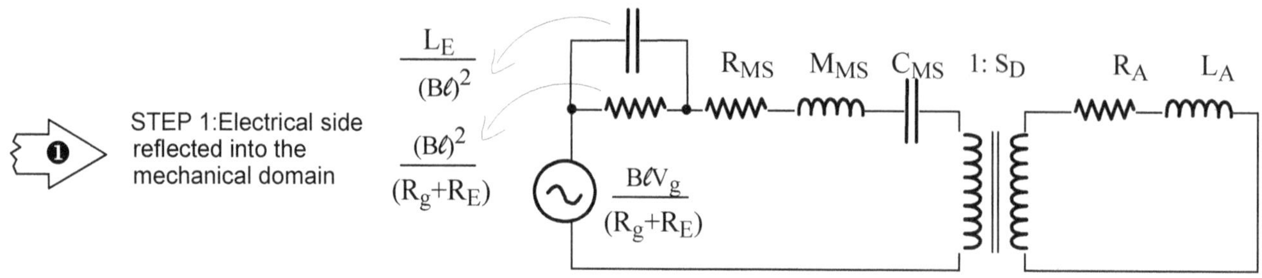

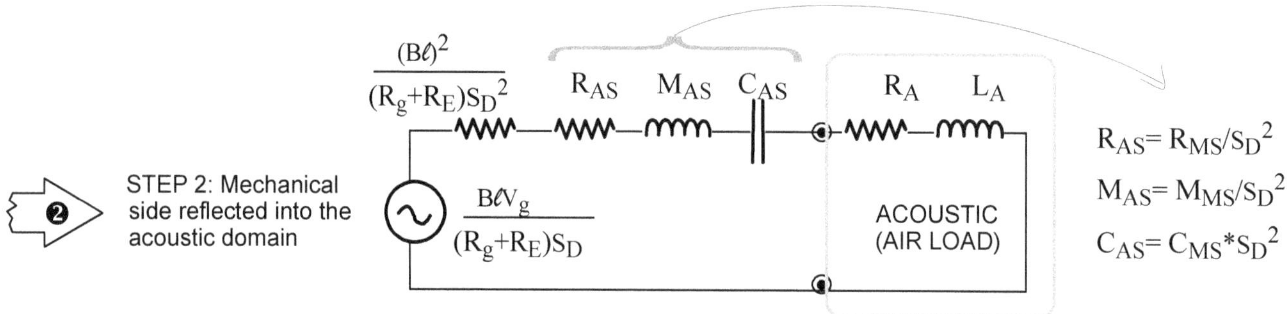

Mechanical » electrical transformation (WARNING: Complex numbers algebra ahead!)

At resonance, the frequency-dependent impedance of the series resonant circuit (R_{MS}-C_{MS}-M_{MS}) is at its minimum. The capacitive $1/(2\pi f_S C_{MS})$ and the inductive $2\pi f_S M_{MS}$ reactances cancel out, leaving only R_{MS}. However, we remember that the speaker electrical impedance (at its terminals, as seen by the amplifier) is at its absolute maximum. How is that possible?

The answer will become obvious once we do the reverse transformation, first from the acoustical to mechanical domain, and then to the electrical equivalent circuit, which is the way the speaker is "seen" by the amplifier.

The mechanical impedance Z_M reflects onto the electrical side as motional impedance $Z_v=(B\ell)^2/Z_M$. If we substitute $Z_M= (R_{MA}+R_{MS}) + j\omega(M_{MS}+M_{MA}) = 1/(j\omega C_{MS})$ and divide the right side of the equation with $(B\ell)^2$ we get

$$Z_v= \frac{1}{\dfrac{(R_{MA}+R_{MS})}{(B\ell)^2} + j\omega\dfrac{(M_{MS}+M_{MA})}{(B\ell)^2} + \dfrac{1}{j\omega C_{MS}(B\ell)^2}} = \frac{1}{\dfrac{1}{R_{ES}} + \dfrac{1}{1/(j\omega C_{MES})} + \dfrac{1}{j\omega L_{CES}}}$$

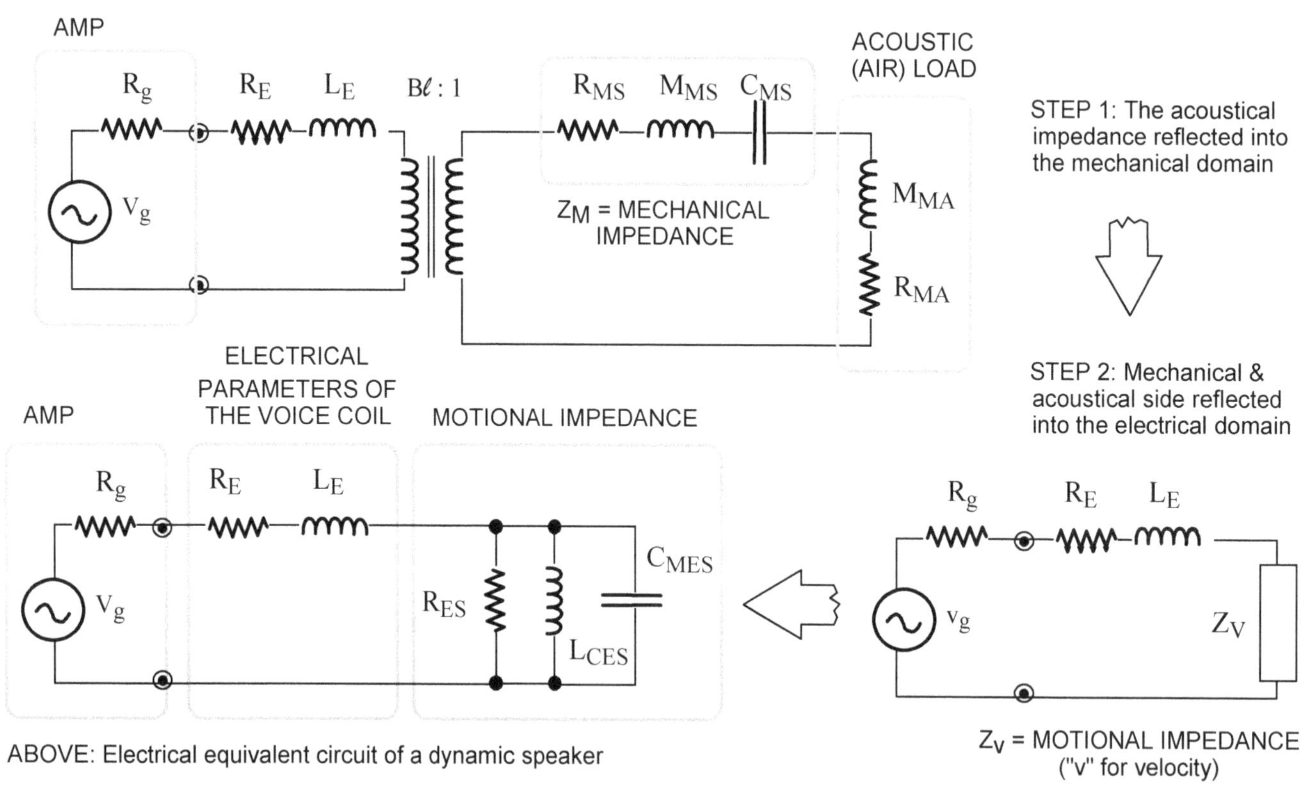

ABOVE: Electrical equivalent circuit of a dynamic speaker

We have a reciprocal (one over something) of the sum of three reciprocal parameters, and we recognize that as the total impedance of three components in parallel (just as with the resistance of three parallel resistors)! The three components are

$$R_{ES}=\frac{(B\ell)^2}{(R_{MA}+R_{MS})} \qquad C_{MES}=\frac{(M_{MS}+M_{MA})}{(B\ell)^2} \qquad L_{CES}=C_{MS}(B\ell)^2$$

While resistances are converted into resistances in this reflection from the mechanical into the electrical world, notice that the inductance (inertia) in the series mechanical model becomes a capacitance in the parallel electrical model, and vice versa: the capacitance (compliance) in the series mechanical model becomes the inductance in the parallel electrical model. that is the logic behind the otherwise confusing suffixes: the "M"in C_{MES} and "C" in L_{CES} indicate such inversion.

At low (bass) frequencies for which these models were developed and are valid (they are not valid for higher frequencies) the reactance due to L_E is very small and can be neglected.

At resonance, the reactances of L_{CES} and C_{MES} cancel out, so we are left with a simple resistive circuit with R_E and R_{ES}.

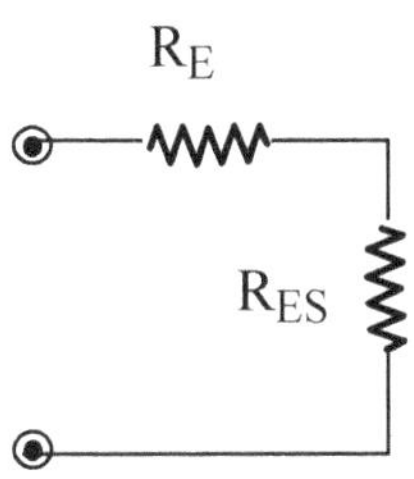

LEFT: The electrical equivalent circuit (pure resistance) of a dynamic speaker at its resonant frequency f_S is the sum of its DC resistance R_E and motional resistance R_{ES}

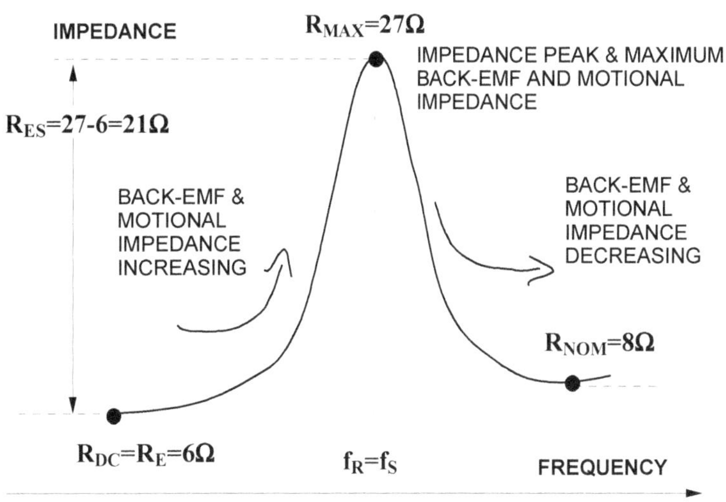

THE SAME MODEL SPEAKERS BUT OF DIFFERENT NOMINAL IMPEDANCE - HOW IS IT DONE, AND ARE THERE ANY SONIC DIFFERENCES BETWEEN THEM?

You are building a speaker cab, to be used with your tube guitar head. The amp's designer valued flexibility, so its output transformers has three secondary taps, connected to three output jacks, marked 4, 8 and 16 Ω. You have heard a speaker (driver) whose tone you like and want to use it in this cab, but it comes in three versions, again, the impedances differ. Which speaker should you buy, the 4Ω, 8Ω, or 16Ω version?

It doesn't matter, most amp, speaker designers would say, as would most guitar players. Perhaps, but there will be subtle sonic differences. Plus, this is supposed to be an educational book, and this discussion will hopefully help you understand a few other important concepts and aspects of speakers and speaker cabs.

A fewer turns of large diameter (or cross section) wire, or many more turns of thinner wire?

The same tubular voice coil former (bobbin) is used for different impedance versions of the same speaker by the dudes in a speaker factory. Actually, the coil winding people (for audio transformers, speaker voice coils, and pickup coils) are more likely to be ladies, since they generally have more patience and better eye-hand coordination, but that is not the point here.

The voice coil "A" consists of five turns of square wire with cross sectional area "S" cm^2. Voice coil "B", wound on an identical VC former, has 20 turns of a thinner wire with four times smaller cross section (S/4).

R_B, the resistance of coil B will be 16x higher than the resistance of coil A, since its total wire length is 4x longer, yet its wire's cross section is 4x smaller. The resistance is proportional to the length and inversely proportional to the cross sectional area, thus 4x4=16! So if R_A=4Ω, then R_B=64Ω.

Assuming the same power rating of both coils, with the same voltage input from an amplifier, the current through coil B will be 4x lower. Power is proportional to current squared multiplied by the coil's resistance, and when you take a square root of the whole equation, a square root of 16 is 4.

So, assuming an identical magnetic structure, cone and suspension, which speaker "motor" will be stronger?

Since the mechanical force on the moving part (VC and cone) is $F=B*I_S*\ell_C$, the current I_S through coil A is 4x higher but its "ℓ_C" (the total length of wire in the coil) is 4x shorter (20 turns versus only 5 turns).

Since B is the same (identical magnetic structure), the force on the motor's moving part will remain the same - both motors will be equally strong!

The only difference (in this case only, without considering the other factors) between the two speakers will be their impedance.

WIRE CROSS SECTION "S"

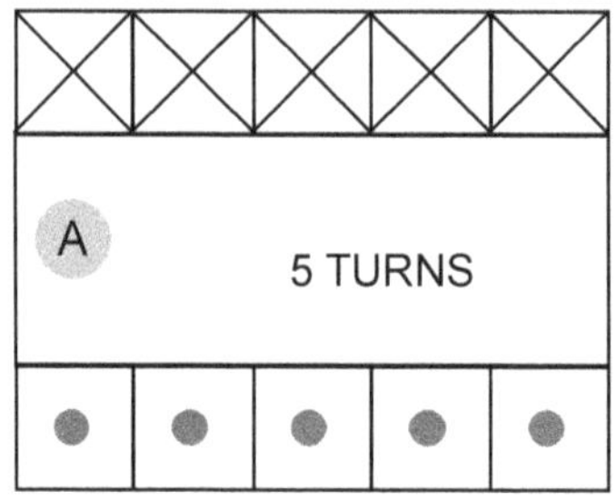

WIRE CROSS SECTION "S/4"

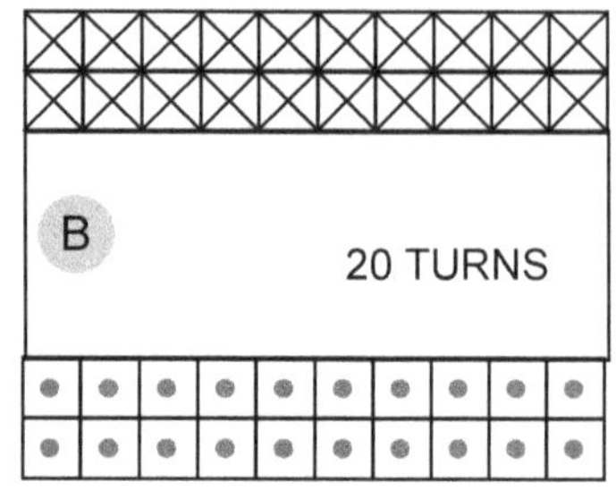

ABOVE: The Force Factor of the two speakers will not be the same (speaker "B" will have a higher $B\ell$ factor), but the force on the speaker's cone will be equal and the motors will be equally strong.

CASE STUDY: The insights from the speaker's spec sheet

To illustrate the difference between the same brand and model speakers but of different nominal impedance (8 ohm and 16 ohm versions), we'll again use Jensen P12R, a 12" Alnico guitar speaker, as an example.

Looking at the spec sheet (below), we see that the 16Ω version has a slightly lower resonant frequency (1), which is determined by the moving mass M_{MS} (2), which is significantly lower (19.8 grams versus 25.2g), and by the mechanical compliance C_{MS} (3), which is higher.

The higher compliance (a more flexible suspension) means that its equivalent air volume inside a closed box V_{AS} (4) will be higher, as it is (72.6/54= 34% higher). Another consequence of looser suspension is a much higher R_{ES} (5), meaning that both the impedance of its resonant peak and its counter-EMF produced will be much higher.

The 16Ω version also has a higher $B\ell$ parameter, indicating a stronger motor (6). This makes intuitive sense, since the magnetic flux density B is most likely the same for both versions, but since the 16Ω version has many more turns of a thinner wire in the magnetic gap, its "ℓ" is much higher. Many more turns automatically means that the 16Ω version's inductance L_E (7) will be higher, as it is, 0.84mH compared to 0.5mH for the 8Ω version.

Finally, it is interesting to note a much lower (31%) efficiency (8) of the 16Ω version, despite which it still manages to (slightly) outperform the 8Ω version in terms of sensitivity (9). Notice a lower X_{MAX}, meaning much shorter maximum excursion limits (25% lower) for the 16Ω version, 0.75mm versus 1.0mm.

Perhaps the reason is a longer VC former (the tube on which the VC turns are wound) used in the 16Ω version, so that all those additional turns could be accommodated? Obviously, a longer VC former would hit the top and bottom plate sooner than a shorter one.

Are these two versions of nominally the "same" speaker going to sound different? Definitely! How much of a sonic difference are we talking about here? Hard to say, it depends on other factors not discussed here, mainly the speaker cabinet and the amplifier driving it. The same speaker can sound significantly different when driven by different amps. More on that important interaction between the amp and the speaker soon.

PARAMETER	SYMBOL		8 Ω	16 Ω	UNIT
Voice coil DC resistance	R_E		6.05	12.0	ohms
Resonance frequency	f_S	❶	80	77	Hz
Mechanical Q factor	Q_{MS}		13.07	12.53	
Electrical Q factor	Q_{ES}		2.67	2.67	
Total Q factor	Q_{TS}		2.1	2.2	
Moving mass	M_{MS}	❷	25.2	19.9	g
Mechanical compliance	C_{MS}	❸	0.157	0.214	mm/N
Force factor	$B\ell$	❻	5.53	6.68	Wb/m
Equivalent acoustic volume	V_{AS}	❹	54	72.6	liters
Maximum linear displacement	X_{MAX}		+/-1	+/-0.75	mm
Reference efficiency	η_0	❽	1.56	1.19	%
Losses electrical resistance	R_{ES}	❺	39	58	ohms
Voice coil inductance @ 1kHz	L_E	❼	0.5	0.84	mH
Sensitivity @1W, 1m		❾	93.7	94.0	dB

ABOVE: Jensen P12R guitar speaker's tabulated Thiele-Small and other important parameters, as published in Jensen's data sheet. There is a 4Ω version as well ...

SPEAKER TONE SECRETS

4

A speaker cab is an instrument, even more so than an electric guitar or bass guitar.

Get Eric Clapton, Mark Knopfler, Carlos Santana or any other legendary guitarist, give them a vintage guitar that now costs more than a new Lexus, but hook them up to a crappy speaker cab, and that's all you will get - virtuoso playing technique but a shitty tone.

In this chapter we look at the main factors that determine the sonic signature of guitar speaker cabs.

- TWO DIFFERENT WORLDS: HI-FI SPEAKERS AND GUITAR SPEAKERS & CABS
- DESIGN & CONSTRUCTION ASPECTS THAT DEFINE GUITAR SPEAKER'S TONE
- BREAKING IN SPEAKERS - MYTH OR REALITY?

TWO DIFFERENT WORLDS: HI-FI SPEAKERS AND GUITAR SPEAKERS & CABS

Guitar amps and speakers versus hi-fi or "audiophile" amplifiers and speakers

Guitar amps and speakers operate on the same physical principles as the hi-fi (high fidelity) or "audiophile" amplifiers and speakers, but their features and sonics differ significantly.

A hi-fi tube amp is a reproduction device that amplifies the low voltage, low power input signal from the source (turntable, CD player, D/A converter, music streamer, etc.) into a high current, high power electrical signal needed to drive the electromechanical transducer (loudspeaker).

Both should be as "neutral" as possible, meaning they should add as little of their sound coloration to the recorded sound as possible. Real amps and especially loudspeakers are never completely neutral or transparent; they always add their "personality" or voicing to the ultimate result.

The basic job of a guitar amp is also to amplify the weak signal from a guitar pickup and drive the loudspeaker. A guitar amp is also an instrument, not just a transducer, since it adds its own voice to the overall result. In other words, a guitar amp should produce a tone or a variety of tones that guitar players are looking for, distorting and "coloring" the incoming guitar signal in specific desirable ways.

In addition to tone and distortion controls, a guitar amp may include additional effects, the most common being tremolo (amplitude modulation of the signal) and reverb or echo.

HIGH FIDELITY SPEAKERS

- main task: sound reproduction
- stationary, setup critical for optimal results
- neutrality, low distortion and minimization of cabinet resonances are the main design aims
- a wide range of drivers used (many different brands and technologies)
- a wide frequency range is paramount - apart from those using "full-range" drivers, all hi-fi speakers use crossovers and multiple drivers: 2-way, 2.5 way, 3-way designs being most common
- designs & topologies vary widely - open baffle, bass reflex, sealed box, horn, D'appolito's arrangement, transmission line, etc.
- expensive, generally $5-10,000, up to $180,000 per pair

GUITAR SPEAKER CABS

- main task: sound creation
- portable, sound adjustable by positioning and slanting of the cab
- distortion welcome, at least the desired kind
- the speaker cabinet is part of the instrument - its distortion behavior and voicing are of paramount importance
- designs are very similar and predictable
- a few major driver manufacturers dominate the field
- (Celestion, Jensen, Eminence, Electro-Voice, JBL, Peavey, Vox), plus a few smaller ones (Warehouse, Weber, Fane, ToneSpeak)
- limited frequency range - very few guitar cabs use tweeters and crossovers, although PA and some bass and keyboard speaker cabs do
- relatively affordable ($300-$2,000)

Guitar amps, being musical instruments, tend to include various controls, of which gain, master volume, tone (bass, middle and treble), "overdrive," "bright," and power attenuation are the most common. Then there are manufacturer- dependent tonal controls named "presence", "resonance", "texture", "contour", "crunch", "fuzz", "Tubescreamer", "Vaporizer" and heaps of others.

Just as guitar amps, guitar speakers and speaker cabs are part of the instrument; their main function is sound creation, and that is the fundamental difference in their design ("voicing") and manufacture. Almost everything undesirable in hi-fi drivers and speaker enclosures is what makes a guitar speaker cab sound a particular way!

Due to the nature of their use, guitar speaker cabs are generally smaller and lighter than their hi-fi brethren, since portability and mobility feature at the top of the list of musicians' requirements. For the same reason, which implies often rough handling and harsher environments, guitar speaker cabs must be mechanically stronger and the drivers' cones tend to be better protected from physical damage.

In the esthetics department, while hi-fi speakers use beautiful lacquered veneers and even polished solid mahogany and other exotic timbers, guitar speaker cabs have their modest plywood, MDF or solid pine innards covered by vinyl covering, pejoratively called "Tolex".

Continental Tire the Americas, LLC, trademarked their flexible and waterproof vinyl material in 1945 under the name Tolex. The trademark expired 50 years later, in 2005, so it can now be used with impunity. A sanded and lacquered tweed fabric is also a popular covering choice for guitar combos and speaker cabs.

Paper cones rule the guitar speaker world

Manufacturers of speaker drivers used in hi-fi and for bass & PA speakers have been experimenting with various materials for their cones. "Exotic" materials such as polypropylene, Kevlar, carbon fiber, aluminium or fiberglass are preferred by some; others even go so far as to use two or more materials in a "sandwiched" construction.

In the guitar speaker world, paper pulp cones have ruled for almost a century. The bravest departure so far has been the trial of hemp cones, with very promising results. While these cones are very light, resulting in the high efficiency and sensitivity of guitar speakers, they are also softer than any of the other materials mentioned.

Such flexing and cone breakup effects are behind the complex overtones and different speaker "personalities". Guitar speaker makers manipulate the mechanical and electrical parameters of their designs to achieve a desired combination of such flaws, which then becomes their feature, their sonic signature.

I call it the "Flaunt Your Flaws" principle: "If you cannot fix the physical problems of electrodynamic speaker drivers and the acoustics of their cabs, make them into features!"

Cone breakup, multiple resonances, partial cancellation of the front and rear waves and a myriad of other problems that hi-fi speaker designers spend thousands of hours to minimize or control, guitar speakers suffer from quite badly. However, nobody bats an eyelid - in fact, such flaws are welcome, since they change the spectral characteristics of the cab (frequency distribution) and its voicing, thus adding to its character, and making it suit particular guitar players and types of music more than others.

The "Flaunt Your Flaws" principle of guitar speakers and cabs

If you cannot fix it, feature it!

Guitar speakers versus hi-fi speaker drivers

Unlike a guitar speaker which must reproduce the whole range of tones produced by an electric guitar, a hi-fi speaker (and in this context we mean a woofer), works only over a relatively narrow range of lower frequencies. Crossovers make sure tweeters and midrange drivers (if any) then take over higher up. This minimizes distortion and ensures that the overall SPL-Vs-frequency characteristic is as flat and smooth as possible.

Lateral vibrations and standing waves are damped as much as possible by the use of soft rubber or foam surrounds. This makes hi-fi speakers very flexible or "high compliance", so they must rely on the stiffness of air inside the enclosure to prevent excessive cone excursions.

Hi-fi woofers have a low resonant frequency, since their incorporation into a sealed or vented enclosure will raise the resonant frequency of the speaker-enclosure combination, and that is clearly undesirable. Ideally, a hi-fi speaker should cleanly and faithfully reproduce frequencies as low as 20 Hz. Although most only manage 40-50 Hz, which is still an octave lower than the lowest frequency demanded of the guitar cabs (80 Hz).

Guitar speakers are generally stiff, or "low compliance". Their suspension, usually a doped cloth accordion fold surround, is designed to prevent over-excursion, so they don't need the help of the trapped air inside the enclosure to achieve the required stiffness; they were designed to work in "free air", as in open-back cabinets.

Guitar speakers generally have a much higher resonant frequency than hi-fi speakers, often way higher than the mentioned 80 Hz ballpark. That is fine in open back cabinets (remember that SPL drops rapidly below the resonant frequency), but sealed and vented enclosures will raise it even further, thus impacting the upper bass registers.

Guitar speakers are generally much more efficient (higher sensitivity) than their hi-fi cousins. Using multiple drivers in the same cab, as we will see soon, will raise the overall SPL levels even further (due to their mutual acoustic coupling phenomenon). Thus, driven by the same amp power, that will make them sound louder, perfect for concerts and larger venues, as compared to much smaller listening rooms in audiophile situations.

Their exposure to vibration and rough handling during transport also demands guitar speakers to be more robust, not just in terms of their structural aspects, but also electrically, since they are much more likely to be subjected to more frequent and more prolonged overload periods.

LEFT: A rubber surround of a hi-fi speaker (1) in contrast to the stiff folded paper or cloth suspension of a guitar speaker (2).

Most hi-fi drivers were designed to be mounted into a recessed opening (3), while the felt "gaskets" of guitar speakers indicate their mounting from the rear.

Finally, notice a deeper corrugated cone of the guitar speaker (5), compared to a smooth and shallow cone (6) of a hi-fi woofer.

Outer suspension ("surround") types and their properties

Both the material used for the outer suspension and its shape (profile) have a profound impact on speaker's parameters, physical properties, sonic character and suitability for certain applications.

Loosely suspended hi-fi speakers have soft rubber or foam surrounds, glued between the. outer edge of the cone and the metal frame. This results in "high compliance" and allows for long cone excursions. The "half-roll" and the "inverted half-roll" are the most common profiles.

Their low stiffness in both axial and radial directions places additional stringent demands on the centering spider, which must prevent lateral movements and coil rubbing.

Guitar speakers' suspensions are either separate rings, made of impregnated fabric, of integral, where the folds are stamped or molded into a one-piece cone and are thus made of the same material as the cone, usually doped paper.

Such suspensions are stiffer so their cones' linear excursions at bass frequencies are much shorter.

Surrounds with fewer folds are less linear than those with fewer folds, but reduce edge resonance effects and provide a more stabile cone motion and reduced off-axis cone movements (better vibrational equilibrium of the cone).

They also maximize the active (radiation) area of the cone and move more air.

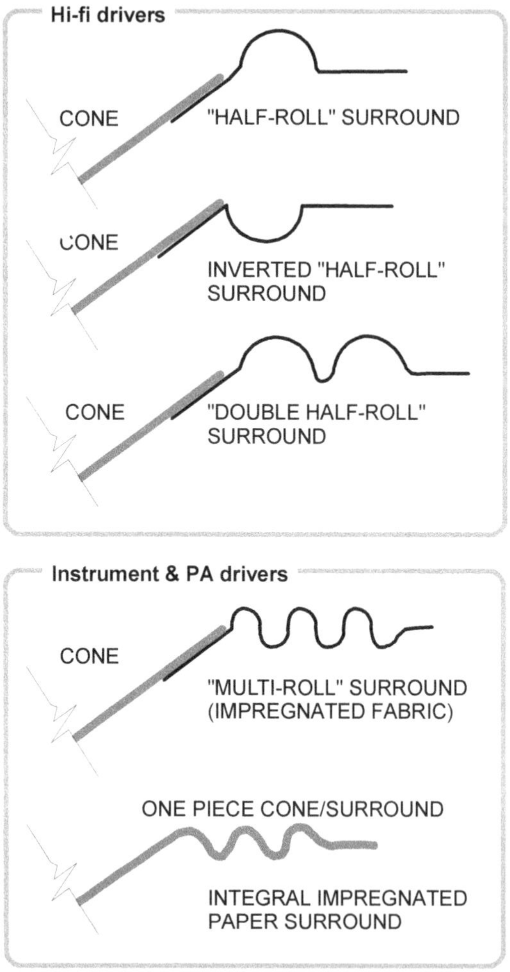

RIGHT: Various types and profiles of outer suspensions

DESIGN & CONSTRUCTION ASPECTS THAT DEFINE GUITAR SPEAKER'S TONE

Guitar amps and speakers compared to bass guitar amps and cabs

An electric (solo, lead or rhythm) guitar produces fundamental tones across three octaves, from 196 Hz (G1) to 1,568 Hz (G3). Of course, the harmonics of these fundamental tones extend much higher, so most guitar speakers and cabs are capable of producing frequencies up to 6-9 kHz.

In short, a guitar is a midrange instrument, there is no need to reproduce very high audio frequencies, those above say 8 kHz. In fact, such high-order harmonics sound shrill and unpleasant (irritating), the reason almost no guitar speaker cab includes tweeters (high frequency speaker drivers), relying on the natural drop in the main driver's frequency response to attenuate such frequencies.

A bass guitar's frequency range ranges from 41 Hz (E3) to 261 Hz (middle C), so this is where deep bass requirements come into play, the same demands imposed on hi-fi speakers. Interestingly, bass guitar cabs also benefit from a wide frequency range, another similarity with hi-fi speakers.

This is due to the (impulsive or transient) nature of bass tones, which have fast rising times and prolonged sustain, making their waveforms closer in shape to square waves than to sine waves. Square waves are very rich in (odd) harmonics, so their reproduction (in hi-fi) or production (in bass guitar cabs) requires a wide frequency range.

If you mike a bass drum and observe its waveform on a storage oscilloscope (which captures fast and/or non-periodic waveforms and stores them in memory so they can be digitally and visually analyzed), you'd see its percussive, a.k.a. impulsive nature.

A spectrum analyzer is an instrument that display's a harmonic structure of any analog signal. The fast-rising transient of a drum would be displayed on its screen as comprising of mid- and high frequency harmonics, and such transient would be followed by the bass wave, which has a much slower rise time.

Without an amp and a speaker cab with an accurate midrange and wide frequency extension, not just on the bass side, but also at the high frequency end, you cannot have an accurate or realistic bass!

I call it "The Loudness Law". I'm sure you realize that such "laws" aren't the laws in the legal or scientific sense, but simply rules-of-thumb, generalizations that apply often, but by no means always!

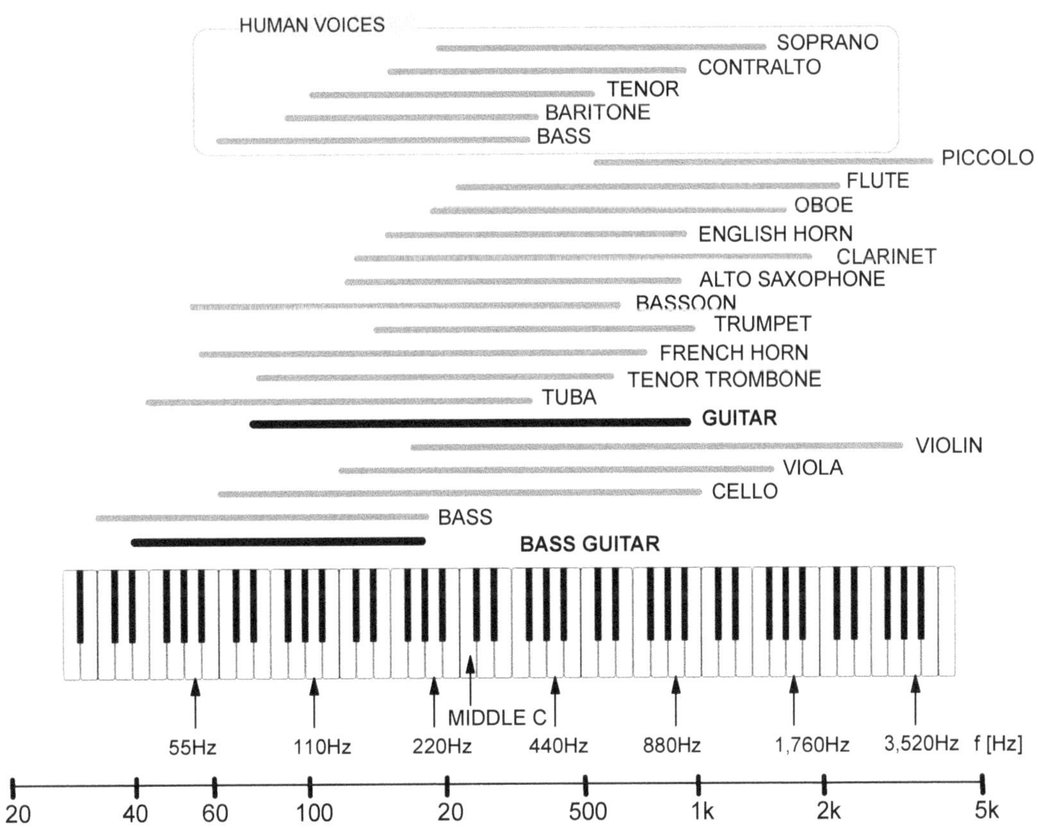

ABOVE: The frequency range of various instruments and human voices compared to the piano scale. The range refers only to the fundamental tones each instrument produces, not its overtones (harmonics), whose frequencies extend way beyond the highest fundamental illustrated here, often even into the ultrasonic band (beyond human hearing limits. Just ask your dog or cat, they can hear such ultrasonic tones ...

The "Loudness Law"

No speaker will sound equally good at all loudness levels; decide upfront which type you prefer, those that sound better at low volumes and SPL levels, or those that come alive when pushed to their limits!

The American vs. the British speaker tone

Generalizing always has some truth behind it, but never the whole or complete truth. Classifying so many different speaker types, made of so many different materials, into only two broad categories is a questionable business, but cultural biases and preferences permeate all products to some degree, and speakers aren't immune to them.

It can be argued that historical musical developments have been even more instrumental (no pun intended) to the development of such tendencies and preferences than the intangible "cultural" inclinations. Simply put, a pop or rock band reaches stardom playing one brand or even one type of a guitar amp (Watkins, Selmer, VOX, Marshall, Hiwatt, Laney and Orange come to mind on the British Isles, while Gibson, Fender, Sunn, Ampeg and Peavey play on the opposite team across the pond), and that sound and tone is forever associated with them or their country of origin.

We can see such perceptions and impressions with guitar amps, and, to a somewhat lesser degree, with guitar speakers. The British speaker tone is considered to be more colored and more aggressive, with a prominent mid- and upper-mid registers, with more "growl" and bite, resulting in thicker and more complex harmonic structure.

The American-voiced speakers generally have more subdued mid range ("scooped-up") with a tight and solid bass, and a more sparkly or chimey treble. Overall, the tone is warmer and cleaner, and the spectrum is more balanced.

Straight and curved cones

Compared to a standard straight cone, a curvilinear cone (a cone with a curved profile) can push slightly more air and reduce resonances. Likewise, shallow cones are also more efficient radiators and have wider dispersion of the mid and high-frequencies.

A straight cone's construction involves sharp folds at its inner and outer perimeter, where it is attached to the voice coil and the frame (if the surround is the integral part of the one-piece cone) or the surround (if a separate surround is used). A curved cone is shaped to provide a smoother transition at those boundaries, thus reducing the severity of the standing wave reflections at these interfaces.

The curvilinear profile also helps significantly reduce the circumferential (bell) mode resonances, prominent in straight cones. Curvilinear or "trumpet" cones present a high mechanical resistance to the bell modes so they cannot develop. Both of these "flatten" and smooth out the multitude of SPL peaks and dips that the straight cones exhibit, thus making the frequency response of curved cone speakers smoother and more uniform.

The curved shape helps to decouple the outer section of the cone at higher frequencies, while the inner section continues radiating, thus improving the directivity. The only drawback of curved cones is their higher cost.

BELOW: Two most common speaker cone profiles, straight (left) and curvilinear (right)

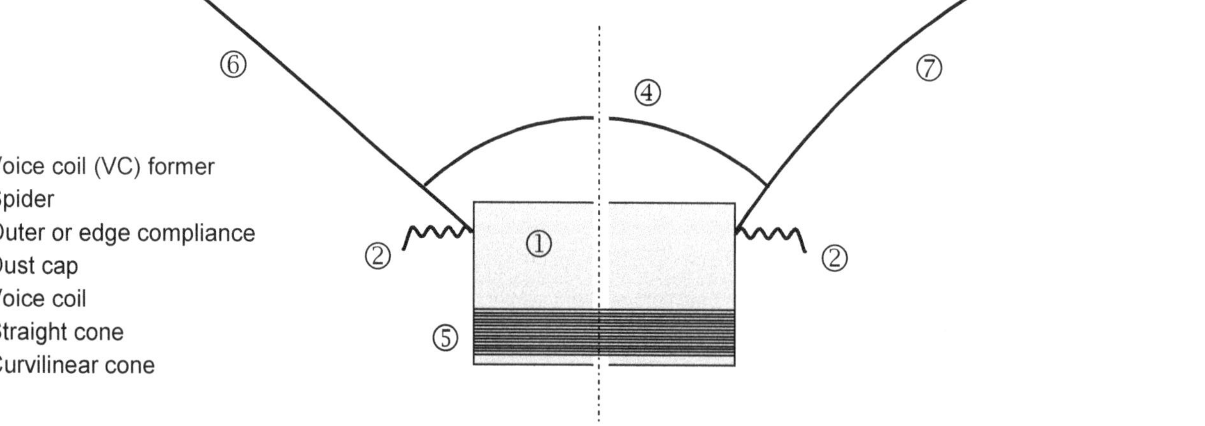

1. Voice coil (VC) former
2. Spider
3. Outer or edge compliance
4. Dust cap
5. Voice coil
6. Straight cone
7. Curvilinear cone

Shallow cones with large diameter voice coils

The fact that their cones are driven from an apex with a relatively small perimeter but the air load is distributed over a much larger surface area, is one of the major drawbacks of dynamic loudspeakers.

Being imperfect and far from the desired rigid piston, their cones flex and buckle, causing uneven frequency response, the already mentioned peaks & dips due to the cone breakup.

One way of reducing this effect is to reduce the disparity in the diameter of the voice coil (and the attached cone's inner diameter) and the cone's outer diameter by increasing the diameter of the voice coil.

Such motor's more uniform thrust results in lower distortion and smoother frequency response.

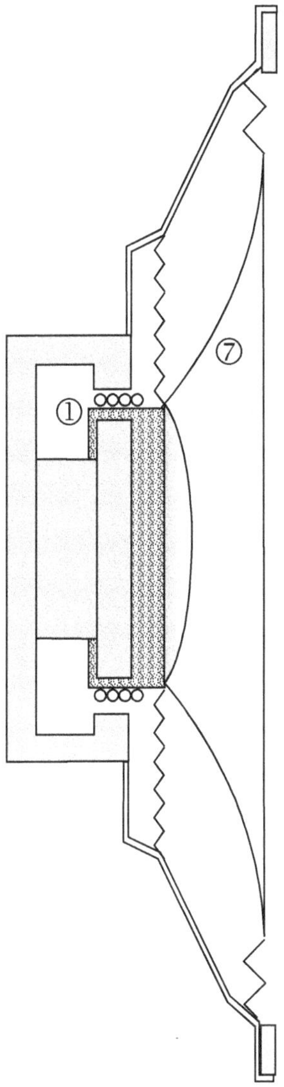

RIGHT: A large diameter voice coil (1), coupled to a stiffer cone with a shallow curvilinear shape (7), distributes the force more evenly and minimizes the cone breakup at higher frequencies

Corrugated or ribbed cones

Introducing concentric corrugations into an otherwise smooth cone does two things. - it makes the cone stiffer for the bell modes (radial modes) of vibration (reduces compliance), while at the same time decreasing stiffness (increasing elasticity) for the concentric resonant modes.

Even more importantly (at least for guitar speakers, where distortion isn't a concern), corrugations act as a series of mechanical filters, progressively decoupling the peripheral sectors of the cone with an increase in signal's frequency. The effective radiating diameter of the cone is thus gradually reduced with rising frequency,

Due to the flexing of the cone, the outer parts of the cone receive a reduced driving force, and the high frequency performance of the driver is improved.

Of course, the number of corrugation, as well as their depth, width and spacing must be optimized, they cannot be simply placed willy-nilly.

The voice coil former material

Paper is light but has low strength and cannot handle high power levels or temperatures, so it was used mostly in the vintage 19560s and 60s speakers.

Aluminum is also cheap and lightweight, but also mechanically/structurally strong. However, it is a conductive metal so the induced eddy currents waste energy and act as a magnetic brake, negatively affecting the dynamics and transient response.

Titanium has an excellent thermal performance (a high melting point and low thermal expansion coefficient), is strong yet lightweight, a perfect combination for VC formers.

Kapton is a nonconductive polyimide film that withstands very high temperature, and is the material of choice for most modern speakers. Nomex is a similar flame-resistant meta-aramid material.

Resonant peaks, phase cancellation dips and cone breakup - the personality traits of a guitar speaker

A speaker cone should be light (so it can move fast and reproduce fast signal transients better) yet rigid (so it doesn't flex and distort), meaning it should have a high degree of internal damping. As it happens almost always in electronics and acoustics, these are contradictory requirements.

Speaker designers and builders (especially of hi-fi speaker drivers) have been trying various materials, almost all of which are still in use - paper (cut and seamed), molded paper pulp, metal, plastic (polypropylene, for instance), and various types of composite materials and layered (sandwiched) hybrids.

No matter how large a cone of a dynamic driver may be or what material it's made of, all cones are driven by a relatively small area of the voice coil and not through their whole surface. At low frequencies (signal wavelengths much larger than the cone size), that is not an issue; a cone moves as a rigid whole (a "piston").

As the signal frequency increases and wavelengths become smaller than the cone, the cone reaches the limits of its rigidity. It starts to act as a mechanical transmission line with distributed parameters (mass, resistance, and compliance).

Such flexing is comprised of a multiplicity of mechanical waves ("ripples") traveling outward as pictured (1) and then reflecting back onto themselves with little attenuation. This effect is known as the "cone breakup."

Cone breakup results in standing waves in various directions, for instance, around the circumference and along the flared edges, as illustrated below.

When the cone radius equals one wavelength (or its multiple), the peaks and dips outward and reflected waves coincide to produce a seemingly stationary ripple or undulation of the cone, also known as a standing wave, so called because the peaks and nulls (zeroes) are always at the same distance from the cone's center, thus appearing as the wave isn't moving at all.

The transverse waves don't just make the speaker's SPL curve peaky; such motion cannot stop immediately after the music signal stops, it continues to be radiated by the cone, and such stored mechanical energy is released as spurious sound that takes time to decay, thus smudging the sound and producing a delayed distortion.

For an 8" speaker, whose effective diameter is around D=17.5 cm (6.75" or 0.175 m), the lowest standing wave frequency is $f_1=v/\lambda = v/r = 2v/D = 2*344/0.175 = 3{,}931$ Hz, or approx. 4 kHz. Its harmonics are then found at $f_2=2f_1=$ 8 kHz (7,862 Hz), and $f_3=3f_1=$ 16 kHz (15,793 Hz).

For a 10" speaker, with effective diameter D = 21cm (0.21m), the lowest standing wave frequency is $f_1= 2v/D = 2*344/0.21 = 3{,}276$ Hz, with harmonics at 6,552 Hz and 9,828 Hz.

A 12" speaker, with effective diameter D = 25cm, will breakup earlier, at $f_1= 2v/D = 2*344/0.25 = 2{,}752$ Hz, with harmonics at 5,504 Hz and 8,256 Hz.

Finally, a 15" driver will start exhibiting cone breakup symptoms earlier still, at 2,293 Hz, followed by 4,586 Hz and 6,879 Hz!

BELOW: Typical sound pressure level versus frequency response ("the SPL curve") of a guitar speaker. Standing waves in various directions, around the circumference "bell" modes) and along the flared edges (concentric or nodal modes), result in resonant peaks of the SPL curve. One or more midrange dips are also often present, caused by the surround moving the opposite way to the cone at some frequencies (f_D)

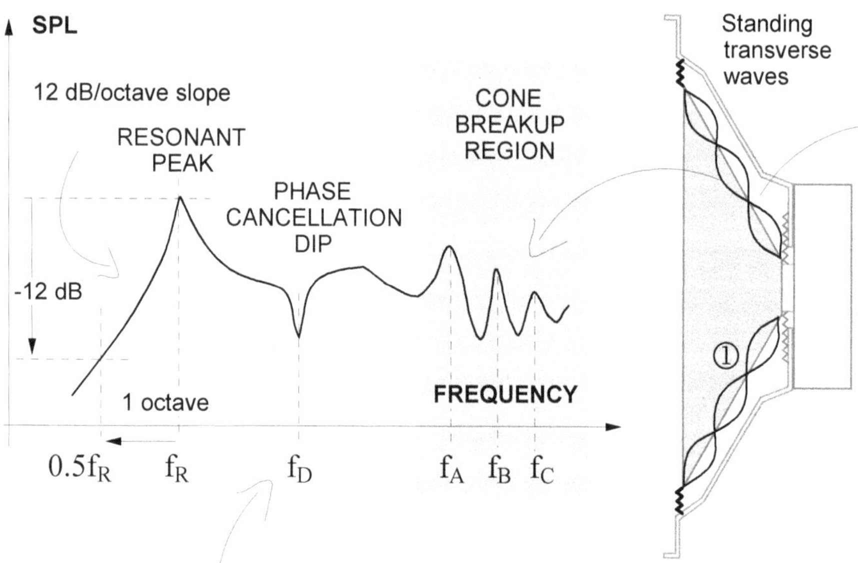

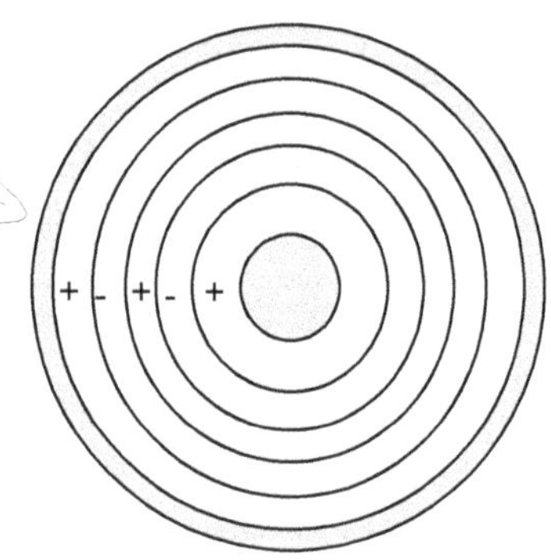

ABOVE: one of many concentric or nodal modes

BELOW: one of many radial, circumeferential or "bell" cone modes

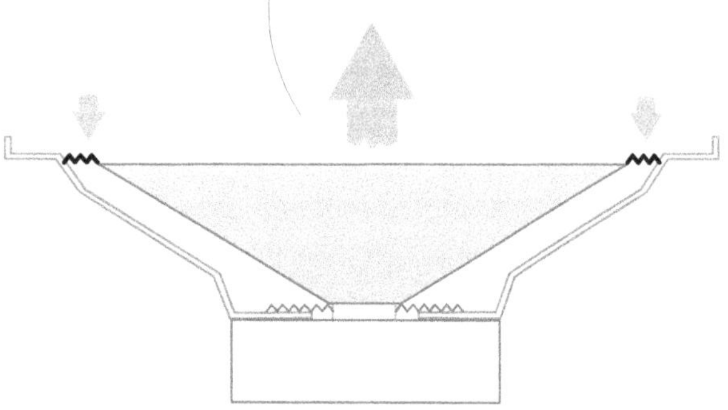

LEFT: The surround moves the opposite way to the cone at some frequencies. This is especially prominent when the surround has many rim folds, causing the cone to vibrate severely off its axis.

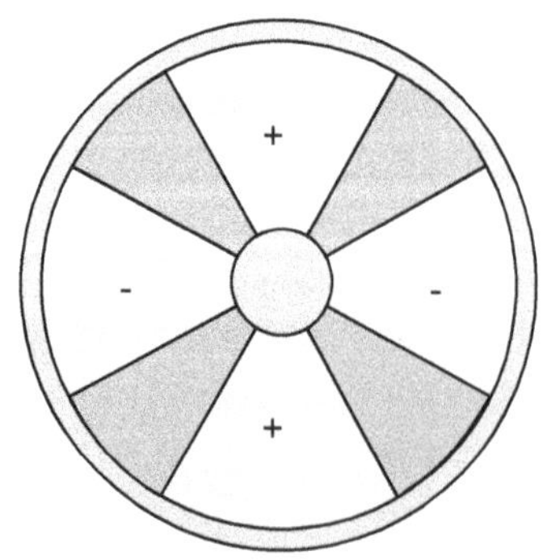

Speaker cones also suffer from radial or "bell" flexure modes (similar to the resonances that makes church bells produce sound), at frequencies where the wavelengths of the sound wave are an integral multiple of the cone's outer circumference.

Both transversal (concentric) and radial standing waves happen simultaneously, reinforcing each other at some frequencies and attenuating or canceling each other at others.

In hi-fi speaker drivers, a considerable effort is made to increase the cone surround's ability to absorb the ripples, but in guitar speakers they are considered one aspect of the speaker's character and sonic signature. The rubber or foam surrounds, which are superior absorbers compared to paper or cloth surrounds, are common in hi-fi speakers, but are never used for guitar speakers.

BREAKING IN SPEAKERS - MYTH OR REALITY?

Mechanical break-in aspects

Most audiophiles believe that all audio equipment and even individual electronic components (on the micro-level) change their sonic signature after a few weeks or months of use.

Car analogies are very useful in explaining audio phenomena, and one such analogy is in order here. Most brand new cars take a few months to "settle down" and generally drive better, smoother, and quieter after 3-5,000 km. The primary issue of engine break-in is the settling of piston rings against the cylinders' walls. Still, the brakes, the hydraulic systems, air-con compressors, and many other mechanical and hydraulic devices also perform better after the initial settling period.

By extension, electromechanical transducers with moving parts, such as loudspeakers, are also subject to the settling-in syndrome and thus benefit from a burn-in period.

When brand new, the speaker's moving parts have never moved before. While the cone relaxes and changes when subjected to the initial stresses and strains during the first hours of use, the speaker's suspension will undergo more radical changes.

The spider (lower suspension) and the cone surround (upper suspension), are initially stiff, and the breaking-in process is simply a change in their materials when exposed to repeated flexing. Both the fibers in their cloth material and the epoxy or phenolic resins used to bind and stiffen them will suffer microscopic breaks and cracks and will "settle in", increasing the flexibility of both suspension components.

This makes the speaker more "compliant", lowering its compliance C_{MS} (one of the Thiele-Small parameters), so its resonant frequency will be lowered significantly, at least 5-10%, but in many cases (depending on the speaker, its construction and the materials used), even up to -20%! So, a speaker with f_S=90Hz will change into one with f_s between 72 and 81 Hz.

This will have a an effect on most other TS parameters of the speaker: V_{AS} will increase but both Q_{MS} and Q_{ES} will go down, so the total Q (Q_{TS}) will be reduced quite a lot (say from 0.4 to 0.28).

Some speakers need to be broken-in again after a prolonged period of being left unused (a week or more). This is most likely due to the resins in their spiders/suspensions that harden again if not in continuos use.

Ultimately, in terms of its sonics, a broken-in speaker will undergo a subtle but noticeable change in its tone. The initial harshness will be reduced, smoothing-out the treble frequencies and increasing the overall warmth. The bass will be deeper and "fatter".

RIGHT: A loosened-up, more compliant suspension, resulting from a break-in period, will result in a few easily measurable changes in speaker's impedance curve and thus its tone:

1. lower resonant frequency
2. higher resonant impedance peak
3. slightly lower inductance, resulting in smaller impedance rise at high frequencies

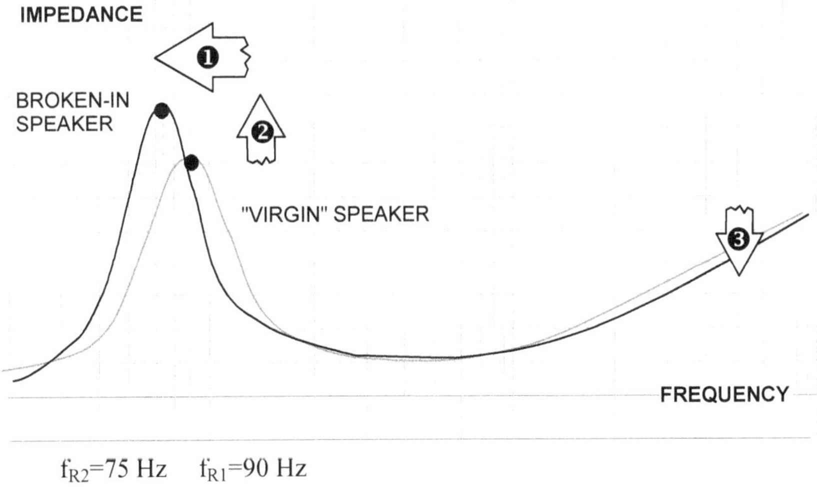

THE SONIC VARIANCE BETWEEN SEEMINGLY IDENTICAL COMPONENTS

Complex components of a mechanical nature (with many interrelated parts), especially if built by human hand (phono cartridges, vacuum tubes, loudspeakers, and headphones), can significantly vary in sound quality, even when made in the same factory. Auditioning one such component (well broken in) and buying another, still wrapped and "virginal," has resulted in the disappointment of many guitar players.

Electrical break-in aspects

While even the hard line absolutists or reality deniers (those fanatics who claim that only measurements count, that sonic changes that can be heard by most but not measured simply did not happen) would agree that a mechanical break-in makes sense, the real debate is about electronic components and systems, those seemingly without any moving parts.

Do electronic components (cables, tubes, transistors, capacitors, inductors, transformers, etc.) sound different after fifty or a hundred hours of usage? I believe they do. Such changes in sonics can be detected even when changing resistors or capacitors but are very noticeable with complex audio equipment such as amplifiers, speakers, and CD players.

There are many possible reasons for the change in sonics over time. Continuous cycles of heating & cooling (thermal expansion and contraction) are just one possible explanation in the case of both transistors and (especially) vacuum tubes inside amplifiers, but also voice coils inside speaker "motors". Subtle changes in the structure of their materials (copper or aluminium wire, but also the voice coil formers) are another.

Capacitors in amps and speaker crossovers, transformers, chokes, and speaker magnets exhibit "memory" effects (hysteresis, for instance), where the state of the magnetic and dielectric circuits depends on their previous operating conditions. If that isn't a temporal change, I don't know what is.

Matching the speaker's character with your personality and other factors in play

Matching our personalities? Dr. Iggy, are you crazy, is this a pseudo-psychological dating book or what?!? If that was your first thought after reading the subtitle above, bear with me for a few seconds more ...

Each speaker (both the driver itself and the box it's mounted in) has a personality, a character. That even applies to high fidelity speakers (which are meant to be as sonically neutral as possible, since they are "*re*producers" of the recorded sound), but even more so to instrument speakers.

A speaker in the guitar and bass context is a sonic producer; what it suppresses and obscures versus what it emphasizes, what it draws to players' and listeners' attention to, is its essence, it's personality, its "soul". Just are there is no perfect guitar amp that will be so versatile that it work equally well will all kinds of music (blues, rock, heavy metal, ...), no speaker will sound "appropriate" in all situations and for all music genres!

By "situation" I mean first and foremost the size and the acoustic properties of the room or venue it is in, but also how well it "matches" the amplifier that drives it.

Some speakers sound good at low volumes and SPL labels, but break up early and turn into harsh, glassy and/or brittle-sounding pest repelling devices when cranked up. Others sound dull and anemic when you try them at low volumes in a music shop or when you practice at home, but come alive when cranked up and pushed close to their mechanical and electrical limits.

NOTES

ENCLOSURES GALORE - TYPES OF SPEAKER CABINETS

5

A speaker (driver) without some sort of cabinet or enclosure would sound tinny and nasal. Due to the mechanical forces acting on its cone, it would also jump around like a stock market during its most spastic fluctuations.

Also, without the structural protection of a cab, its delicate cone and precisely-centered voice coil mechanism (motor) wouldn't last long in the rough & tumble music world.

Many types and designs of speaker enclosures have been devised in the last hundred or so years, but only a few main types are currently used commercially, namely open back, closed back (sealed box) and bass-reflex.

- TWO MOST COMMON TYPES OF GUITAR CABS: OPEN BACK AND SEALED BOX
- BASS-REFLEX ENCLOSURES
- DOUBLE-BAFFLE PORTED DESIGNS OF YESTERYEARS
- OTHER TYPES OF SPEAKER ENCLOSURES - PASSIVE RADIATORS, TRANSMISSION LINES, AND HORNS
- RESONANCE EFFECTS IN SPEAKER ENCLOSURES

TWO MOST COMMON TYPES OF GUITAR CABS: OPEN BACK AND SEALED BOX

The ultimate loudspeaker problem: how to faithfully reproduce bass frequencies

Most loudspeaker enclosure problems fall into the bass and low midrange frequency bands, typically below the frequency at which the speaker's cone's circumference equals the wavelength of the sound wave.

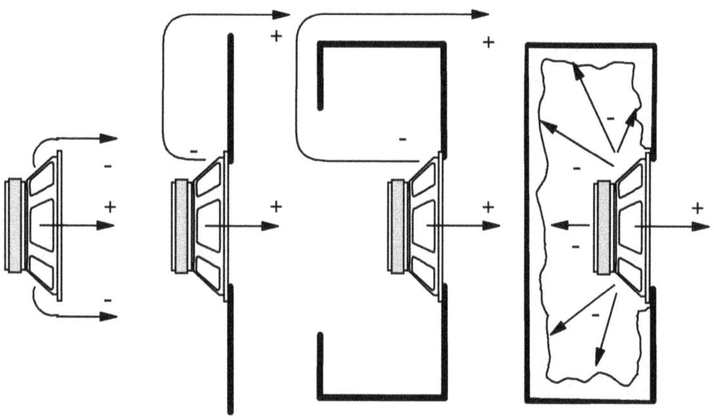

ABOVE: The progression from "naked" driver, through flat baffle and open back cabinet, to the closed (sealed) box

For a 10" (25.4 cm) speaker with a 21 cm effective cone diameter, the circumference is $C = D*\pi = 66$ cm or 0.66 meters.

Using the fundamental acoustic formula $\lambda = v/f$ (v is the speed of sound in air, 344 m/s, λ is the wavelength, and f is the frequency of the sound wave), we can calculate that critical frequency as $f = v/\lambda = 344/0.66 = 521$ Hz.

Using a "naked" driver, just a loudspeaker suspended on some sort of support, the bass frequencies (below say 500 Hz in the example above) would be severely attenuated. The front and rear sound waves would be of equal magnitude but opposite phase, effectively canceling each other. The speaker, no matter how large, would sound "small" and "treble-heavy".

Mounting a speaker on a flat baffle

To properly reproduce low midrange and bass frequencies, an enclosure of some sort is required in order to "separate" the front and rear sound waves produced by the moving cone, so they don't cancel each other.

Mounting a driver onto a flat baffle reduces the cancellation of front and rear sound waves, but it's still far from a feasible or practical arrangement. For instance, to reproduce a modestly low frequency of 50 Hz (generally considered insufficiently low for top-quality hi-fi systems and a bare minimum for bass guitar cabs), the baffle would need to be at least 4 x 4 meters in size!

Flat baffles sound better than is generally believed. Unlike all other types of speaker enclosures, they are free from cabinet resonances and associated colorations and, in case of poor designs, a "boxy sound" character. For this and other reasons, expensive flat baffle speakers are currently a rage in high-end audiophilia. Some use a single full-range drivers and no crossovers, while others opt for 2- or even 3-way designs with crossovers. However, for obvious reasons (size and lack of driver protection), flat baffle designs are not nor will ever be used for guitar and instrument speakers.

Open back cabinets

Folding the baffle a few times into a cabinet open at the back gets us closer to the goal of a reasonably sized loudspeaker enclosure, an approach used in many guitar amplifiers. The rear sound wave's path to the listener's ear is much longer than the direct, frontal path, yet, compared to large flat baffle that would need to be used, the overall cabinet dimensions are much smaller.

An electric guitar amp and the cab don't have to amplify and reproduce frequencies below 80 Hz, so open-back cabinets are fine.

Open back cabinets are never used in hi-fi loudspeakers since they suffer from multiple resonances (standing waves). The open box cab acts as a resonant tube or cavity, boosting frequencies around its audible resonance, resulting in a dominant "one-note bass" or "boombox" character.

To make matters worse, as the frequency of reproduced sound increases, the cab exhibits a series of minor resonances evident as peaks and dips on the cab's frequency response curve.

ABOVE: Although the back panel covers most of its rear, this combo amp is still classified as an open back cabinet

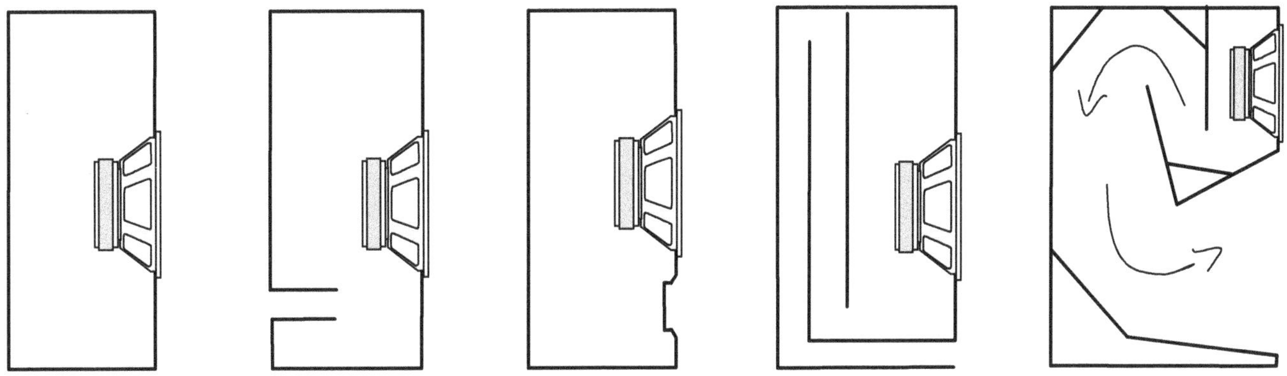

ABOVE: A few of the most commonly used enclosure types using dynamic speaker drivers.

Closed-back cabs (sealed box)

Bass guitar amps, on the other hand, require closed cabinets for proper bass extension, as do their hi-fi brothers. Closed cabinets eliminate all the radiation from the driver's rear, but solving one problem causes a few new ones.

The trapped air behaves like a spring, so its "springiness" or "compliance" (a fancy word for elasticity, the opposite of stiffness) increases the resonant frequency of the sealed speaker box above the resonance of the driver itself.

As the frequency of reproduced sound increases, the speaker box exhibits a series of minor resonances evident as peaks and dips on the speaker's frequency response.

These can be minimized (but never completely eliminated) by padding or stuffing thc cnclosurc intcrnally with fiberglass, long fiber wool, polyester filling (Foam Fill, Poly Fill, Fairy Floss Fill, Acousta-Stuf, etc.), or bonded acetate fiber (BAF) wadding.

The effect of such stuffing is also an increased apparent size (volume) of the enclosure.

Of course, the physical size of the enclosure does not change, we are talking about an acoustic size, the size the speaker "sees".

The air trapped in a sealed enclosure resist the movement of the loudspeaker's cone, acting as an acoustic stiffener, reducing the compliance.

The effect of this stiffening is the increase of the resonant frequency of the speaker in such an enclosure. The smaller the box, the higher the stiffening effect and the higher the resonant frequency is raised!

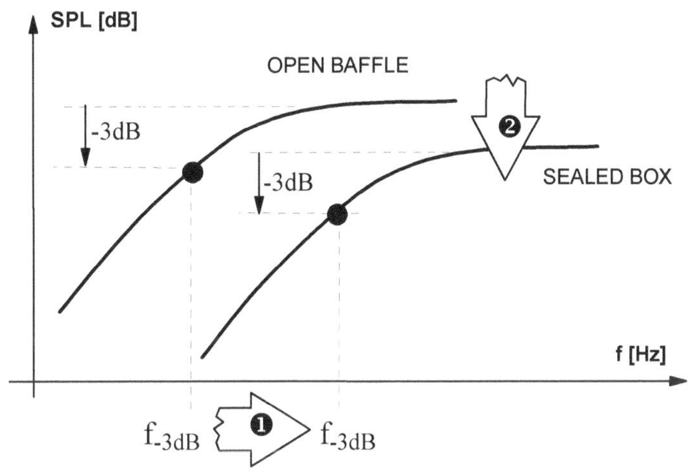

ABOVE: Compared to the same driver on an open baffle, once it's enclosed in a sealed box, the resonant frequency goes up, and so does the -3dB frequency (1), meaning lower bass is attenuated. The efficiency and the sensitivity are also reduced (2).

BELOW: The smaller the enclosure, the higher the rise in both the resonant frequency and the Q (quality factor) of the enclosure

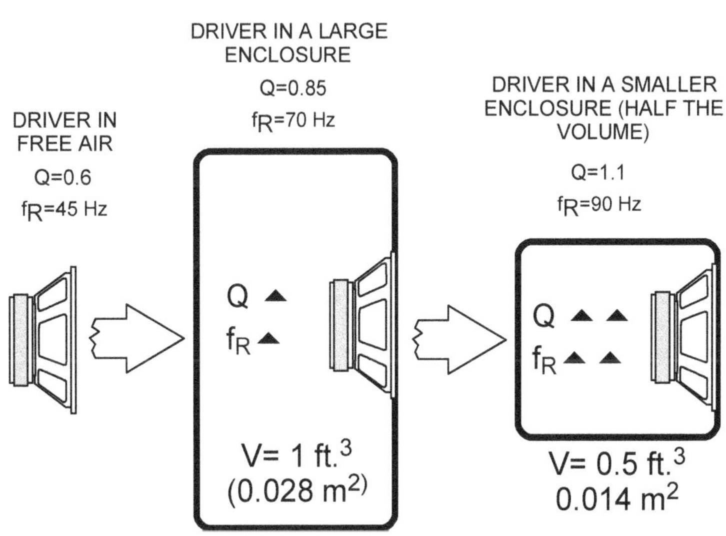

Hofmann's Iron Law

Named after Josef Anton Hofmann (1924-2010), an audio engineer and speaker designer of the KLH fame, this "law" is a simple observation that many of his notable contemporaries in the audio business also made, but for some reason his name remained associated with it.

The realization was of an inevitable tradeoff between three principal aspects of speakers mounted in enclosures - deep bass (low frequency reproduction), small enclosure size, and high sensitivity (and the related parameter, efficiency). A designer may choose to optimize two of these three parameters, but in doing so, will compromise the third one. For instance, you can obtain a deep bass extension (low -3dB or cutoff frequency) and high efficiency and sensitivity (SPL levels), but only with a very large enclosure.

Likewise, I have seen and heard many highly acclaimed bookshelf hi-fi speakers (small size), producing copious amounts of bass, but their sensitivity was atrocious (extremely low), meaning they were notoriously hard to drive, ruling out all low powered tube amps, even many higher-power ones!

A typical example is the famous family of LS3/5 hi-fi speakers, made over decades for BBC as studio monitors by Rogers, Harbeth, Spendor, and a few other companies, with a sensitivity (on a good day) of only 82 dB/W/m.

A formula for the efficiency of a sealed enclosure describes Hofmann's "law" in a single line. Efficiency is proportional to the internal volume of the box V_B, but by far the most significant factor is the -3dB or cutoff frequency f_{-3dB}, since it is cubed in that formula.

Doubling the size of the box doubles efficiency, but to half the cutoff frequency (lower it by only an octave) will cause the efficiency to drop $2^3 = 8$ times!

EFFICIENCY OF A SEALED BOX SPEAKER

$$\eta = k * f_{-3dB}{}^3 * V_B$$

k= constant

f_{-3dB} = -3dB (cutoff or corner) frequency

V_B = enclosure (box) internal volume

Bipolar versus dipole speaker cabs - what's the difference?

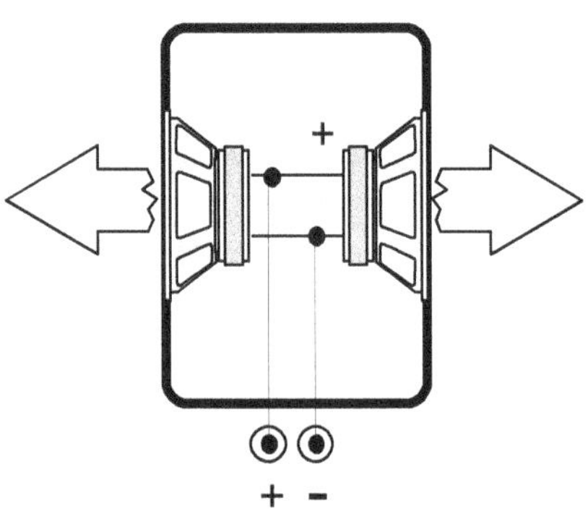

A BIPOLAR SPEAKER CAB

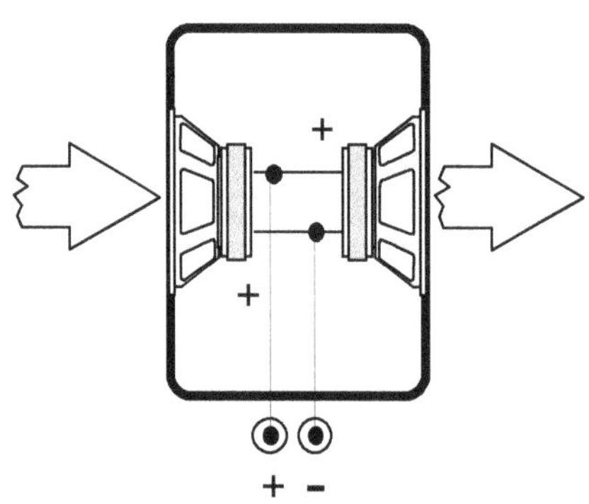

A DIPOLE SPEAKER CAB

A bipolar ("two poles") speaker is a bi-directional *in-phase* arrangement of two drivers sharing the same enclosure) facing in different directions. At an extreme, the two drivers can face in the opposite directions (a "back-to- back" placement).

A dipole speaker (or enclosure) is also a bi-directional arrangement of two drivers sharing the same enclosure, but the drivers are *out-of-phase* (or of the "opposite phase").

The impedance & SPL characteristics of Marshall 4 x 12" cab before & after modifications

Marshall MC412A-E is an "entry-level" angled 4x12" full-size cab. Rated at 200 watts mono (4Ω) or 2 x 100 watts stereo (8Ω each), it features four 12", 16Ω 50 watt Celestion Rocket 50 drivers. Its retail price in Australia in 2023 was AU$699, but we got it at a fire sale for AU$450.

MC412A is a sealed cabinet, yet there is no internal padding at all, meaning standing waves and multiple internal reflections and interference of the cones' back waves will be prominent, resulting in severe peaks and dips in its frequency response.

Notice how two left drivers are connected in parallel, their output (7) hooking up to one side of the stereo-mono switch (located on the back cover - not shown), and how the two drivers on the right hand side are connected in the same way, output (8). A cross-brace (9) is also prominent, one self-tappered screw attaches the rear cover to it.

We measured this cab's impedance and SPL characteristics in its original state, both against frequency, of course. Then, we covered the two sides and the bottom panel with acoustic padding. The back cover was braced using four parallel braces (1), and then padded with the same material (2).

The same two measurements were repeated on the modified cab. Notice I didn't say the "improved cab", but I do believe the improvement was obvious, both on paper (graphs on the next page) and during the actual play-test by our professional guitarist friends.

This wasn't a proper, scientific "blind" test, we had only one cab. For a proper blind test we would need two cabs, one in its original state and the other modified, so the guitar players evaluating the cabs' sonics would not know which is which, so there would be no bias of any kind.

ABOVE: The two drivers on the left are paralleled (16 Ω || 16Ω = 8 Ω), as are the two drivers on the right side. When the back switch is in "MONO" position the two 8 Ω pairs are paralleled again, resulting in a 4 Ω cabinet.

LEFT: Marshall 4x12" cab after modifications, parallel bracing (1) and acoustic padding of the back cover (2) and the three sides (3). Obviously, the acoustic padding (2) was stapled on top of the bracing after the photo was taken.

The original cab had an impedance peak at around 105 Hz, of just under 11Ω (1), which seemed pretty innocuous. After the two mods, the resonant peak was slightly lower, around 10Ω (2), but the resonant frequency shifted downwards, from the original 105 Hz to 90 Hz, which was quite a significant improvement.

The original cab had a prominent SPL peak at around 120 Hz (4), which resulted in a bass boost and boomy character of the lower registers. The modification lowered the average SPL levels, which, together with the lowering of the resonant frequency, makes us think that the original cabinet was "leaky", not truly behaving like a sealed cab.

The lower f_R indicates that the drivers "see" the modified cab as a larger volume enclosure, with an increased compliance. Remember, the smaller the apparent acoustic volume of an enclosure, the more the enclosure will raise the resonant frequency of the driver(s).

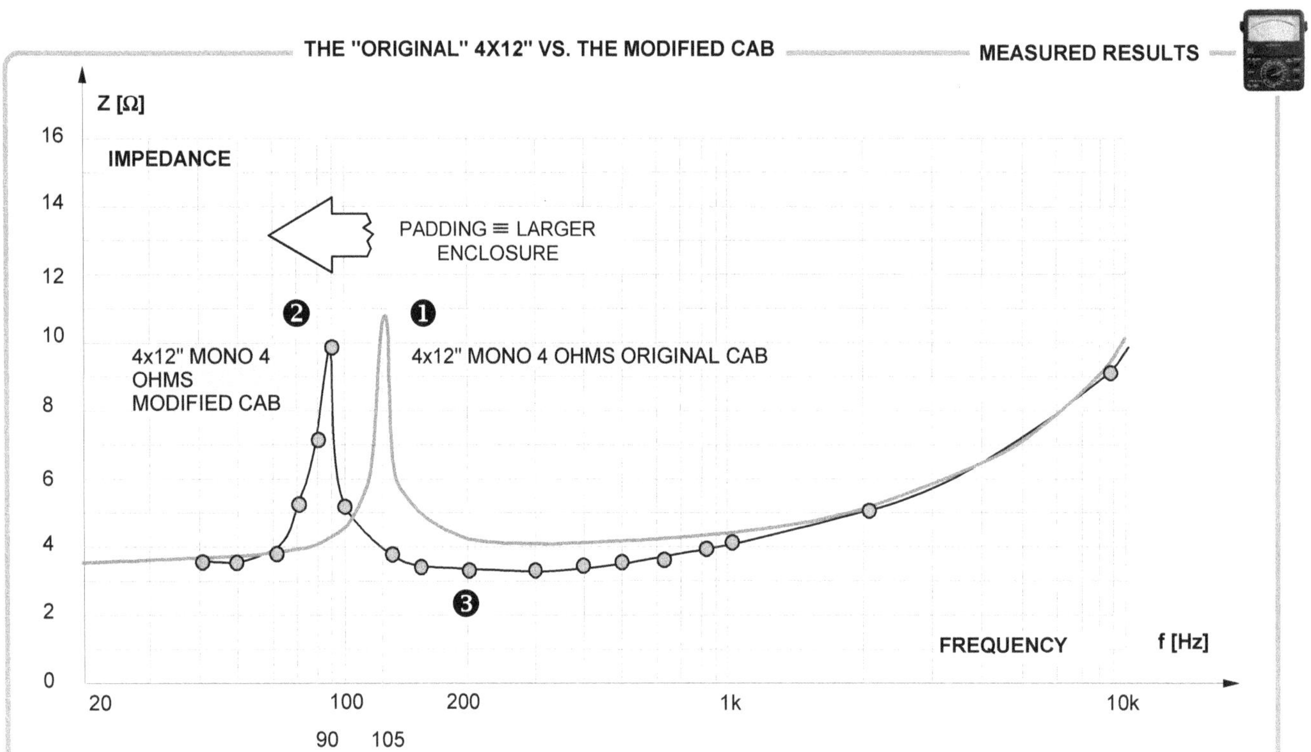

ABOVE: The impedance curve of the original 4x12" cab before & after bracing and acoustic padding

BELOW: The SPL (sound pressure level) curve of the original 4x12" cab compared to the SPL curve of the same cab after bracing and acoustic padding

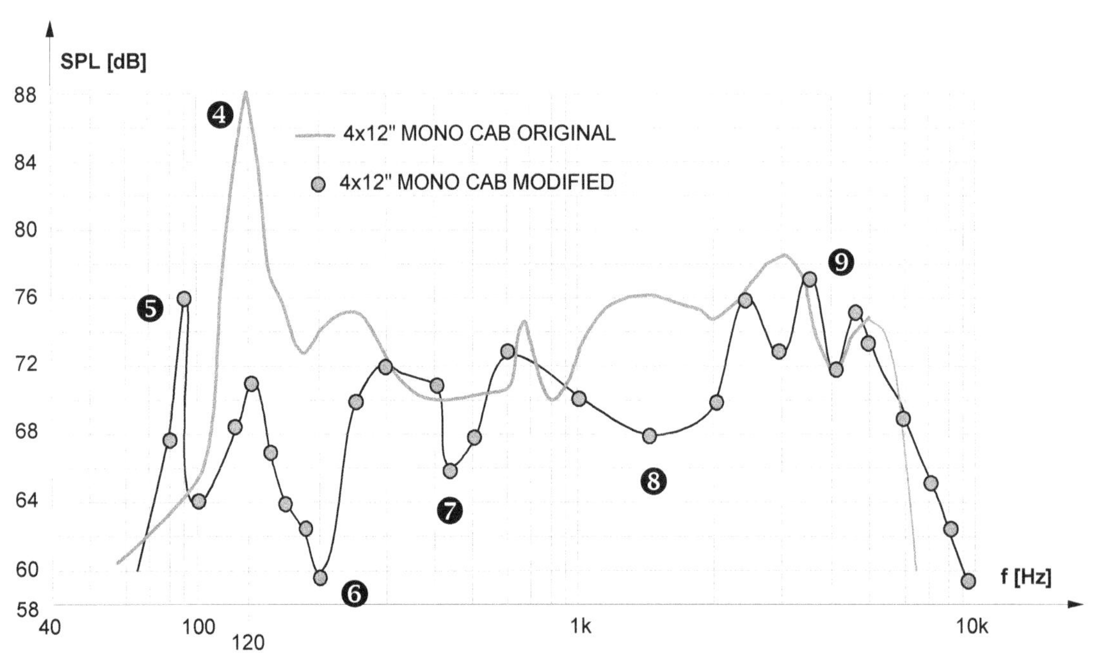

Also, a leaky enclosure (behaving almost like an open-back cab), will produce higher SPL levels than the properly sealed box, which would have a reduced sensitivity, as was definitely the case here.

The modified cab had a much milder SPL peak at a much lower frequency (5), 76 dB at the new resonant frequency of 90 Hz, followed by another smaller peak at 120 Hz, and a significant dip at 200 Hz (6), which was audible during play tests.

Another dip at 420 Hz followed (7), but what really defined the sonic character of the modified cab was the "scooped-up" midrange, from about 600 Hz to 2.2 kHz, centered around the serious dip at 1.5 kHz (8). A couple of peaks and dips in the 2-5 kHz range, most likely the results of cone breakup effects were somehow more prominent than with the original cab (8). After 5 kHz, there was no notable difference between the two cabs.

BASS-REFLEX ENCLOSURES

How they work

Instead of trapping the woofer's rear sound waves and dissipating their energy into heat, a bass reflex cabinet utilizes a duct or "port" to delay the back waves enough to invert their phase. They exit the enclosure in phase with the main, front-radiated sound waves, thus reinforcing them and producing higher SPL levels than an equivalent sealed box speaker. For that reason these designs are also known as phase inverter cabs.

The port can be facing forward ("front-firing") or rear ("back-firing"). Its opening can be round, rectangular or any other shape.

To achieve the required volume of air in the port, smaller (mostly hi-fi) enclosures, use ports of a small cross-section but deep (a long tube).

Larger (musical instrument) cabs often have a large rectangular port spanning the whole width of the cab, but there is no added tube as such - the front baffle thickness is the depth dimension of the port.

As for the impedance curve, the cab no longer has a single resonant peak at f_R; Instead, two lower amplitude peaks appear, about an octave above and below it, one corresponding to the in-phase motion and the other resulting when the two oscillatory components move out of phase.

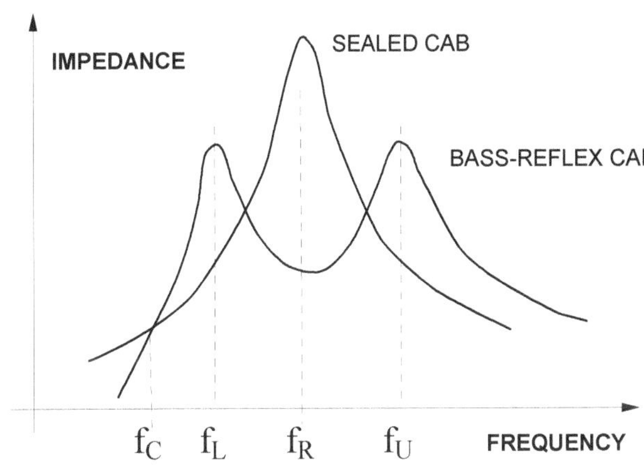

ABOVE: Typical impedance versus frequency response of tuned port ("bass reflex") enclosure and the sealed box at bass frequencies

The Helmholtz resonator

Named after German physicist Hermann von Helmholtz, this resonator consists of a cavity with rigid walls and volume "V," with a neck area "S" and neck length "L." The acoustic pressure in the cavity provides the stiffness element (acting as a spring), and the air in the tube (neck) provides the resistance, the mass element.

At resonance, the mass of air in the neck swings back and forth, and the mass of air in the cavity V acts as a spring, resisting that movement. The mechanical equivalent of such a resonant system would thus be a mass suspended on a spring.

The fundamental resonance occurs at the angular frequency ω_R (formula on the right). The resonant frequency f_R is $f_R = \omega_R / 2\pi$.

The port of a bass reflex speaker cabinet is the neck, with "S" its opening area and "L" its depth or length., while "V" is the internal volume of the enclosure.

"Tuning" the port involves adjusting "S", the area of the port, or "L" (the depth of the neck or port opening), often both. The enclosure volume "V", once chosen, is a given and cannot be changed.

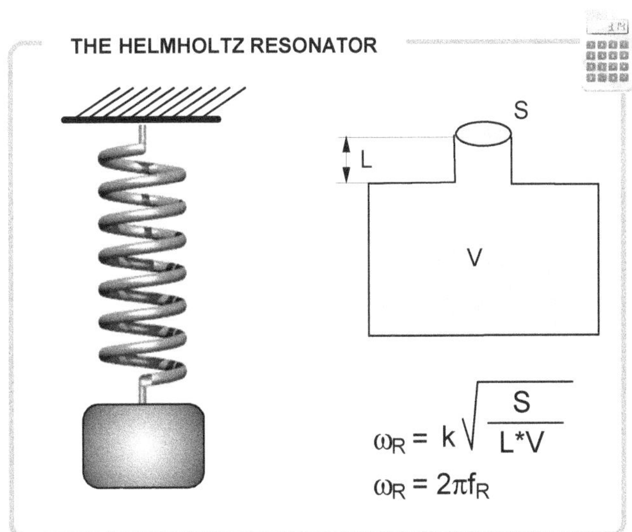

The Q or quality factor, is an indicator of the resonator's behavior at the resonant frequency. A higher Q factor means a more efficient resonator (lower losses) and higher resonant amplitude.

On the right is an SPL curve of a typical bass reflex speaker around the tuned frequency (70Hz).

The port is a narrow bandpass filter, whose characteristic is almost of an inverse shape to the cab's. As the port's tuned frequency is approached (from either side), the contribution of the cab diminishes, and the SPL is maintained by the increasing contribution of the column of air oscillating inside the port.

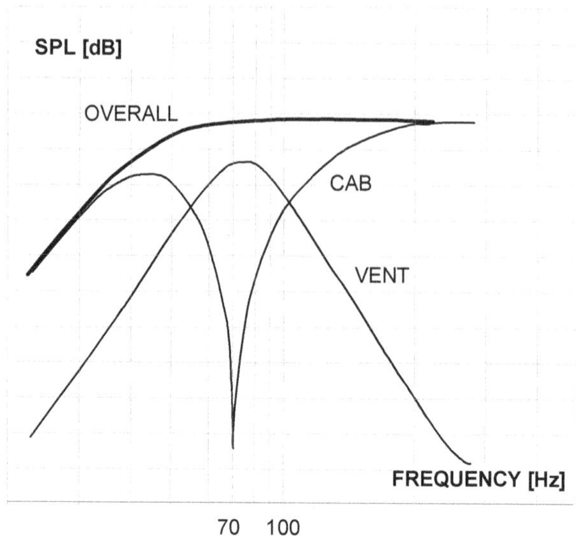

RIGHT: The relative contributions of the cab and the port around the tuned frequency of the port

Port tuning on bass-reflex speakers: controlling the bass response of a vented cab by its port

As an oscillating system, the bass-reflex speaker has two components coupled together, the driver (woofer) itself and the air from the port, making it a tuned system. The tube's size and length determine the system's Q-factor (quality factor) and its resonant frequency.

The tube's cross-sectional area and its length are chosen to extend the lower bass so that the resonant frequency of the enclosure (a Helmholtz resonator) equals the resonant frequency of the speaker (f_0), hence the term "tuning" of the port. A wider or larger port (cross-sectional area) raises the resonant frequency of the enclosure, while a longer neck (tube) lowers it.

When properly tuned, the two oscillating components become so strongly coupled (in the acoustic sense) that the resultant SPL-vs-frequency characteristic features two equal, symmetrical peaks. A smaller or nonexistent lower peak indicates that the port is too small or only partially open.

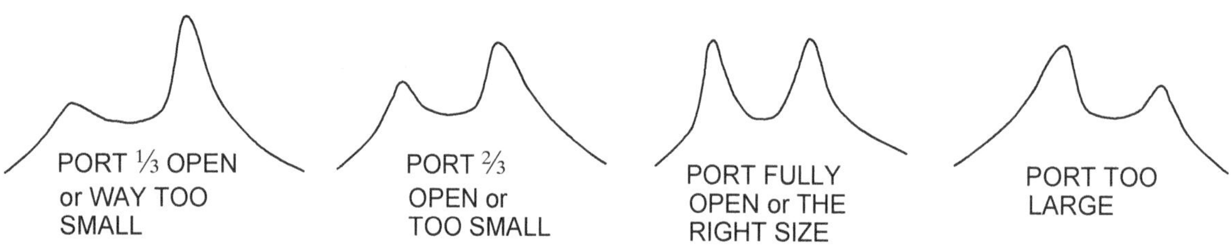

ABOVE: The impedance of a bass-reflex cab in the lower bass region will indicate if its port is of the optimal size (two equal peaks), too small (two cases illustrated) or even too large (lower upper peak).

BELOW RIGHT: How to determine the required port area (in square decimeters) for a given cabinet volume (in cubic decimeters or liters) and the driver's resonant frequency (in Hz)

What would happen if we blocked the port of a bass-reflex enclosure? How much of a difference would that make to its bass response?

These comparison curves (next page) will shed some light on the issue. Keep in mind that the differences in SPL in decibels and the frequencies at which dips and peaks happen will depend on the actual speaker and the enclosure, we are looking at the principles and general conclusions here.

For this driver and this enclosure the blocking of its port had no effect above 200 or so Hz (5).

As we went deeper into the bass territory, the SPL differences between open and closed port became more pronounced.

The lowest peak in the response was at 50 Hz (1), the next one around 85 Hz (3).

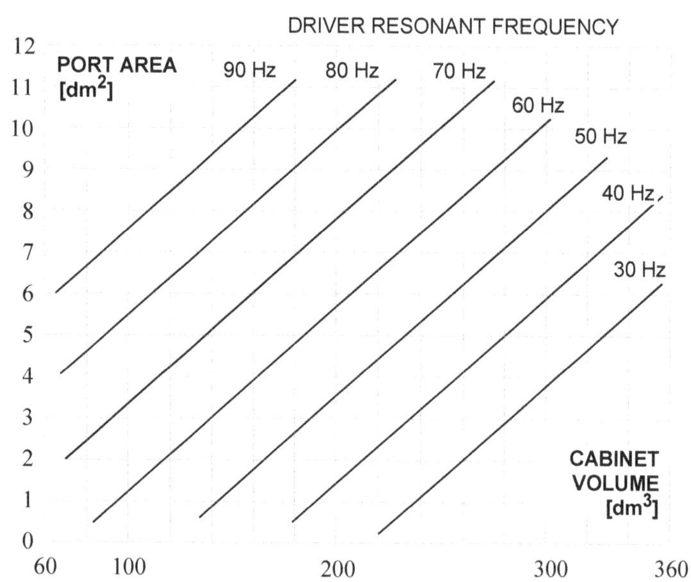

The two dips were at 72 Hz (2) and 92 Hz (4). Between 50 and 100 Hz, blocking the port reduced SPL 4-5 dB, which would be noticeable.

So, if you feel your vented cab's bass is too boomy, try blocking its port fully or partially. This is much easier in the case of a small (usually circular) opening of a ducted port.

Just get a piece of dense foam and cut a cylindrical shape out of it. Unless you want only a partial blocking of the port, it should fit snuggly into the tubular port.

A large and shallow (the thickness of the front baffle) rectangular port is a bit more difficult to block. Cut a suitably-sized plywood piece and screw it on top of the opening, making sure the seal is airtight.

JVC VS-5323 are typical 1970s hi-fi speakers in every sense (the veneer finish, the 3-way design, the drivers used), all except one; they included a continuously tunable bass-reflex port (photo below). This unusual front baffle-mounted control was used to vary the bass response of the cabinet.

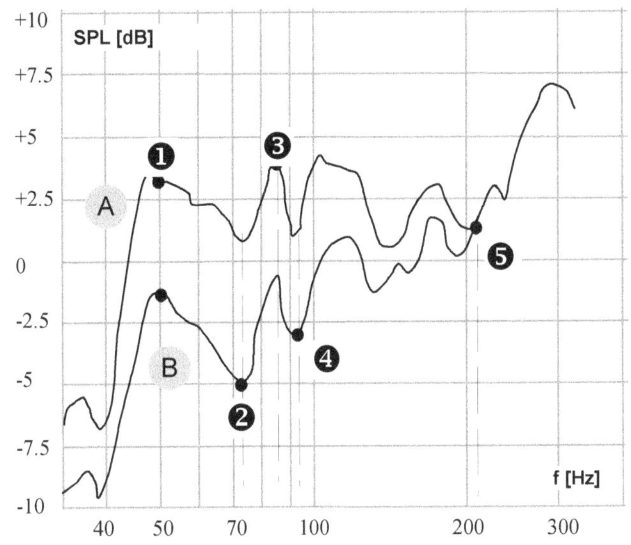

ABOVE: The low frequency (below 400 Hz) response of a 10" driver in a 75 liter bass-reflex enclosure: A) port open B) port blocked

From the "0 dB" position, turning the dial CCW to +2dB (1) opened the circular port, making it a tuned bass-reflex enclosure. Rotating the dial in the other direction closed the port, so in the extreme "-2dB" position (2), the port was fully closed and the cab worked as an air suspension enclosure and the low frequencies were damped.

Moving the dial further left or right from these two settings repeated the behavior of the aperture system, with another "0dB" setting directly opposite the top one (3).

Marantz models, HD-66, HD-77 and HD-88 (1976) included the "Vari-Q" feature where the user could insert or remove a foam plug (with an aluminium front and handle attached) from the circular port.

From their promotional literature: "By removing the acoustical plug to change the system to a tuned port design, the speaker output is greater (3 to 4 dB) in the 30 to 75 Hz region. In this frequency range the speaker's power requirement is cut in half. However, extreme low frequency response (below 50 Hz) will be reduced when the plug is removed. Again, experimentation and personal taste will determine the ideal arrangement."

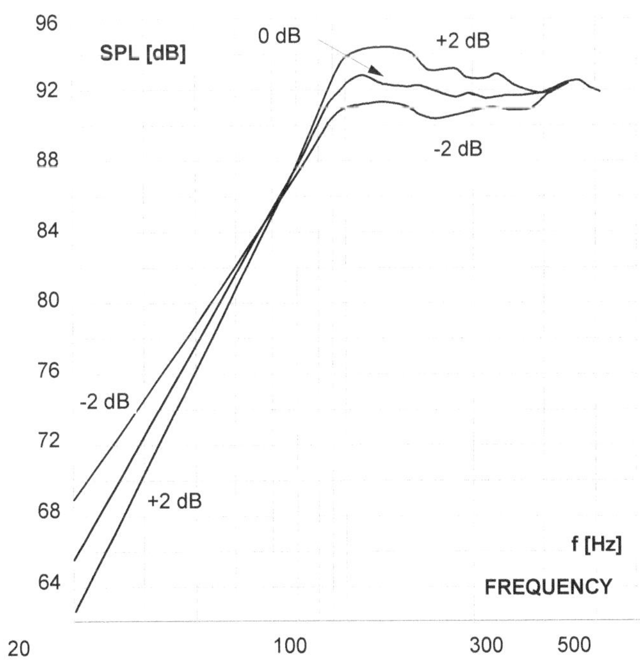

ABOVE LEFT: Another Tone Tweak from the last century. Back in the 1970s, JVC VS-5323 high fidelity speakers had a tunable bass-reflex port

ABOVE RIGHT: The LF response of JVC VS-5323 speaker at three settings of its tunable bass-reflex port

What happens when we stuff a sealed or bass-reflex cabinet?

Stuffing the enclosure with damping material absorbs some of the natural cabinet resonances and, since it minimizes multiple reflection paths, it also prevents others from developing. By pushing and rubbing the fibers against each other, the driver's rear radiation energy is converted into friction and thus heat, which is then dissipated.

Long-fibre wool and long-staple polyester are two of the best damping materials, followed by cotton batting. Fiberglass and foam bats are less effective and impart a drier tone to the cab.

Most guitar speaker cabs come without any inside lining or stuffing. Perhaps their manufacturers are trying to save every penny, or perhaps they deliberately aim at achieving a peaky response.

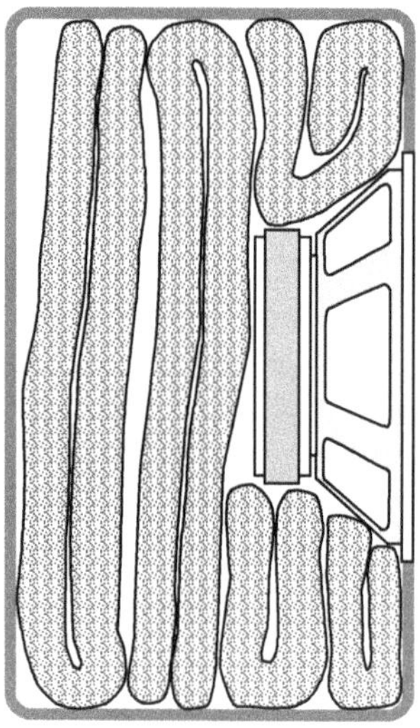

As we have seen, the pronounced peaks and dips in the produced SPL are due to the rear waves bouncing around between the walls, enforcing each other at some frequencies (wavelengths), while canceling or attenuating at others.

The easiest modification is to line all internal walls with absorbent sheets and staple them securely. This will reduce the tangential and oblique resonant modes (standing waves), but for axial resonant modes, the sound absorbent material is most effective not at the internal surfaces but at points of maximum velocity (the speed of the sound waves), which is always away from the walls.
Open any good quality hi-fi speaker - its internal walls will be bare (unlined), but the whole internal cavity will be stuffed with an acoustic material. Fiberglass used to be the traditional material of choice, but its glass fibers and the chemicals used during the fabrication process cause dermatitis and a highly irritant to the eyes, throat and lungs as well.

Better and healthier choices are polyester batts ("Dacron"), polyurethane, bonded cellulose acetate fibers, and the best of all, long fiber wool. However, long fiber wool is the most expensive and can harbor moths, so polyester and other man-made materials are much more practical.

How to increase the compliance and the efficiency of an existing cabinet while lowering its resonant frequency and thus getting a deeper bass from it

Another profound change happens when a speaker cab is filled with an acoustic material. The compression of the trapped air then stops following the adiabatic law (remember your high school thermodynamics classes?) and changes to an isothermal mode, where the temperature stays constant.

In other words, in an "empty" sealed box, the speaker is compressing the trapped air (changing it volume), thus raising the air's pressure and temperature.

Stuffing adds a large thermal mass which absorbs that heat that is generated by the system. It causes both the internal pressure and sound velocity to drop, the effect of which is a rise in the compliance (elasticity) of the air in the box, just as it would happen if we increased the physical size of the enclosure. And there's the beauty of it - the enclosure stays the same physical size, but the speaker is "fooled" into "seeing" (and "feeling") a much large internal volume, up to 15–25% larger, depending on the stuffing material used.

The efficiency of the enclosure is also increased substantially and its resonant frequency is lowered. Nice!

How bass-reflex speakers compare to the sealed cabs

In comparison with the sealed box, bass-reflex speakers have higher overall sensitivity, meaning they sound marginally louder when driven by the same amp (at the same input power level). Bass reflex enclosures are the most flexible choice because their size and tuning capability makes them useable with to many different loudspeakers, unlike the sealed cabs which don't always result in a good loudspeaker/enclosure match.

Sealed enclosures are preferred by most bass players due to their cleaner sound (lower distortion) and superior transient response. However, while their lower peak boosts specific bass frequencies, those below the lower peak are attenuated faster (24dB per octave) since the speaker's frequency response curve now drops at a higher rate (steeper curve). So, while a bass-reflex enclosure may provide a bit more bass at frequencies immediately below the original peak at f_R, the opposite is the case at deeper bass frequencies, below the frequency f_C (see page 85).

For those reasons (and due to their more complex and more expensive construction), bass reflex designs are not used for guitar speaker cabs, where such a marginal improvement in lower bass makes no sonic difference at all.

It must be emphasized that, just like sealed cabs, bass-reflex cabs must also be completely sealed - there may not be any air leaks anywhere. The only "leak" is the strictly controlled and tunes radiation from the port!

DOUBLE-BAFFLE PORTED DESIGNS OF YESTERYEARS

Ampeg's "Portaflex" cabinet

Ampeg's B15 speaker cabinet was a double baffle version of a bass reflex enclosure, with eight 6"x ¾" rectangular ports around the perimeter of the inner baffle. The inner baffle was set back from the outer baffle by a precise distance, found by experiment to yield the best and smoothest bass extension. Earlier units used Jensen drivers, replaced by CTS drivers on later ones, with an optional upgrade to JBL D140. It was in production until 1966.

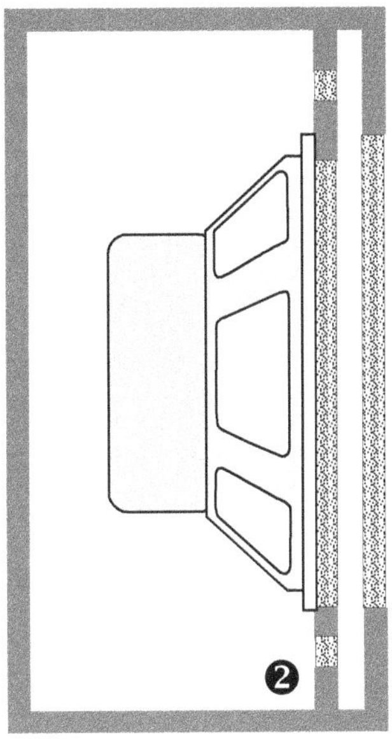

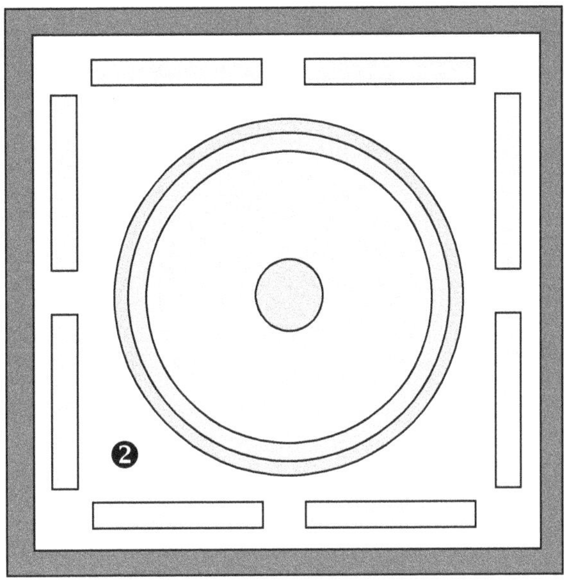

FAR LEFT: The side cross-section of Ampeg Portaflex enclosure

1) Front baffle

2) Inner baffle

LEFT: Front view of Ampeg Portaflex enclosure's inner baffle, showing eight narrow slots (distributed bass reflex ports)

Fender's "Projector Ring" cabinet

This must be Fender's most original idea of all times, their take on Ampeg's "Portaflex" enclosure. It also used a double-baffle front, (1) and (2), but instead of multiple distributed ports, the bass-reflex port is a circular gap around the driver, achieved by the use of a steel "projector ring" (3).

This completely sealed enclosure, as a bass-reflex cab should be, had padded inner surfaces (4), and parallel bracing of the back cover (5).
The commercial life of this design was short lived as it was time consuming and thus costly to build.

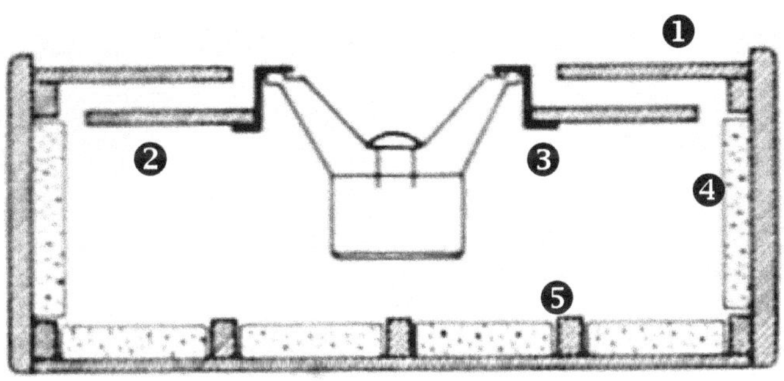

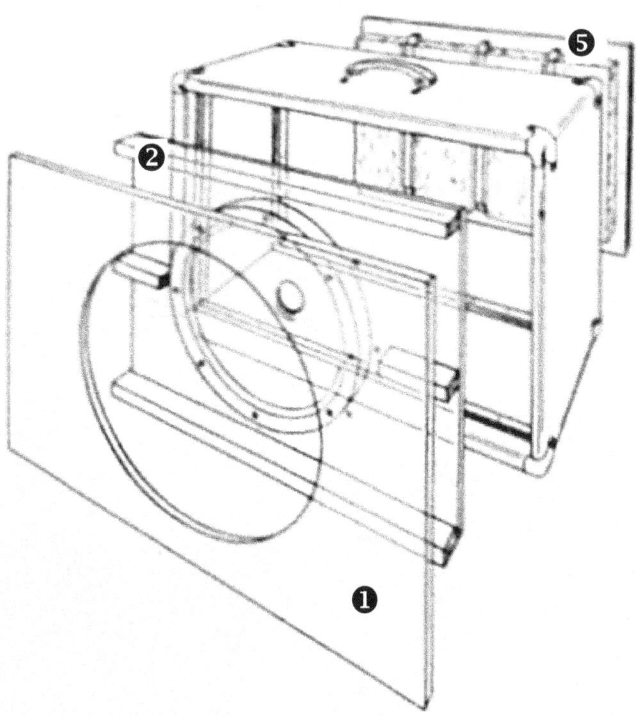

RIGHT: A cross-sectional and deconstructed view of an enclosure using a "Projector Ring" mounting adaptor (ABOVE), from Fender promotional literature
© Fender Musical Instruments

The Onken enclosure

The Onken enclosure was (and still is) revered in some circles of the audiophile community. Depending on the driver used, detailed drawings for 12" and 15" versions can be found online.

The total area of the two vertical bass-reflex ports (one on each side) should equal or be slightly larger than the area of the speaker's cone.

A 1" thick high quality plywood must be used (or marine grade ply), lined with 15mm thick felt lining.

Despite many theoretical disadvantages of rear speaker mounting, a rear-mounted speaker sounded better.

Although it may seem so at a first glance, the equivalent circuit analysis of this enclosure shows that it isn't a bass-reflex design, it is a dampened low-pass filter and not a resonant circuit!

The RJ enclosure

While the Onken enclosure had relatively large dimensions (this smaller version for a 12" speaker was 850 mm high, 650 mm wide and 450 mm deep), the RJ enclosure was truly petite, barely larger than the speaker used. There were 8" and 12" versions of the enclosure.

A 12" speaker was mounted over a smaller 10.5" hole in the internal baffle, which is spaced away from the front baffle, thus forming a ducted and ported output for the back waves to exit through the front..

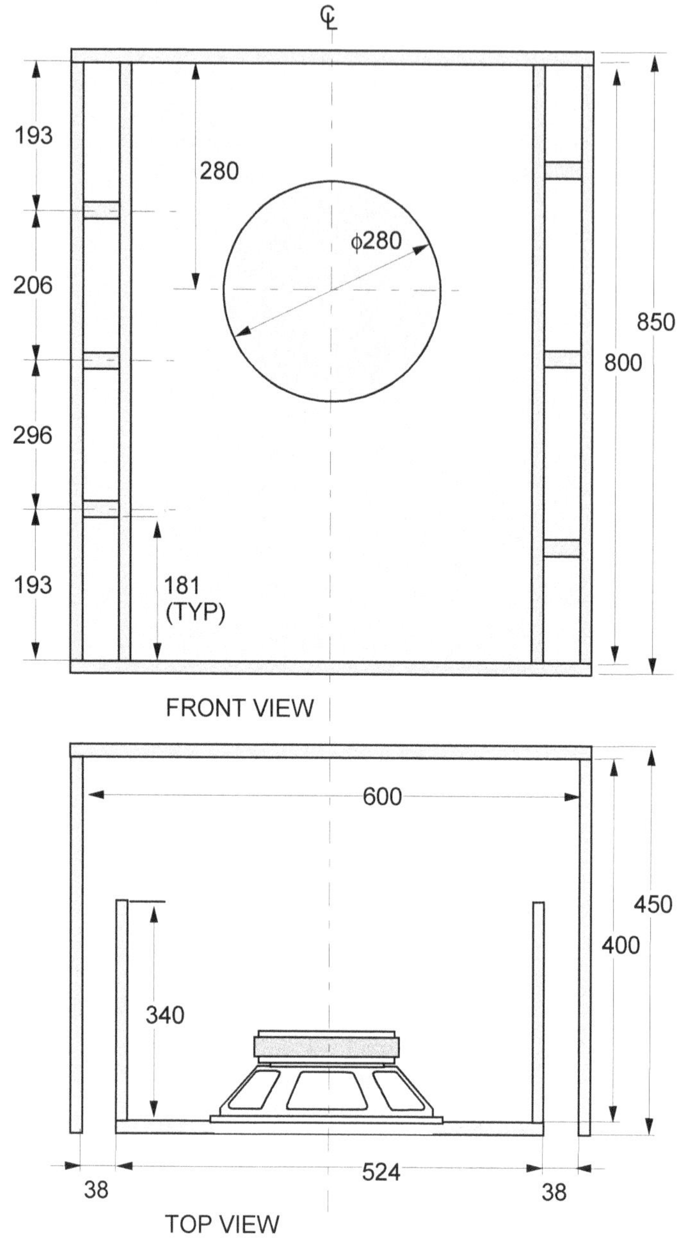

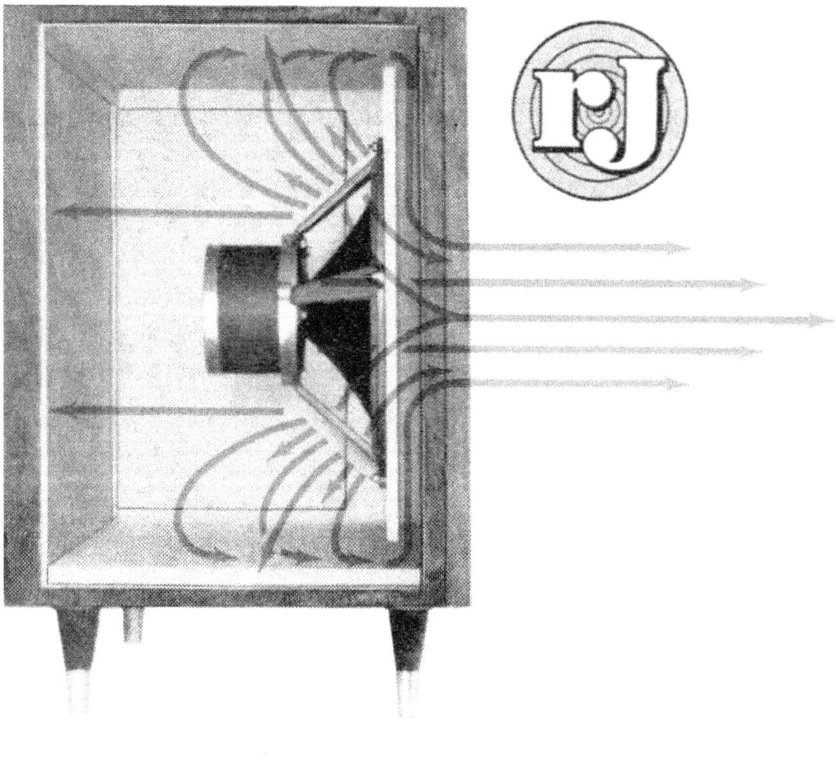

ABOVE: The Onken enclosure introduced vertical bass-reflex slots on both sides. Dimensions (in mm) are for a 12" driver and a smaller version, called "Petitte Onken" in French literature.

The area of the front baffle's slot is only 40% of that of the internal, circular opening.

All internal surfaces except the front baffle were lined with 1/2" lining. The 15.75" deep cabinet was constructed using 1/2" plywood sheets, with 3/4" space around the internal baffle and between the two baffles.

The RJ enclosure was, just like all bass-reflex cabs, based on the Helmholtz resonator's principle.

However, such resonating systems have a very high Q (quality factor) and thus a very sharp resonance, resulting in a very high impedance and SPL peak at the resonant frequency.

The RJ design increases the acoustic impedance of its vent channel by adding an appreciable resistive component. Its magnitude could be adjusted by varying the distance between the two baffles, thus changing the damping.

Robbins and Joseph lodged their U.S. patent 2,694,462 ("Acoustic System for Loud-speakers") in 1951. This original design featured a down-firing woofer, through a rectangular aperture, and a separate upper compartment for a compression horn tweeter, a much more complex arrangement than the subsequent iteration, the relatively simple "RJ Enclosure."

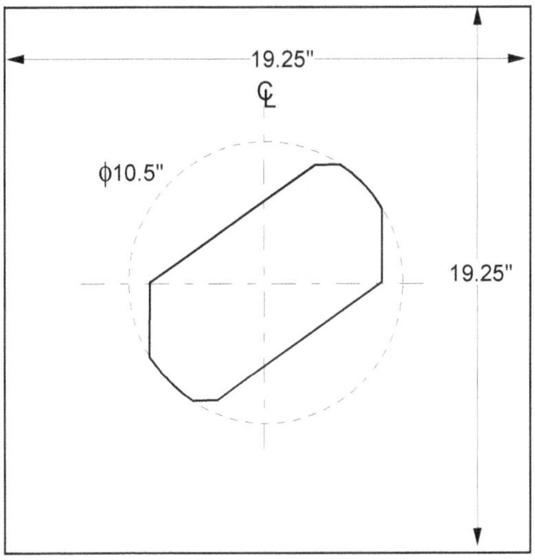

RIGHT: The original design featured a slanted front baffle opening, as illustrated on the right; later versions had a simple rectangular horizontal port.

The Ringmaster enclosure

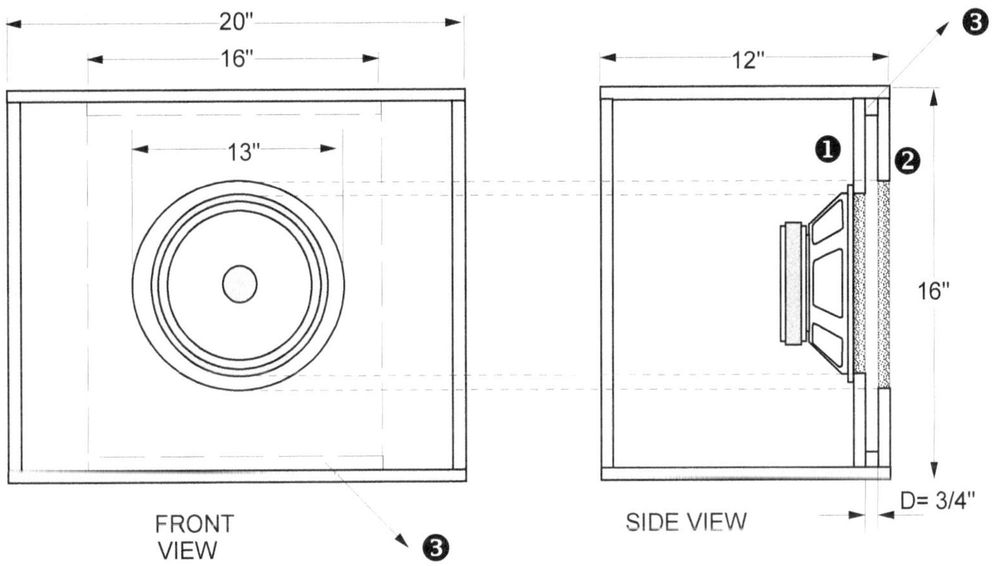

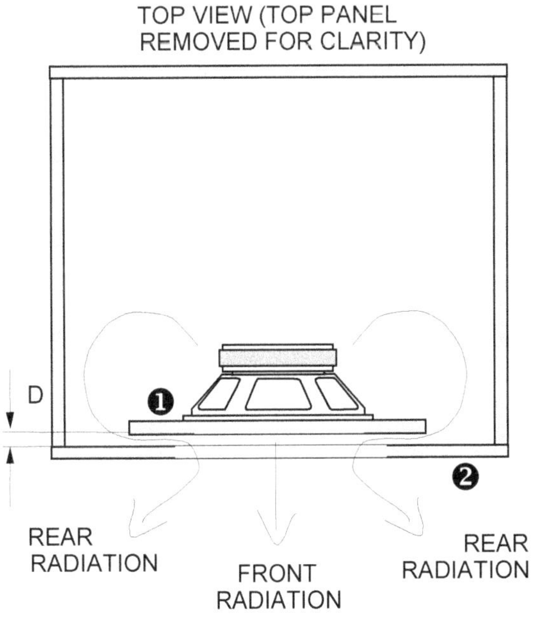

This design for a 12" driver is similar to Fender's "Projector Ring" cabinet, but without the use of the expensive and hard-to-get ring adapter. The internal baffle (1) was spaced back from the front baffle (2) by two 3/4" x 3/4" profiles (3), top- and bottom- mounted.

This design also significantly simplifies the construction. The simple internal baffle was a 16" x 16" square with a 13" circular hole in the middle. In fact, of all designs in this section, this one is the easiest to make.

Looking from the front, the port is circular, or, more precisely, ring-shaped. The duct is rectangular, a space between the overlapping sections of the simple front baffle with a 13" circular hole, and the rear baffle on which the driver is mounted.

Again, just as with all enclosures in this class, the Q and the degree of damping of the resonant peak could be adjusted by varying the distance between the two baffles. For these dimensions, a 3/4" spacing is optimal.

A modern take on an old concept: Jackson Ampworks Dual Ported cabinet

Just like the Onken enclosure, Jackson Ampworks dual ported cabinet features two vertical rectangular ports along its sides, (1) and (2), which, according to Jackson Ampworks, gives you "... the "airiness" of an open back cabinet with the presence and low-end of the best closed back cabinet you've ever heard!"

An internal high frequency diffuser is claimed to completely avoid the beaming that plagues closed-back cabs. While that may be true for the rear waves channeled forward through the side ports, I cannot for the life of me see how would that prevent the beaming of high frequency waves produced by the front radiation, which has nothing to do with the cab's construction.

The cab is also available in kit form for US$295 or US$395 fully built (no speaker, unfinished). The DIY assembly will take you about two hours and save you US$100.

A fully-finished unloaded cab's (no speaker) list price is US$645, which means the finishing (covering the cab with Tolex) costs US$250.

Both the kit and the production version use 3/4" 13-ply Baltic birch plywood, while the speaker baffle is made from 1/2" Baltic birch.

A few more ideas and conceptual ported designs

Both of these designs are very simple to build. The front baffle of design A) has no hole, while the two front baffles of design B) are identical (5). Port tuning in design A) is done by changing the incline angle of the inner baffle or its length above the speaker (2).

Even the height of the front baffle can be adjusted, so this design really lends itself to experimentation. Strictly speaking, it is more of a horn-loaded enclosure than a bass-reflex one.

Design B) achieves an improved horizontal dispersion angle, due to its vertically-stacked speaker arrangement. Changing the depth "D" of the internal shelves (4) that provide depth to the vent, and the height "H" of its rectangular port (3) will change its tuning.

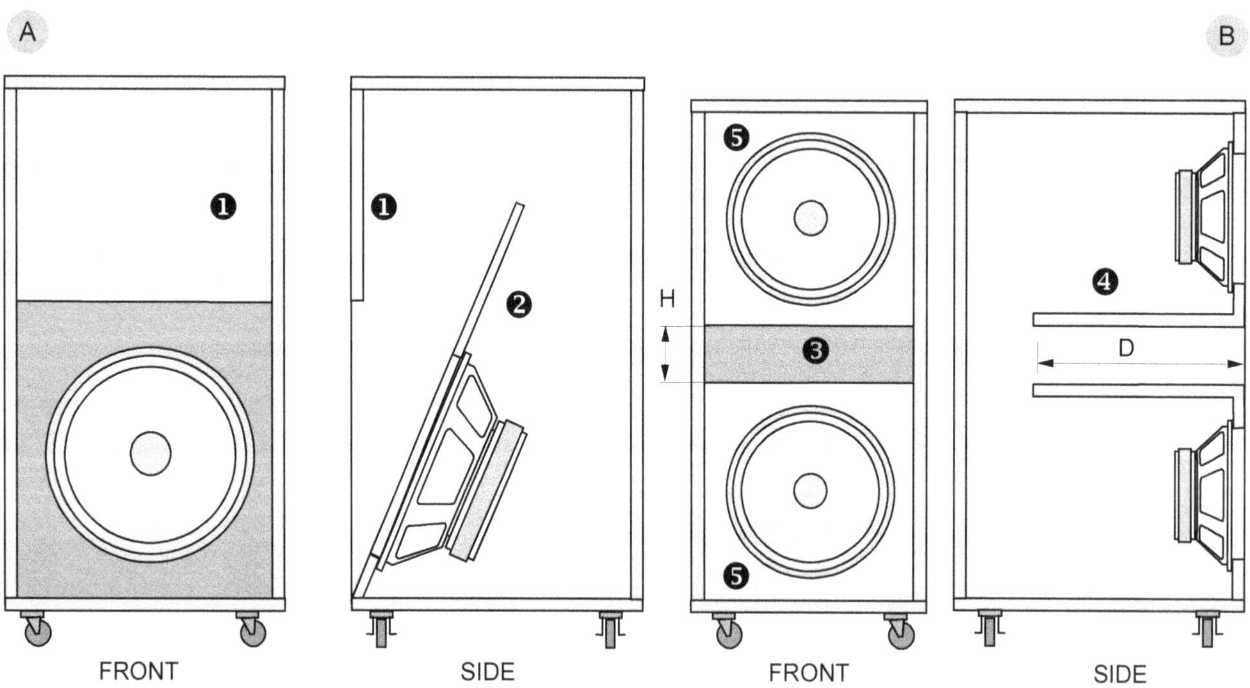

OTHER TYPES OF SPEAKER ENCLOSURES - PASSIVE RADIATORS, TRANSMISSION LINES, AND HORNS

Passive radiators ("drone cones")

A passive or auxiliary bass radiator enclosure replaces the conventional bass reflex port with a moving diaphragm, usually of a standard woofer or large midrange driver. The frame with the cone (often called a "drone" cone) and its suspension (spider) is retained, but the magnet assembly and voice coil are removed.

A passive radiator with a decent cone area and excursion capability helps achieve a tout, low distortion bass. The "chuffing" (wind noise) effect of poorly designed ports is also eliminated, especially at higher volumes. Passive radiators can be used in small or shallow enclosures where the required length of the vent would exceed the enclosure's dimension(s).

While they share vented enclosures' qualities such as lower distortion and increased power capability, on the debit side, passive radiators have a higher cutoff frequency and steeper drop rate than equivalent vented designs.

Instead of using just its cone for acoustic damping, some designers retain the entire auxiliary driver and load it with a resistor across its voice coil terminals. As the "drone" cone moves its voice coil in the magnetic gap, the induced voltage in the coil pushes current through the load resistor. Instead of working as a linear motor in regular speaker duty, it acts as an AC generator.

The current flowing in this simple circuit (voice-coil + resistor) then produces its own magnetic field that tries to oppose the motion of the coil & cone, resulting in electromagnetic damping. The energy is dissipated in the terminating resistor as heat.

If this short description has raised your curiosity, you can easily convert your 2- or 4-driver cab into a passive radiator enclosure. Simply disconnect (electrically) half of the drivers. In a 2x12" cab, disconnect one speaker, and in 4x12" cabs disconnect two of the four speakers. The inactive speakers will act as drone cones.

Of course, that will halve the maximum power rating of your cab and its impedance, but it may be the way of finding that mojo that the original cab lacked. See Chapter 7, page 134.

Ideally, a passive radiator cone should be able to move twice as much air volume than the active driver. That means it should have a higher compliance and a heavier cone.

When modifying an existing cab and keeping the front baffle, you don't have much choice. You could coat the paper cone to make it heavier, or remove the original guitar speakers and install cheap hi-fi speakers with rubber surrounds (higher compliance).

If you are building your speaker cab, cut two 12" openings in the front baffle for drone cones and two 10" holes for the active drivers, and two 15" openings for two 15" drone cone drivers (if you have to space).

RIGHT: Depending on the wiring of your 4x10" or 4x12" cab, it may be easier to disconnect the two top or bottom drivers (especially with stereo-mono switchable cabs wired left-to-right).
Alternatively, use a cross-arrangement, disconnecting one of the top and one of the bottom drivers, as illustrated.

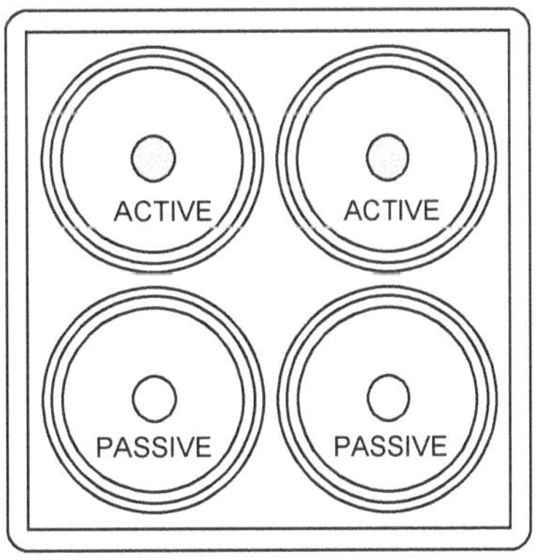

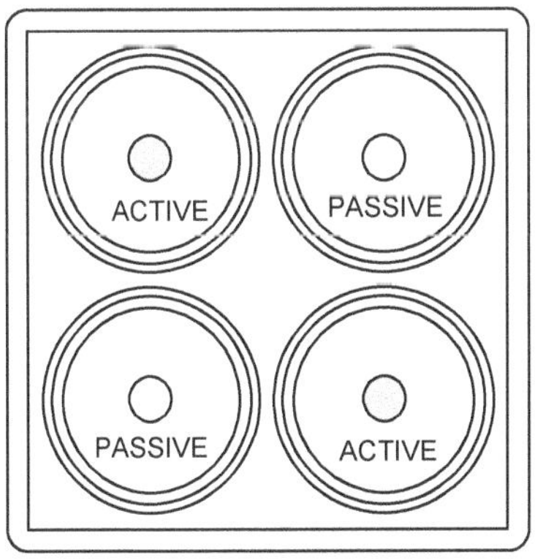

Transmission line speakers

A transmission line acts as an acoustic waveguide. This folded labyrinth at the back of the woofer in transmission line speakers is designed to absorb as much of the driver's rear mid- and high-frequency energy as possible. It allows only the lowest bass signals to pass through and exit the enclosure, making it an acoustic low pass filter.

The total length of the folded transmission line should be equal to $\lambda/4$, a quarter wavelength at the lowest bass frequency the speaker is designed to reproduce. It must be lined with damping material to avoid a series of minor resonant peaks and dips in the frequency response.

Well-designed transmission line speakers have smooth impedance curves and roll of the low bass frequencies more gently than bass reflex designs, thus providing better woofer control and cleaner bass.

We know that low bass frequencies have extremely long wavelengths, so it's the large size and weight of TL cabinets that make them unpopular even in hi-fi circles, and almost unheard of in guitar cab designs.

Horn enclosures

A saxophone is an acoustic horn, a small cross-sectional area (the "throat") opening up through its "neck" (a flared body) into a much larger "mouth." Horn-loaded speaker enclosures works on the same acoustic principle.

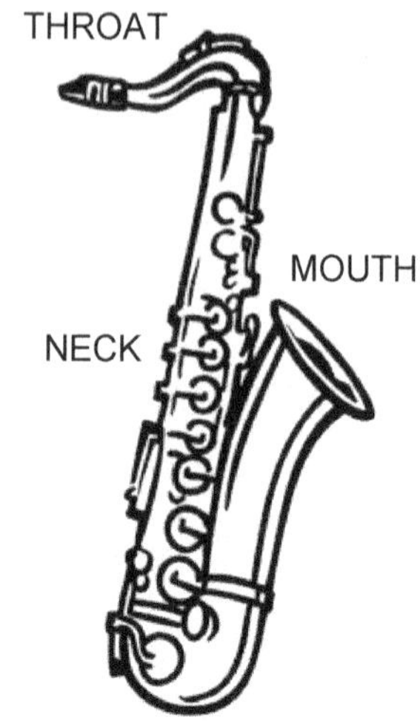

A horn is not an amplifier, but an acoustic transformer; it couples the small area of the throat to a large area of the mouth. The constricted throat area and the resulting high pressure mean high acoustic impedance.

As the sound waves travel through the continually enlarging transmission channel, the high pressure and low volume velocity at the throat are gradually transformed to low pressure and high velocity at its mouth (low output impedance), thus achieving better coupling to the surrounding air and much higher efficiency.

Again, we see the mechanical analogy with cars (coupling high-RPM engine to low-RPM wheels through a gearbox) and an electrical analogy with tube amps, whose output stages must be coupled to low impedance speakers by output transformers.

There are so many types of horn enclosures and variations on that theme that it's impossible to proclaim one design to be a "typical" horn.

In the 1950s and 60s, the golden era of horns, apart from Paul Klipsch and his company, many types of horns were marketed by various well-known brands, Altec Lansing being one of the most prominent.

Many other "horns" are actually hybrids, using a standard enclosure for the woofer, while only the midrange driver and perhaps the tweeter are horn-loaded.

Also, most folded horn cabinets feature straight sections, which are only a crude approximation of a true horn's smooth, exponentially flaring curvature. A smoother flare means less chance of standing waves developing and lower acoustic loss (sound attenuation) along the way.

Various tapers are used, hyperbolic, exponential, and tractrix being the most common. The tractrix contour expands faster than the exponential, resulting in shorter horn lengths.

The diameter of the mouth (or, more commonly, the width of a rectangular cross-section) must be at least $D = \lambda/4$, 1/4 the wavelength λ of the lowest frequency f_C to be reproduced, and since $f_C = v/\lambda$, we get $f_C = v/4D$.

A designer can decide on the overall size of the mouth and then calculate the cutoff frequency or decide on the cutoff frequency and find out how large the enclosure will be.

Full range horns have their faithful followers in a section of the audiophile community, but are too large and cumbersome as guitar or bass enclosures. However, due to higher frequencies and small wavelengths involved (which means smaller and thus manageable dimensions of the horn), many instrument and PA enclosures use horn loaded midrange drivers and tweeters.

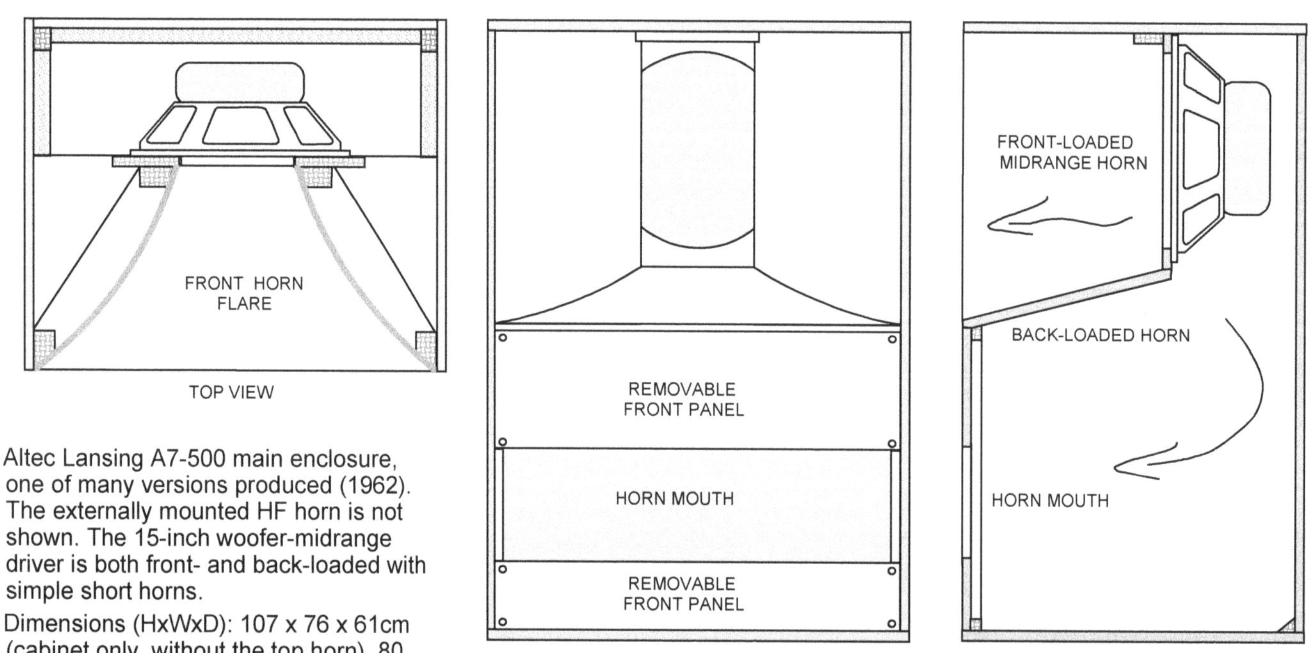

Altec Lansing A7-500 main enclosure, one of many versions produced (1962). The externally mounted HF horn is not shown. The 15-inch woofer-midrange driver is both front- and back-loaded with simple short horns.

Dimensions (HxWxD): 107 x 76 x 61cm (cabinet only, without the top horn), 80 kg, 102.5 dB/W/m sensitivity

RESONANCE EFFECTS IN SPEAKER ENCLOSURES

Standing waves

The sound wave produced by the rear of the cone propagates through the enclosure until it hits one of its four sides or its back cover. As we have seen in the "Acoustics 101" chapter, most of its reflects back at the same angle with respect to the vertical, some of its energy gets absorbed by the material the cab was made of, andsome passes through it and exits the cab.

At certain frequencies, where half the wavelength equals the distance between the two parallel sides, the peaks of the travelling wave always fall in the same spot, so it seems that th wave isn't travelling at all.

Named standing waves for that reason, they are most pronounced between the cab's parallel surfaces, its two sides, top and bottom, and front and back cover (baffle).

The points of minimum vibration, where the incoming and reflected wave (theoretically, assuming a complete reflection off the cabinet's wall) cancel each other, are called "nodes", while those where the vibration is at maximum (where the incoming and reflected wave double up), are the "antinodes",

Since all guitar cabinets are of finite dimensions, back-to-front sound cancellations or reinforcments (dips and peaks) will inevitabily occur at some low frequency and its harmonics.

When the speaker is mounted in the middle of the front baffle, the "up-down" and "left-right" cancellation frequencies will be equal and the peaky response will be exarcebated!

Even worse would be a case of closed cab whose depth equals its other two dimensions, or is exactly or close to half its height and width; in that case, the peaks and dips frequencies would be identical.

Thus, if your speaker cab is large and its centrally-positioned speaker can be moved to one side, you can experiment by making a new baffle with an asymetrically- positioned hole. You'll be rewarded with a smoother, fuller response of the cab.

How to calculate the troublesome frequencies in any cab

For closed-back cabs, the front-to-back standing waves are at frequencies which are a multiple of f_A (corresponding to $\lambda_D/2$), with peaks at f_D, $3f_D$, $5 f_{D,...}$ and dips (cancellations) at $2f_D$, $4f_D$, $6 f_D$, ... and so on.

For open-back cabs, the front-to-back standing waves are also at frequencies which are a multiple of $f_{A,}$ with peaks at f_D, $3f_D$, $5 f_{D,...}$ and dips (cancellations) at $2f_D$, $4f_D$, $6 f_D$, only this time corresponding to $\lambda_D/4$.

You may be asking yourself how is it possile to have a standing wave when the back is open, there is no rear baffle for the sound to reflect from?

This effect is known as a resonant open pipe, and church organs and many other musical instruments work on that principle - that's how their tubular nature produces sound.

To save you from unnecessary therory and algebra, we will not go into the physics of it, this principle is explained in all good physics textbooks.

The lateral standing waves (along the height and width of the cab) are at frequencies which are a multiple of f_H and f_W (also corresponding to half wavelenghts $\lambda_H/2$ and $\lambda_W/2$).

The peaks are at f_H, $3f_H$, $5f_{H.}$ ... and dips (cancellations) are at $2f_H$ $4f_H$, $6 f_H$,... for the height. For the width, the peaks are at f_W, $3f_W$, 5_W,... and dips (cancellations) happen at $2f_W$, $4f_W$, $6 f_W$,... , and so on.

The standing waves along the width of the cab (dimension looking into the paper) have not been not shown on this and following side-view diagrams to preserve clarity.

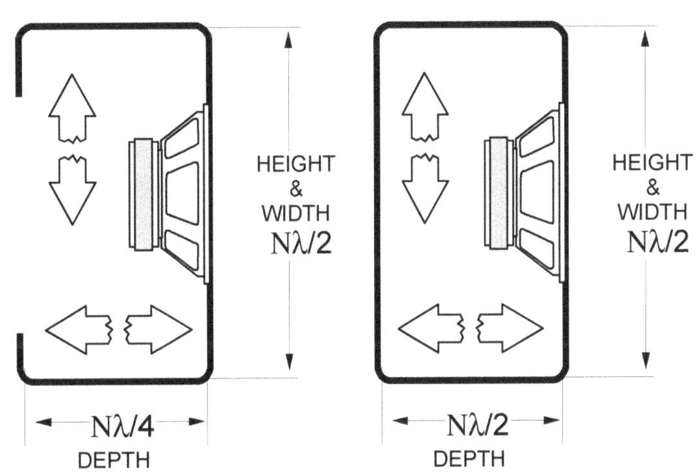

ABOVE: Standing waves in all three cabs' dimensions happen at certain frequencies, when the the cab's internal dimensions is an integer (N) multiple of the reproduced sound wave's half-wavelength, or, just for open-back cabinets' back-to-front reflections, as an integer (N) multiple of the reproduced sound wave's quater-wavelength.

Standing waves in closed and bass-reflex cabinets

This side-view of a typical closed-back (infinite baffle, sealed box) cab may visually illustrate and clarify a few issues (next page). All standing waves are reflections between the three pair of the opposing surfaces, front-back, up-down and left-right. The front-to-back standing waves are at frequencies which are a multiple of f_D (corresponding to $N*\lambda_D/2$), with peaks at f_D, $3f_D$, $5f_D$,... and dips (cancellations) at $2f_D$, $4f_D$, $6f_D$, ... and so on. The "d" in the suffix signifies DEPTH, and N are integer numbers, 1, 2, 3, 4, etc. .

Likewise, the lateral standing waves (along the height and width of the cab) are at frequencies which are a multiple of f_H and f_W (also corresponding to half wavelenghts $\lambda_H/2$ and $\lambda_W/2$).

The peaks are at f_H, $3f_H$, $5f_H$,... and dips (cancellations) at $2f_H$, $4f_H$, $6f_H$, ...for the height. The peaks are at f_W, $3f_W$, $5f_W$,... and dips (cancellations) at $2f_W$, $4f_W$, $6f_W$,... for the width.

The "H" in the suffix signifies HEIGHT and the "W" in the suffix signifies WIDTH.

At the fundamental resonance frequency the antinodes (emphasized with a ●) appear halfway between the opposite surfaces (H/2, W/2 and D/2), which is at the middle of the enclosure. Just as with the fundamental, the harmonics also have a pattern of nodes and antinodes.

The second harmonic's antinodes (points of maximum pressure) are at one-third and two-thirds of the internal dimension of the cab (H/3, 2H/3 for the height H, ditto for W and D). The third harmonic's antinodes (marked with a ○) are at 1/6th, 1/2, and 5/6ths of the height, and again, the same for the depth and the width.

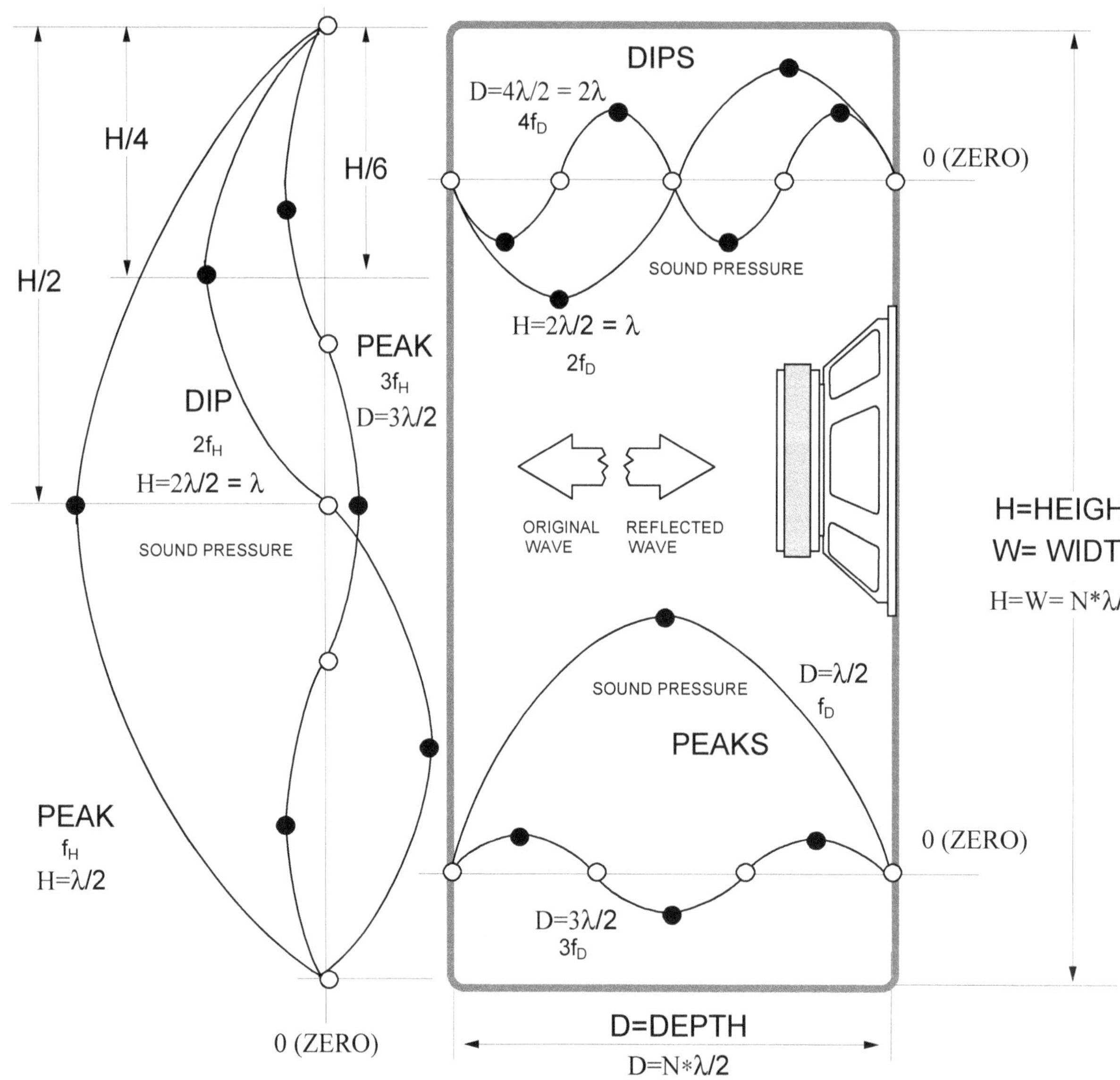

ABOVE: For closed-back cabs, the front-to-back standing waves are at frequencies which are a multiple of D (corresponding to $N\lambda_D/2$), with peaks at f_D, $3f_D$, $5f_D$,... and dips (cancellations) at $2f_D$, $4f_D$, $6f_D$,... N are integer numbers, 1, 2, 3, 4, etc. The same applies to their width W and height H.

Standing waves in open-back cabinets

For open-back cabs, the front-to-back standing waves are also at frequencies which are a multiple of f_D, with peaks at f_D, $3f_D$, $5 f_D$,... and dips (cancellations) at $2f_D$, $4f_D$, $6 f_D$, only this time corresponding to $D= \lambda_D/4$.

With regard to its back-to-front standing waves, the open back enclosure acts as a resonant tube, boosting frequencies around this fundamental resonant frequency (corresponding to $\lambda_D = D/4$), which may result in a prominent "one-note bass" or "boombox" character. To make matters worse, as the frequency of reproduced sound increases, the cab exhibits a series of minor resonances evident as peaks and dips on the cab's frequency response curve.

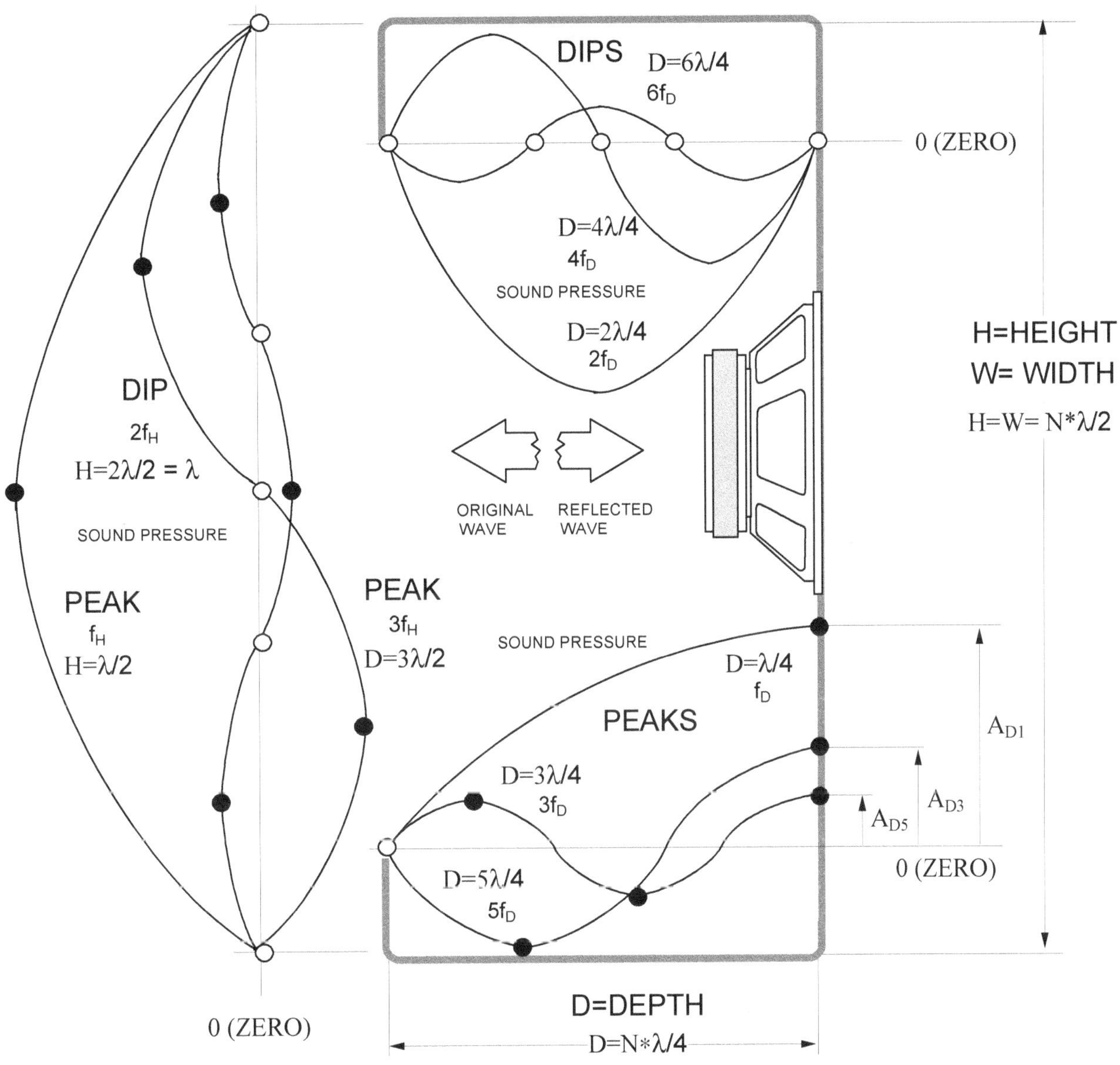

ABOVE: For open-back cabs, the front-to-back standing waves' frequencies are a multiple of f_D (this time related to $\lambda/4$, not to $\lambda/2$ as for the closed-back box), with peaks at f_D, $3f_D$, $5 f_D$,... and dips (cancellations) at $2f_D$, $4f_D$, $6 f_D$,..., etc. The lateral standing waves (along the height and width of the cab) are the same as for the closed-back cabs.

Case study: A 2x12" solid pine speaker cab

As always, a practical example is the best way to explain a concept, so let's look at a typical (in terms of size and its proportions) 2x12" guitar cab.

An old solid pine wardrobe looked daggy, but its timber was first class, obviously seasoned and stable after thity or so years, meaning all the shrinking and warping had been done.

Obviously limited by the existing dimensions of the wardrobe, the internal dimensions we chose for this solid pine cab were 51 cm (height), 73 cm (cab's width) and 16.5 cm depth. Let's do a few quick "back-of-the-napkin" calculations and see how well we managed to spread the resonant frequencies of this cab.

When depth D=16.5 cm = λ/4, the first reinforcing standing wave has a wavelength of λ=4D = 66cm; its frequency is around f= v/ λ = 344/0.66= 521 Hz. The second reinforcing wave has a wavelength of ⅓λ/4 (3x shorter), so its frequency is 3x higher, of 521*3 = 1,563 Hz, the third is at 521*5 = 2,605 Hz, and so on. These were the frequencies where the reflected wave reinforces the original, resulting in peaks in the frequency response.

At frequencies where depth D==2λ/4 (λ/2), 4λ/4 (λ), and so on, the incoming and reflected waves are out-of-phase and cancel each other, resulting in dips in the frequency response.

That happens at λ=2D = 33cm, so the highest frequency of a dip is at f= v/ λ = 344/0.33= 1,042 Hz. The following dips are at 1,042*2 = 2,048 Hz, 1,042*3 = 3,126 Hz, etc.

The lowest enforcing frequencies for the height and width can be calculate as 337 and 236 Hz respectively, meaning that the frequencies 236 - 337 - 521 and their harmonics were quite well distributed.

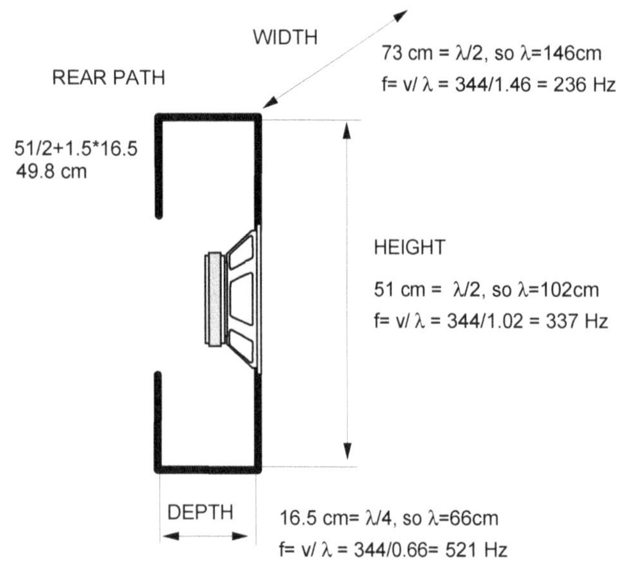

ABOVE: The internal dimensions of our 2x12" open back cabinet and its three dominant resonant frequencies

How to minimize standing waves and cabinet resonances

The worst aspect ratio for a speaker cab is a cube (A), whose three dimensions are identical. The nodes and antinodes of the standing waves in such an enclosure coincide (occur at the same frequencies), so their impact is amplified. The three cabinet dimensions (height H, width W, depth D) should all be different, but not proportional.

Finally, even in enclosures where H≠ W≠ D, the speaker(s) should not be centrally placed (B). The driver should not be centered or equidistant from the top and bottom, or from the left and right edges of the enclosure.

If at all possible, de-center the driver to one side, or to the corner (C). The standing waves will still occur, but the aim is to avoid coincidental standing wave nodes in two or three dimensions. Such resonant frequencies should be as equally spaced as possible.

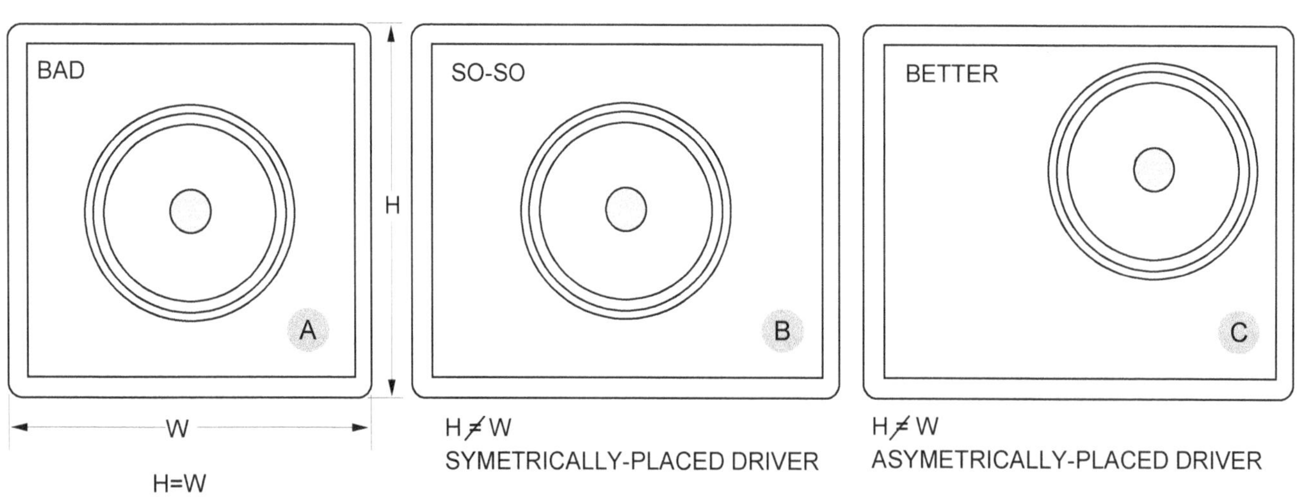

EQUIVALENT CIRCUITS OF SPEAKERS IN SEALED AND VENTED BOXES

How the enclosure changes the acoustic load

In this section we will study how an enclosure changes the acoustic load on a speaker (driver) and its behavior. We will only look at the sealed and bass reflex enclosures (with its close relative, the passive radiator type), since the other types mentioned (apart from the ubiquitous open-back cabinet) are never used as guitar cabs.

The universal acoustic model bellow shows three parallel branches loading the driver and the amp. The $C_{AB}R_{AB}$ branch models the effect of the air trapped inside the enclosure (for all three types). C_{AB} is the acoustic compliance of air (remember, trapped air acts as a spring or suspension in addition to the driver's own suspension compliance), while R_{AB} is the acoustic resistance of enclosure losses caused by internal energy absorption.

The addition of a vent ("port") or a passive radiator membrane introduces a series resonant circuit as a parallel branch to the resistive-capacitive box branch. M_{AP} is the acoustic mass of the column of air inside the port or the acoustic mass of the passive radiator's moving parts, including air load. C_{AP} is acoustic compliance of passive radiator's suspension and R_{AP} is acoustic resistance of the port or passive radiator frictional losses.

Finally, if an enclosure isn't properly sealed, there will be a substantial air leakage, which is a resistive element R_{AL} in its own parallel branch. This does not affect the resonant frequency but it does affect the Q-factor, damping and transient response of the speaker, as well as its efficiency and sensitivity. To preserve clarity and maintain simplicity, we will disregard this factor in our further discussions.

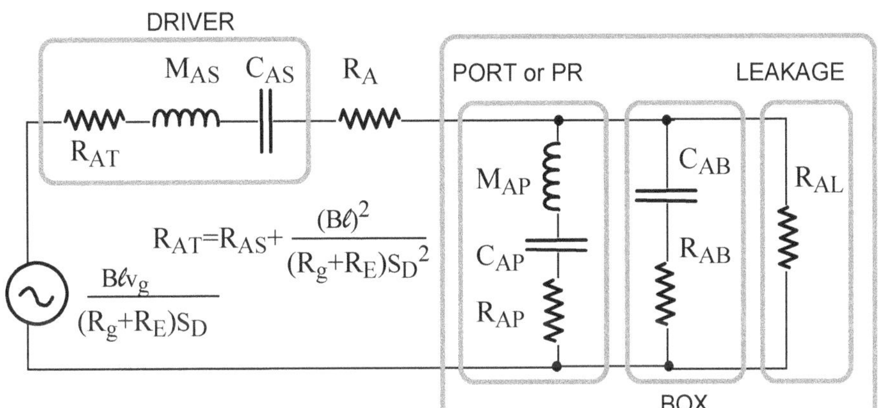

ABOVE: A universal acoustic model of the three basic types of enclosures

- R_{AT} total electrical & mechanical resistance reflected into the acoustic domain
- R_{AP} acoustic resistance of port or passive radiator losses
- M_{AP} acoustic mass of port or passive radiator including air load
- C_{AP} acoustic compliance of passive radiator suspension
- C_{AB} acoustic compliance of air in the box (enclosure)
- R_{AB} acoustic resistance of enclosure losses caused by internal energy absorption
- R_{AL} acoustic resistance of enclosure losses due to air leakage
- R_A radiation resistance

Acoustical model for a speaker in a sealed box

A sealed box is the simplest to design, build and analyze. Assuming there's no leakage (a properly sealed box), R_{AL} is negligible. If the box isn't filled with acoustic material, R_{AB} can be omitted, too, resulting in a simple series RLC circuit, just like that of the driver itself.

The only difference is that the compliance of the trapped air C_{AB} must be added to that of the driver (C_{AS}), resulting in a much smaller overall compliance: $C_{TOT} = (C_{AS}*C_{AB})/(C_{AS}+C_{AB})$. We can say that the box has a stiffening or compliance-reducing effect!

Remember, when two capacitors are in series, the total capacitance (or compliance, its mechanical equivalent, as in this case) is always smaller than the smaller capacitance of the two.

A lower overall capacitance means a higher resonant frequency. You'll remember that we mentioned how enclosing a driver in a closed box raises the resonant frequency, and now you understand why.

Compliance is acoustic capacitance and a larger box has a larger compliance (the air trapped inside is easier to compress). Thus, installing a driver into a larger box will raise its resonant frequency by a smaller percentage than installing it into a smaller box, whose compliance is lower (the air inside a smaller box is stiffer)!

BELOW: The acoustic model (equivalent circuit) of a closed-box speaker cab

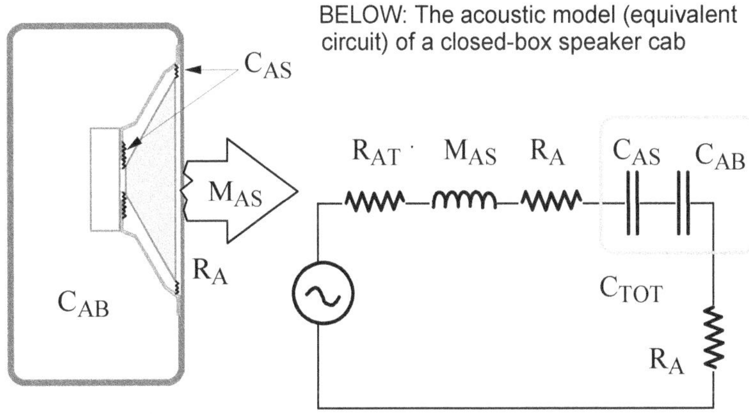

What happens to the Q-factor once a speaker is installed into a sealed enclosure?

So far we have considered the Q-factors of the speaker or driver only. As soon as a driver in mounted in an enclosure, the overall Q-factor will change to Q_{TC} ("C" for cabinet). This parameter takes into consideration the damping of the cabinet as well.

Mounting a driver into a cabinet raises its resonant frequency from f_S to f_{CB} in the same ratio that the overall Q goes up (Q_{TC}/Q_{TS}):

$$\frac{f_{CB}}{f_S} = \frac{Q_{TC}}{Q_{TS}} = \sqrt{\frac{V_{AS}}{V_B}+1}$$

Since the cabinet or "box" volume V_B is in the denominator, that rise in resonant frequency and Q-factor is inversely proportionate to the enclosure's volume - the larger the box, the smaller such an increase will be.

This makes intuitive sense as well. To compress the air inside a smaller box is harder (requires a larger force), than compressing the air inside a larger sealed box. We say that the larger box is more compliant, so it will add less stiffness to the stiffness of the driver itself.

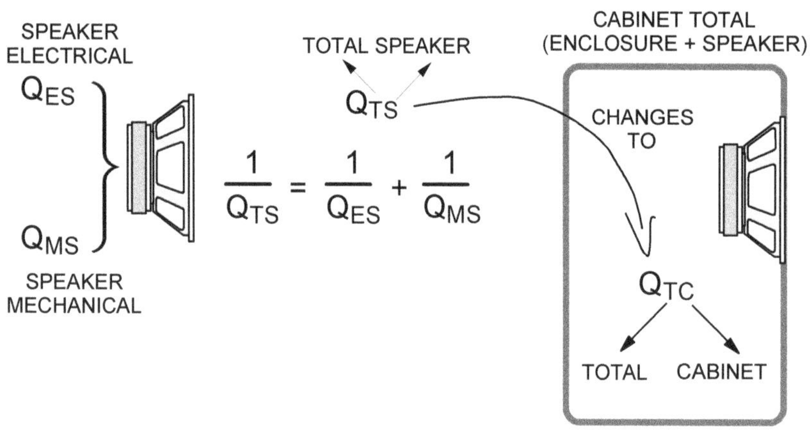

THE Q- FACTORS & RESONANT FREQUENCIES

$$Q_{TS} = \frac{Q_{ES} \times Q_{MS}}{Q_{ES} + Q_{MS}}$$

$$\frac{f_{CB}}{f_S} = \frac{Q_{TC}}{Q_{TS}} = \sqrt{\frac{V_{AS}}{V_B}+1}$$

TRANSIENT RESPONSE OF SPEAKERS IN SEALED ENCLOSURES

Transient response indicates speaker's ability (or inability) to faithfully reproduce fast changing music signals. Reproducing steady-state sine wave test signals is easy; reproducing "spiky" impulses that start and stop abruptly is the real challenge.

It is easy to recognize a speaker cab with a superior transient response; it sounds "faster", more dynamic and "alive", more responsive, and thus more pleasurable to play a guitar through.

What speaker parameters determine its transient response?

We have already established that the mechanical force acting on the speaker's cone is $F=B*I_S*\ell_C$, where I_S is the signal current flowing through the voice coil placed in the uniform magnetic field with flux density B. The third variable, ℓ_C, is the total length of the voice coil wire inside the magnetic gap.

To reproduce fast transients faithfully, the cone and the attached voice coil should be very light, so they can move fast and thus be able to follow rapid changes in I_S, the signal current.

The acceleration (force over mass), $a = F/M$ or $a = B*I_S*\ell_C/M$, should be as high as possible, ideally infinite. Since that is clearly impossible, a speaker's designer has three variables (B, ℓ_C and M) to play with. The name of the game is to minimize the mass of the voice coil and the cone M and to maximize the other two factors.

To maximize B requires a large and strong magnet, but the limits are very quickly reached, B cannot be increased beyond a certain level, so that avenue is quickly exhausted.

Maximizing ℓ_C involves again a large magnetic structure and a thin wire so to enable winding of as many turns as possible on the given size VC former (tube).

Again, very quickly we reach a practical limit. Plus, an ugly contradiction raises its head - the more turns we use and the longer the VC wire, the heavier the VC becomes so M goes up instead of down. The same applies to increasing the diameter of the voice coil - the thicker the tube, the heavier it becomes.

From an electrical point-of-view, the news aren't good either - the more turns we have on the VC, the higher its inductance, and the slower it will react to rapid transients!

The compromise between a fast transient response and a LF extension (deep bass)

The total Q factor of a cabinet Q_{TC} is the best predictor of its impulse response - the lower the Q_{TC} the faster the response. Below we see a few selected waveforms in the time domain. All responses "undershoot", dip below the zero or reference level before settling onto it, it's just a matter how much of the undershoot we will have.

For Q_{TC} values below 0.71 the undershoot is relatively tolerable (1), and there is no "overshoot", which is noticeable with Q_{TC} =1.0 (2), and which grows progressively larger as Q_{TC} increases, as in (3) and (4).

Alas, undershooting and overshooting is not the end of our problems. Notice how the response takes longer and longer (with an increasing Q_{TC}) to settle to zero, oscillating a few times above and below the zero level. Engineers call such responses "underdamped", while the responses for Q_{TC} values below 0.71 are overdamped. The Q_{TC} of 0.71 isn't just a weird choice of a random number, for that figure the response is optimally dampened.

You may be thinking now "Why are we wasting paper and time discussing this issue, let's just make as small as possible and enjoy all its positive consequences (superior dynamic response)!"

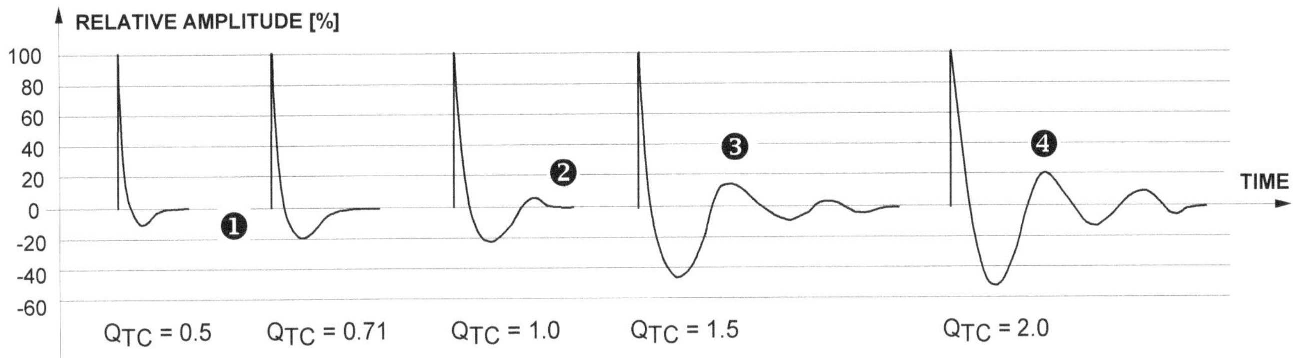

ABOVE: The impulse (sharp transient) response of a sealed loudspeaker cabinet for a few typical values of Q_{TC}

The answer lies in the graph below. Here we see the SPL produced by the same cabinet with different values of Q_{TC}. As Murphy's Law would have it, the better the transient response, the weaker the cab's deep bass!

At some upper bass frequency f_U, all five versions of the cab will produce the same SPL, the one we have marked as the referent or zero dB level. With a lowering of the signal frequency, the response of the cab with Q_{TC}=0.5 will start dropping very early (1), followed by that of the cab with Q_{TC}=0.71, and so on.

Look what happens at a conveniently chosen lower bass frequency f_L. The response of the cab with Q_{TC}=0.5 will be down a whopping -6 dB and the cab with Q_{TC}=0.71 will be down 3dB.

In contrast, the high-Q cabs would actually be boosting the bass, up by 3dB for the cab with Q_{TC}=1.5 and up by 6dB for the cab with Q_{TC}=2.0!

The Q_{TC}=0.5 is called a "critically damped" response. The optimal compromise is usually in the Q_{TC} range of between 0.71 and 1.00.

The former is called "Butterworth alignment", resulting in the "maximally flat" response, while the later results in wider bandwidth at the expense of worsened transient response accuracy.

This issue usually isn't important for guitar speaker cabinets, but is critical in the design of hi-fi and bass guitar speakers. For guitar cabs lower values of Q_{TC} are preferred, since deep bass isn't necessary, and, in fact, can become a distraction, muddying the sound.

Many bass players, especially those who also play an acoustic bass, also prefer low Q_{TC} values (0.5) due to its associated superior transient response. For them a deep, penetrating bass isn't nearly as important as fast & faithful dynamic response and "attack".

Obviously, to achieve such low values of Q_{TC}, the speakers (drivers) used must have their QTS values below 0.5, since the total Q of the cabinet (Q_{TC}) can only go up once the driver is in the enclosure!

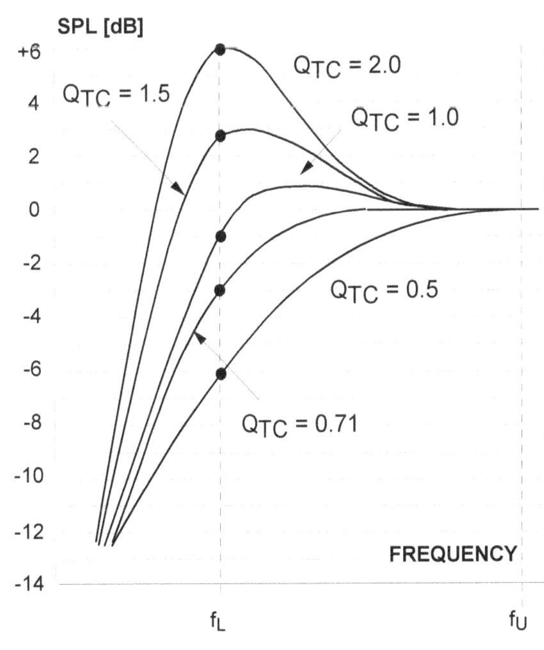

ABOVE: The effect of Q_{TC} on the low frequency response of a sealed loudspeaker cabinet

The acoustical model for speakers in bass-reflex and passive radiator enclosures

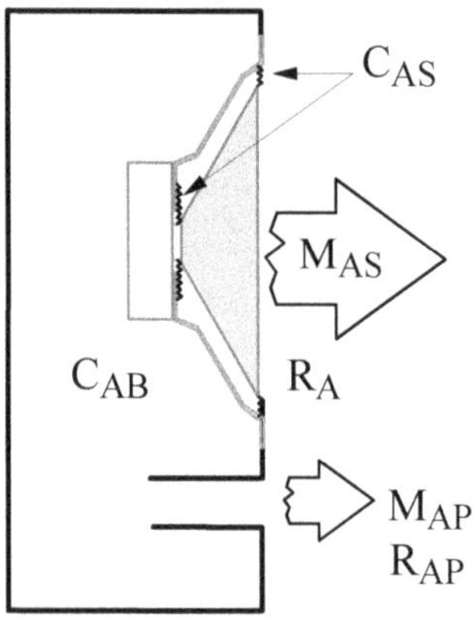

ABOVE: The acoustic model (equivalent circuit) of a bass-reflex speaker cab

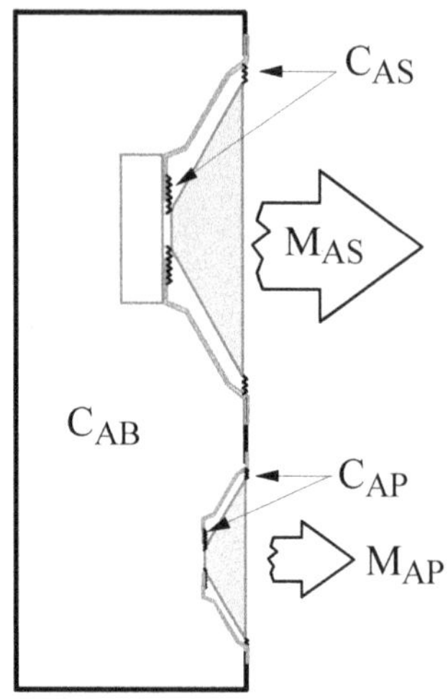

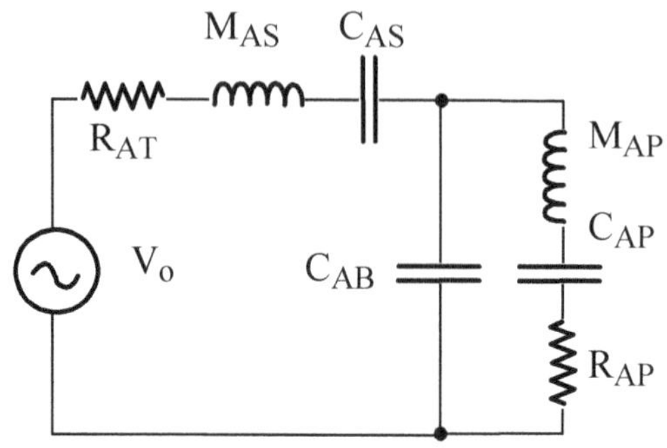

ABOVE: The acoustic model (equivalent circuit) of a passive radiator speaker cab

We have already seen that a vented speaker enclosure behaves as a Helmholtz resonator. The equivalent electrical circuit will help us understand the two-humped camel shaped impedance curve.

When the speaker produces the resonant frequency of the box, the motion of the air in the vent is at maximum and in phase with the motion of the cone, whose velocity is at its minimum. This is the resonance of the parallel tuned circuit, and since its impedance is at its maximum and speaker current is reduced, it is an "anti-resonance".

Below f_R the vent air mass moves in anti-phase to the cone, the cone's motion is mainly transferred by the air in the enclosure to the vent air mass. At those frequencies the dominant parameter is M_{AP}, C_{AB} can be disregarded, and we have a series resonant circuit with three resistances, one capacitance (C_{AS}) and two inductances (M_{AS} and M_{AP}), whose resonant frequency is f_L.

Above the anti-resonance the air mass in the vent becomes too inert to easily move, so the vent is practically blocked and the bass-reflex enclosure turns into the sealed cab. As in all closed boxes, the added stiffness of the enclosed air, the resonant frequency rises above its unbaffled value f_R (driver only) to f_U ("U" for "upper").

In the equivalent circuit this acoustic compliance of air in the box (enclosure) C_{AB} becomes dominant and resonates with the series RCL circuit (R_{AT}-R_A-C_{AS}-M_{AS}) at f_U.

The acoustic model of a passive radiator speaker cab has one more element, C_{AP}, which is the compliance of the PR's suspension.

RIGHT: The equivalent electrical diagram at low frequencies

BELOW: The equivalent electrical diagram around f_L

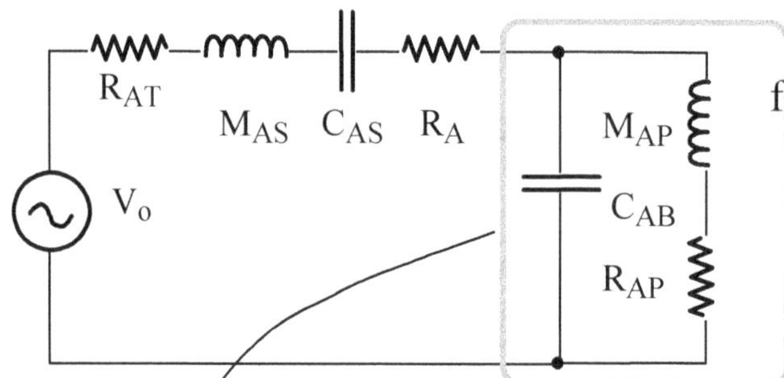

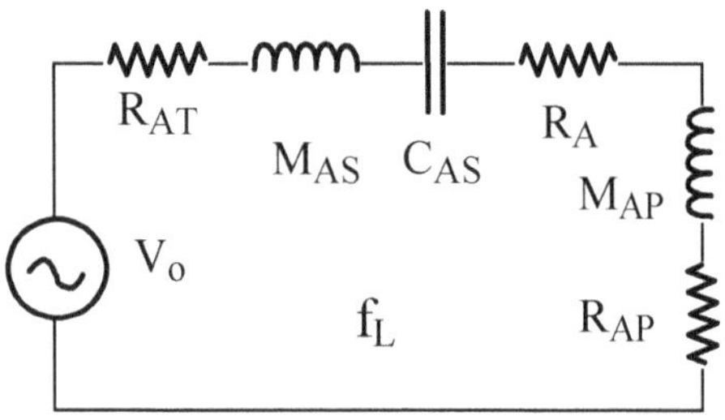

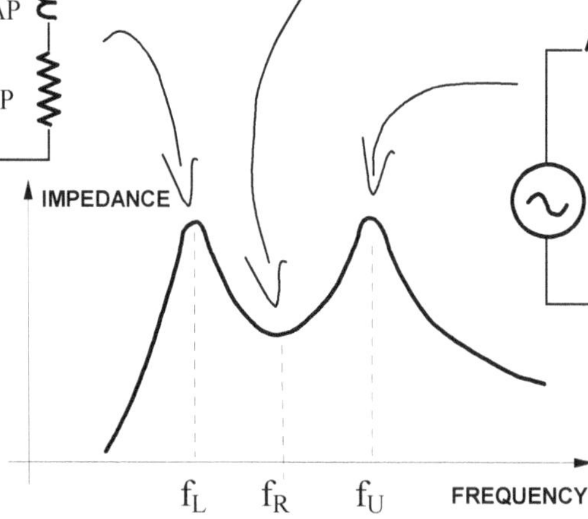

RIGHT: Typical impedance versus frequency response of ported ("bass reflex") enclosure

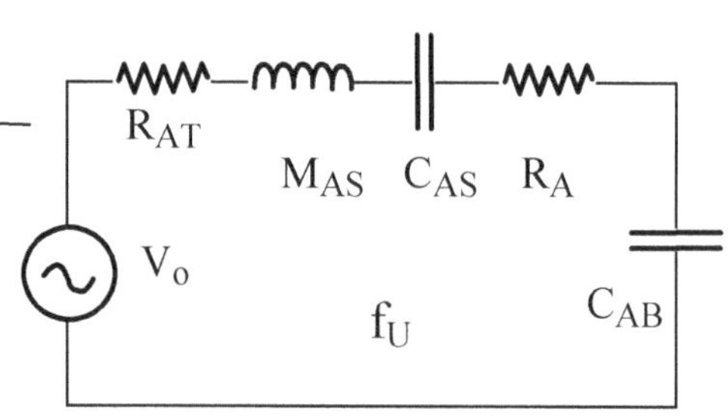

ABOVE: The equivalent electrical diagram around f_U

MATCHING AMPS AND SPEAKERS - OUTPUTS, IMPEDANCES & INTERACTIONS

6

The guitar amp-speaker cab interface and interaction is a complex issue, a minefield of potential problems. Let's start with the top three!

Solid state (transistor) and tube guitar amps behave differently with regard to the varying speaker impedance.

Secondly, there is no standard way of configuring and marking the outputs on guitar amps.

Thirdly, proclaiming a guitar speaker to have a (presumably) fixed impedance (4, 8 or 16 ohms) is gross simplification and can be both misleading and confusing.

Plugging an 8 ohm speaker cab into a 4-ohm or 16-ohm amplifier output will result in a different tone from that of the 8 ohm - 8 ohm connection, and will often sound better. But is it safe to do so?!?

In this chapter, we answer all these questions and clarify all such issues for you.

- 1/4" PHONE JACKS (PLUGS AND SOCKETS) - A CRASH COURSE
- NAVIGATING THROUGH A MINEFIELD - HOW TO FIGURE OUT AMPLIFIER SPEAKER OUTPUTS
- WHAT HAPPENS WHEN YOU CONNECTS HIGHER OR LOWER "NOMINAL" IMPEDANCE SPEAKERS TO A TUBE GUITAR AMP
- HOW ATTENUATORS IMPACT SPEAKERS' IMPEDANCE & SPL CHARACTERISTICS

1/4" PHONE JACKS (PLUGS AND SOCKETS) - A CRASH COURSE

The official or proper name for guitar and speaker cab connectors or "jacks" is 1/4" phone jacks. "Phone" comes from the fact that they were designed in the 19th century for use in manually operated telephone exchanges and switchboards. Thus, they had to be reasonably large (to fit snugly into a palm of the operator's hand) and robust to withstand thousands of plugging & unplugging operations.

1/4" indicates their diameter, around 6.3mm. The symbols for various 1/4" sockets are based on the basic symbol (below). The rectangle represents the "female" sleeve contact, basically a tube that fits snugly around the "male" sleeve. Likewise, the arrow-like male tip clicks into the V-shaped and flexible tip contact.

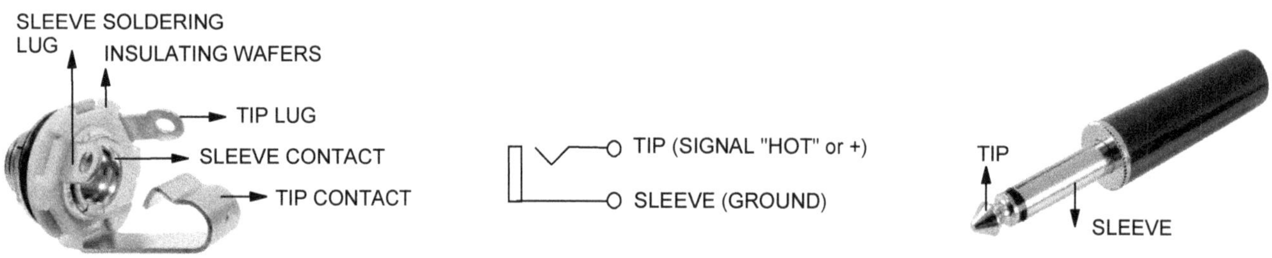

ABOVE: The simplest 1/4' socket (left) and plug (right), also called "mono" or TS jack. T denotes a TIP which is "hot" or + of the signal, S denotes a SLEEVE, which is the negative of the signal or GND (ground).

Open type sockets

There are many other open 1/4" sockets, such as the mono switched socket, unswitched stereo socket, stereo switched socket, etc. All switched sockets operate on the same principle.

When the male plug isn't plugged in, the tip of the shunt contact is touching the tip contact, which is thus called NC or "Normally Closed" contact. This is often used at the input of guitar amps, with a tip shunt lug wired to GND. That way, when a guitar cable isn't plugged in, the tip is grounded, meaning the grid of the input tube is also grounded, so it cannot pick up any stray signals or interference, keeping the amp quiet.

When a guitar cable is plugged in, the tip pushes the tip contact downwards and breaks the mechanical connection between the tip shunt contact and tip lever.

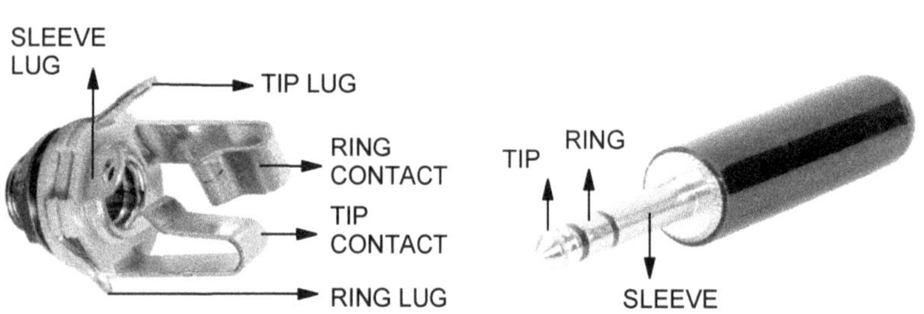

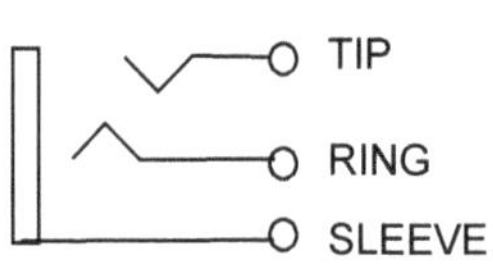

ABOVE & LEFT: Open-type stereo unswitched socket, its symbol and the stereo plug

RIGHT: Open-type switched mono socket with one NC (normally-closed) contact (shunt) and its symbol

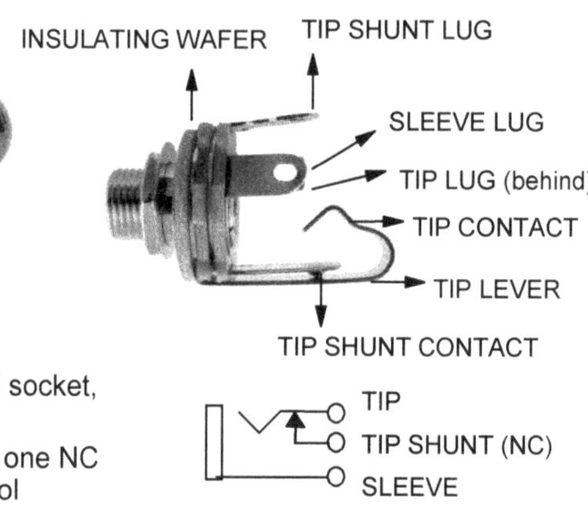

Closed type sockets

Just as its open-type version above, the simplest closed type mono-socket has only two lugs, the tip and the sleeve (GND). The next complexity level is the mono-socket with one NC contact, as illustrated below. Lug 1 is the sleeve; lug 2 is the tip, and lug 3 is the normally closed (to the tip) auxiliary contact.

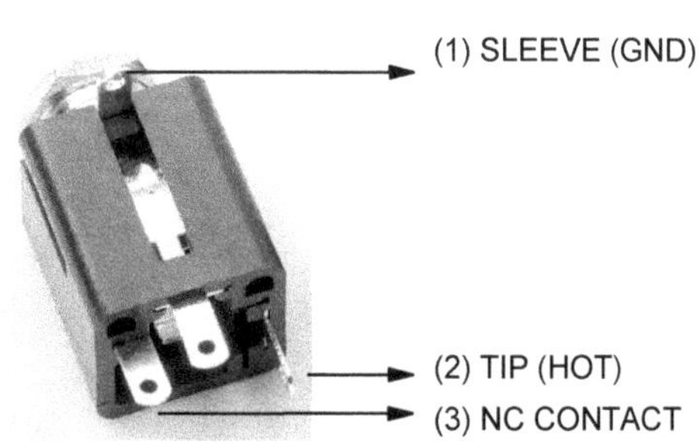

RIGHT & LEFT: 6.35mm 2PST ("two-pole, single throw") mono socket, its symbol and dimensions (in mm)

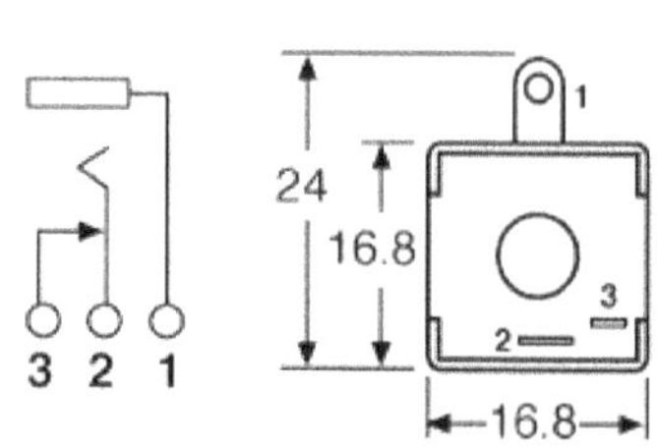

In both open and closed sockets the switching contact can be normally-closed to either the TIP or the SLEEVE. That way it can serve different purposes, short-circuiting something to ground or common if connected to the sleeve, or to the signal input (hot or +).

The stereo socket illustrated below has two C/O (Change Over) contacts. Apart from lug 9 (sleeve or GND), all lugs are at the back. Lugs 4 and 8 are the tip and the ring, two "hot" or signal + lugs.

The three lugs at the top (1, 2 & 3) and the three at the bottom (5, 6 & 7) are the two auxiliary switches. Contacts 6 & 7 are normally closed and are broken apart (opened) by the insertion of a stereo plug, which then pushes contacts 5 and 6 together, so they close. Ditto for contacts 2&3 and 2&1 on the other switch.

Remember that in this case none of the six switching lugs or contacts ever come into contact with the actual tip or ring (the audio signal); they are only mechanically actuated by them.

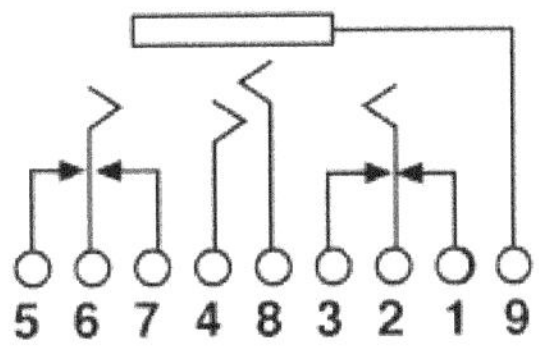

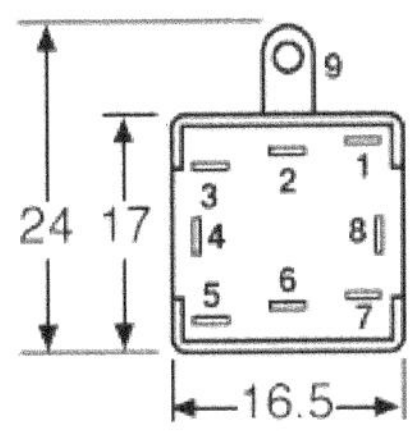

LEFT: 6.35mm stereo socket with two changeover auxiliary contacts, its wiring diagram, and rear view with dimensions (in millimeters) and lug numbers

Both open and closed type sockets can be insulating or non-insulating. That refers to the relationship between the mounting chassis in amplifiers (almost always metal) and the socket's sleeve.

In non-insulating sockets, the metal sleeve ring touches the chassis, so there is a galvanic connection between the two. This may be undesirable if the amp builder does not want to use the metal chassis as the ground bus and uses a star grounding system instead. An insulating socket would then ensure that there is no double grounding, a situation that can cause a ground loop and result in an annoying hum at the amp's output.

These types of sockets are seldom used in speaker cabinets, and since the metal plate they are mounted on isn't connected to anything, the insulating- versus non-insulating issue isn't usually relevant in their speaker cab applications.

Finally, there is a third socket type, illustrated below, easily recognizable by its open-type contacts mounted on top of the socket, bridging from one to the other side of its body.

Each active element has its own normally closed (NC) contact or shunt, which opens when the plug is inserted. A mono-socket has two switchable contacts; one is normally closed to the tip, the other to the sleeve, designated by DPST (double pole single throw), DP meaning the two poles or contacts are separate (no galvanic or mechanical connection between them).

A stereo socket has three pairs of lugs or legs and three "ST" or "single throw" contacts, thus 3PST in its catalog name.

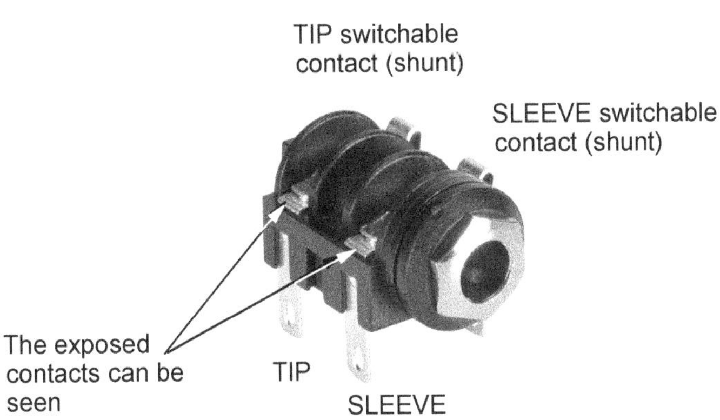

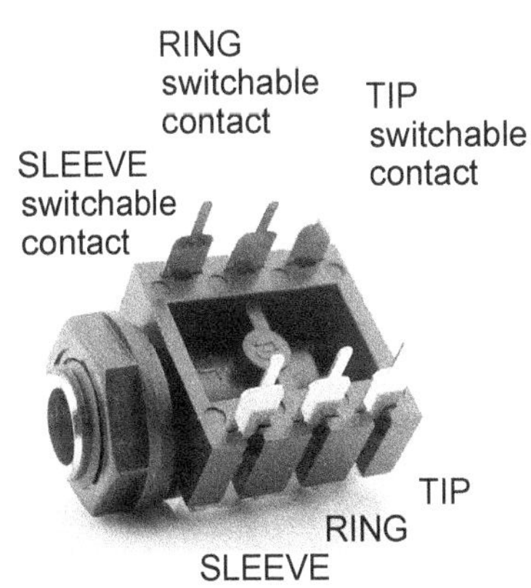

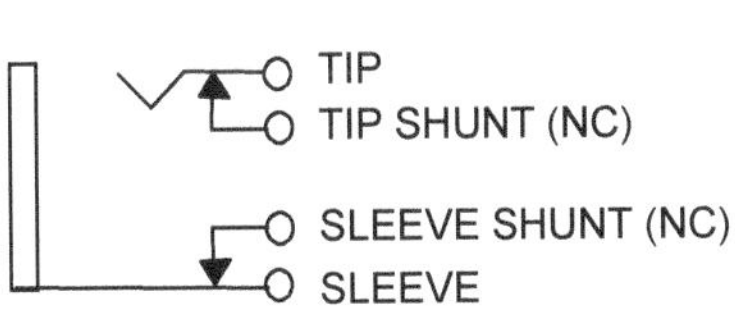

LEFT: 6.35mm DPST non-insulated mono socket, its symbol and contacts

RIGHT: 3PST non-insulated stereo socket, its symbol and contacts

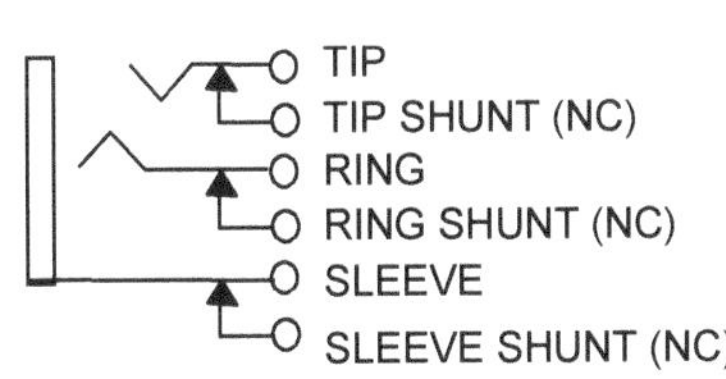

NAVIGATING THROUGH A MINEFIELD - HOW TO FIGURE OUT AMPLIFIER SPEAKER OUTPUTS AND CABINETS' INPUT CONFIGURATIONS

Marshall DSL50 and DSL100 speaker connections

Marshall DSL50 and DSL100 feature three output jacks. Two are marked "4 & 8 Ohm PARALLEL JACKS" and one is marked "16 Ohm ONLY". There is a note saying "16 Ohm Output Jack negates use of others", the meaning of which may not be obvious at the first glance. There is also a "SELECT" switch marked "4 Ohm/8 Ohm".

Only after studying the circuit diagram (the relevant portion redrawn on the right) we understand the thinking behind Marshall's designers' choices.

Notice how the negative terminals of both "4 OR 8 Ω PARALLEL OUTS", (1) and (2), are wired through the auxiliary contact of the "16 Ω ONLY" jack (3) to the COM terminal on the output transformer (4).

As soon as a speaker cab is plugged into the "16 Ω ONLY" jack, these COM terminals are disconnected and it does not matter what is plugged into one or the other (or both) "4 OR 8 Ω PARALLEL OUTS", the "16 Ω ONLY" speaker overrides both!

With the "IMPEDANCE SELECTOR SWITCH" (5) in the 8Ω position (as drawn), both jacks are in parallel. With one 8Ω cab connected to either jack, all is well.

However, "4 OR 8 Ω PARALLEL OUTS" doesn't really say much, so a guitarist should not be blamed for plugging in two 8Ω cabs together. That would effectively mean a 4Ω load into an 8Ω amp output, a gross mismatch! A similar mismatch is possible with the "SELECT" switch in the 4Ω position (a 2Ω load into a 4Ω amp output.

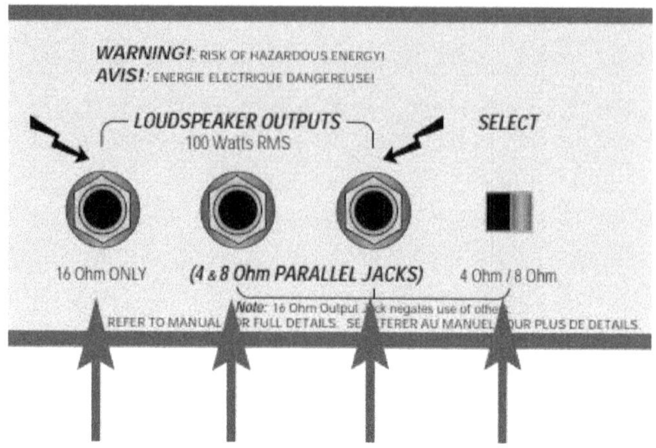

ABOVE: Marshall DSL50 and DSL100 speaker outputs ▲

BELOW: How the output jacks are wired ▼

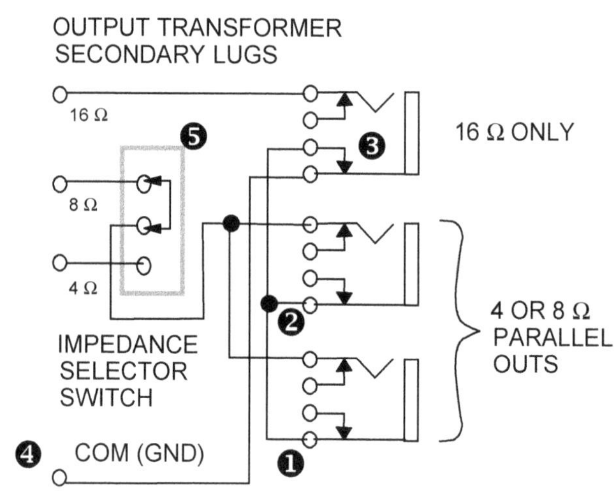

Vox AC30C2 speaker connections

AC30C2 is a combo amp with two internal 8Ω speakers connected in series, for a total of 16Ω.

Apart from a 16/8 ohm impedance selector switch, there are also two jacks for external speakers not marked in terms of the required impedance, only saying that one mutes (disconnects) internal speaker, while the other doesn't.

Notice the clever use of different words, "external" for external speaker only (internal speakers disconnected), versus "extension" for an external speaker in addition to the internal speakers!

Can you see any possible issues with this arrangement?

With ISS in the 16Ω position and no external speaker plugged in, all is well. If we plug in an external speaker in the bottom jack (B), its impedance is now in parallel with the still connected internal speakers.

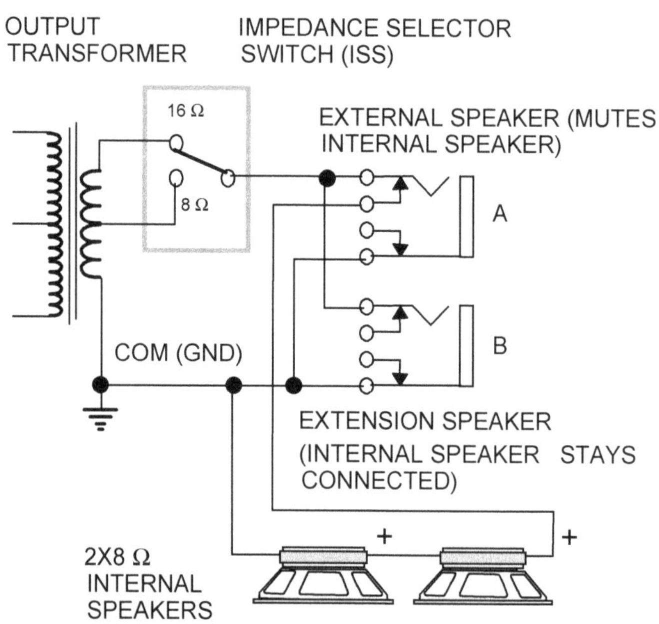

ABOVE: VOX AC30C2 output circuit ▲

With a 16Ω external cab we must flick the ISS into the 8Ω position (16Ω||16Ω=8Ω). What if we forget to do that or if we plug in an 8Ω or a 4Ω cab?!? We have a problem, a mismatch that could burn the external speaker out since it is of lower impedance than the internal speakers so it will draw a proportionally higher current from the amp!

Also, if we now unplug the external speaker but forget to flick the ISS back into the 16Ω position, we again have an impedance mismatch, 16Ω internal speakers connected to the 8Ω tap on the output transformer. So, again, another simple but far from a foolproof circuit.

Fender Hot Rod Deville 212 speaker connections

Let's see if things are any different across the pond, and if they are, are they better or worse ...

The output of Fender Hot Rod Deville 212 relies on two mono speaker sockets. The "MAIN SPEAKER" socket has a single, auxiliary switch which breaks contact (1) when a speaker jack is inserted. The main speaker (presumably its internal speaker) is then hooked up from the "hot" tip (4), thought the "EXTERNAL SPEAKER" tip and contact (2) to the 8Ω tap of the OPT.

Plugging an external speaker (with the internal speaker still plugged in) opens the AUX contact (2) and closes the AUX contact (3), thus connecting both speakers' "hot" tips together and to the 4Ω tap, which is correct, since two 8Ω speakers in parallel represent a total amplifier load of half that value or 4Ω!

Plugging in just an external speaker (with the internal speaker unplugged) opens the AUX contact (2) and closes the AUX contact (3), thus connecting the 8Ω speaker to the wrong (4Ω) tap.

However, notice that circuit is also closed through the MAIN SPEAKER "hot" tip and its AUX contact (1) to ground (GND) and the common OPT terminal (COM). This short-circuits the 4Ω tap to ground, creating a short-circuit at the amp's output!

The same happens when no speaker is plugged into either of the amp's outputs. Since contact (3) is open, the 4Ω tap is disconnected and inactive, but the 8Ω tap is short-circuited to GND through (2) and (1), a situation just like the previous one.

That is why the amp's manual states "D. EXTERNAL SPEAKER - Plug-in connection for an external speaker. This jack is wired in parallel with the main speaker jack and must ONLY be used in conjunction with the main speaker jack. If this jack is used without the main speaker jack, there will be no sound."

Although the OPTs have taps for three load impedances (2-4-8Ω), there is no ISS; each amp is either wired for 4Ω (taps 2Ω & 4Ω used, 8Ω unused) or 8Ω (taps 4Ω & 8Ω used, 2Ω unused).

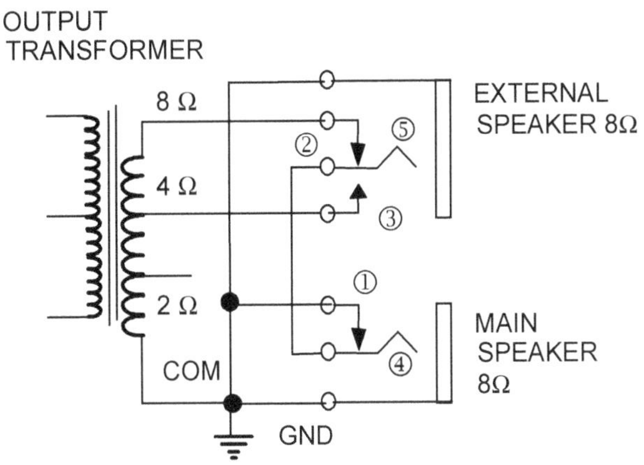

ABOVE: Fender Hot Rod Deville 212 output wired for 8 ohm speakers

BELOW: Fender Hot Rod Deville 212 output wired for 4 ohm speakers

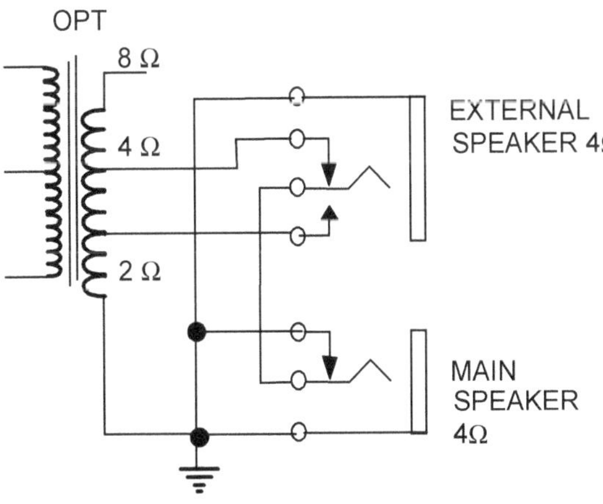

The Good, the Bad, and the Ugly - a few more examples of speaker connections on guitar amps

Let's briefly discuss a few more examples of speaker connections on commercial guitar amps (next page).

The first configuration is Fender's, a "4 OHM - 8 OHM" switch and two jacks in series. "Always observe proper load impedances" means it's up to the user to figure out whet those impedances are.

This is a vintage amp, when many hi-fi lovers and guitar players had a high level of technical prowess, and were designing and building their own amps and speaker cabs. They knew that two 4Ω speakers in series would yield 8Ω, two 8Ω speakers in series would yield a 16Ω load, etc.

These days the technical knowledge levels among users (of anything) are so low that car battery makers feel necessary to include a prominent warning not to drink the acid from them!

Mohave Plexi 45 configuration isn't foolproof, but is simple and easy to understand - one 16Ω jack and two paralleled 8Ω jacks that can be used with a pair of 16Ω cabs.

Laney CUB Super 10 combo only has one speaker jack, into which, Laney claims, any speaker of 4-16Ω impedance can be plugged in. However, both the internal speaker (permanent wiring) and the external speaker (plugged in) are connected to a single 8Ω tap on the OPT. Goes to show that Laney's designers think a 2:1 load impedance ratio mismatch is perfectly fine

Victory's 212-VC cab has a jack marked "INPUT" and another one marked "LINK", whatever that means (most likely for daisy-chaining another cab).

The warning (the capitals are theirs) reads "DO NOT CONNECT TWO AMPLIFIERS TO THIS CABINET OR BOTH MAY EXPLODE AND RUIN YOUR ENTIRE DAY."

Their cheeky sense of humor aside, we are still none-wiser - would the second cab be connected in series or parallel? Without opening the cab and doing some continuity (ohmmeter) investigation we will never know ...

Ibanez 4-8-16Ω outputs are grouped that way, and no switching is used.

A single 16Ω jack is self-explanatory, and the other two pairs are marked in the same way.

The bottom jack is for a single 4Ω or 8Ω speaker, while both vertically aligned jacks are for paralleled speakers of twice the bottom impedance, 8Ω and 16Ω each cab, respectively.

I prefer this "no words" approach to long, awkwardly-worded sentences that instead of clarifying things, leave you even more perplexed.

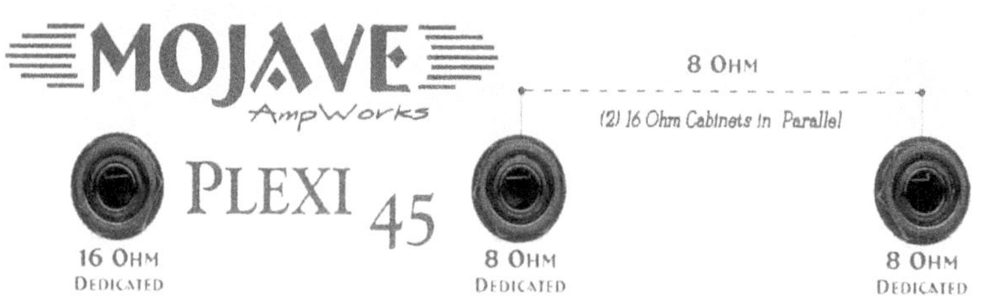

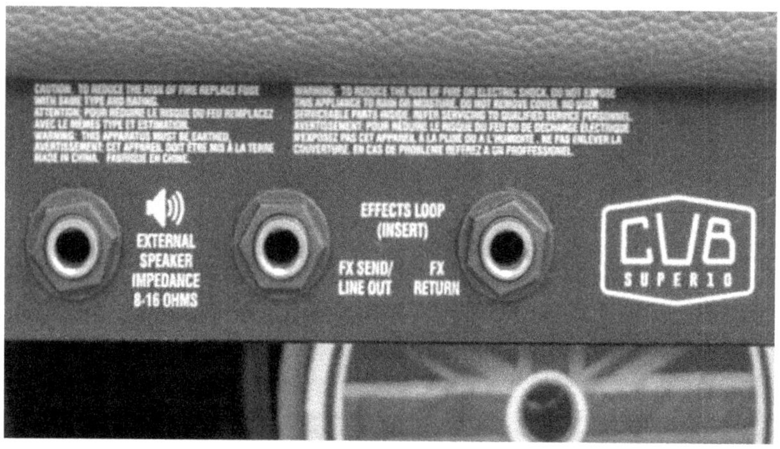

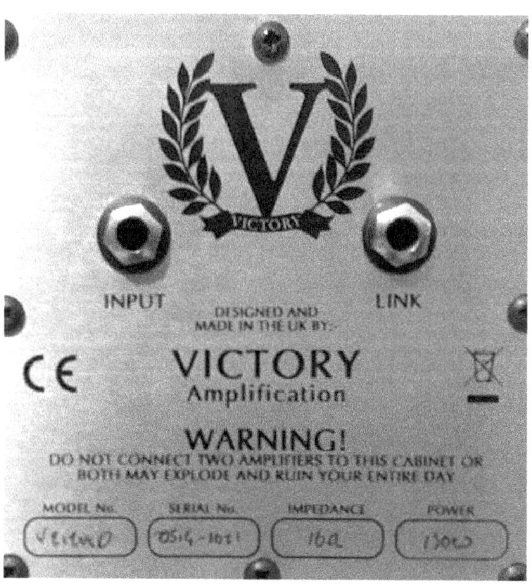

Series internal cab wiring?

The Kemper Kabinet has two jacks, marked "IN" and "OUT" and, in tiny letters, "SERIES WIRING". Their manual only says "Two or more KEMPER Kabinets can be safely daisy-chained by the additional jack, as the built-in circuitry has been designed to run multiple Kabinets in series automatically."

What does that really mean and how could the cab's jacks be wired ("the built-in circuitry")?

We have never had a Kemper Kabinet on our test bench, so this may or may not be the way its jacks are actually wired.

In any case, if you wanted to daisy-chain cabinets in this fashion, what kind of jacks do you need and how should they be wired?

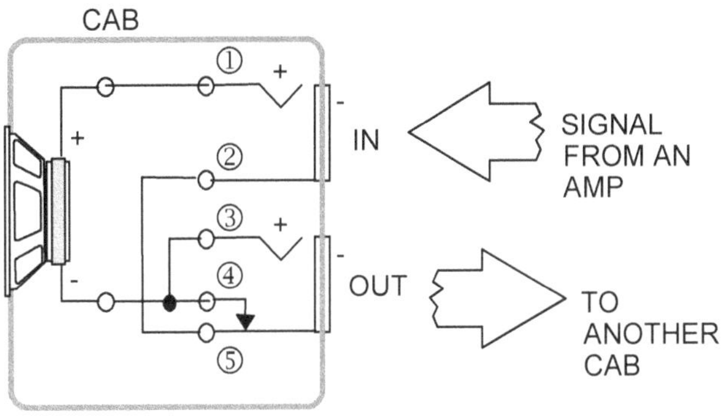

ABOVE: One possible way the jacks inside the Kemper Kabinet could be wired

The simplest solution is one non-switched jack, with plus terminal (1) and negative terminal (2), and a similar jack with the positive terminal (3), negative terminal (5) and one switched terminal (4), normally-closed to the negative terminal (5).

As drawn, nothing is plugged into the "OUT" jack (no second cab used), so the + from the AMP goes into the + terminal of the cab's speaker.

The negative terminal of the speaker is connected internally to both the + terminal of the "OUT" jack and to its switching (auxiliary) terminal (4). That terminal is now connected to negative terminal (5) of the "OUT" jack and he negative terminal (2) of the "IN" jack.

The + and - terminals of the "OUT" jack are short-circuited (connected together) by its switching AUX contact (4), but that does not matter, since nothing is plugged in (yet) into that jack.

As soon as another cab is plugged into the "OUT" jack, the AUX contact (4) opens, breaking its connection with terminal (5). The - lug of the cab's speaker is permanently connected to the + terminal of the "OUT" jack and the + lug of the extension cab's speaker, whose - lug is permanently connected to the - lug of the "IN" jack and the amp's - output. Thus, the desired + - + - series connection is achieved with only one AUX contact!

Main speaker, external speaker, extension speaker, series speakers ... oh man!

We have already learned that an "extension" speaker refers to an additional speaker, added either in series or in parallel to an existing, already connected speaker, which can be internal or plugged in externally. Usually, for an extension speaker to work (produce sound), the main speaker must be plugged in.

Referring to the drawing on the next page, in (A) the extension speaker jack is connected in the usual way and plugging a speaker in there would produce sound, but notice that if nothing is plugged into the "MAIN SPEAKER" jack, the AUX contact short circuits the two parallel connected tips to ground (GND) and thus shorts circuits the amp's output.

The situation in (B) also requires the main speaker to be plugged in for the extension speaker to work. Assuming that is so, the COM (negative) side of the "MAIN SPEAKER" jack is connected to the + (hot) tip of the "EXTENSION SPEAKER" jack, but also to its AUX contact.

When nothing is plugged into the "EXTENSION SPEAKER" jack, the (negative) side of the "MAIN SPEAKER" jack is connected to GND through that normally closed (NC) AUX contact.

As soon as the extension speaker is plugged in, that AUX contact opens, and now the (negative) side of the "MAIN SPEAKER" jack is connected to the + (hot) tip of the "EXTENSION SPEAKER" only, so the two speakers are now in a series connection.

In (C) we see a series connection of two jacks where a single speaker can be plugged into either jack, there is no more main jack and extension jack.

Just as in (A), with no speaker connected, the normally closed AUX contacts on each jack bypass those jacks, so the amp's output is shorted.

As soon as one speaker is plugged in, its jack's AUX contact opens and the short across that jack is thus removed and in its place we have the speaker's terminals. For the circuit to work, the AUX contact of the upper jack (whose + or tip is connected to the + amp's output) must be NC to the negative side of the jack, while the AUX contact of the lower jack (whose - is permanently connected to GND) must be NC to the plus (tip) of that jack.

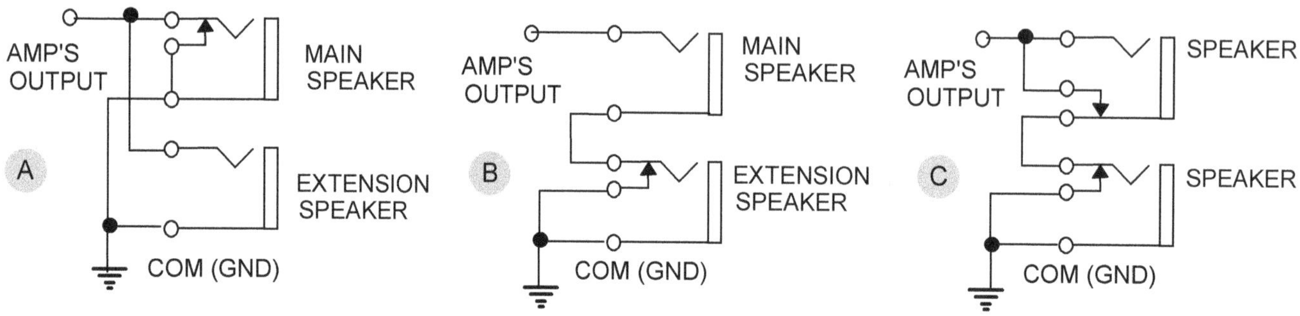

A) The MAIN SPEAKER must be connected for the EXTENSION SPEAKER to work, otherwise the amp's output is shorted. The two speakers are in parallel connection.

B) The MAIN SPEAKER must be connected for the EXTENSION SPEAKER to work, but the amp's output is left open. When both connected, the two speakers are in series.

C) Two speaker outputs of equal priority (no more MAIN or EXTENSION speaker). When both connected, the two speakers are in series.

CAN I CONNECT A HIGHER OR LOWER IMPEDANCE SPEAKERS TO MY AMP?

Over the years I have seen dozens, perhaps hundreds of online discussions on this topic. The prevalent opinion is that you should not use a lower impedance cab connected to a higher impedance tube amp output, for instance, a 4Ω cab with an 8 ohm amp output or an 8Ω speaker cab with a 16Ω output. That prevalent consensus is wrong, there is no danger of using a lower impedance cab than the "nominal"! And this is why ..

The impedance of a speaker or a whole cab is almost never equal to its nominal impedance (an arbitrary figure proclaimed by its manufacturer), for most of the audio frequency band it is much higher. So, the impedance of your 4Ω cab is actually 8Ω or higher for most of the frequency components (harmonics) of the guitar signal.

Secondly, and now we have to get very technical, a tube guitar amp is not an ideal voltage source, meaning it does not keep the output voltage constant with changing load impedance. Transistor amps (solid state amps) are much closer to that ideal.

You may have seen specs for transistor amps, hi-fi or guitar, it does not matter, saying "Output power 100 Watts into 8Ω, 200 Watts into 4Ω". Providing the amp's power supply and its output stage is capable of supplying the required current (the current doubles into 4Ω compared to an 8Ω load), the transistor amp will keep the output voltage constant and double the current.

Of course, that cannot continue indefinitely, halving the load again to 2Ω will not double the output power again (to 400 Watts) but will most likely blow a fuse or destroy the output transistors!

Tube amps do not behave that way. A couple of graphs (next page) will illustrate the point. The output characteristics are of EL84 pentode for a single-ended stage, but the discussion is also valid for push-pull stages.

Such characteristics are published by tube manufacturers for a few selected screen grid voltages, in this case V_S = 250V (1), meaning the screen grid was held at a constant DC voltage of 250V with respect to the cathode.

All the voltages marked, the control grid DC voltage ("bias voltage") and the anode voltage are also in reference to the cathode voltage, whatever that voltage may be.

The characteristics show the complex relationship between anode current (2) as a function of the anode voltage (3) and the input signal voltage entering the control grid, which is biased (zero signal) around the bias voltage (4).

The curves are published for each 1V of bias voltage (negative DC voltage on the control grid, where the signal enters the tube), starting with V_G=0V, and the last one is for V_G=-13.5V.

We also see a quiescent (no signal) operating point Q, positioned at the chosen bias voltage of -7.3V_{DC}, and right on the curve of the maximum dissipation, P_A=12 Watts, which means it is 12 watts for that particular tube (EL84). Such positioning by the amp's designer maximizes the output power of the amp.

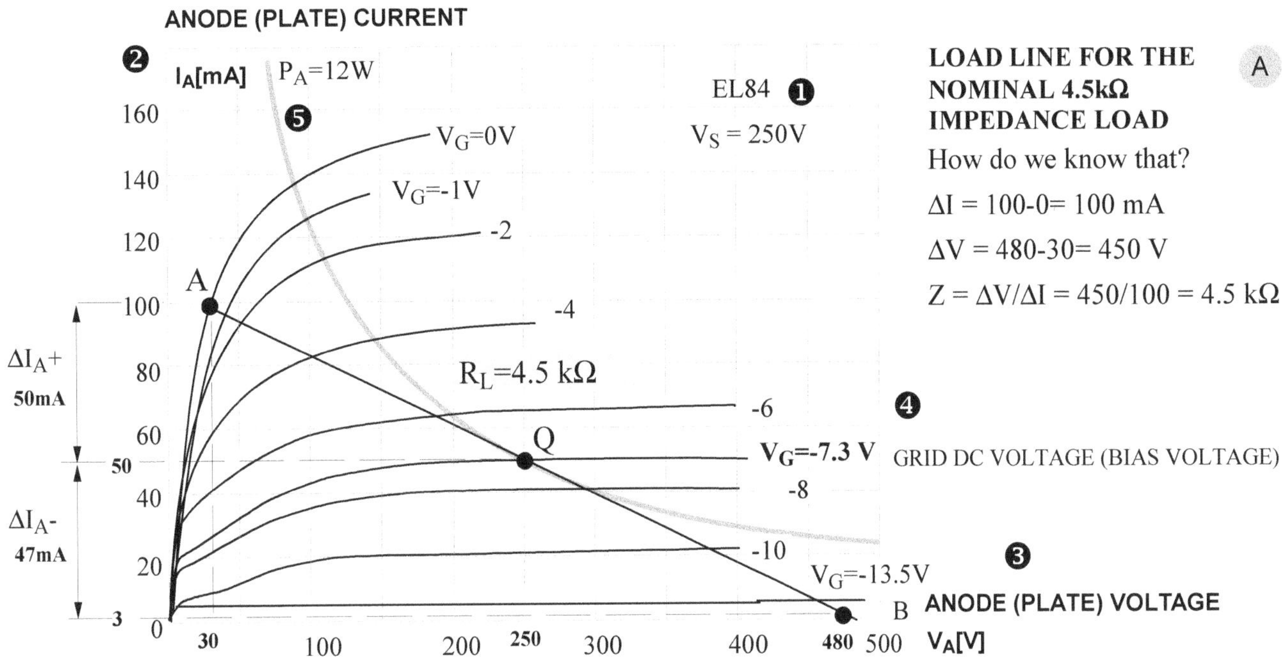

Now the designer has to choose the optimal impedance of the output transformer's primary winding. In this case, 4.5 kΩ was chosen. If we draw a straight line with that slope (remember, impedance or resistance is voltage divided by current) on this graph (which has current on the vertical and voltage on the horizontal axis) though the operating point Q ("Q" for "quiescent"), we get the following situation.

For no input audio signal on the control grid, there is no AC current or voltage swing, the Q current is 50 mA, and the Q voltage is 250V. Imagine now a sinusoidal test signal at the grid input, zeroed around the bias voltage (V_G=-13.5V); what is the maximum voltage swing that this output stage will take?

The intersection of the operating load line with zero bias voltage (maximum current through the tube) is A, and the max. anode (plate) current is 98 mA. The grid voltage cannot become positive if a coupling capacitor is used (as it is on almost all guitar amps), so that is our limit.

On the other side, for the negative input voltage swing, the anode voltage will swing 7.3V from Q down to a curve for V_G=7.3+7.3=14.6V. There is no such curve published, the lowest one is for grid bias of -13.6V, but it will be just under it. So we estimate a current of a few mA only, let's say 3mA (the tube is almost in a cutoff state).

The positive voltage swing is 480-250V−230V, the negative swing is 250-25 = 225V. The current swings are +50 and -47mA. Both the current and voltage swings are almost equal so the distortion should be relatively low.

Notice also that the load line never crosses the "forbidden" territory, that above the parabola of the maximum anode dissipation (5).

The metal box (plate) of the anode can only get so hot (the current through a tube is a stream of electrons that hit the anode and thus heat it up) and dissipate so much of the heat into its surroundings, to achieve a thermal equilibrium. That is the meaning of the maximum dissipation.

Let's see now what happens if we halve the load on the power tube, from 4,500Ω to 2,250Ω (2.25 kΩ). This is what happens when instead of a nominal 8Ω speaker cab, you connect a 4Ω cab to the same amp. Now, instead of reflecting an 8Ω speaker impedance back to the primary side as 4,500Ω, the output transformer will reflect a 4Ω speaker back as a 2,250Ω load.

If we draw that load line through Q (next page), it now crosses into the "forbidden" territory, above the curve of maximum plate dissipation (6), but only for the portion of the signal (roughly 1/2 half of the total swing).

With the same biasing point Q, the maximum input (grid) signal swing this tube can take is again Q-C or 7.3 Volts, but on the other side the swing is much shorter.

We don't know exactly the grid voltage at the cutoff point D, but we can estimate that it's just below -13.5V (the lowest curve published), say -14 Volts, making the maximal grid signal swing 14-7.3V= 6.7 Volts.

That doesn't seem too bad, but look what happened to the current swing: QC=130mA-50mA=80mA, but QD=50-0=50mA only. Likewise, the voltage swing QC is now smaller than before, 250-65= 185V (we had 225V before), and the QD voltage swing is even shorter, only 365-250= 115V!

185V versus 115V means that the positive and negative halves of the signal will not be amplified equally, resulting in huge distortion! So what does all of that mean in practical terms?

It can be shown (we will not do it here) that the maximum obtainable power in this case, compared to the previous case (load impedance matched to the amplifier's output impedance), is lower. We have already shown that the distortion will be significantly higher.

But will the output tube or the output transformer be damaged? No! While the tube's dissipation will be exceeded in the QE segment (6), the dissipation in the QD segment (7) will be so much lower that the average power dissipation (what really matters) may actually go down!

While 2:1 impedance mismatch ratio is OK, this conclusion cannot be extrapolated indefinitely. You should not connect a 4Ω cab to a 16Ω output or a 2Ω cab to an 8Ω output! The output tube(s) will work much harder and may overheat (depending how hard the amp is driven and how loud it's played), and, eventually, fail prematurely.

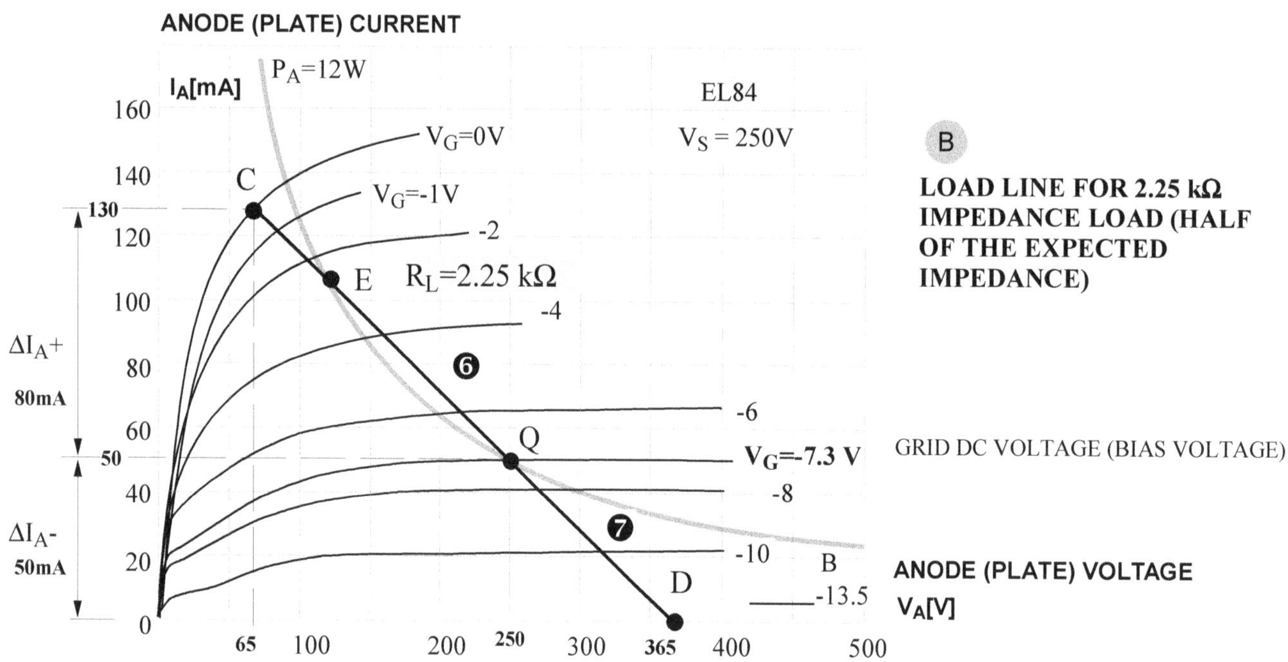

Question: Can I connect a higher impedance cab to a lower impedance tube amp output?

To answer this question, let's draw the 4.5kΩ x 2 or 9 kΩ load line. We see an opposite situation (next page), the QF swing is much shorter than the QG, let alone the QH swing. The graph doesn't extend beyond 500V on the anode, but we can extrapolate the load lone to point H (where it crosses the zero anode current line).

The load line is in the forbidden territory (above the maximum dissipation line) from Q to G (by pure coincidence, I assure you), and then it reverts back to the lower dissipation area between G and H (9).

Since the current swing is much narrower, the maximum obtainable output power will be much lower than the "matched" impedance situation, but the distortion will (initially at least) be somewhat lower, providing that input signal isn't too high (beyond point G).

Cranking the "GAIN" on the amp will, after that point, cause the output stage to start distortion significantly!

Again, just is the previous "downward" mismatched case, while the 2:1 impedance mismatch ratio is OK, going any higher could be disastrous! Draw the line for a 16Ω cab on a 4 ohm amp output case (a mismatch of 4:1) - it will be almost horizontal and a huge section of it will be above the curve of max. plate dissipation (5).

However, remember that a speaker will generate a large "back EMF" or "flyback" voltage. A speaker is an inductor, and when the magnetic field in the inductor drops or collapses, the inductor will try to fight this change and will generate (self-induction) a huge voltage spike that will be reflected by the OPT back into the anode circuit of the power tube multiplied by the OPT's voltage ratio.

With a higher load impedance (double the nominal), that flyback voltage spike will also be double than it would normally be. So, a higher load impedance is actually more likely do damage either the output transformer, the output tube or both, than the lower load impedance!

So, there you have it. Risk-wise, a 2:1 mismatch either way should be fine, but don't push your luck beyond that! Let's summarize these findings with an illustration.

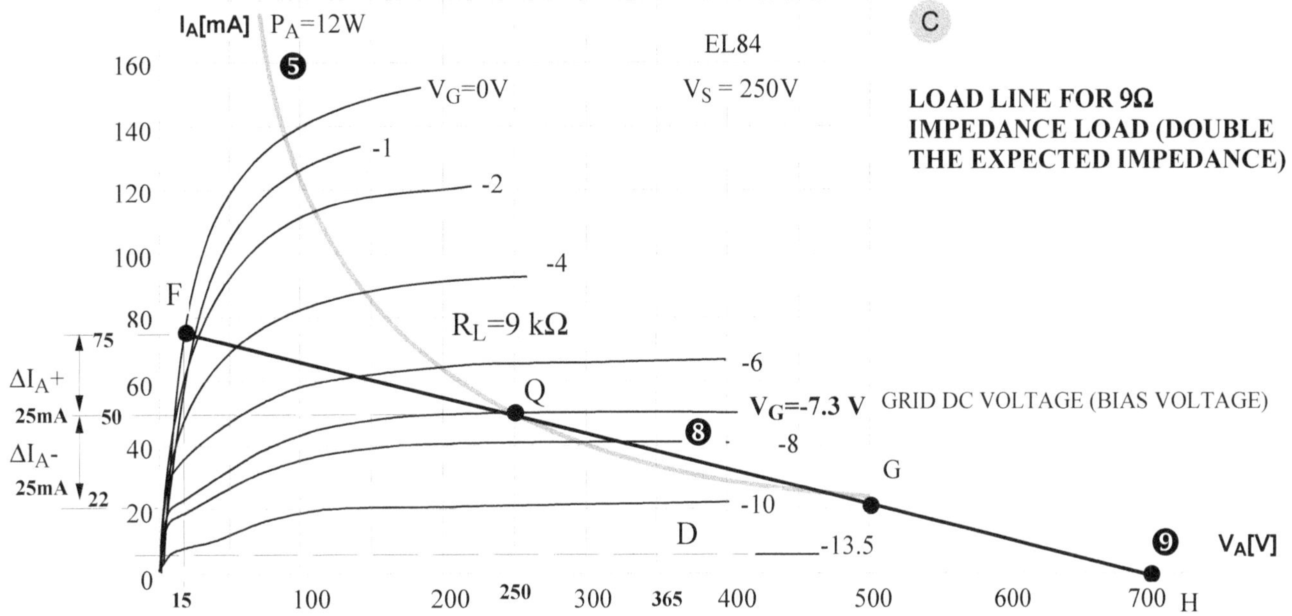

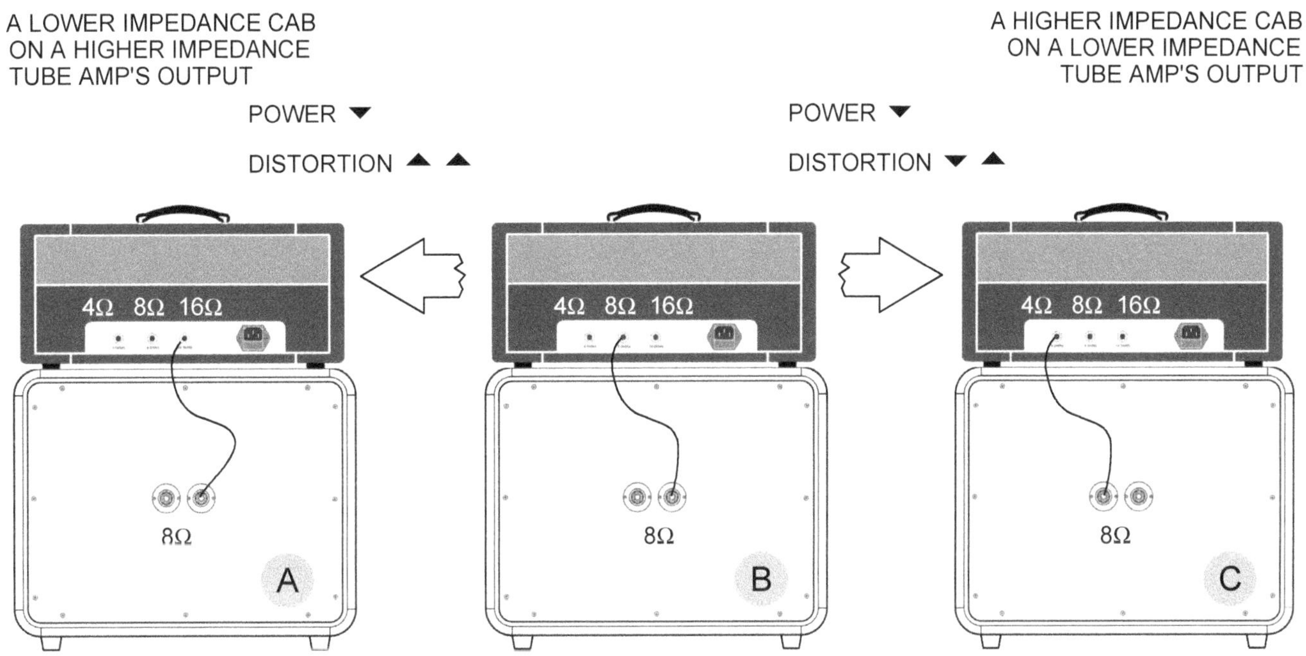

ABOVE: What happens to the maximum available output power and distortion when the speaker load is matched to the amplifier's output (MIDDLE), compared to when the speaker load is half of what the amplifier "expects" (LEFT), and double to what the amplifiers connection says (RIGHT).

Is it possible to connect two dissimilar speaker cabs to one tube guitar amplifier?

The (qualified) answer is yes, it is! To understand the implications of such a hookup, let's briefly review the interaction between the output transformers (OPTs) used in tube guitar amps and speakers (speaker cabs).

We will assume an amplifier (NEXT PAGE) with two OPT secondary outputs (4Ω and 8Ω), but the same logic applies to amps with 8/16Ω outputs and even amps with a triple output, 4-8-16Ω. The primary impedance of the OPT does not matter, we have used 8kΩ as an example (A).

Connecting a 4Ω speaker to the 4Ω output (B) or an 8Ω speaker to the 8Ω output (C) both result in the "proper" loading of the output tube. Both speaker impedances are reflected onto the primary side as an 8kΩ load on the power tube(s). The discussion applies to both push-pull and single-ended output stages.

Disregarding possible power ratings and overload issues (in this discussion we are only interested in impedances), in (D) each tap is loaded with a cab of its own impedance (4Ω speaker to 4Ω tap, and 8Ω speaker to 8Ω tap).

The OPT will dutifully reflect each back to the primary side as 1 x Z_P (nominal primary impedance), in this case as 8kΩ, but the total primary impedance will now be two of them in parallel or half of the Z_P (nominal primary impedance), 4kΩ!

In (E) each OPT tap is loaded with twice the expected impedance (8Ω speaker to the 4Ω tap, and 16Ω speaker to the 8Ω tap). The OPT will reflect each to the primary side as $2Z_P$ (double the nominal primary impedance), in this case as 2x8kΩ, so the total primary impedance will be $2Z_P$ in parallel or Z_P, the nominal primary impedance of 8kΩ!

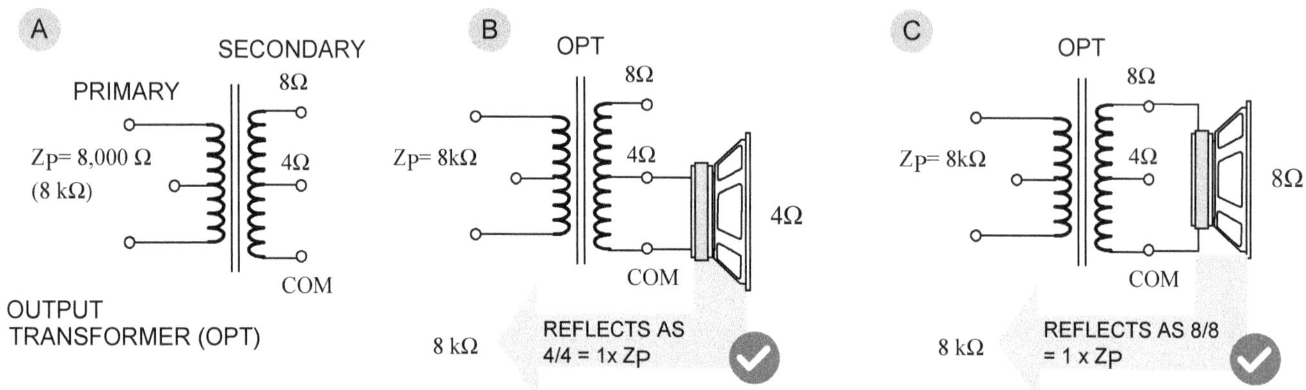

ABOVE: Connecting a 4Ω speaker to a 4Ω tap (B), or an 8Ω speaker to an 8Ω tap (C) reflects their impedance to the primary side as 1 x Zp , the nominal primary impedance (in this example 8kΩ plate-to-plate)

D): Connecting two speaker cabs of dissimilar impedances to their corresponding amplifier outputs (4Ω speaker to the 4Ω tap, and 8Ω speaker to the 8Ω tap) results in a loading of the output stage that is only half of the expected (nominal)!

E): When both 4Ω and 8Ω output taps are loaded with speakers of twice the expected (nominal) impedance 8Ω speaker to the 4Ω tap, and 16Ω speaker to the 8Ω tap), the output stage of the amplifier "sees" the expected impedance.

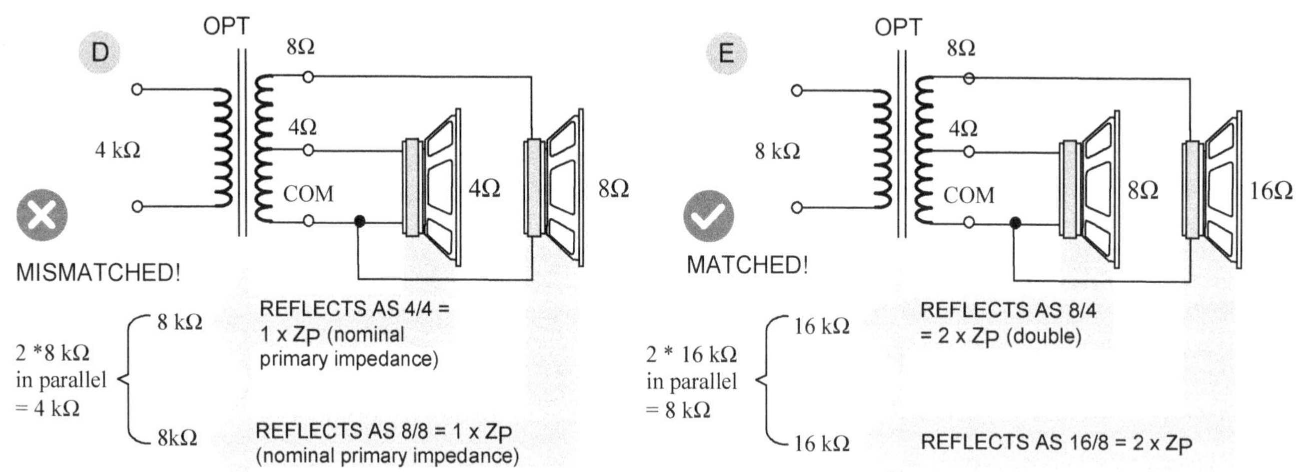

How speakers' reactance impacts its tone and amplifier distortion levels

We have already learned that real speakers are not a purely resistive load (1) on the amplifier. Dynamic speakers are mostly of inductive nature (frequency bands "A" and "C", illustration on the next page)) and in the first approximation they can be modeled by a series RL circuit, the impedance being $Z_L=R_L+X_L$.

Instead of the operating point moving up and down along the straight load line AB as with a purely resistive load (only possible during testing with a dummy resistive load - a high power resistor), now the operating point follows an elliptical trajectory. For capacitive loads (a dynamic speaker reproducing guitar frequencies in the "B" range and electrostatic speakers), it rotates counter-clockwise, for inductive loads it moves in the clockwise direction.

If you draw a horizontal line through any value of anode current, it will intersect the ellipse in two points. There are two solutions to the loadline equation, for each current there are two voltages that will produce such a current. The same applies to anode voltages, if you draw a vertical line for any anode voltage between you will get two different plate currents.

Intuitively, without going into complex maths, the larger the reactive component of the load (inductive or capacitive reactance), the wider or "fatter" the ellipse. The smaller the reactance, the flatter the ellipse, and the more such an ellipse will approach a straight line.

However, the reactive component of the load isn't constant - it changes with the frequency of the audio signal the speaker is trying to reproduce. The higher the frequency of the signal, the larger the inductive reactance $X_L=2\pi fL$ will be and the fatter the ellipse.

So, with a complex signal comprising of many different components of various frequencies, one ellipse will turn into dozens. If you observe that on an oscilloscope, the load "line" will be a messy, fuzzy, filled "ellipse" of ever changing size and shape.

I put even ellipse in quotation marks, because the tangled mess cannot be called an ellipse any more, unless a single sinewave signal of fixed frequency is being reproduced, which is never the case with music, only during testing.

A few other important issues are immediately noticeable. The fatter the ellipse (the load "line'), the deeper it crosses into the tube overload territory (2), above its parabola of maximum anode power dissipation, which is P_A=12 Watts in this case of the EL84 power tube (3).

Finally, notice that the anode current drops to zero and remains there between points (4) and (5), meaning the tube is in cutoff state and does not amplify at all. During that short period of time no sound will be produced by that amp-speaker combination!

However, due to its extremely short duration and its physics, delayed signals and similar phenomena, such tiny breaks in sound (re)production are not always noticeable.

What is noticeable, though, are the extremely high levels of distortion that such elliptical load lines imply. The "fatter" the ellipse (the more inductive or capacitive the load), the earlier such cutoff situation will happen and the higher the distortion will be!

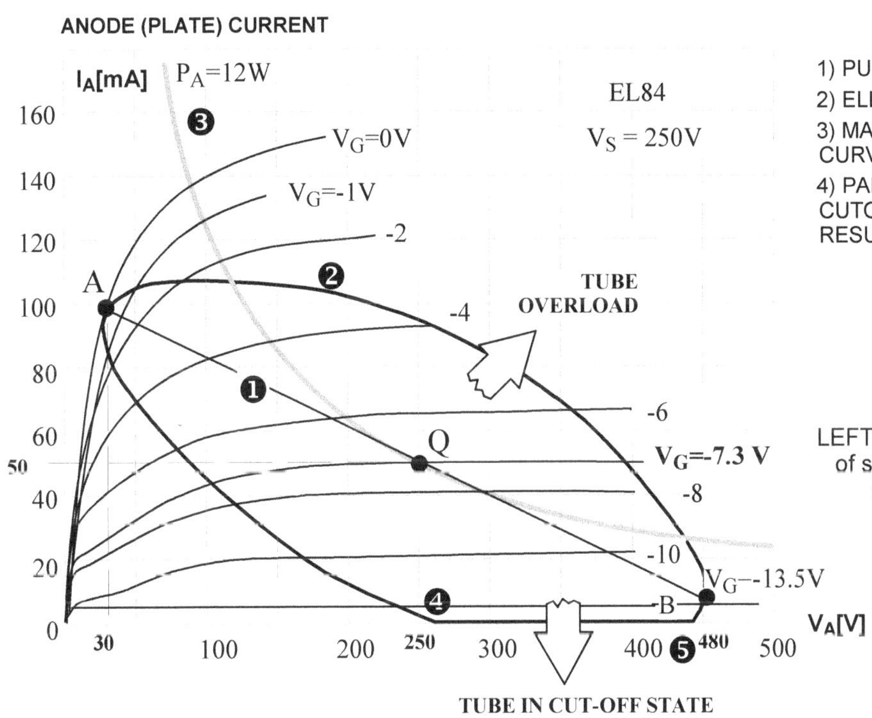

1) PURELY RESISTIVE LOAD LINE R_L=4.5 kΩ
2) ELLIPTICAL LOAD LINE
3) MAXIMUM ANODE POWER DISSIPATION CURVE (in this case 12 watts)
4) PART OF THE ELLIPSE MISSING (TUBE IN CUTOFF STATE, NO CURRENT FLOWING), RESULTING IN HIGH DISTORTION LEVELS

LEFT: An elliptical load "line", a consequence of speakers' inductance and capacitance, is one of the main causes of distortion and other undesirable effects

A tale of two speakers

Let's compare two speakers (next page), both of the same "nominal" impedance. Speaker #1 has a resonant frequency of 90 Hz, while speaker #2 has a much lower f_R or 60 Hz. That means that with no guitar frequencies below 80 or so Hertz, speaker #2 will never cross into Region A (the lower inductive range), while speaker #1 will be crossing between the lower inductive (A) and the capacitive region (B) often, which is undesirable. The boundary between regions A & B is different for speaker #1; the shaded region is for speaker #2 only.

More importantly, the impedance peak of speaker #2 is much lower (6) than that of speaker #1 (7). In fact, its whole impedance curve is flatter, its impedance rise with increasing frequency is also much lower (8).

The load ellipse of speaker #2 will be much thinner than the fat load ellipse of speaker #1, so its distortion will be lower and the tone cleaner.

You may be thinking, but wait Dr. Iggy, guitarists love distortion, what's the big deal about getting some more?!? Remember, the distortion we are talking about here is not the pleasant, harmonic or somehow desirable distortion guitar players crave, but a harsh, shrill, unpleasant distortion of the worst possible kind.

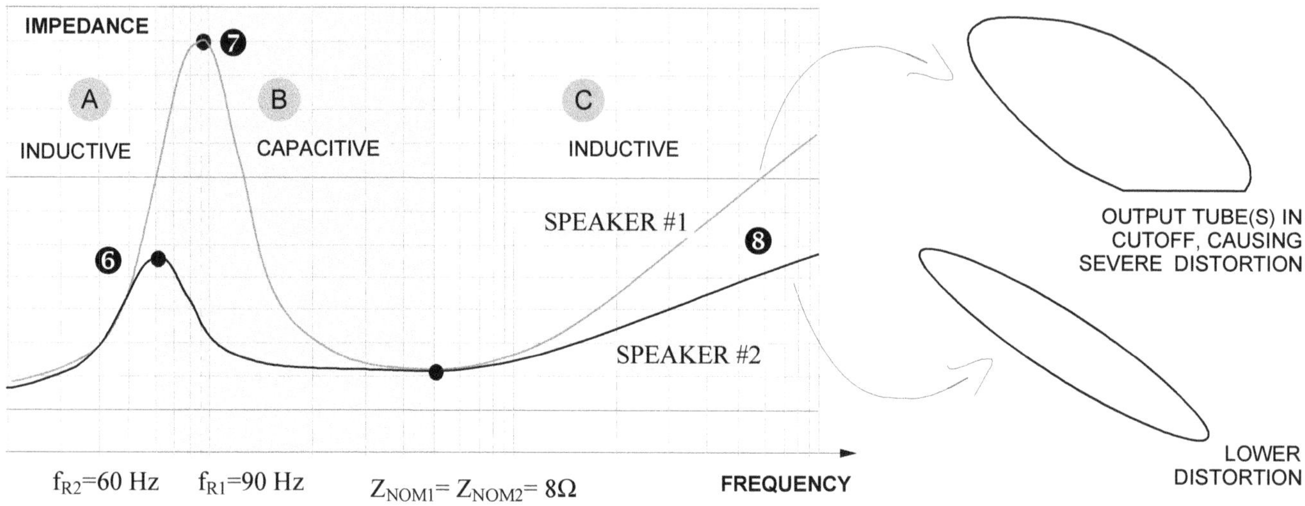

ABOVE: The flatter the speaker's impedance curve, the better. A "fat" elliptical load "line" of speaker #1 (high impedance peak) is much more likely to take an output tube into a cutoff mode and cause severe distortion!

The importance of speakers' sensitivity

Consider two guitar amp/cab combo amps, a low power (10 Watts) tube amp with a speaker of 98 dB/W sensitivity and a high power (100 Watts) combo whose speakers have a low 83 dB/W sensitivity. Which one will sound louder in the same room?

If we express the ratio of power outputs of the two amps as a ratio we get $P_2/P_1 = 100/10 = 10$ or in dB terms $10\log P_2/P_1 = 10\log 10 = 10$ dB.

I've deliberately chosen these friendly & easy (but very realistic) numbers. Those who understand decibels could have quickly figured that result out since ten times higher power of the 100 W amp is +10 dB above the output level of the low-powered amp.

However, the higher power amp's speaker has 10 dB/W lower sensitivity, so to get the same SPL (sound pressure level) as the low-powered amp with high-efficiency speaker, the amp must have 10 dB higher amplification, which it has. Therefore, both setups will sound equally loud. In the second case, what we gained in amplifying power we lost by using a very inefficient speaker!

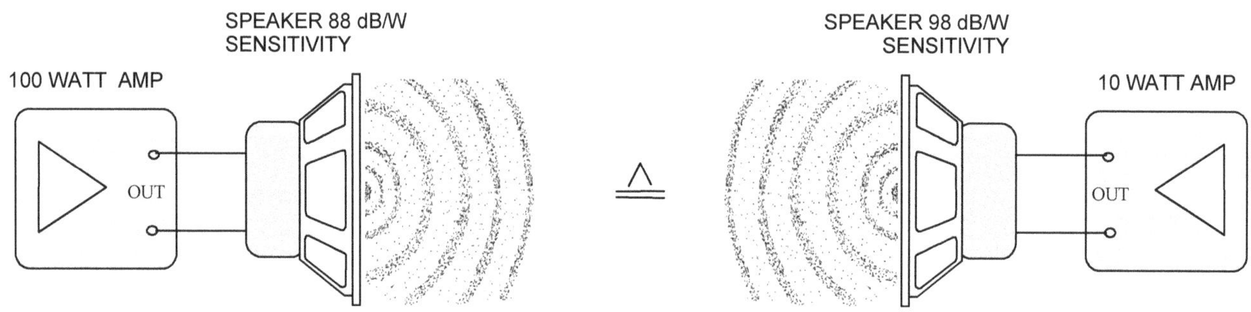

ABOVE: Since we don't listen to amplifiers alone, but through loudspeakers, the maximum power of an amplifier does not matter on its own; the loudspeaker sensitivity also has to be taken into account. The two combo amps illustrated produce the same SPL and will sound equally loud.

COMPARING APPLES AND ORANGES (SPEAKERS' SENSITIVITY USING VOLTAGE AND POWER FIGURES)

Apart from tube-rolling in their amps, many guitarists try speaker-rolling in their cabs, replacing existing speakers with (usually expensive so called "boutique") speakers of higher sensitivity, to get a better tone but also a louder sounding (a few dB) rig. Some even go in the other direction, replacing relatively efficient speakers in their cabs with lower efficiency speakers, to get similar tone (distortion) but at lower volume levels.

In many cases, they do not achieve their goal and get disappointed, since the sensitivity figures declared by the replacement and original speaker makers may not be directly comparable!

HOW ATTENUATORS IMPACT SPEAKERS' IMPEDANCE & SPL CHARACTERISTICS

CASE STUDY: Weber MASS Lite attenuator

MASS Lite comes in 50 Watt and 100 Watt versions, with three optional features: balanced line out, headphones output, and footswitch bypass. Our unit had the first two options installed. The basic 100 Watt Mass Lite sold for US$276 in 2024, with the headphone option an additional $30. The unit pictured is the older version; the current model has different cosmetics, but we believe it is practically identical inside.

The "Lows-Mids" rheostat acts as the primary volume control, while the "Mids-Highs" knob provides treble control. Its action, however, is always dependent on the "Lows-Mids" control. The higher the "Lows-Mids" attenuation, the less effect "Mids-Highs" or treble compensation will have on the sound.

With two main rheostats set at maximum attenuation, fully CW in "10" position, MASS Lite can be used as a dummy load for amplifier testing.

Since some amps don't have the headphones or the line outputs, this versatile device also adds those two features to its primary attenuation purpose.

As with any attenuator, the heart of the unit is a couple of 50W rheostats (rated at 100W on a 100W version).

Rheostat (1) controls "Lows-Mids" frequencies, and rheostat (2) "Mids-Highs." There is also a third rheostat (3) at the back, labeled "HP Volume," which controls the headphone output volume.

The "Low" position of the RANGE switch adds 6db of attenuation to the setting of the attenuator control.

In the center (4) is a voice coil assembly (VCA). There is no metal basket or paper cone, so the VCA doesn't really behave like a "real" speaker, more like a choke, but it does add some counter-EMF to the response.

Weber is one of the rare attenuator makers that includes VCAs in their designs. Most guitar players would agree that simple resistive attenuators significantly change the sound of an amp, and that seems to be due to many factors, the most important being the absence of the speaker-amp interaction.

With resistive loads, there is no induced back EMF (Electromotive Force) that a dynamic speaker feeds back into the output transformer's secondary. How successful or otherwise has Weber been in mitigating this problem? We will leave it to guitar players and their personal opinions.

The newer model has no numerical markings or scales on its rheostat controls, which have been changed to "Bass (Main Volume)" and "Treble". Pity we had already sold the pictured 50 Watt version, otherwise we could have measured the attenuation-versus- rheostat position curves for each controls. With the 100 Watt version we had nothing to use as a position indicator!

Tucked in between the "Line out" and "Speaker" jacks is a small audio transformer (5) that converts the single-ended (unbalanced) signal to a balanced output. This is an optional feature. The transformer's primary is fed from the attenuator's input via a resistive voltage divider, two film resistors joined in a messy fashion and simply hanging in the air (6).

LEFT: The internal view of Weber MASS Lite attenuator (50W version)

Notice two NP (Non-Polarized) electrolytic capacitors (7), the kind used in speaker crossovers. This means there is some kind of filtering and some frequency-dependent manipulation of the signal, which will enable a degree of control over the tonal characteristics of the output signal. One of the caps is in parallel with two 135Ω 15 Watts power resistors (8).

The top switch (9) is the bypass switch, which bypasses the whole attenuator when engaged, feeding the input signal straight to the output.

The headphone and line out stay active even in bypass mode. The HP and line-out circuits are independent of the attenuator.

The line out signal on the primary side of the transformer is 1/(1+10)*100 = 9.1% of the input signal from the amp's output. Since power is $P=V^2/R, V=\sqrt{PR}$. For instance, with 5 Watts coming in and an 8Ω amp output, the RMS voltage is $V=\sqrt{(5*8)} = 6.3V$ and the line out is 6.3*0.091 = 0.57V.

When closed, the Range switch bypasses the two paralleled 135W resistors (62.5W total) and thus shorts points "Y" and "Z" together, increasing the output level. With the Range switch open (as shown), the 62.5Ω resistance is in series with the load (speaker), and the volume is somewhat reduced.

ABOVE: The voice coil assembly of Weber MASS Lite 100W attenuator: 1) ceramic magnet 2) central iron slug 3) voice coil windings 4) voice coil former (bobbin) 5) cloth suspension, glued to the bottom of the enclosure

The reactance of the 4μ7 parallel capacitor will decrease as the frequency rises, so its bypassing effect will be more pronounced. With rising frequency, the series 62.5Ω resistance will be gradually taken out of the circuit. HF attenuation will be reduced, thus providing a mild HF boost. This cap was most likely included to prevent the attenuator from sounding "dull." Without it, the "Lows-Mids" attenuator would be a simple resistive affair.

The *"Mids-Highs"* rheostat's wiper feeds the outgoing signal through a capacitor, which forms a simple low-pass (CR) filter with the load's resistance. Assuming a closed *Range* switch and an 8Ω load, the lower -3dB (half power) frequency of that filter would be $f_L=1/(2\pi RC)$ = 904 Hz, meaning that "Mids-Highs" refers to frequencies above approx. 1kHz.

Finally, a handy feature of Weber attenuators is that there is no need for separate units for each speaker or amplifier output impedance, say 4, 8 or 16Ω. These units can be used with any of those load impedances.

However, no matter how well designed or intricate an attenuator may be, they always change the sound of the amp-speaker combination since they change the amp-speaker interaction, frequency response, and tonal balance.

In electronics, just as in life in general, there is no such thing as "free lunch"; there is always a price to be paid!

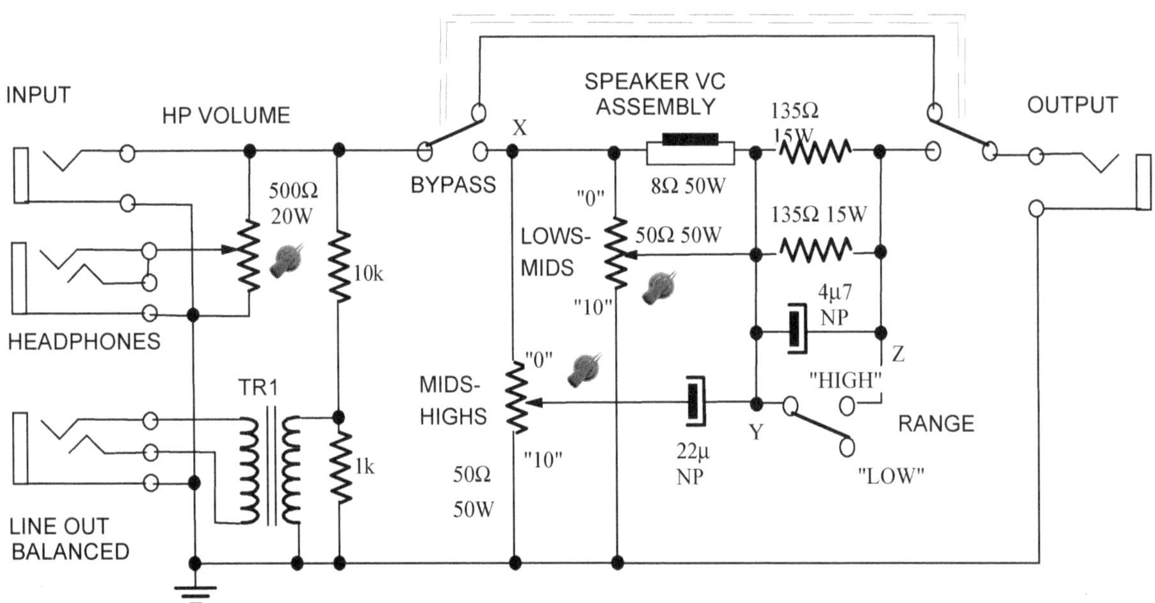

ABOVE: Weber MASS Lite 50W attenuator circuit diagram, © "Ted Weber's Famous Loudspeakers"

The output impedance of the attenuator and the amplifier

In this experiment we hooked up a Vox Night Train amplifier head to Peavey 6505 speaker cab, first directly and then through Weber Mass Lite 100 attenuator. This is made easy by the fact that the attenuator has an inbuilt bypass switch. We kept the amplitude of the signal constant, the tone controls on the amp were all in their mid (presumably but not necessarily neutral) positions.

Two True-RMS multimeters were used, one as an AC mA-meter in series with the load (speaker), the other as an AC voltmeter, in parallel with the load.

At higher frequencies (1), the SPL curve of the amp - attenuator (bypassed) - cab combination was practically identical to that of the cab driven directly by the function generator ("cab by itself"), meaning the amplifier-speaker interaction did not impact the impedance at high frequencies.

In the 1 - 2 kHz region there were noticeable differences between the cab's own curve (thick gray line) and the cab+amp curve (2), and then between 300 and 800 Hz the curves coincided again (3). Below about 350 Hz the cab by itself (4) was much louder than the amp-attenuator-cab combination (5).

This is most likely primarily due to the limited and sharply dropping output of the VOX amp below 300 Hz.

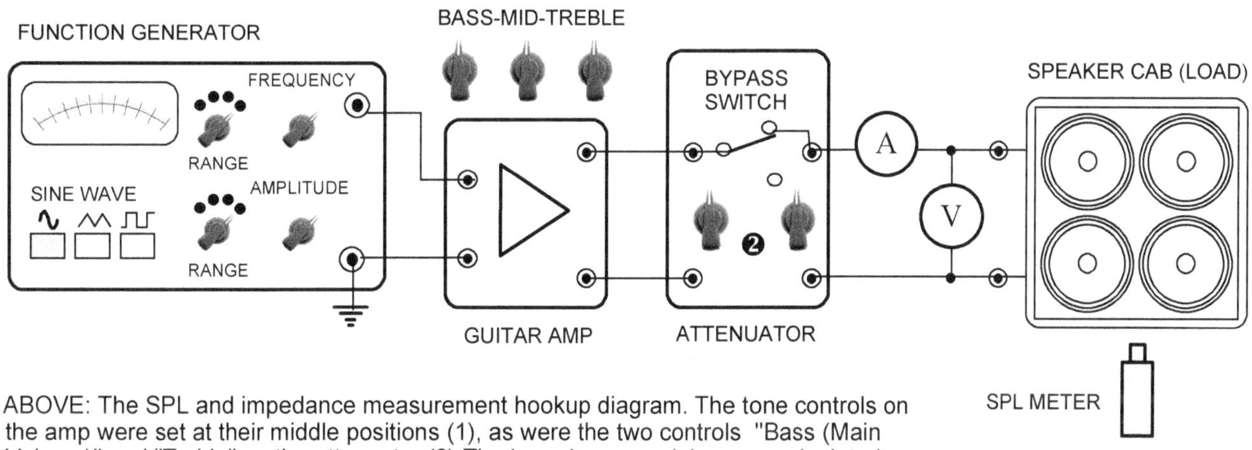

ABOVE: The SPL and impedance measurement hookup diagram. The tone controls on the amp were set at their middle positions (1), as were the two controls "Bass (Main Volume)" and "Treble" on the attenuator (2). The impedance modulus was calculated as the ratio of the measured AC voltage V and AC current (A for ammeter).

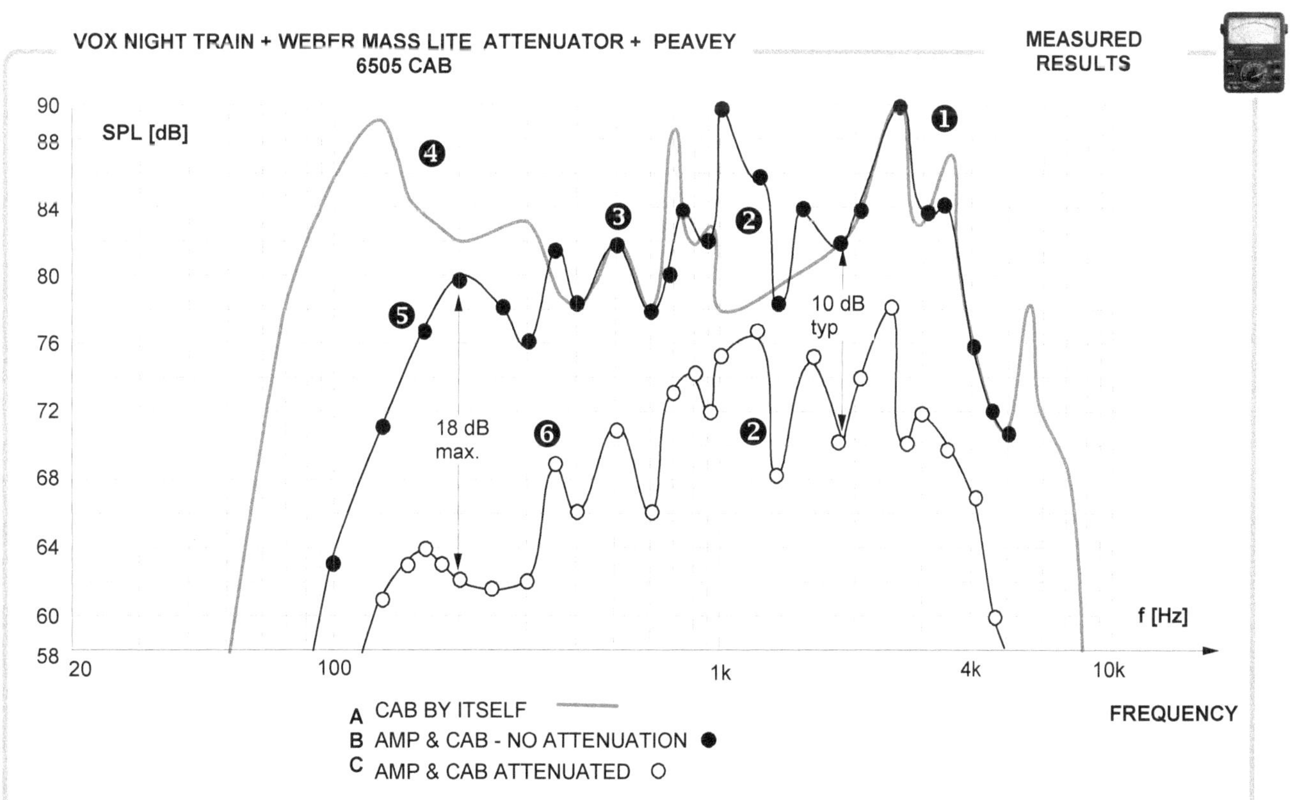

The attenuated SPL curve follows the upper SPL curve (attenuator bypassed) all the way down to about 300 Hz, (6), with slightly larger attenuation in the 150-300 Hz band of 18 dB. Otherwise, with both attenuator controls around their middle positions, the attenuation was fairly constant, 10-11 dB.

Of those 10-11 dB, 6 dB were added by the "Low" position of the RANGE switch, so it seems that in their upper (middle) positions, the two attenuator's control (rheostats) attenuate by 4-5 dB.

Also, notice that the impedance in the 300 Hz - 5 kHz range as lower than that of the cab by itself. At low and high frequency extremes, the impedance was much higher than that of the cab by itself.

With or without attenuation, the impedance modulus curve (next page) had a shape of a bathtub; gone was the impedance peak of the cab itself (1). The flat section of the curve extended further into midrange frequencies (2), the impedance rise started later but was steeper (3) . The flat section also dropped lower (4). The impedance rise in the bass region was even steeper (5), reaching 318Ω at 60 Hz (without attenuation) and a whopping 1,733Ω with the attenuator!

Notice that the vertical scale for the impedance (next page) with the attenuator is different (on the right), five times higher than without the attenuator in the circuit. For instance, without the attenuator, the measured impedance 1 kHz was 10.3Ω, but with the attenuator is jumped to 38Ω!

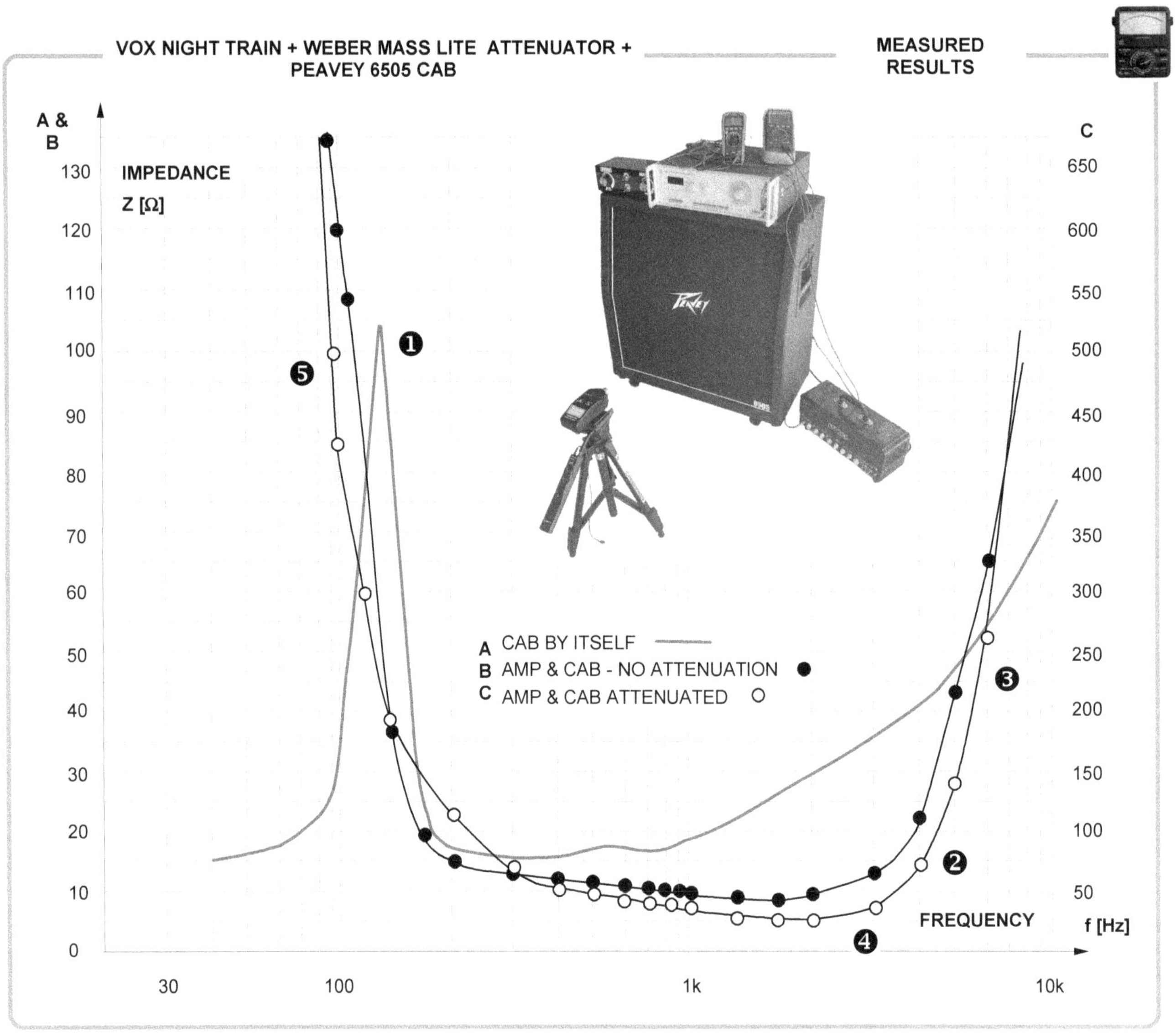

ABOVE: The SPL and impedance measurement setup: function generator (sine wave), two multimeters (one as a mA meter, the other as a mV meter), Weber Mass Lite attenuator, Vox Night Train amplifier head, Peavey 6505 quad cab, and an SPL meter on a tripod

MULTI-SPEAKER CABINETS

7

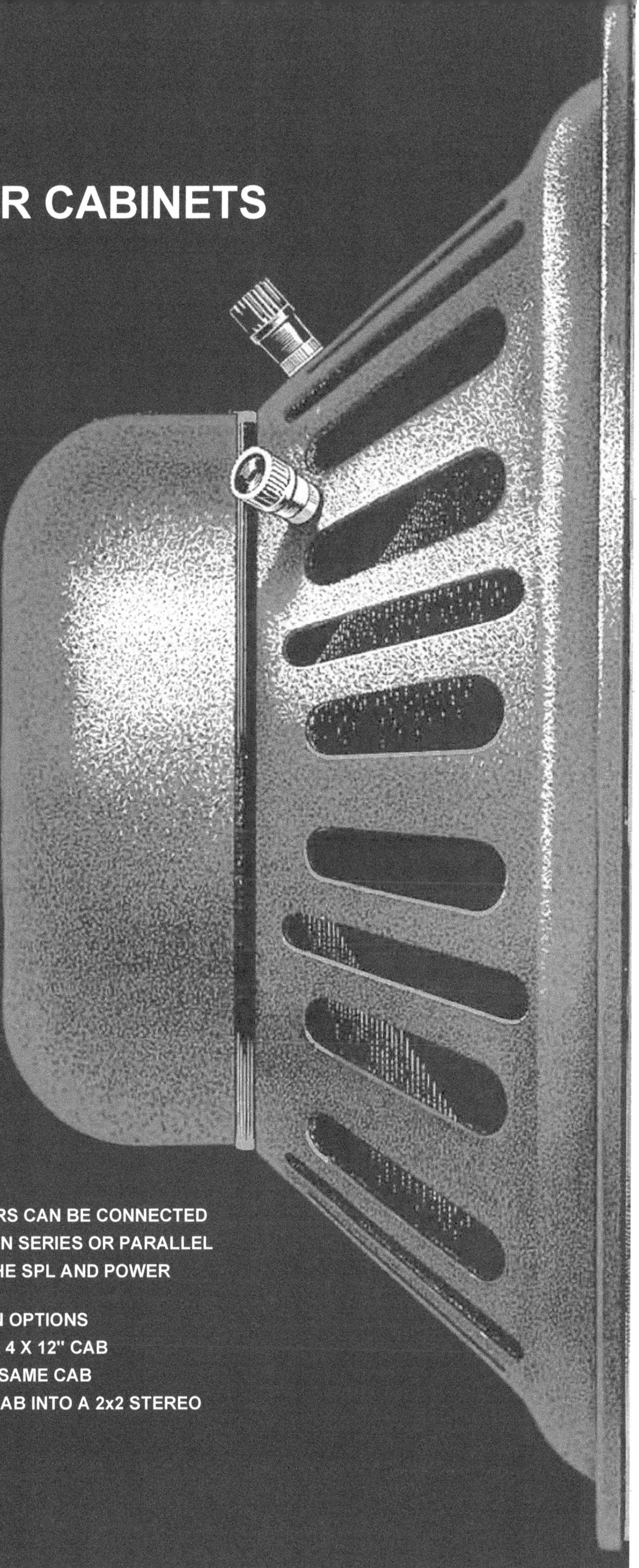

Apart from small combo amps, which have only one speaker, stand alone cabs usually feature two, four or even more drivers. That increases their power handling capability but also boosts the overall SPL levels due to the mutual coupling effect.

There are various ways multiple drivers can be hooked-up inside a cab, some better than the others.

In this chapter we also take a typical 2x12" cab and modify it in various ways. Open vs. closed cabs, passive radiator, detuned (bass reflex), and dissimilar drivers are just some of the mods at your disposal.

Before you rush out to buy a new cab, try some of these hacks on your existing cab first and spend your hard-earned money on more important things such as beer, women, holidays, or "all of the above"!

- THE THREE WAYS MULTIPLE DRIVERS CAN BE CONNECTED
- HOW TO CONNECT EXISTING CABS IN SERIES OR PARALLEL
- HOW MULTIPLE DRIVERS IMPACT THE SPL AND POWER HANDLING OF A CAB
- 4 x 12" or 4 x 10" CAB MODIFICATION OPTIONS
- PRACTICAL TESTS ON A MARSHALL 4 X 12" CAB
- USING DISSIMILAR DRIVERS IN THE SAME CAB
- HOW TO TURN YOUR QUAD MONO CAB INTO A 2x2 STEREO CAB

Now that we understand how speakers work and what their limitations are, the time has come to see if those limitations and contradictory requirements can be overcome, or at least minimized.

The first and arguably the most important is the limited power that can be achieved by a single driver. It is also the easiest problem to solve, and the solution is to combine two or more drivers in the same speaker enclosure.

As always, we will start with simple and easy-to-understand arrangements and move onto more complex driver combinations later on.

THE THREE WAYS MULTIPLE DRIVERS CAN BE CONNECTED

In our discussions to follow, we will assume a quad cab, four drivers in the same enclosure, but the principles are universal and valid for any "N", or number of drivers. Thus, from now on, our "N" will equal four (N=4).

Also, we will limit our analysis to identical drivers in the same cab. Later on we will investigate what happens when we mix drivers of different sizes, power ratings, impedances, sensitivity, and resonant frequencies.

Two drivers can only be connected in series or parallel. Again, assuming four or more drivers, there are three possible arrangements of multiple drivers, which can be connected in series, in parallel, and in some mix of series and parallel connections.

The series arrangement

The current through all drivers is the same, while the voltage across each driver is determined by their impedance at that particular frequency.

In the simplest case of identical drivers, the voltage across each driver will be the same and will equal the total voltage across the cab's terminals (X-Y) divided by the number of drivers.

In the case of different drivers in the same cab, the total voltage at amplifier's and cab's terminals will be distributed in such a way that the voltage drop will be the highest across the driver with the highest impedance, and the lowest across the driver with the lowest impedance: $V_{TOTAL} = V_1 + V_2 + V_3 + ...$

From a practicality and reliability aspects, if one driver's voice coil burns out, the series circuits is broken. All remaining drivers will lose power and stop producing sound, so there will be no sound coming out of the cab.

For two speaker cabs, the flexibility is great, the two most common options being two 4Ω drivers in series for a total of 8Ω, and two 8Ω drivers in series for a total of 16Ω.

Since only 4-8-16 Ω drivers are made, and amplifier's are designed for the same three load impedances, there's only one practical option in four speaker cabs using series arrangement, to use four 4Ω, drivers for a total of 16Ω.

The combined inductance of drivers in series is the sum of all the inductances in the loop (series circuit).

If a typical 12" driver has a VC inductance (at a certain test frequency) of 1.2 mH, the inductance the amplifier will "see" is gonna be for times that, or 4.8 mH.

We have seen how the inductive reactance of a dynamic speaker drivers rises with the rising frequency, so adding them up will exacerbate that effect, which is clearly undesirable (it will significantly attenuate the treble, making the cab sound dark and dull).

SPEAKERS IN SERIES

- Impedance: N*Z, where Z is the impedance of each identical driver and N is the number of drivers
- Power rating of the cab is N*P, where P is the power rating of each identical driver and N is the number of drivers
- Inductance: N*L, where L is the inductance of each identical driver and N is the number of drivers
- Capacitance: C/N, where C is the capacitance of each identical driver and N is the number of drivers

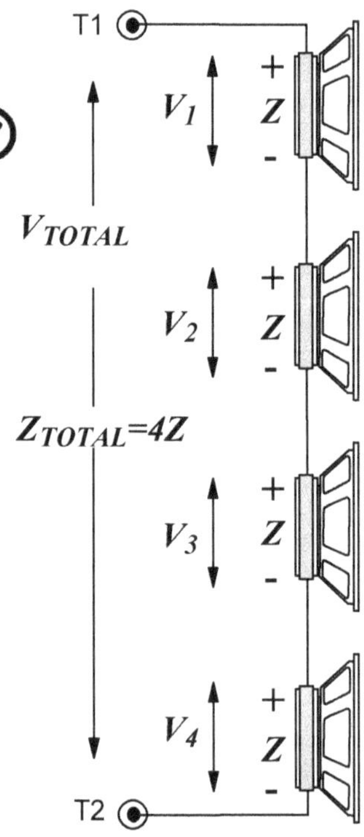

ABOVE: Four identical speakers (impedance Z each) in a series arrangement

The same happens with the inevitable low frequency resonant peak. Some guitar speakers' impedance peaks at 30-40Ω, but others peak at 100-200Ω. Imagine four such drivers in series, at that bass frequency (usually between 50 and 120 Hz), the impedance peak will shoot up to 400-800Ω! Another strike against the series connection.

Last, but certainly not least, cabs with series-connected drivers suffer from inadequate damping, which results in the dreaded hangover effect and poor transient response.

This is most obvious with 2x12" or 2x10" cabs. If you've ever compared two different cabs and found one to be fast, accurate and responsive, while the other was "slow" and "muddy", I bet you the slow and "inaccurate" one had drivers connected in series!

Damping, "hangover" effect and muddy bass

To explain the interrelated concepts of damping, counter-EMF and hangover, we must enter the dangerous waters of electrical engineering. This is purely voluntary, if you are math challenged, you can safely skip this section. All you need to remember is "Don't connect your speakers in series!' and move on.

We have seen that a dynamic speaker behaves as a "load" or "motor" only at certain frequencies, while at others it behaves as generator, feeding the counter-EMF (CEMF) back into the amplifier.

The amp is the source of the signal voltage V_A and has its own internal or output impedance R_A. When a single speaker is acting as a load, or "motor" driven by the current the amp is pushing through its voice coil, the amp "sees" the speaker's impedance Z (which varies with frequency).

For simplicity sake, we will treat the speaker impedance as a pure resistance R_S, which, although oversimplified, will help us to keep things simple so we don't get bogged down in endless high-level maths.

Damping factor DF is defined as the ratio of R_S and R_A: $DF=R_S/R_A$.

The lower the amp's output impedance, the higher the damping factor and the faster and tighter the amp-speaker combination will sound. Hi-fi buffs (a.k.a. audiophiles) pay special attention to that issue in their endless pursuit of perfection in sound reproduction.

Generally, solid-state or transistor amps have much lower internal impedances and thus achieve higher damping factors than tube amps and faster/tighter/cleaner bass.

So, the damping factor tells us how well the amp "controls" the speaker. When the driving signal suddenly stops, as happens all the time with musical tones and their transient nature (as compared to smooth and unchanging sine wave test tones), the speaker's cone will continue to move and produce sound due to inertia. This is pejoratively called "hangover".

When the speaker acts as a generator, it's voltage or counter-EMF (induced in its voice coil's windings) feeds the current back into the amplifier's output stage. The speaker "sees" only the amp's internal resistance R_A as its load.

If that resistance is very low (below say 1Ω for a tube amp or in solid-state amps below 0.1Ω), that is practically a short-circuit and that signal quickly "dies down", to use a very technical term. In other words, it is well damped.

Now, let's see what happens with four identical drivers in series. Each driver, acting as a generator, now "sees" not only the amp's internal resistance R_A as its load, but also the impedance of the other three drivers in series.

The modulus of that impedance, as we have seen, can vary between say 4Ω and 200Ω at resonance. Its phase also swings back and forth, revealing the speaker's dual inductive and capacitive character.

The counter-EMF signal from each of the four drivers isn't short-circuited any more by a very low amp's internal impedance but drives the other three speaker voice-coils, meaning the damping is practically nonexistent.

RIGHT: A simplified electrical model of an audio amplifier driving a single speaker, plus a reversed situation when the speaker acts as a voltage generator and drives CEMF (self-induced voltage) back into the amplifier's output.

The lower the amp's output resistance R_A, the better the damping and the tigher the bass!

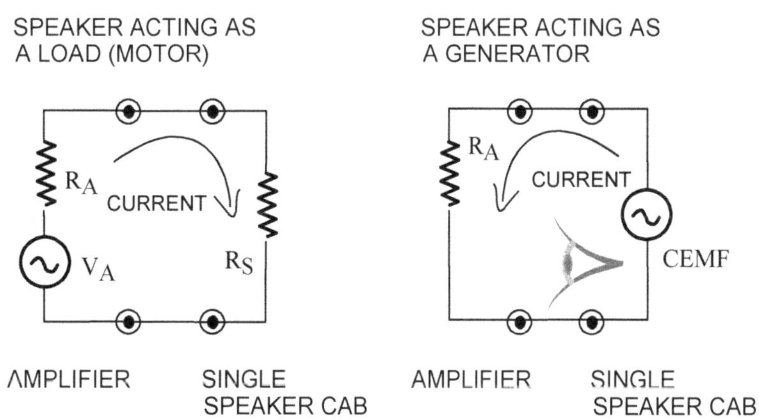

SPEAKER ACTING AS
A LOAD (MOTOR)

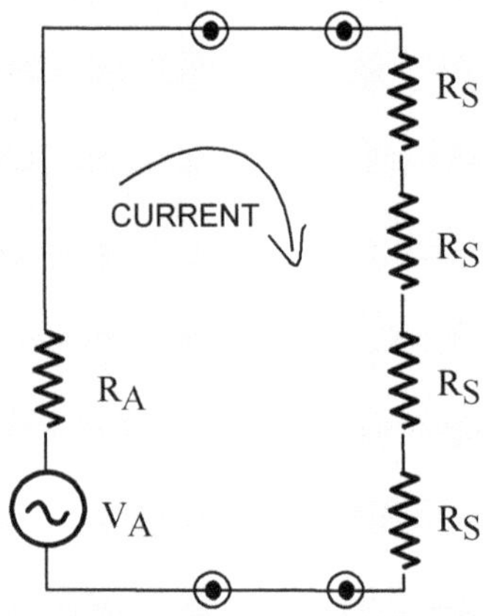

AMPLIFIER QUAD CAB

LEFT: A simplified electrical model of an audio amplifier driving a speaker cab with four identical drivers in a series connection

RIGHT: A reversed situation when one driver acts as a voltage generator and drives back EMF into the amplifier's output through the other three series-connected speakers.

The total damping resistance each driver "sees" is $R_{TOT} = R_A + 3R_S$

No matter how low the R_A may get (even zero), the extremely high factor $3R_S$ means the damping is practically nonexistent!

SPEAKER ACTING AS
A GENERATOR

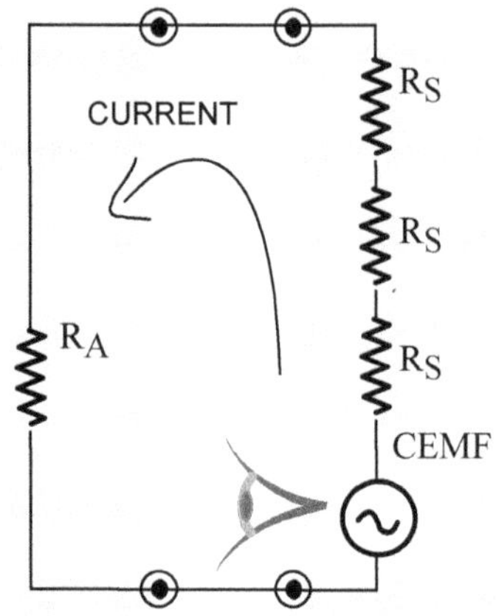

AMPLIFIER QUAD CAB

The parallel arrangement

Connecting drivers in parallel reduces the overall impedance, increases the capacitance the amplifier sees, and reduces the overall inductance across the amplifier terminals.

The total capacitance is multiplied by N, the total inductance is divided by N (assuming identical drivers).

From a practicality and reliability aspects, if one driver's voice coil burns out, the other drivers won't be affected and will keep working.

Let's say each driver in a 4x12"cab is rated at 50W, a total of 200 watts and one fails with an amp cranked to its maximum, which, let's keep ti simple, is also 200 watts. Now each of the three remaining drivers is dissipating 200/3= 66.7 watts, an overload of 66.7/50 = 1.333, or 33.3%.

So, unless the fault is noticed and the situation rectified, the other three drivers may also fail. This, of course, assumes that the overload situation persists, that we keep using the amplifier driving the cab at its maximum output power, which may exceed the rated power of the cab.

Just as with the series arrangement, for two-speaker cabs the most common options are two 16Ω drivers in parallel (for a total of 8Ω), and two 8Ω drivers in parallel for a total of 4Ω.

In four-speaker cabs, the parallel arrangement is predominately used with higher impedance drivers, most often 4x16Ω, resulting in a 4Ω total, and occasionally but very rarely with 4x8Ω, resulting in a 2Ω cab.

SPEAKERS IN PARALLEL

- Impedance: Z/N, where Z is the impedance of each identical driver and N is the number of drivers
- Power rating N*P, where P is the power rating of each identical driver and N is the number of drivers
- Inductance: L/N, where L is the inductance of each identical driver and N is the number of drivers
- Capacitance: N*C, where C is the capacitance of each identical driver and N is the number of drivers

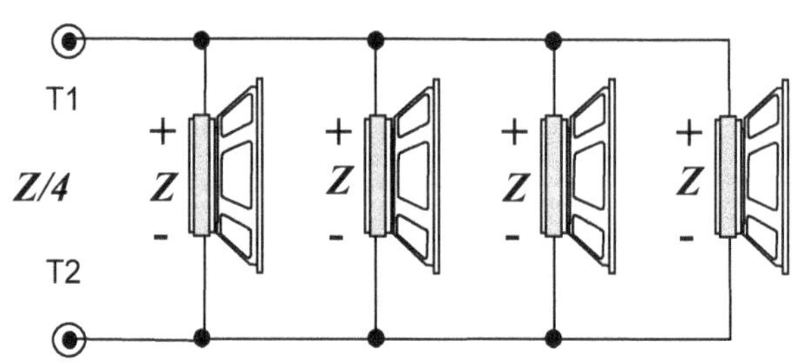

ABOVE: The parallel arrangement (four speakers shown)

Each driver, acting as a generator, now "sees" not only the amp's internal resistance R_A as its load, but also the impedance (resistance) of the other three drivers in parallel.

You may have realized by now that no matter how high or low these three impedances may be, since R_A is much lower, the overall impedance each of the four drivers sees is always lower than R_A! Remember, current always takes the path of the least resistance, which in this case is R_A.

This is an almost identical situation to a single driver back feeding the amplifier, as discussed recently. Thus, the parallel connection is way better (more optimal) than the series arrangement from that point-of-view (damping and hangover) as well. Thus, especially for bass guitar cabs, never use series-connected drivers!

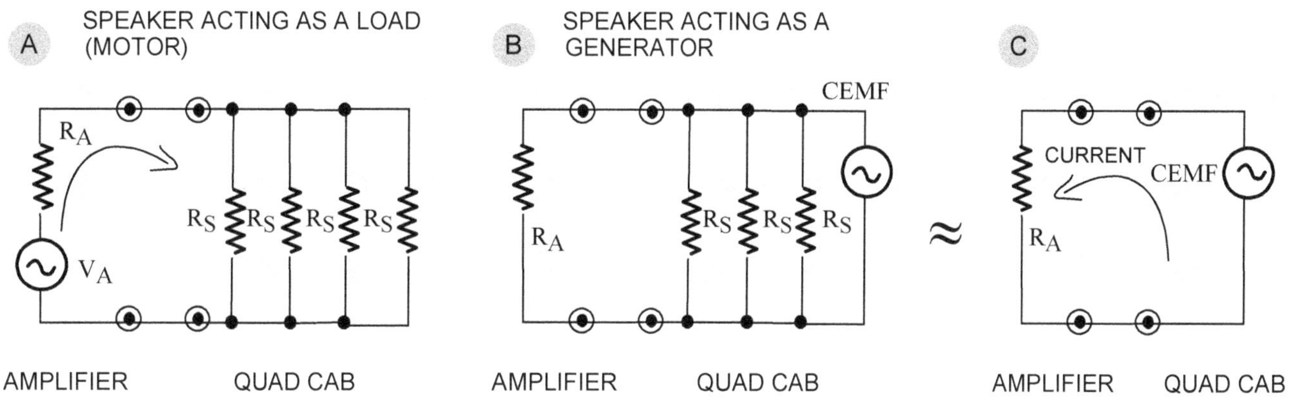

ABOVE: An electrical model of an amplifier driving a speaker cab with four identical drivers in parallel, and a reversed situation when one driver acts as a voltage generator and drives CEMF into the amplifier's output. The total damping resistance each driver "sees" is $R_{TOT} = R_A \parallel R_S/3$. Since R_S is much higher than R_A, the three R_S in parallel can be omitted and the model simplified, so the superior damping (campared to the series connection) of this configuration becomes obvious.

The series-parallel and parallel-series arrangement

The mix of series and parallel connections gives us the highest degree of flexibility as far as the driver impedances are concerned. There are two possible arrangements, a series connection of two paralleled driver pairs (the series-parallel arrangement), and a parallel connection of two pairs of drivers, each pair connected in series, called, you've guessed it, a parallel-series arrangement.

If the impedance of each of the four identical drivers is Z, the overall impedance of the quad cab will also be Z, which is very neat and easy to remember. So, an 8Ω cab will have four 8Ω drivers, a 16Ω cab will have four 16Ω drivers, and a 4Ω cab will have four 4Ω drivers.

What would happen if one of the drivers fails, burns out due to temporary overload. Using driver #1 as an example, in the series-parallel arrangement, all the current from the amp, which was divided between drivers #1 and #2, will now flow through driver #2, so it could be overloaded too.

The overall impedance of the 3-driver cab is now Z (driver #2) in series with Z/2, or 1.5Z; the failure of driver #1 caused the cab's impedance to increase 50%. This will reduce the current through the cab and thus make the failure of driver #2 less likely. The failure will be audible, the SPL level of the cab will drop.

In the parallel-series arrangement, the branch with drivers #1 and #3 is now be open, thus protecting driver #3 from failure. Now only drivers #2 and #4 are working, and the cab's impedance would be doubled, from Z to 2Z.

Assuming a sealed enclosure, the other two drivers (#1 and #3) will work as drone cones (passive radiators), so there may not be any drop in loudness, or it may be frequency dependent and thus not easily noticeable. With an open back cab, the loudness level will be noticeably lower.

What about damping, CEMF and hangover? Again, we will assume four identical drivers in the cab, each with impedance R, and, for simplicity's sake, treat each driver as a resistor, ignoring for a moment its complex, frequency dependent impedance. Indeed, at one particular frequency, that impedance can be treated as a fixed resistance. Phase shifts (is that impedance capacitive or inductive) are of no interest to us here.

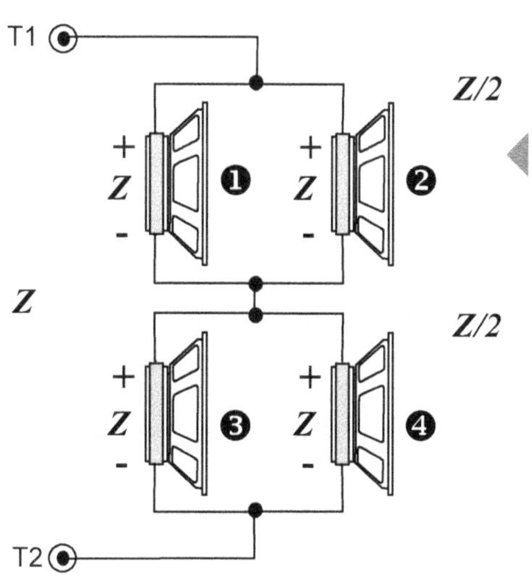

LEFT: Series-parallel arrangement (two pairs of paralleled speakers in series)

RIGHT: Parallel-series arrangement (two pairs of series-connected speakers in parallel)

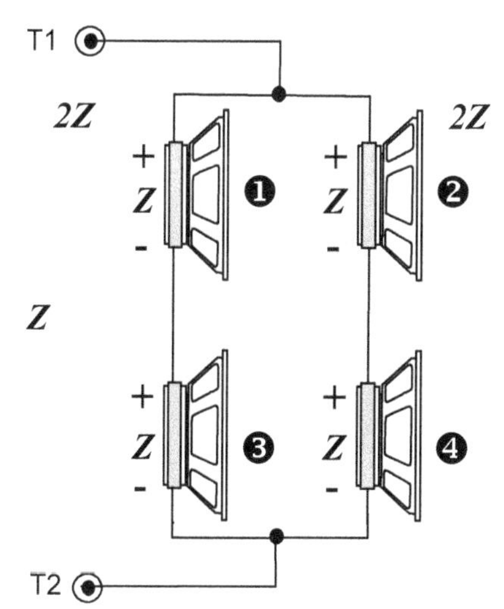

Looking at the series-parallel arrangement first, let's choose the driver between T1 and T3. The driver "sees" two of its fellow drivers in parallel ($R \| R = R/2$) in series with the amplifier impedance R_A, all paralleled by the driver connected to the same terminals T1 & T3.

Let's assume $R=8\Omega$ and $R_A=4\Omega$ (a tube amp with a relatively high output impedance). The impedance each of the four drivers sees is thus $Z = R \| (R_A+R/2) = 8 \| (4+4) = 8\|8 = 4\Omega$, or half it's own impedance.

Even if the amp driving the cab had an ideal, zero ohms output impedance (($R_A=0$), the driver will still "see" an impedance of $R\|R/2 = 8\|4 = 2.67\Omega$, or a third of it's own impedance.

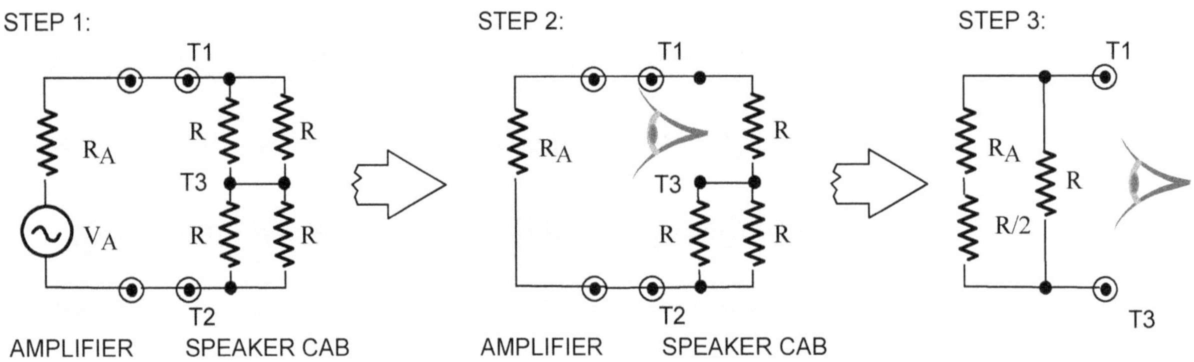

ABOVE: What impedances each driver "sees" in the series-parallel arrangement

In the parallel-series arrangement, the driver between terminals T1 and T3 "sees" two of its fellow drivers in series ($R+R= 2R$) and in parallel with the amplifier impedance R_A, all in series with the driver connected in series with our active driver in the cab.

Again, let's assume $R=8\Omega$ and $R_A=4\Omega$. The impedance each of the four drivers sees is thus $Z=R+ (R_A \| 2R) = 8 + (16 \| 4) = 8 = 3.2 = 11.2\ \Omega$.

In an ideal case, the amp driving the cab would have zero output impedance (($R_A=0$), and then ($R_A \| 2R$) would also be zero, since any impedance in parallel with a short circuit is a short circuit or zero impedance. However, our "active" driver will still "see" an impedance of $R = 8\Omega$, which is the impedance of that driver it is connected to in series, the driver between T2 and T3; that is the absolute minimum attainable.

In conclusion, we see that the damping impedances are much lower in the series-parallel arrangement's case, which will thus generally sound superior to the same cab and drivers in the parallel-series arrangement.

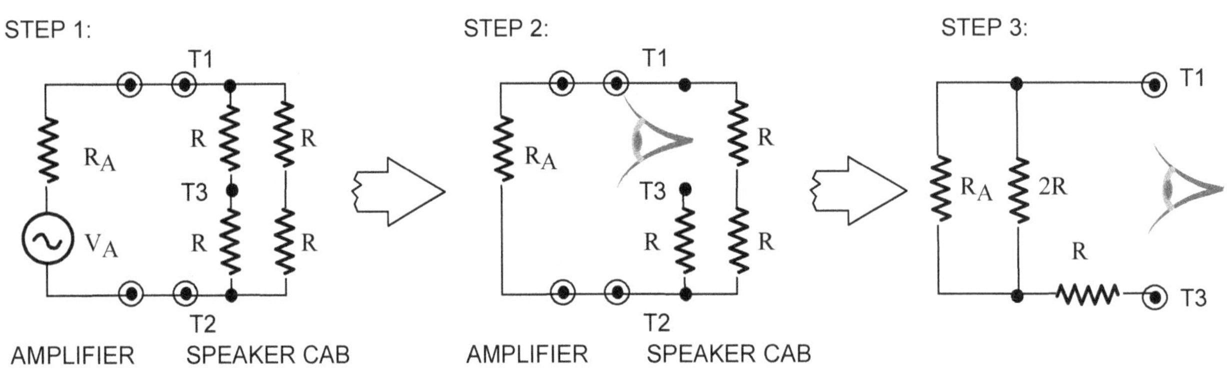

ABOVE: What impedances each driver "sees" in the parallel-series arrangement

HOW TO CONNECT EXISTING CABS IN SERIES OR PARALLEL

Connecting two cabs in parallel

To connect two or more cabs in parallel, the internal wiring of each cab stays as it is ("standard"), meaning the driver (or all drivers) have their + connections connected to the + of the socket (1), and all negative speaker terminals to the negative tab on the socket (2). A standard (+ to + and - to -) speaker cable is assumed.

The speaker cab with two paralleled sockets we could call "Master", while the added cab with a single socket is the "Slave". It does not matter which of the sockets on the Master is used for the amp and which for the cab B connection (they are in parallel).

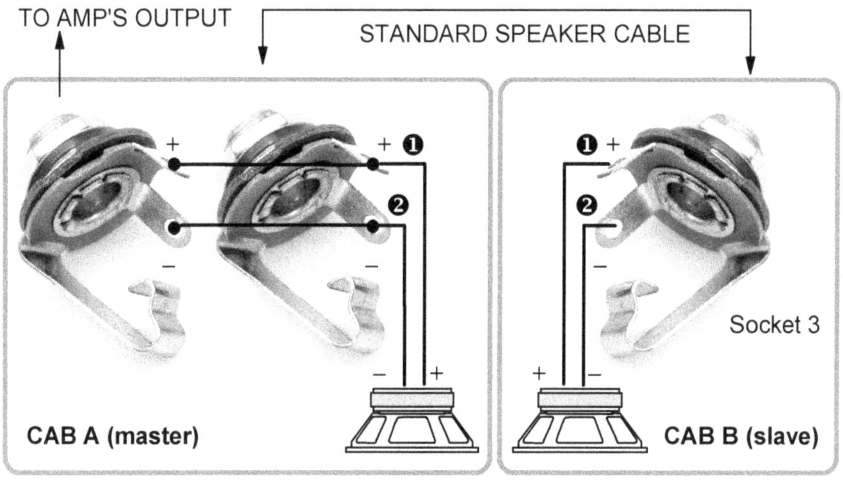

ABOVE: The wiring diagram of the two cabs in parallel

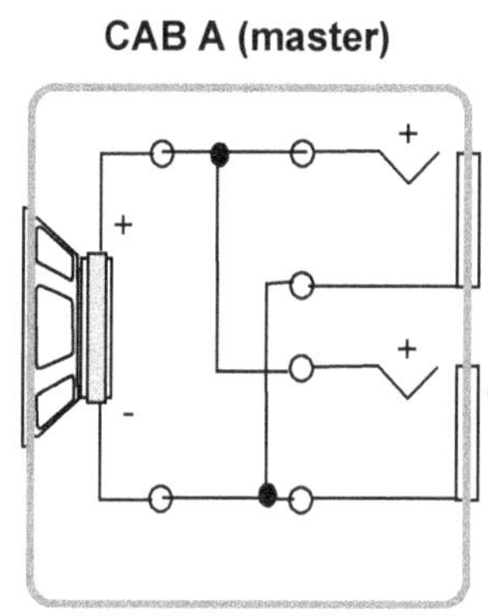

ABOVE: The schematic of Cab "A" (Master)

Connecting two cabs in series

To connect two or more cabs in series, again using a standard speaker cable, the internal wiring of the Slave (cab B) must be reversed, the + of the driver has to be connected to the negative lug of the socket, and the negative lug of the driver to the positive lug of the socket.

The two sockets on the Master cannot be wired in parallel any more. The new wiring is illustrated below.

Due to the clever wiring used, it does not matter which of the sockets on the Master is used for the amp and which for the cab B connection. With the amp plugged into socket 1, the signal goes through Cab A driver first and Cab B driver second; if plugged into socket 2, the signal goes through Cab B driver first and Cab A driver second.

The wiring diagrams like those just illustrated can be a tad confusing, so to dispel any doubts and traces of confusion, go back to the series and parallel circuit diagrams, a.k.a. schematics on page XX.

The first issue with this arrangement is that the cab A would not work by itself. Notice that the + of the driver is wired to the + of one jack, but its - is wired to the - lug of the other jack.

The way to solve this problem is to use a shorting plug on the unused socket, a plug that would short-circuit its + and - connections.

The second issue is that cab B is wired out-of-phase, so if used on its own, it may be out-of-phase with the other cabs/amps in the band.

The only solution is an added inversion switch inside cab B, which would in one position connect + (driver) to + (jack) and - to - (standard wiring), and in the other connect the wiring as illustrated above, + to - and - to +!

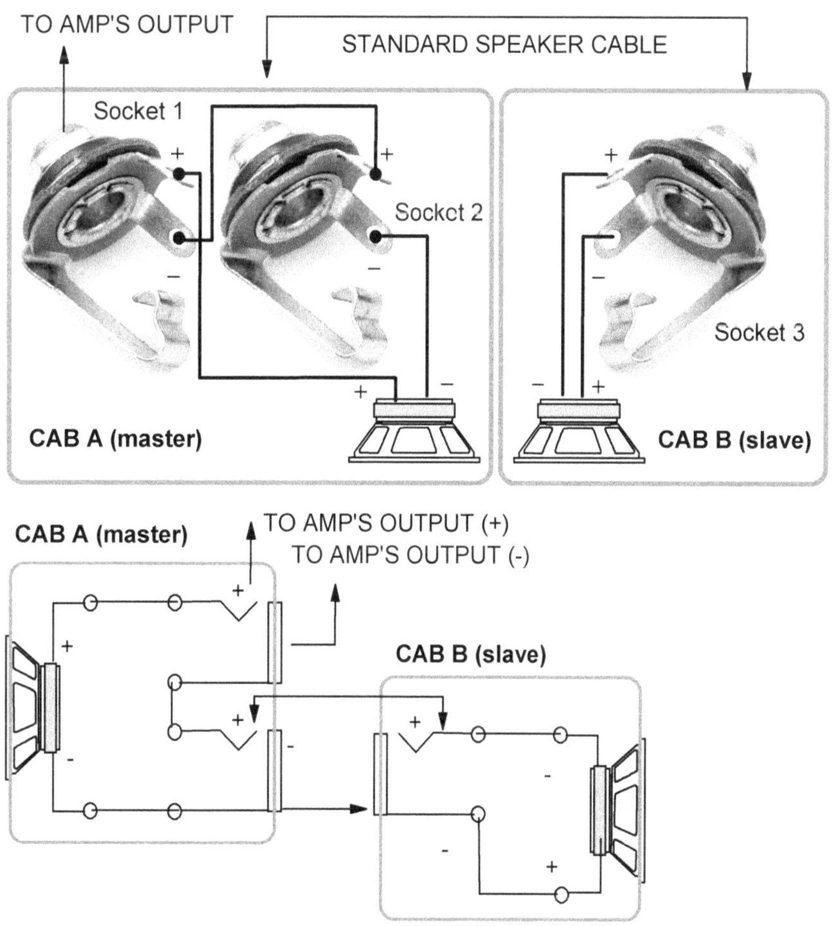

RIGHT: The wiring diagram and the schematic of the two cabs in series

HOW MULTIPLE DRIVERS IMPACT THE SPL AND POWER HANDLING OF A CAB

The way it used to be: Danelectro Commando

In production from 1954 to 1960, Danelectro Commando was a 30 watt combo guitar amplifier. Four 6V6 lovelies in the push-pull output stage and two lots of four 8" Rola Alnico speakers, four firing forward and four facing the rear, in a suitcase-type enclosure.

I have never come across a combo guitar amp with more speakers than The Commando - eight! To gain access to the jacks and controls, the hinged two-part enclosure had to be opened in use. This improved the cooling of the tube amp and also enabled the guitarist to adjust the angle of the two halves and thus control the cab's dispersion. Total awesomeness!

Multiple drivers in the same cab and the resulting increase in SPL

Compared to a single driver, the sensitivity of three identical drivers acting in unison (due to the mutual coupling effect) is 10log3 = 4.8 dB higher, which is a noticeable increase in loudness. Four identical drivers, as in each of the two quads used in the Danelectro Commando would be 10log4 = 6 dB louder!

Of course, using four drivers instead of one also quadruples the power handling and increases the SPL level by the same factor, 10log4 = 6 dB! So, 1 watt of amplifier power into a quad of guitar speakers of unremarkable sensitivity, say 95 dB/Watt@1m, would produce a sound pressure level of 95 + 6 = 101 dB!

Thus, four identical drivers working electrically in parallel and acoustically operating as a closely-coupled cluster could be 2x6 = 12 dB louder than a single driver, an incredible boost. This assumes that the cab is driven to its maximum power rating by an amp.

The mutual coupling effect applies primarily to wavelengths λ greater than twice the distance "d" between the cones ($\lambda_{MIN} = 2d$), in which case the driver outputs will be substantially in-phase at any point in the room. In a typical 4x12" cab the drivers are about 32cm apart so we get $f_{MAX} = v/\lambda_{MIN} = v/2d = 344/(2*0.32) = 537$ Hz.

Thus, above 500Hz or so, the mutual coupling effect diminishes since the individual waves from each driver becomes more and more out-of-phase. Regions of in-phase and out-of-phase interference will develop.

SPL INCREASE BY USING MULTIPLE DRIVERS
(N = NUMBER OF DRIVERS)

ACOUSTICAL COUPLING EFFECT: $P_A = 10\log N$ [dB]

ELECTRICAL POWER HANDLING: $P_E = 10\log N$ [dB]

4 x 12" or 4 x 10" CAB MODIFICATION OPTIONS

If you already have a closed 4x12" or 4x10" cab, you have a flexible platform that can serve as a starting point for many awesome modifications and improvements. In literally five minutes you can convert it into a "detuned" cab, a passive radiator cab or an open-back cab. Or, you can remove the back cover and try it out as an open-back cab. All you need is a screwdriver and some courage to experiment. So go ahead, make my day ...

Option 1: Disconnect two drivers, thus making them work as drone cones (passive radiators)

Remove the back cover of your cabinet (assuming a sealed design) to gain access to the drivers and connections. Unplug two of the four drivers. In this case, due to symmetrical wiring (capable of a stereo connection) you must disconnect one driver from the left vertical pair and one driver from the right pair.

Due to the way our particular cab was wired (Marshall MC412A-E), the easiest was to unplug the two drivers at the bottom.

Replace and bolt the back cover into place. The box is still sealed so the paper cones of the two bottom drivers will be acting as drone cones or passive radiators.

The power rating (the maximum amplifier power the cab can handle) is halved (in this case from 200 watts to 100 watts), but the actual loudness (the sound pressure levels) may not be. The experiments and measurements that follow will shed more light on that issue, so keep reading!

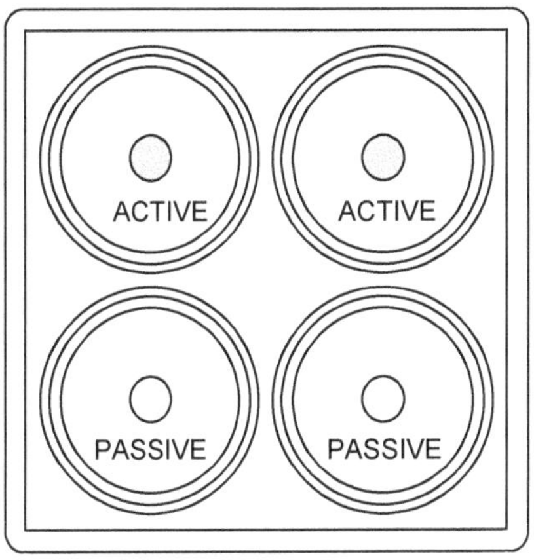

Option 2: Completely remove two drivers, thus making the cab work as a bass-reflex enclosure ("detuned cab")

Removing two out of four drivers (one from the left and one from the right side) effectively converts a sealed cabinet into a bass-reflex design. Each opening acts as a port of the same size as the effective speaker cone area. Some call this a "detuned cab".

Obviously, just as in the previous option, the power rating (the maximum amplifier power the cab can handle) is halved (in this case from 200 watts to 100 watts), but the actual loudness (the sound pressure levels) may not be. In theory, the sensitivity of a vented (bass-reflex) cab should be higher than of the original sealed design, so nothing should be lost in terms of the overall loudness.

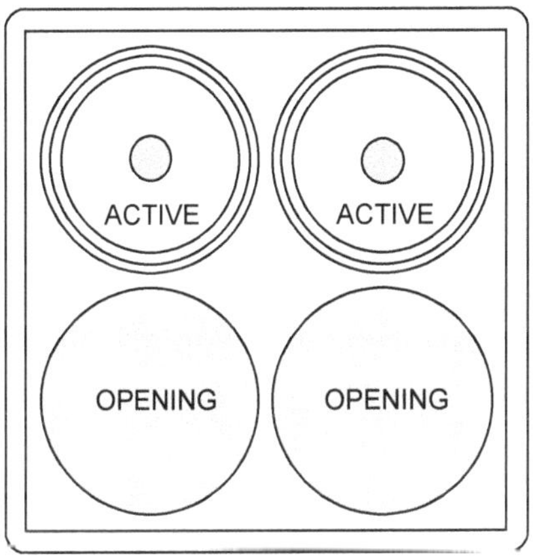

Option 3: Install dissimilar drivers

While the two previous options only require ten or so minutes of your time, this configuration requires a financial investment. You need to purchase at least one driver, or if unhappy with the drivers in your cab, you need to replace three or even all four of them.

While all four should be of the same physical size, you can combine speaker drivers of different power ratings, impedances and characteristics (resonant frequencies f_R, Q_{TS} and other parameters), all in order to get a superior mix or blend of their sonic signatures.

You can mix ceramic and Alnico drivers, drivers with low and high resonant frequencies, and even throw in an unusual driver or two, such as "full range" drivers with whizz-cones or integral tweeters.

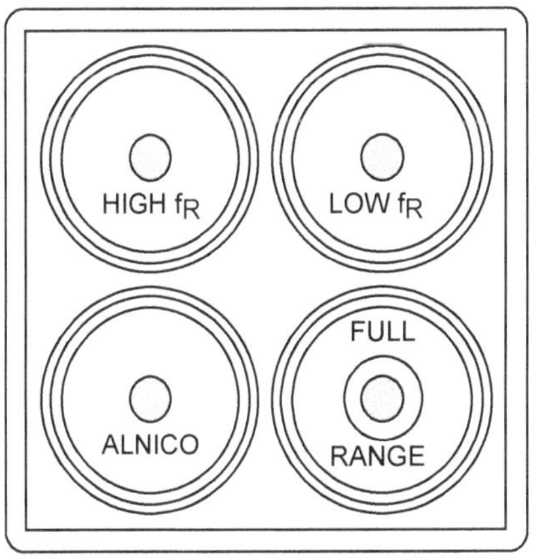

Option 4: Conversion into an open back cabinet

Turning closed-back (sealed) cabinets into an open back design is the easiest modification of all - simply remove the back cover and you'll change the cab's sonics.

In this experiment all four drivers have been retained, connected in parallel, resulting in 200 watt, 4Ω mono open-back cabinet. For discussion purposes the SPL and impedance graphs will be superimposed on those of the original 4Ω closed back configuration.

PRACTICAL TESTS ON A MARSHALL 4 X 12" CAB

The aim of these experiments is to compare the impedance-versus-frequency and the SPL-versus-frequency characteristics of a typical 4 x 12" speaker cab in its original state (sealed box, four drivers in parallel) with the same cab after modification options #1, #2, #3, and #4 were performed (previous page).

The test circuit

This point-by-point test requires a function generator with a calibrated display of its output signal's frequency (1), an SPL meter (2) on tripod, and a True-RMS multimeter capable of measuring signals with frequencies up to at least 10 kHz (5), not shown on the photo.

Most quality function generators keep the output voltage steady, regardless of the load. If your doesn't, you will need a second multimeter V_1 (AC voltmeter) to measure the output voltage at each frequency (3).

The value of the series resistor R_S is not critical (4), it should be at least 220 ohms; we used a 270 ohms resistor.

Its function is to convert the constant voltage sinewave signal V_{OUT} at the output of the function generator to a constant current signal to drive the speaker (cab) under test. With constant current (not dependent on the speaker's impedance), the voltage drop across the speaker is directly proportional to its impedance modulus.

Starting with 20 or 40 Hz frequency signal, you measure two parameters, the AC voltage V_2 (across the input of the speaker cab) and the SPL meter's indication (sound pressure in dB).

Up to 200 Hz we measured at every 10 Hz, then from 200 to 1,000 Hz (1 kHz) every 100 Hz, and from 1kHz to 10 kHz every 500 Hz.

ABOVE: The sound pressure meter on a tripod, speaker cab under test and function generator. The multimeter and the series resistor are not shown here (see the photo on the next page).

BELOW: The interconnection diagram for the impedance and SPL measurements using the two-voltmeters method

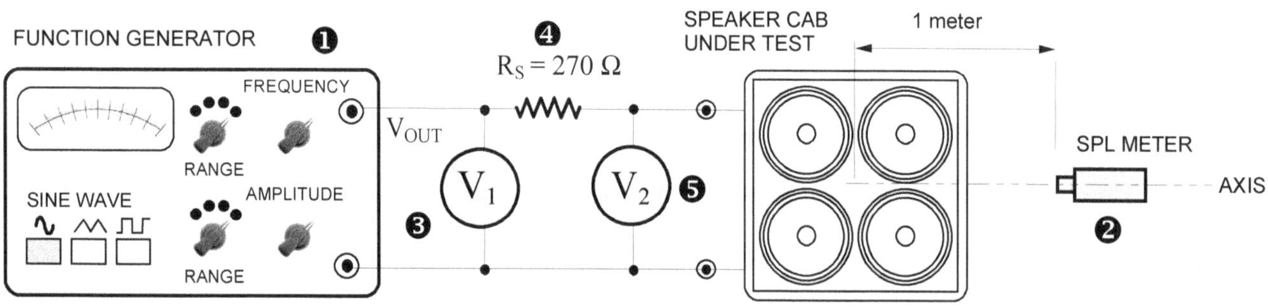

To facilitate the measurements and minimize the possibility of poor contacts or a loss of contact, we soldered the series resistor (5) directly to a pair of banana plugs (4), which plug into the output of the function generator, and to a mono speaker jack (6), which plugs into the speaker cab.

The multimeter is then connected across the cab's input (two wires soldered to the jack) via short test leads (7) and (8), terminated with crocodile clips.

When using only a single multimeter, should you wish to make sure that the output voltage of the function generator is constant, simply move one crocodile clip, from the cab-side of the series resistor to its other leg, its generator side.

Once you write down the voltmeter and SPL meter readings for all those selected frequencies, transfer them onto a spreadsheet and calculate the impedance using the formula on the right.

Again, V_1 is usually constant and V_2 is the AC voltmeter (multimeter on AC volts range) reading. R_S is the value in ohms of the series resistor.

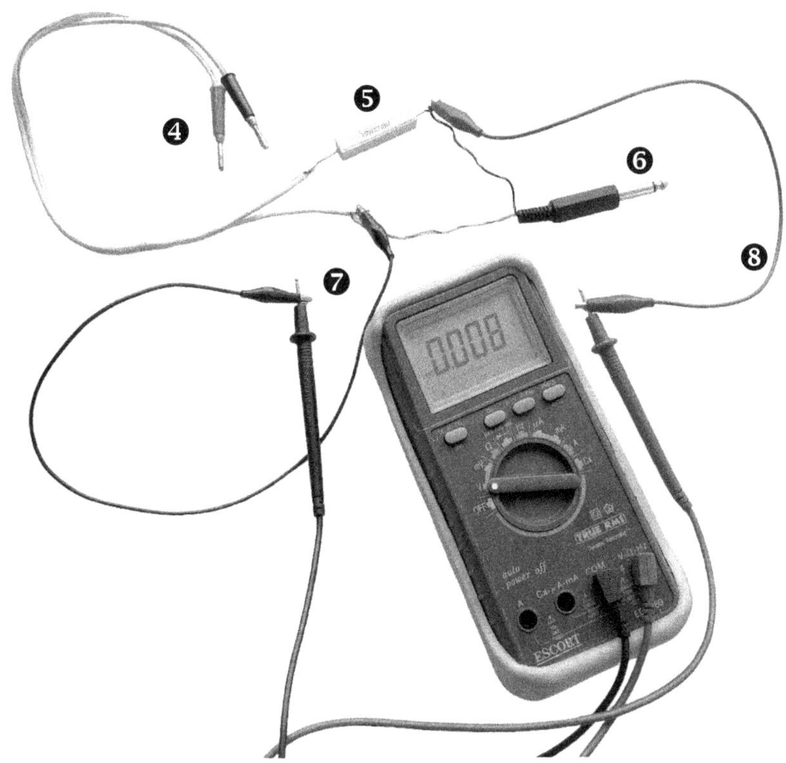

ABOVE: The series resistor and multimeter (AC voltmeter) hooked up to the test jig. The two banana plugs plug into the output sockets of the function generator

SPEAKER IMPEDANCE MEASUREMENT BY TWO VOLTMETERS METHOD

$Z(f) = R_S V_2(f)/[(V_1(f)-V_2(f)]$

$V_1(f)$ and $V_2(f)$ are measured results at various frequencies, Z(f) is the calculated impedance modulus at each test frequency "f".

The testing platform - Marshall MC412A-E 4x12" slanted cab

Marshall MC412A-E is an "entry-level" angled 4x12" full-size cab. Rated at 200 watts mono (4 ohms) or 2 x 100 watts stereo (8 ohms each), it features four 12", 16 ohm Celestion Rocket 50 drivers, each rated at 50 watts.

MC412A is a sealed cabinet, yet there is no internal padding at all, so standing waves and multiple internal reflections and interference of the cones' back waves will be prominent, resulting in severe peaks and dips in its frequency response.

Should you wish to experiment and alter its tone, try covering at least one side and the bottom of the cabinet with acoustically absorbing material. Do the same for the back cover.

The two left drivers are connected in parallel, their output (7) hooking up to one side of the stereo-mono switch located on the back cover (not shown in the photo). The two drivers on the right side are connected in the same way, output (8).

One self-tappered screw bolts the rear cover to the cross-brace (9).

BELOW: The two drivers on the left are paralleled (16 Ω || 16Ω = 8 Ω), as are the two drivers on the right side. When the back switch is in "MONO" position the two 8 Ω pairs are paralleled again, resulting in a 4 Ω cabinet.

SPL MEASUREMENTS

The sound pressure level measurements in this book are not standardized. Different SPL meter distances and input signal amplitudes were used with different speakers, so we don't disturb the neighborhood and cause panic among the local feline and canine population.

You cannot directly compare the graphs of different cabs in terms of loudness and absolute sensitivity, which wasn't our goal anyway. The objective was to get a visual indication of the frequency characteristic of each speaker or speaker cab, it's shape, peaks & dips, bandwidth, and behavior at the frequency extremes (bass and treble).

IMPEDANCE MEASUREMENTS

Unlike the SPL figures, the impedance graphs in this book are standardized, so you can compare the shapes, peaks and average (nominal) impedances of different speakers or speaker cabs.

Options 1 & 2 impedance measurements using Marshall 4 x 12" cab

The Z-curve of the original 100 watt, 4Ω closed back mono configuration (1) and with two drivers disconnected (2) have an identical shape above 200 Hz, but two issues are immediately obvious. The "passive radiator" impedance curve (2) rises steeper than that of the original cab's (1).

Also, while the single resonant peak of the sealed cab was around 105 Hz (3), the "passive radiator" impedance curve has two resonant peaks, at around 90 and 120 Hz. In complete agreement with the theory, the two peaks are equally-spaced in relation to the original resonant frequency, 105+15 =120 Hz and 105-15 = 90 Hz.

You'd imagine that the total impedance of the four identical drivers in parallel, with around 4Ω midrange value (4), would be half of the impedance of only two active drivers in the "passive radiator" option.

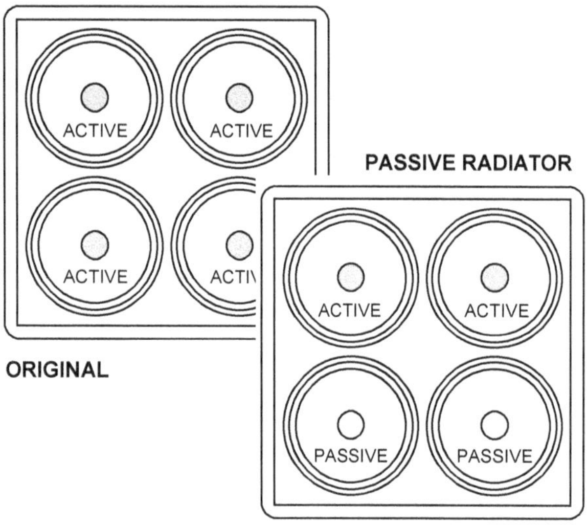

THE "ORIGINAL" 4X12" VS. THE PASSIVE RADIATOR — MEASURED RESULTS

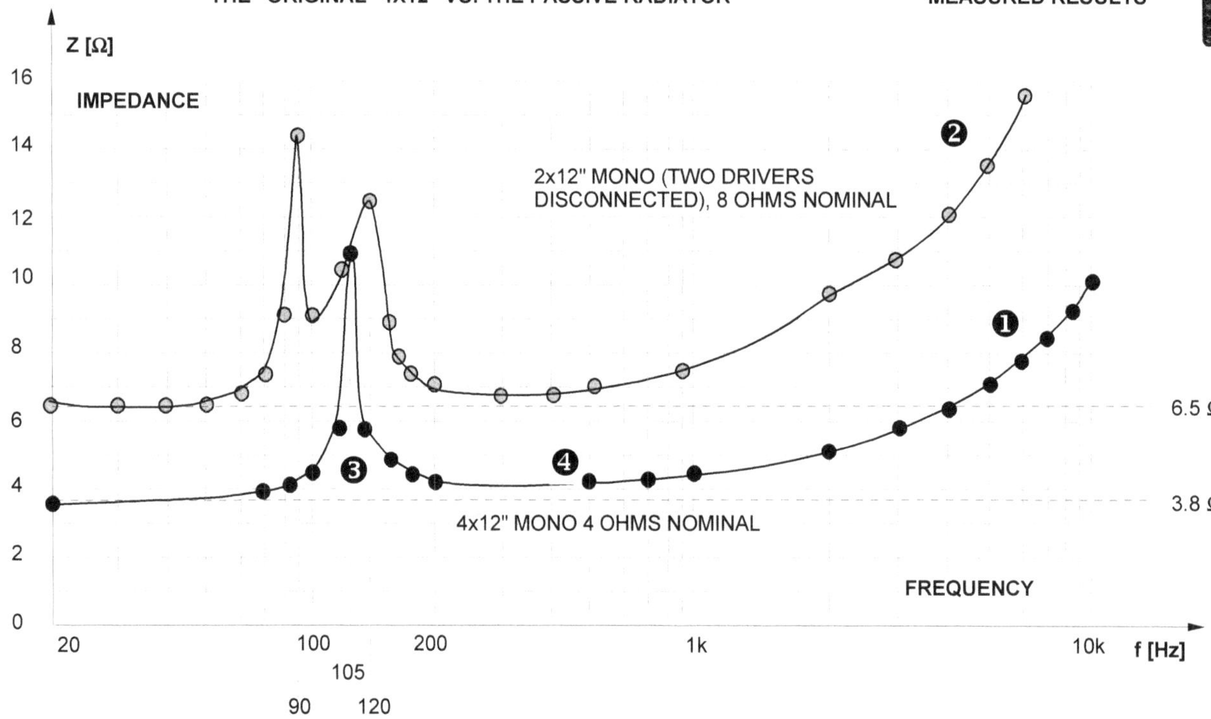

ABOVE: The impedance curve of the original 4x12" cab (1) in comparison with the impedance curve (2) of the same cab with two drivers disconnected electrically, but still working acoustically as passive radiators ("drone cones")

Alas, that was not the case! The DC resistance increased from 3.8Ω to 6.5Ω, while the midrange AC impedance increased from around 4 Ω to 6.8 Ω! In other words, neither had doubled in value.

With the same two drivers now removed from the cabinet (and the back cover replaced to make it a sealed cab), theoretically we have a detuned bass-reflex cabinet. Thus, its impedance curve should change to a typical double-peaked shape.

However, that didn't happen; the impedance curve retained its single-peak character, but with a significantly lower resonant frequency, 75 Hz compared to 105 Hz with all four drivers in a sealed box.

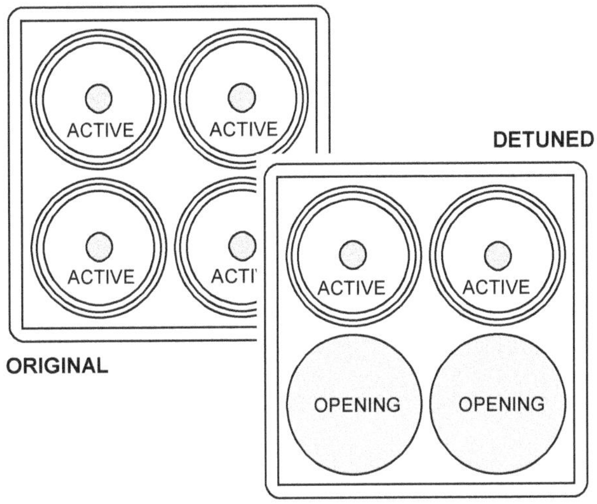

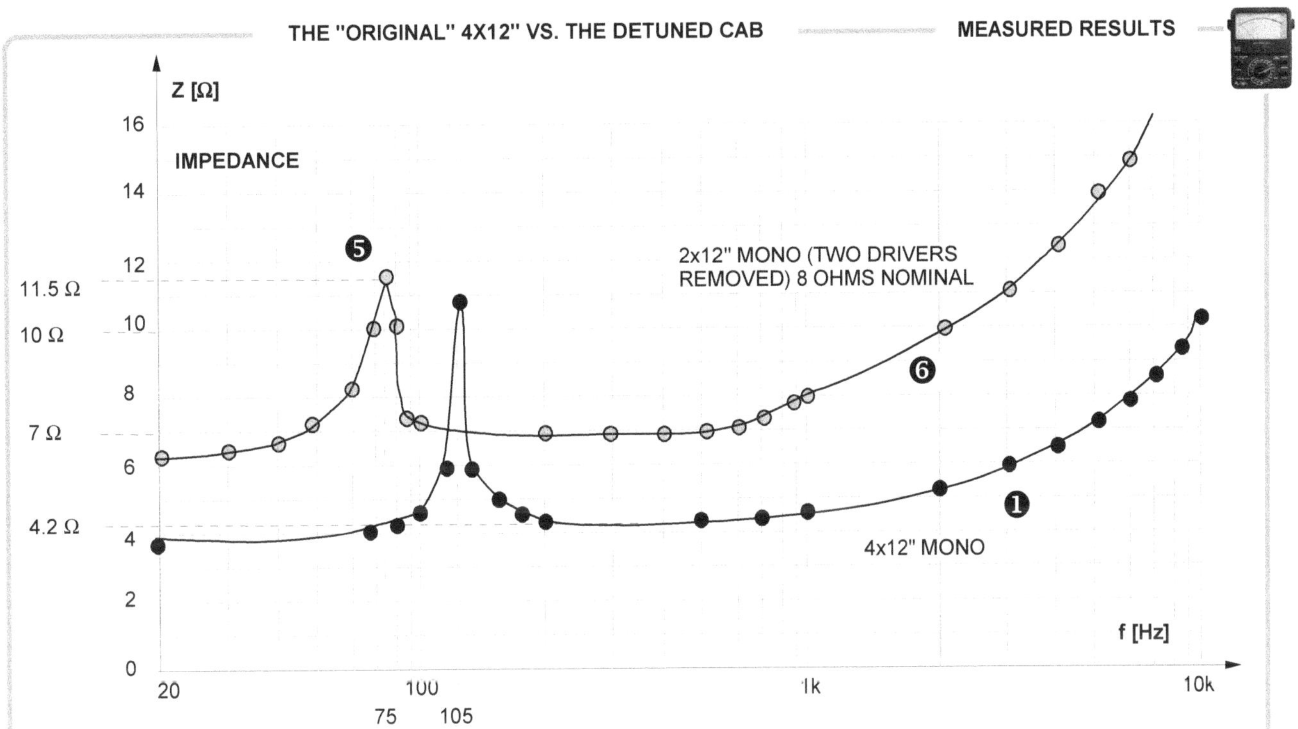

ABOVE: The impedance curve of the original 4x12" cab in comparison with the impedance curve of the same cab with two drivers removed, working as a "detuned" cabinet

Again, just like in the previous case, the total impedance was not doubled, and the resonant impedance peak was relatively speaking (in percentage terms) lower than the resonant peak of the original 4x12" sealed box. From around 7Ω in the midrange, the impedance rose to 11.5Ω at resonance, a 64% rise, while in the original 4x12" box the resonant peak rose from 4.2Ω in the midrange to 11Ω at resonance, a 262% increase!

Options 1 & 2 SPL measurements using Marshall 4 x 12" cab

Below 200 Hz, the SPL levels of the original sealed cab and the passive radiator cab (next page) were almost identical. The small differences could be due to measurement errors and variations, which in our case could not be ruled out. The SPL peak at around 120 Hz did not change a bit.

The passive radiator cab's loudness levels fluctuated more - a dip (7) at around 400 Hz, a peak at 600 Hz (8), another dip at 2 kHz (9), and finally a small peak at 2,400 Hz.

If we (somewhat arbitrarily) choose 73 dB as the referent 0 dB level, we look for the points where the curve drops 3dB from that level. These -3dB points (also called "half-power" points) are around 105 Hz in the bass region, a relatively poor result (the lower half-power frequency is too high), and around 5,800 Hz in the treble region, a relatively decent result (compared to similar cabs by other manufacturers).

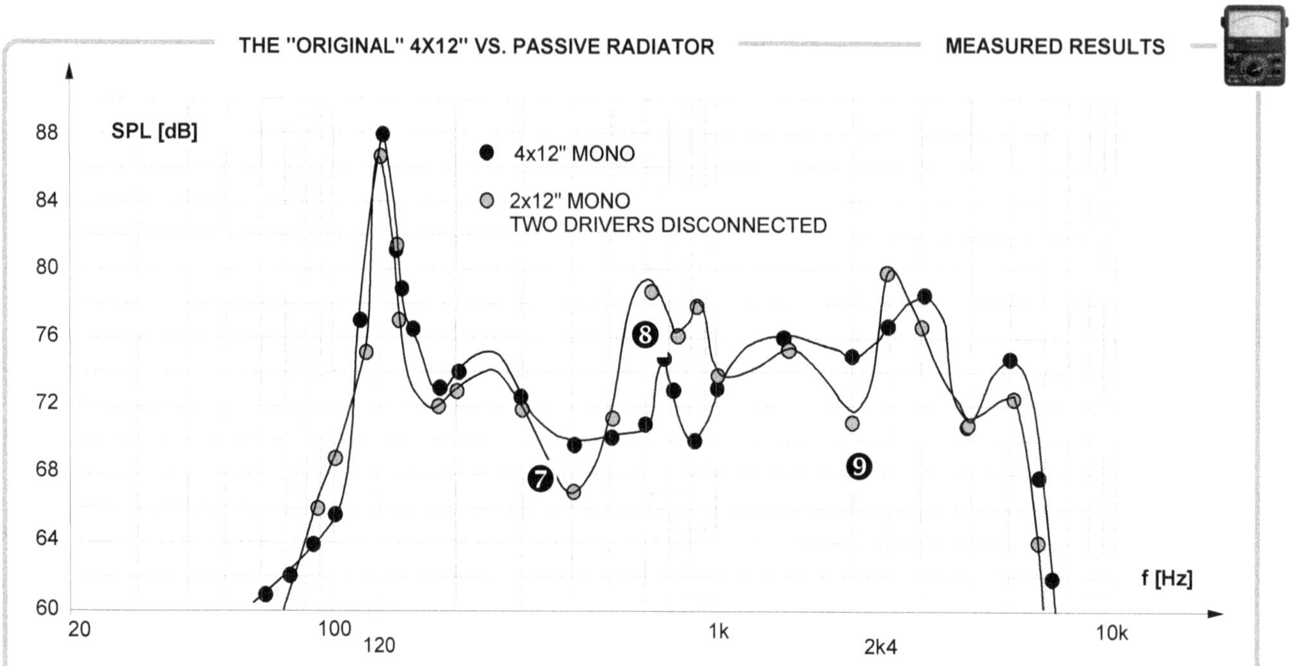

ABOVE: The SPL (sound pressure level) curve of the original 4x12" cab compared to the SPL curve of the same cab converted to a passive radiator ("drone cones") arrangement

Although in both configurations the peak at 130 Hz remained (1), the response of the original cab dropped sharply, from 84 dB @ 130 Hz to 70 dB @100 Hz and only 60 dB @ 70 Hz (2). The response of the detuned cab followed the same drop down to about 100 Hz, but then rose again, peaking at 70 Hz (3), dropping to 60 dB at 50 Hz. This means an improved and more prominent deep bass of the detuned option.

The response of the detuned cab was flatter in the octave between 160 and 320 Hz (4), but suffered from the same dip around 400 Hz (5).

While the SPL level of the original cab recovered faster between 400 and 600 Hz (6), the SPL of the detuned cab significantly exceeded its levels above 900 Hz, all the way up to 3 kHz (7), where it suffered a more pronounced fall down to 66 dB at 4 kHz (8).

It peaked again around 5 kHz (9), thus extending the upper reach of the detuned cab about 2 kHz beyond the frequency limit of the original cab.

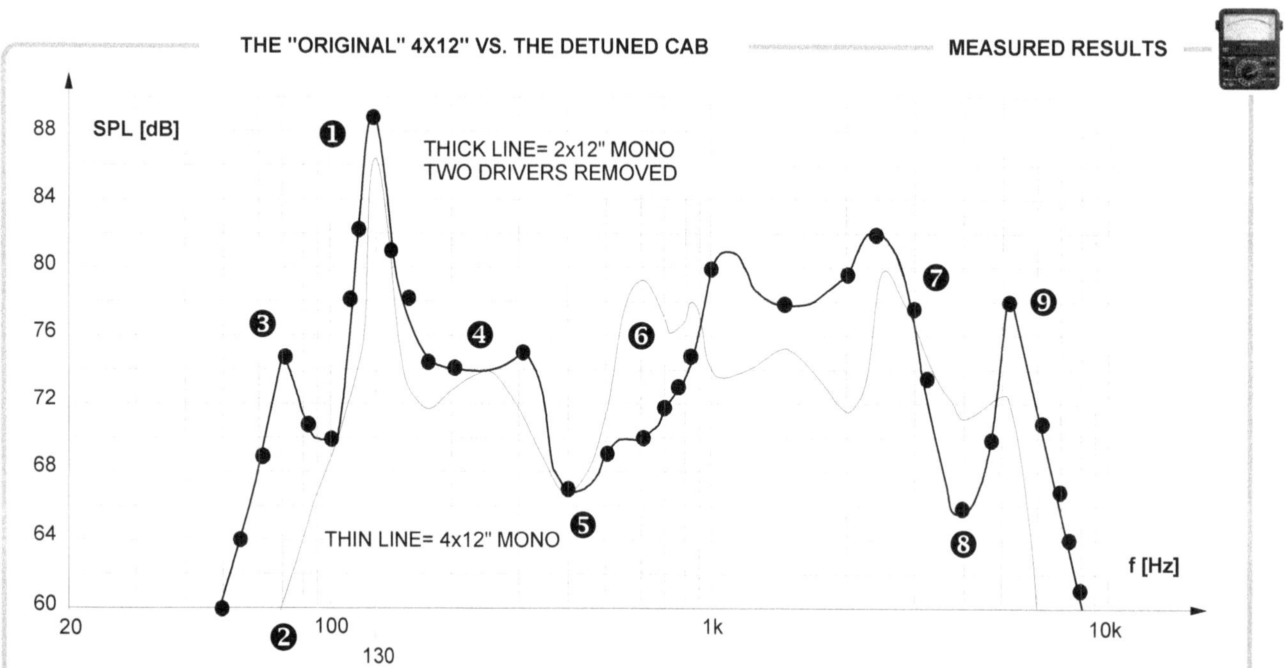

ABOVE: The SPL (sound pressure level) curve of the original 4x12" cab compared to the SPL curve of the same cab with two drivers removed ("detuned" cab)

The average sensitivity of the original cab

As a small digression at this point, lets have a quick look at a highly technical technique of guesstimating a cab's average SPL rating.

If we (admittedly, somewhat arbitrarily) choose 73 dB as the referent 0 dB level, we look for the points where the curve drops 3dB from that level. These -3dB points (also called "half-power" points) are around 105 Hz in the bass region, a relatively poor result (the lower half-power frequency is too high), and around 5,800 Hz in the treble region, a relatively decent result (compared to similar cabs by other manufacturers).

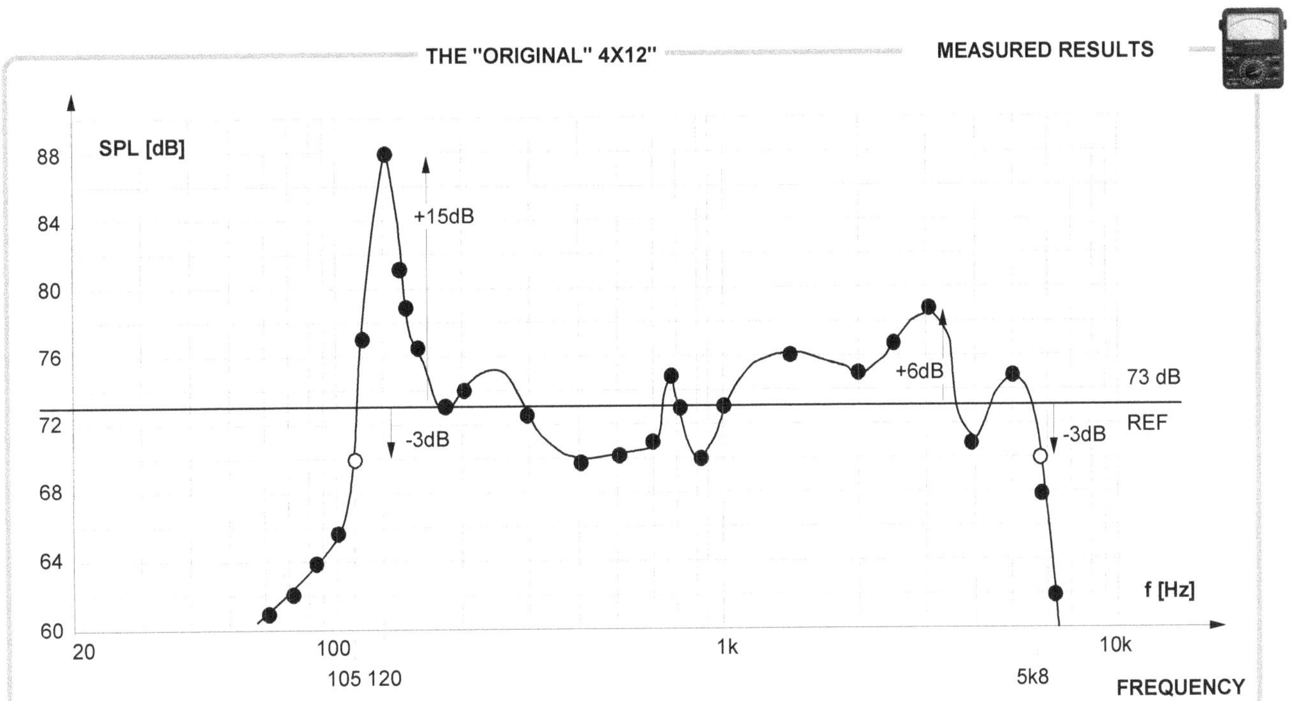

ABOVE: The SPL (sound pressure level) curve of the original 4x12" cab

Off-axis measurements of the original Marshall 4 x 12" cab

As far as the conventional wisdom goes, the higher the frequency, the more pronounced the beaming and narrowing effect, and the higher the drop in the SPL levels of the highly directional quad cabs. Alas, just as the so called "common sense" doesn't seem to be very common, we had a sneaking suspicion that the conventional wisdom may not be "conventional" at all, so we decided to perform a couple of simple measurements.

Using the original Marshall 4x12" cab, we repeated the SPL measurements at 30° and 60° off axis.

Again, the conventional wisdom claims that frequencies below a few hundred Hz would not be affected, so we started this test at 200 Hz and went up.

The two curves almost coincide between 400 and 900 Hz; the peak (1) was slightly shifted in frequency, but, overall, there was no major difference below 1 kHz or so.

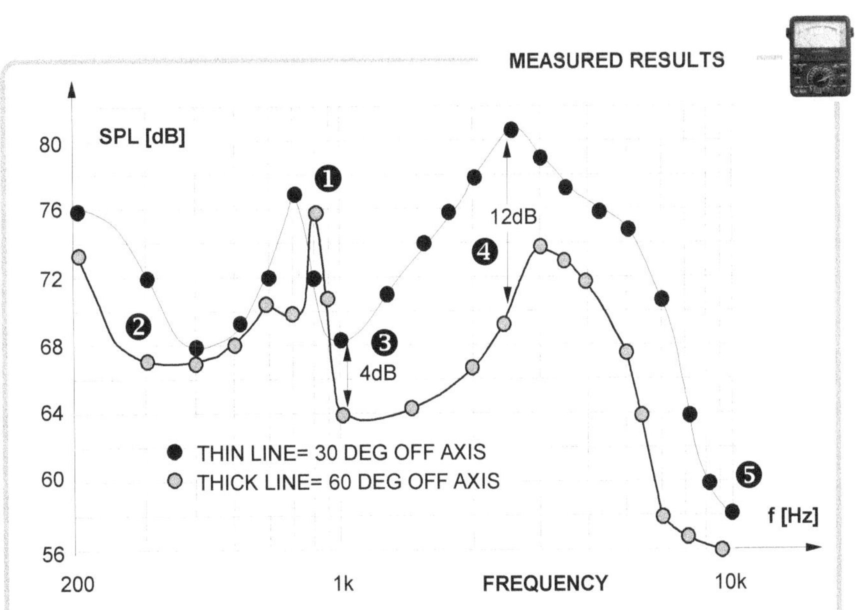

ABOVE: The SPL curve of the original 4x12" at 30 and 60 degrees off axis

However, notice that, unexpectedly, in that deeper octave between 200 and 400 Hz the SPL levels 60° off axis were also lower than 30° off axis. Above 1 kHz, the SPL differed between 4 dB at 1 kHz (3) and 12 dB at 2.3 kHz (4).

At higher frequencies the difference remained around 4-6 dB, but then, again unexpectedly, it actually narrowed down to no more than 3 dB above 9 kHz (5).

Which is louder, a detuned (DT) or a passive radiator (PR) cab?

In conclusion to this discussion of options #1 and #2, let's compare the SPL curves of the 4x12" cab with two drivers removed ("detuned" cab) versus the same two drivers electrically disconnected, working as passive radiators.

Wile the two resonant peaks happened at almost the same frequency (90 Hz versus 94 Hz), the loudness peak (1) of the PR cab was 97 dB, compared to only 84 dB for the DT cab.

Its response below the resonant frequency was also superior (2). Thanks to its second peak, the DT cab was louder (3) between 110 and 130 Hz, and between 160 and 175 Hz (4).

So, if for some reason you crave deep bass, or, indeed if you are a bass player, go for the drone cone modification; simply disconnect two of the four drivers and Bob is your uncle!

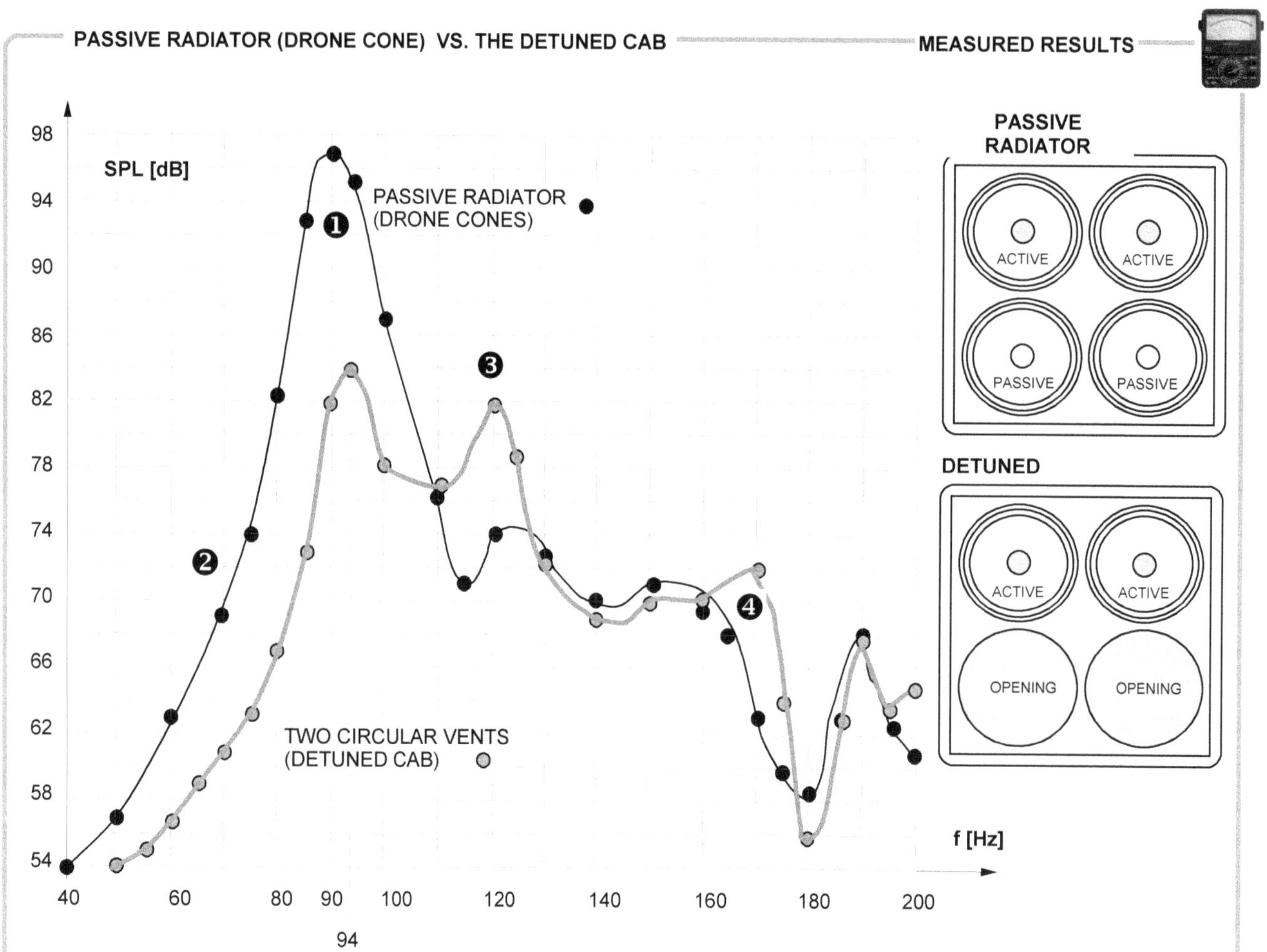

ABOVE: The SPL (sound pressure level) curve of the 4x12" cab with two drivers removed ("detuned" cab) versus the same two drivers electrically disconnected, working as passive radiators

Option #4: SPL and impedance measurements using Marshall 4 x 12" cab (open back)

In this experiment all four original Celestion drivers have been retained, connected in parallel, resulting in the 200 watt, 4Ω mono *open-back* cabinet. Again, for discussion purposes, the SPL and impedance graphs will be superimposed on those of the original 4-ohm *closed back* configuration.

The impedance curve (next page) for the open and closed back have an identical shape, but two issues are immediately obvious. While the resonant frequency of the sealed cab was around 105 Hz (1), with its back panel removed, its f_R was significantly lower, around 71 Hz (2).

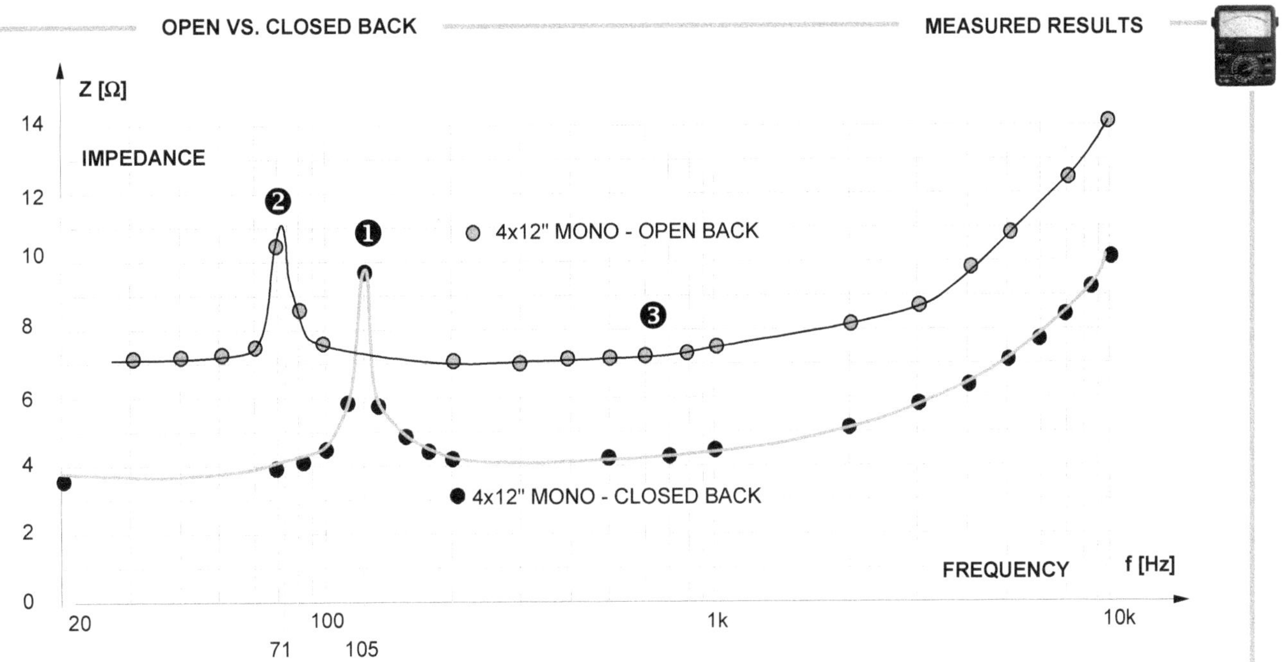

ABOVE: The impedance curve of the original 4x12" cab with the closed and open back

The second phenomenon, which I am unable to explain, is that the average or "nominal" impedance of the open cab almost doubled, from around 4Ω to 7-8 Ω! In the 200 Hz-1kHz midrange band (3) it hovered around 7-7.5 Ω. Just to make sure we didn't make a mistake connecting the drivers in parallel, we double-checked, and our trusted multimeter indicated a DC resistance of 4 ohms.

As for the SPL differences, the closed back cab had a prominent peak around 120 Hz (why not 105 Hz?!?), below which its SPL dropped rapidly (5).

While the SPL levels of the same cab with an open back were lower in the upper bass region where the closed back cab had a huge peak (4), they were both higher and smoother (more constant) in the 70-150 Hz region (6). So, again, if deeper bass is your poison, an open back has your name on it!

The open back's upper bass response is more scooped-up (7), as is its midrange (8), its response being around 3-4 dB lower in that critical (the peak of human hearing) 1-3 kHz frequency band.

Finally, the mid-treble of the open back cab was louder in the 4-6 kHz region (9), dropping abruptly thereafter, much earlier than the more extended treble of the closed back cab!

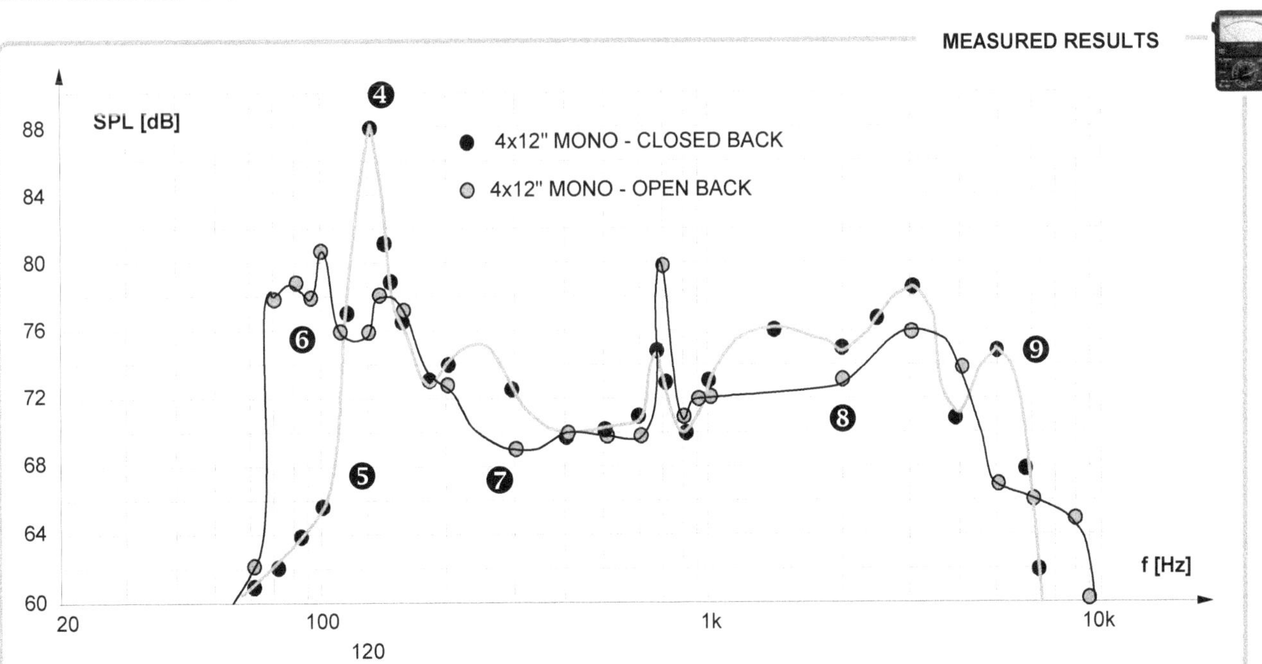

ABOVE: The SPL (sound pressure level) curve of the original 4x12" cab with the closed and open back

USING DISSIMILAR DRIVERS IN THE SAME CAB

Paralleling dissimilar speaker drivers

We have already seen how the widely varying driver impedance (often shooting way above 100 ohms at the resonant frequency) causes an irregular load on the driving amplifier and an SPL (loudness) that also varies with frequency, boosting some frequencies and attenuating others.

A series connection of two (illustrated here) or more dissimilar drivers results in an impedance curve with multiple peaks in the two-octave frequency band (50-100-200 Hz) where such resonant frequencies fall.

An often forgotten benefit of paralleling dissimilar speakers is a significant reduction in the resonance peak. In simple terms, with N identical paralleled drivers (N=2,3,4,..) whose resonant peaks happen at the same frequency and are of the same magnitude, the resulting peak is n times lower.

For instance, when connected in parallel, four Jensen C12-N drivers with the impedance peak of just above 80 ohms (3) will behave like a single driver with an impedance peak of 80/4 = 20 Ω. This simplified rule-of-thumb assumes a certain degree of consistency that should be expected but isn't always present. Manufacturers have their own tolerances and allowable variations in speaker parameters, such as the impedance maxima and the resonant frequencies.

Let's take things one step further and assume three dissimilar speakers with the same resonant frequency but different values of Z_{MAX}, say 80, 250 and 260 Ω. The resulting peak would be $1/Z=1/Z_1 + 1/Z_2 + 1/Z_3 = 1/80 + 1/250 + 1/260 = 0.026$ Siemens, so $Z=1/0.026 = 38.46$ Ω. What an improvement!

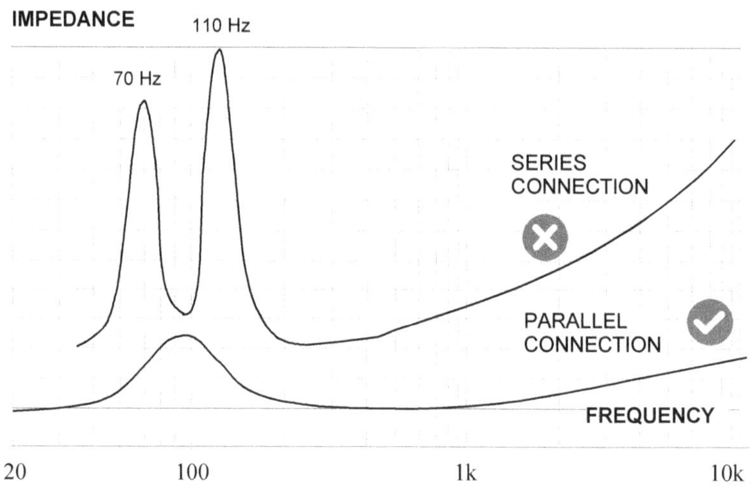

ABOVE: A qualitative sketch showing what happens to the composite impedance curve when two speakers with different resonant frequencies and different impedance peaks are connected in series and parallel

BELOW: The impedance curves of three 16 ohms, 12" drivers by different manufacturers, superimposed on the same graph

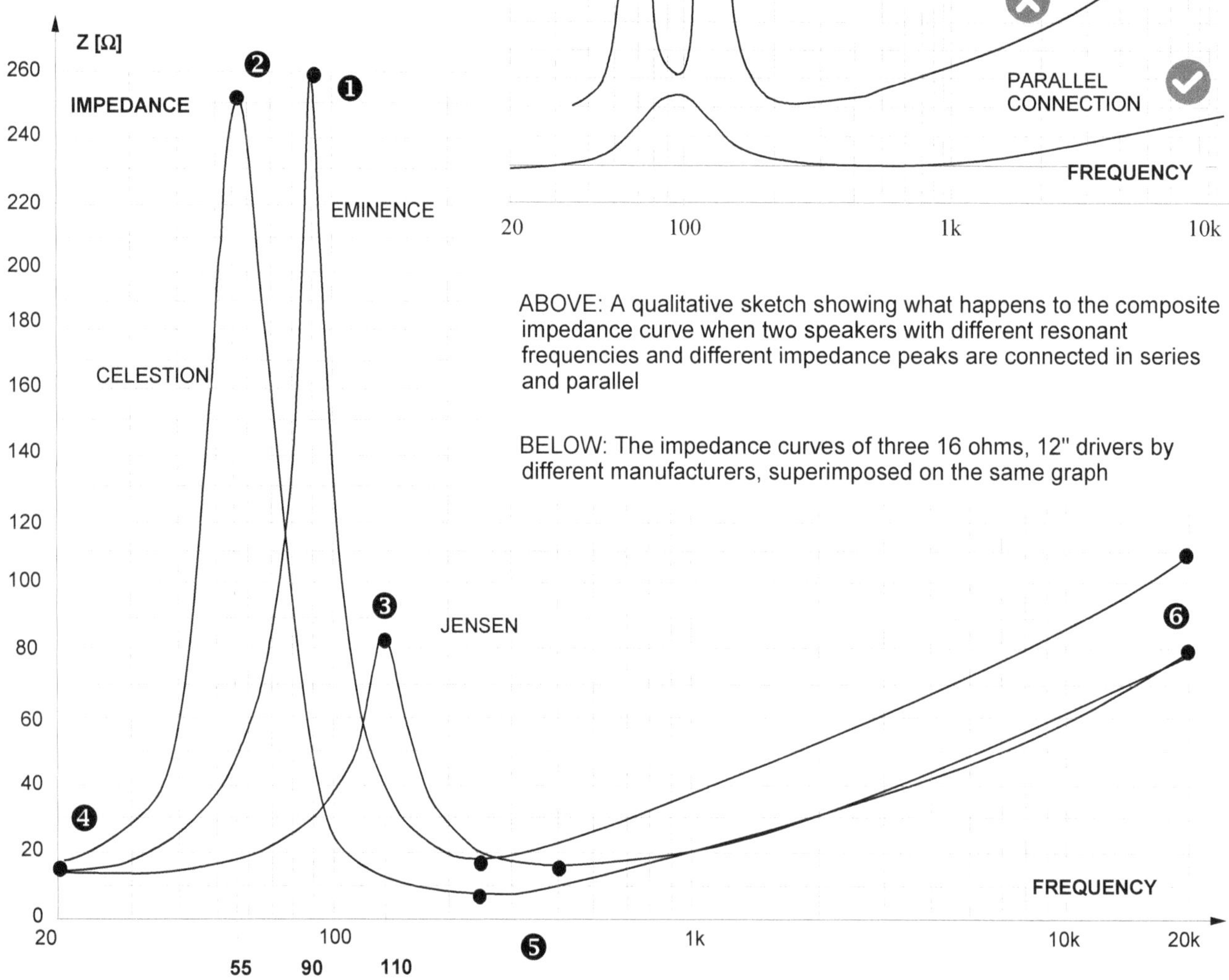

However, the main benefit occurs if both the impedance maxima and the resonant frequencies vary as much as possible. I didn't just pull the impedances used in the previous example out of my ... hat, they are the actual values of three currently produced 12" drivers, Jensen C12-N, Celestion G12H(55) "Greenback", and Eminence Texas Heat 16. Their resonant frequencies and impedance peaks are 110 Hz, 55 Hz (2) and 90 Hz (1) respectively.

SIDE-BY-SIDE	Jensen C12-N	Celestion G12H(55)	Eminence Texas Heat 16
Power rating	50 W (100 W)	30 W	150 W
Resonant frequency	110 Hz	55 Hz	91 Hz
Sensitivity [dB/W@1m]	97.5 dB	100 dB	100.2 dB
Magnet Weight	29 oz - 810 g	50 oz - 1.42 kg	38 oz - 1.06 kg
Voice Coil Diameter	1.5" - 38 mm	1.75" - 44.5mm	2" - 50.8 mm
DC resistance	12.6 ohms	12.11 ohms	14.7 ohms
Coil/Former/Cone	Copper/Kapton/Paper	Copper/Aluminium/Paper	Copper/Polyimide/Paper

All three speakers feature a ferrite (ceramic) magnet, pressed steel baskets and paper cones, but differ in most other aspects, most noticeably in their power rating (50, 30 and 150 Watts). The sensitivity figures of the Celestion and Eminence speakers are almost identical, around 100 dB/W@1m.

You may be thinking why on earth would we parallel drivers of such a wide range of power ratings. Wouldn't that mean the cab's maximum power would be determined by the lowest powered driver while the higher powered drivers would not be fully utilized? It certainly would, but I chose these drivers due to their widely varying resonance frequencies and impedance peaks, to illustrate the point I'm just about to make, and that is a much more desirable impedance curve without sharp and high peaks.

Indeed, the calculated composite impedance curve looks much more benign, with only a single broad impedance peak of 16.5Ω at around 80 Hz (7).

As for the "nominal" impedance, paralleling three 16 ohm drivers should result in a 16/3 = 5.33Ω cab, but notice that at almost all frequencies the impedance is way above that "nominal" value.

In fact, I would proclaim this to be an 8 ohm cab, and even then, the impedance is way above 8 ohms in the 40 - 150 Hz bass region (8), and in the 1.5 kHz- 20 kHz treble region (9).

BELOW: The impedance curve of the three 16 ohms, 12" drivers (Jensen C12-N, Celestion G12H(55), and Eminence Texas Heat 16) in parallel

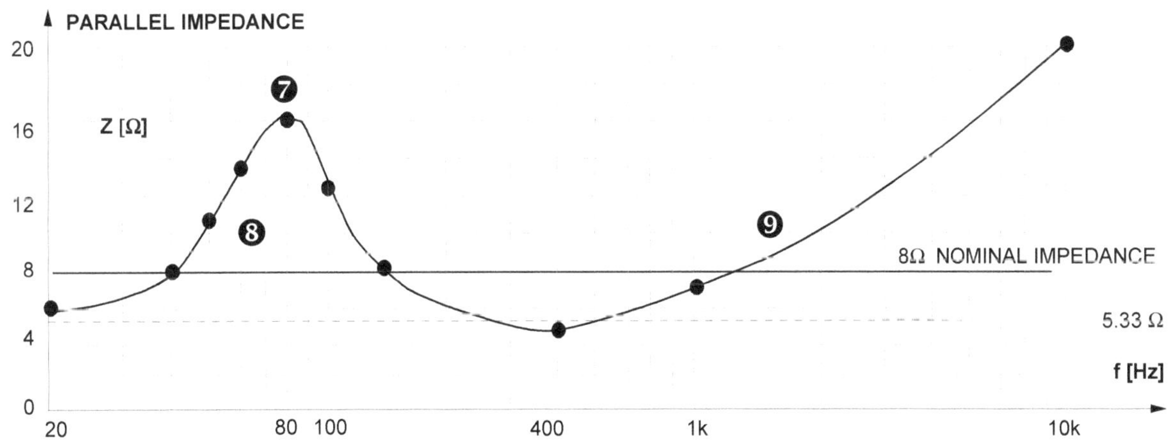

Dissimilar speakers in Marshall 4 x 12" cab

Now that we understand the theoretical background and the principles behind this approach, let's see how would paralleling dissimilar drivers work in a real quad cab. We had two 16 ohm 12" vintage Australian-made Etone lovelies, one with a dual cone, so we substituted them for the two bottom Celestion drivers in the Marshall cab.

The singular impedance peak moved from 105 Hz down to 90 Hz, while its amplitude was also lower, 8Ω (2) compared to 11Ω previously (1).

Interestingly, the new overall impedance was more than 1 ohm higher in the low bass region (3), but almost 1 ohm lower in the upper bass and lower midrange band (4), from about 110 Hz to 1 kHz.

While the impedance curve was only marginally changed, the SPL curve (next page) was drastically altered. The deep bass region below 100 Hz was now 2-3 dB louder (5). The two Etone drivers with lower f_R (resonant frequency) and much more powerful magnets than the el cheapo Celestion ones obviously could reproduce lower bass noticeably better and deeper.

MEASURED RESULTS

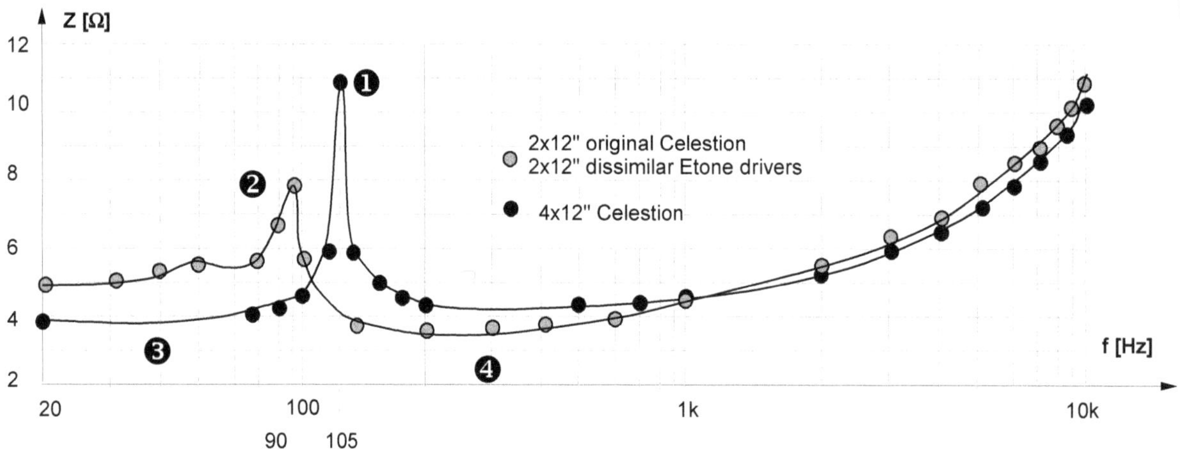

ABOVE: The impedance curve of the original 4x12" cab ("MONO") in comparison with the impedance curve of the same cab with two Celestion drivers replaced by two dissimilar Etone drivers

BELOW: The SPL (sound pressure level) curves

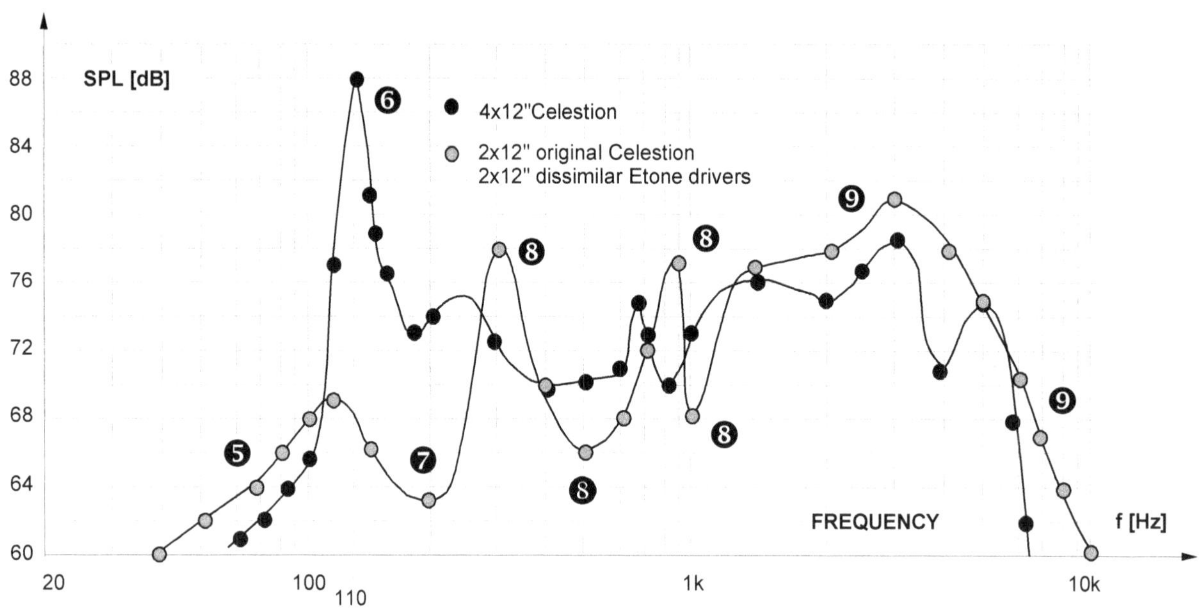

The huge loudness peak at 100 Hz (6) was gone, and a significant dip at around 200 Hz (7) was introduced. The new SPL curve then peaked and dipped throughout the midrange (8), with two major peaks at 300 and 900 Hz, and two dips at 500 Hz and 1 kHz.

Treble frequencies above 2 kHz were louder in the new arrangement (9), both throughout the midrange (1.5 kHz-5 kHz), and above 6 kHz. This is most likely due to the presence of the whizzer cone on one of the newly-substituted Etone drivers.

Peavey 6505 slanted cab

Peavey 6505 4x12" slanted cabinets loaded with Peavey Sheffield drivers sell for US$999, so when this one in perfect, as new condition came up for sale locally for AU$300 (US$180), we snapped it up.

The impedance curve dips below the nominal 16Ω impedance in the mono mode (two pairs of paralleled 16Ω drivers connected in series) only below 50 Hz (where there are no guitar produced tones) and around 300 Hz; at all other frequencies it is much, much higher. The resonant peak at 110 Hz reaches a whopping 105Ω (1).

From our experience, drivers in series sound inferior to the same drivers connected in parallel. Also, many tube guitar amps only have 4Ω and 8Ω taps, 16Ω outputs are relatively rare. For those two reasons, and, most importantly, to tame such a high impedance peak, this cab should be rewired so that in the "Mono" mode, all four drivers are in parallel. That would bring the nominal impedance to 4Ω, but in reality higher than that, between 4Ω and 8Ω.

If we arbitrarily choose 81 dB as the referent 0 dB level, we see that the cab's frequency range extends from 75 Hz to only around 3.8 kHz. The -3dB points are not much different, at 70 Hz and 4 kHz. Selecting a higher referent level, say 82 or 84 dB, would shrink that relatively narrow range (especially on the upper side) even further.

A series of equal dips follows at 400, 600 and 1,000 Hz (2), with two more peaks at 700 and 900 Hz (3). The sensitivity of the cab then rises and peaks at around 2,600 and 3,600 Hz (4), only to fall very early and steeply to a minimum at 5kHz (5). There's one more peak (below the referent level) at around 5k5 Hz (6), followed by the final sharp fall.

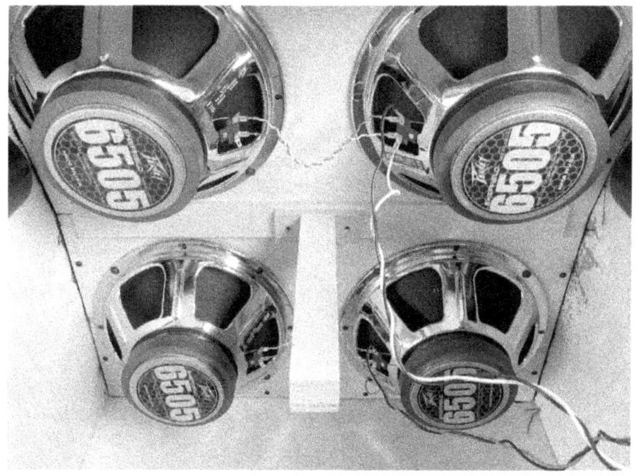

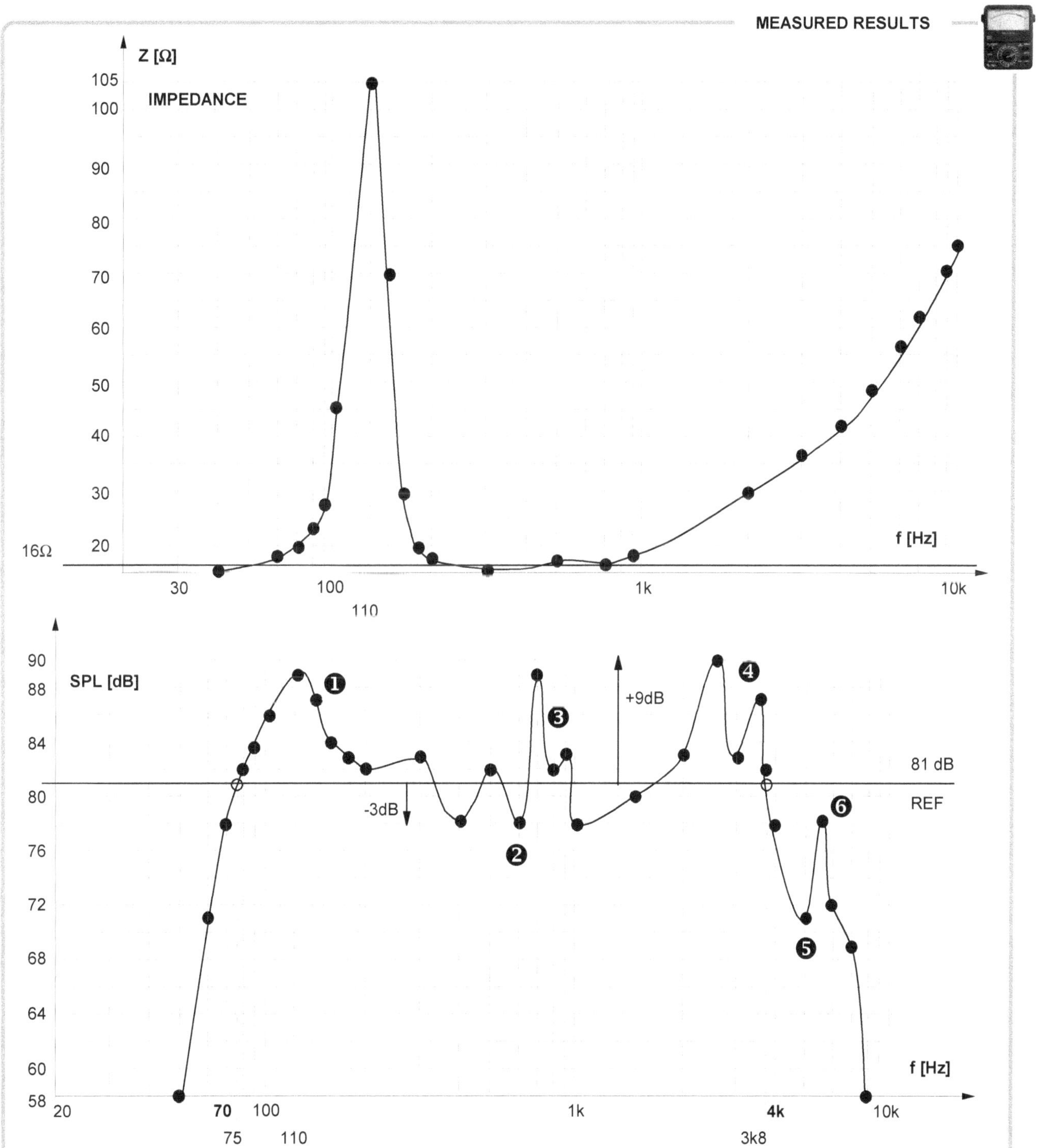

ABOVE: The impedance and SPL (sound pressure level) curve of Peavey 6505 slanted 4x12" cab

HOW TO TURN YOUR QUAD MONO CAB INTO A 2x2 STEREO CAB

Option 1: Simple but fixed wiring

Converting a quad box into a stereo speaker cab can be achieved in various ways. Assuming you are absolutely sure you will never need to operate it again as a mono cab, this simple conversion involves only adding a second mono jack and a simple rewiring of the four drivers.

Speakers "A" and "B" remain connected in parallel, just as they were in the original mono 4Ω cab (assuming that all four 16Ω drivers were in parallel), and so do drivers "C" and "D". Each pair is then wired to its own jack.

A vertical pairing is illustrated here, for better horizontal dispersion. You can wire the top two drivers together in one and the bottom two drivers in the other channel if you prefer.

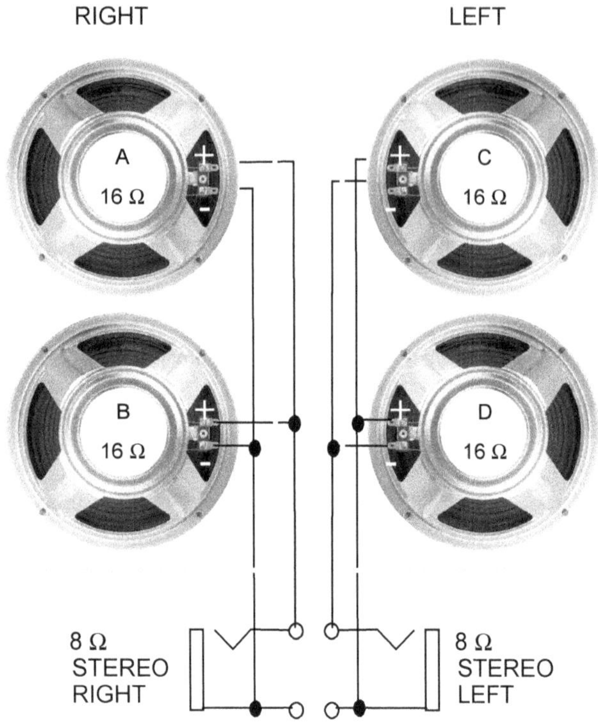

Option 2: Switchable "STEREO 2x8Ω - MONO 4Ω" wiring

MONO-STEREO switching can be done using the simplest mono jacks (no auxiliary contacts). However, only one MONO configuration is then available, either 4Ω (all four drivers in parallel) or 16Ω (drivers A&B in parallel connected in series with the parallel drivers C&D).

This way of wiring the changeover switch provides a 4 ohm mono output at the "8Ω STEREO RIGHT" jack only. In the "MONO" position, nothing is connected to the + terminal of the "8Ω STEREO LEFT" jack. (2).

The negative terminals of all four speakers are connected together at (3) and to the jacks at (4). The + of speakers A & B is wired to the central terminal #2 of the left switch, while the + of speakers C & D is wired to the central terminal #5 of the right switch.

In the "STEREO" position, terminal #2 is connected to terminal #1 and thus to the + of the "8Ω STEREO RIGHT" jack. Likewise, terminal #5 is connected to terminal #4 and thus to the + of the "8Ω STEREO LEFT" jack.

In the "MONO" position, terminal #2 is now connected to terminal #3, which is permanently wired to the + of the "8Ω STEREO RIGHT" jack, which is now acting as the "4Ω MONO" jack.

Terminal #5 is connected to terminal #6, which is strapped to terminal #3, and to the + of the "8Ω STEREO RIGHT" jack. Thus, the + terminals of speakers C & D are now also connected to the + of the "8Ω STEREO RIGHT" jack, so all four speakers are now in parallel.

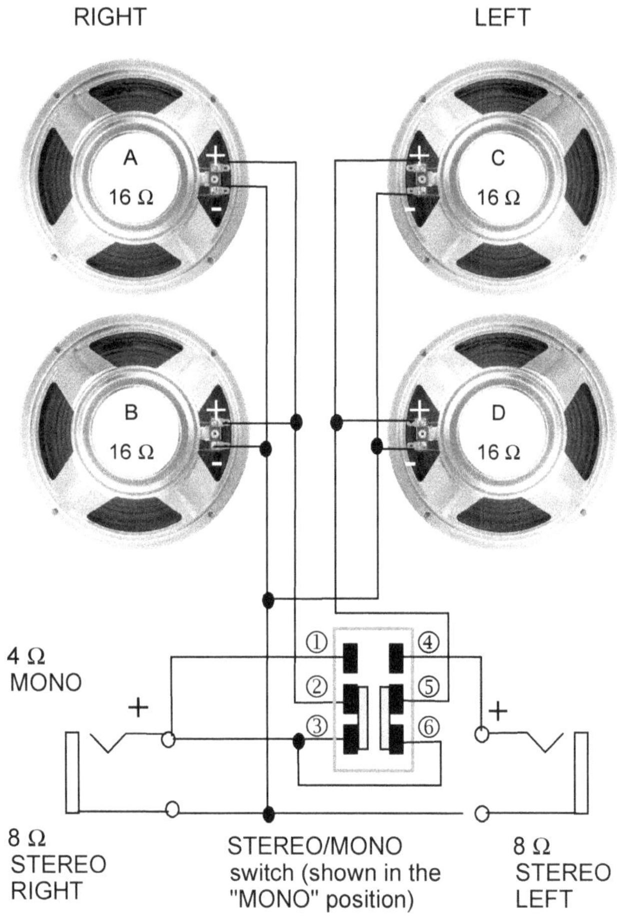

ABOVE: The "Stereo/Mono" switch gives you the flexibility of turning your 4x16 ohm mono cab (4 ohm total) into 2x8 ohm stereo cab while still retaining the option of using it as a 4 ohm mono cab at a flick of a switch

Option 3: Switchable "STEREO (2x8Ω) - MONO 4Ω - MONO 16Ω" wiring

This clever arrangement uses the auxiliary contacts to provide both 4Ω and 16Ω mono options at the two jacks.

You need is a pair of mono jacks with two auxiliary switching contacts and a DPDT (double pole double throw) switch of some kind - toggle, sliding, rotary, ...

Keep in mind that in the MONO mode, only one jack can be used at any time. The whole setup works providing that the amplifier's output is plugged into only one of the two jacks.

If you like a challenge, try to trace the connections on the diagram (RIGHT) for either 4Ω MONO or 16Ω MONO jack and see how the switching is achieved.

Keep in mind that the AUX contacts are connected ("made" or closed) when there is no male jack plugged in (as drawn here) and that they are open or disconnected once a male jack is inserted!

Looking at the RIGHT pair of speakers (A&B), their negative terminals are connected together and then to the lug #1 of the switch and also to the negative terminal of the 4Ω mono jack. Their + terminals are both connected to the upper terminal (#4) of the other pole of the double switch and to the + terminal of the 4Ω mono jack.

The sliding contact of the switch connects terminals #4 and #5, thus strapping the positive lugs of the LEFT pair of speakers (C&D) to positive terminal of the 4Ω mono jack, together with the positive lugs of the RIGHT pair of speakers (C&D).

Likewise, the left sliding contact of the switch connects terminals #1 and #2, thus connecting one of the AUX contacts of the 16Ω mono jack to the - terminal of the 4Ω mono jack.

Since that AUX contact is normally connected to the - terminal of the 16Ω mono jack (as pictured, without anything plugged into the 16Ω mono jack), that effectively connects the two - jack contacts together, and makes the left 8Ω parallel pair paralleled with the right 8Ω parallel pair, or 4Ω in total.

Now, assuming nothing is plugged into the 4Ω mono jack and an amplifier output plugged into the 16Ω mono jack, how is the 16Ω connection achieved?

Tracing the circuit from the + terminal of that jack, through terminals #4 and #5 of the switch we reach the two + terminals of the RIGHT speaker pair (A&B), whose - terminals, through switch contacts #1 and #2 are connected to the AUX contact of the 16Ω mono jack.

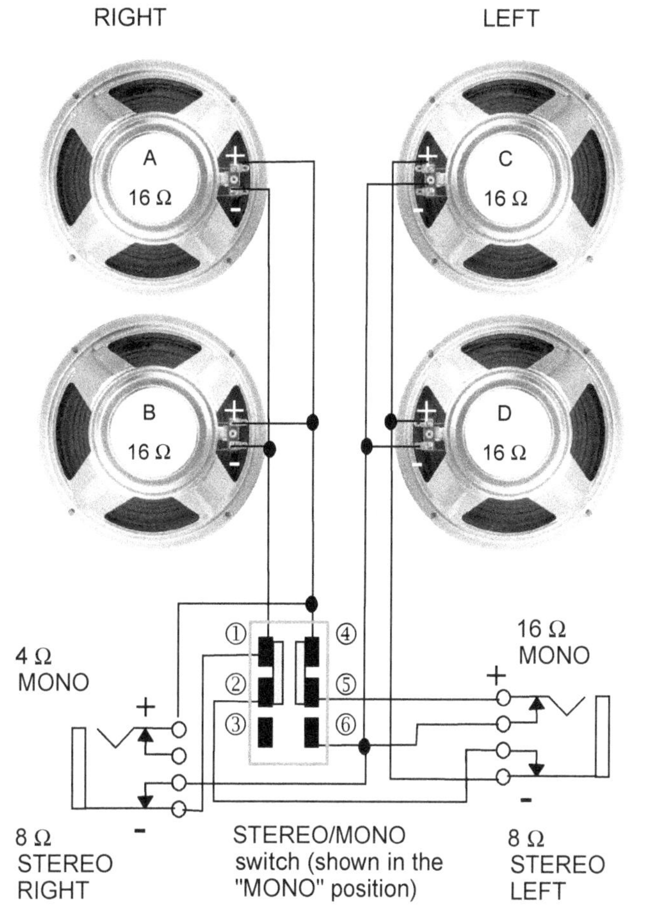

BELOW: The same mod, but this time the "Stereo/Mono" switch is in the "STEREO" position.

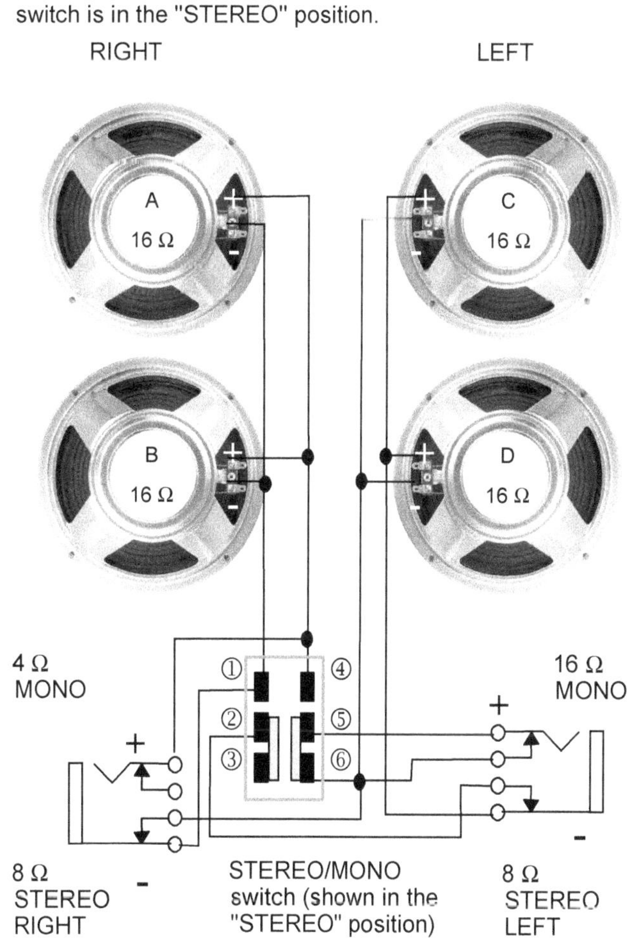

However, once the speaker cable is plugged into it, that contact is disconnected from the - terminal of the jack and thus does nothing.

Terminal #1 of the switch is also connected to the - terminal of the 4Ω mono jack, which makes contact with its AUX terminal, wired to terminal #6 of the switch and the + terminals of the LEFT pair of speakers (C&D).

It is also connected to the upper AUX contact, but that is now disconnected from the + terminal of the 16Ω mono jack and is thus inactive.

Finally, the - terminals of the C&D speaker pair are directly connected to the - terminal of the 16Ω mono jack, thus placing the right and left 8Ω pairs in series, thus achieving a 16Ω mono connection.

Unless your head is spinning now and you are cursing both me and the twisted logic behind this ingenious wiring example, lets try to figure out how the STEREO connections are achieved.

Keep in mind that there are mono jacks plugged into both sockets, thus disconnecting all four AUX contacts from their + or - terminals.

To make our analysis a bit easier, I have redrawn the sockets with those four AUX contacts disconnected.

Looking at the RIGHT paralleled pair of speakers (A&B), their - terminals go to switch terminal #1 and to the - lug of the socket.

The + goes to switch terminal #1, which isn't connected to anything now, but the + is also permanently wired to the + lug of the socket.

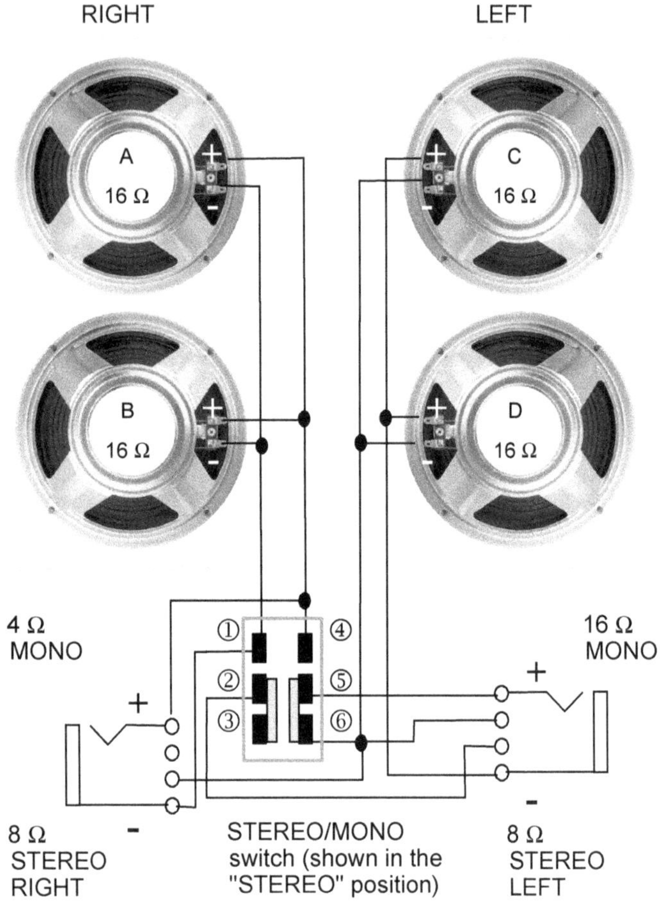

ABOVE: The wiring with the STEREO/MONO switch shown in the "STEREO" position (and both amplifier channels plugged into both sockets). The four "open" AUX contacts have been removed for clarity.

The - terminals of the LEFT pair of speakers (C&D) are permanently wired to the - lug of the "8Ω STEREO LEFT" socket, while their + terminals get connected to the + lug of the "8Ω STEREO LEFT" socket through switch terminals #6 and #5, thus completing the circuit.

NOTES

DIY CORNER - HOW TO DESIGN AND BUILD YOUR OWN SPEAKER CAB

8

Unhappy with the bland, boring, black boxes?

Want to stand out from the crowd not just in terms of the awesome looks of your guitar cab, but also by its wicked tone?

Have the basic woodworking skills?

If you've answered YES to all three questions, this chapter will give you a few ideas of what is possible on a low budget and a couple days of your time ...

Happy woodworking!

If you've answered "no", read the stuff anyway, you may change your mind and give it a go!

- SPOILT FOR CHOICE - DECISION TIME MADE (A BIT) EASIER
- A FEW DESIGN & CONSTRUCTION ISSUES: BASS, BAFFLES, AND BRACING
- THE QUAD - MODULAR CABS FOR EVERY CONCEIVABLE STAGE SITUATION
- TIME & MONEYSAVING HACK: THE NEW YORKER CAB
- DIY PROJECT FOR THE INTREPID: THE ANNIHILATOR
- DOUBLE TROUBLE - THE DIPOLE ENCLOSURE
- DISSIMILAR DRIVERS WITH A SLOT OPENING AND DISTRIBUTED PORT LOADING

SPOILT FOR CHOICE - DECISION TIME MADE (A BIT) EASIER

Why should you consider building your own speaker cab

Musicians make their cabinets for different reasons. Some are unhappy with the unimaginative commercial offerings, which either fall into the cheap & nasty category, with lousy tone, or into the very expensive "boutique" class. These "hi-end" cabs may be thicker or use better drivers, but are usually still of the bland & boring variety.

Others want unique-looking cabinets, those that will reflect their personality, preferences and prestige, and a black 4x12" box doesn't do that for anyone.

Finally, we have the tinkerers, those who enjoy building their amps and/or speaker cabs possibly even more than playing them. The cost (savings) could be a factor for some, but most don't usually care even if their handmade beauty ends up costing twice as much as the priciest commercial option.

The speaker cab design decision tree

Commercial speaker cab makers are limited by what they think an average guitar or bass player wants, or by what their marketing departments tell them their buyers want. When you design & build your own cab the only limit is your imagination.

To help you think about issues you probably wouldn't consider, I've tried to visualize these design & build factors as a decision tree. I'm sure you will think of other aspects and options, but let's use it as a starting point.

To build your own or modify and existing cab is the highest level question. Many cheap (usually solid state) guitar amps have decent cabs that can easily be modified, open back can be converted to closed back, they can be covered in a different color Tolex or tweed, etc. Shelves and other furniture pieces can be used as frames; just add a front baffle, and you have a speaker enclosure. All of these options are illustrated in this book.

The issue of expansion possibilities isn't usually considered. For ultimate flexibility you can build two, three or even four identical cabs, or cabs that look the same but have different tone (different drivers and construction).

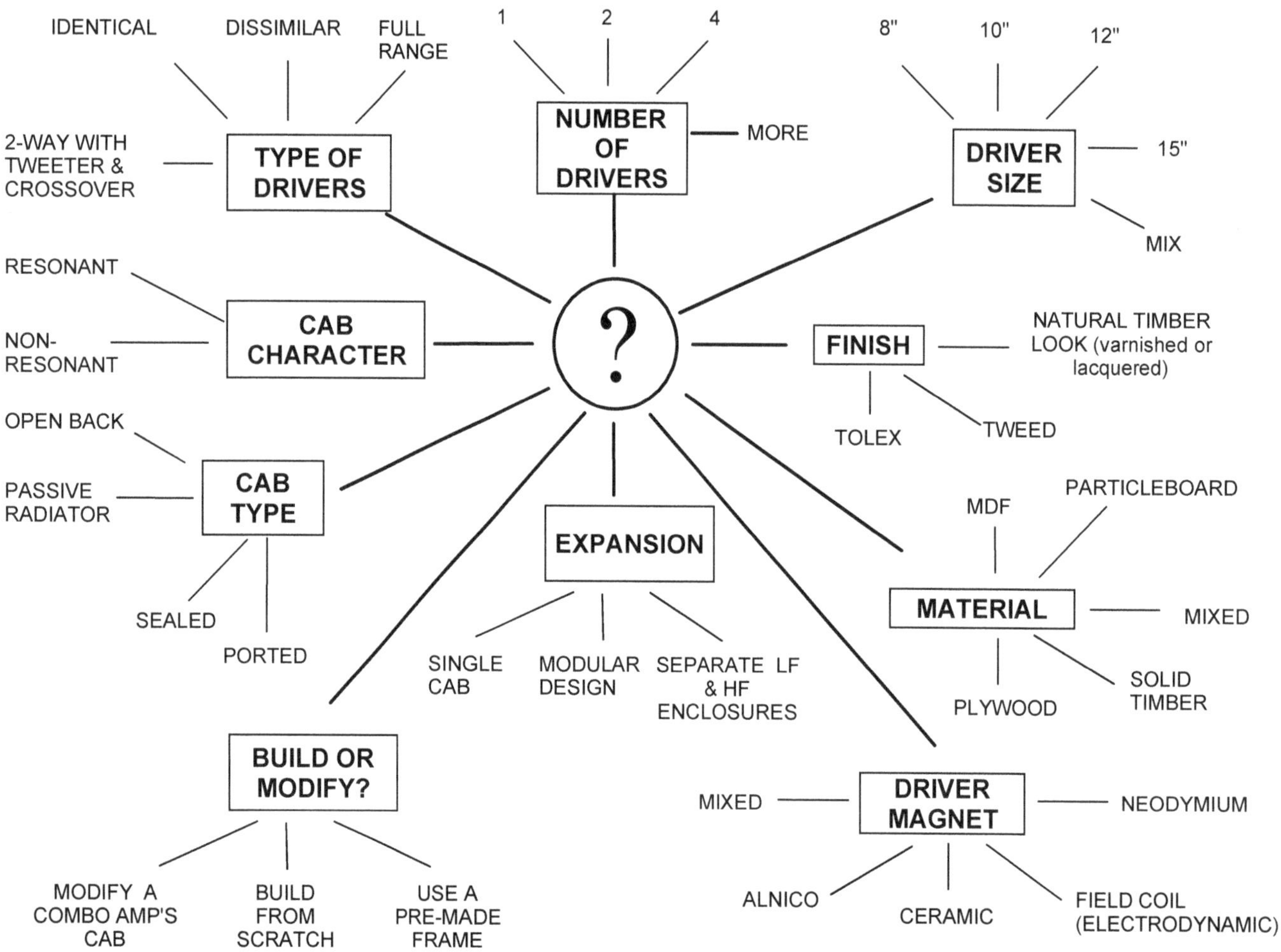

ABOVE: a plethora of cabinet design & construction possibilities - so many decisions to be made

Different MATERIALS sound different, so you can experiment to see which you prefer. This decision should be considered together with the choice of FINISH.

Some options, such as solid timber frame and natural timber finish (varnished or lacquered) obviously go together. Others don't make much sense, for instance varnishing a particleboard cab, which should be covered in Tolex, a much more sensible option.

Guitar speaker cabs are not hi-fi speakers, which should be as a neutral and non-resonant as possible. If you are happy with some frequencies being more prominently reproduced than others, experiment with resonant designs, not just in terms of a resonating front baffle, but also a non-braced, flexing frame and (if used) back cover.

Bass guitar speaker cabs should always be of the non-resonant variety.

Just as different timber types impact the cab's tonal character, so does the choice of magnets for its drivers. While ceramic magnets dominate, Alnico is making a comeback, and smaller and more powerful neodymium (rare earths) magnets are gaining acceptance.

Even some electrodynamic speakers (those without a permanent magnet) are being manufactured again, and, although relatively rare, vintage ones regularly come up for sale online.

While open back and sealed cabs predominate, you should experiment with passive radiator (drone cone) and ported (bass reflex) options. Quad cabs (with four identical drivers) are a great platform to try all three options. We covered such experiments in Chapter 7.

While we usually choose the driver size (diameter) based on the limitations of the cabinet size and the number of drivers we want to fit in, some go the opposite way, choosing the desired driver size (and even brand and model) and number, and then figuring out the smallest cab dimensions that would accommodate that many drivers.

A FEW DESIGN & CONSTRUCTION ISSUES: BASS, BAFFLES, AND BRACING

Internal bracing

Minimizing enclosure walls' vibrations is of paramount importance in hi-fi and bass guitar speaker cabinets, but should also be one of the aims while designing and building guitar cabs.

Vibrating cabinet walls introduce distortions harmonically unrelated to the musical signal, which are unpleasant and discordant or dissonant-sounding), "hangover" effects (delayed and sustained sound radiation after the signal has stopped), peaks and dips in the cab's frequency response and other unwanted effects.

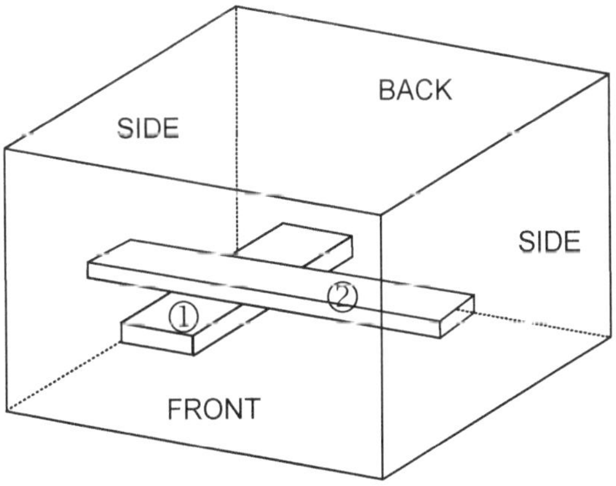

ABOVE: Bracing the opposite cabinet walls, front-to-back (1) and side-to-side (2).

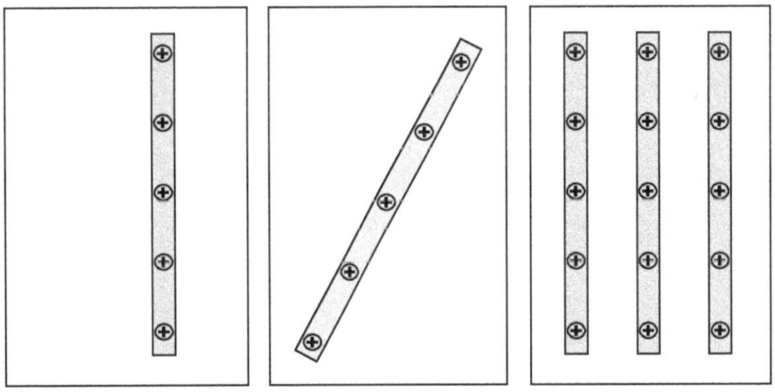

ABOVE (L--R): Back panel off-center brace, back panel diagonal cross-brace, and back panel multiple parallel cross-braces

Front baffle - floating or rigid?

While the frame of a speaker enclosure should always be as rigid and acoustically inert as possible, the front baffle can be approached two ways.

Non-resonant enclosures use thicker baffles, firmly attached to the frame on all four sides to minimize the baffle's vibration. The overall "color" and tone of the cab is predominately determined by the speaker.

In resonant enclosures, a thinner baffle is attached on two sides only (usually left and right sides, but top & bottom can be fixed instead). The baffle now acts as a soundboard in an acoustic piano, reinforcing the speaker vibrations by its own and sympathetically vibrating together. Compared to "tighter" sounding and more focused character of their non-resonant brethren, resonant enclosures generally sound warmer and more "lively".

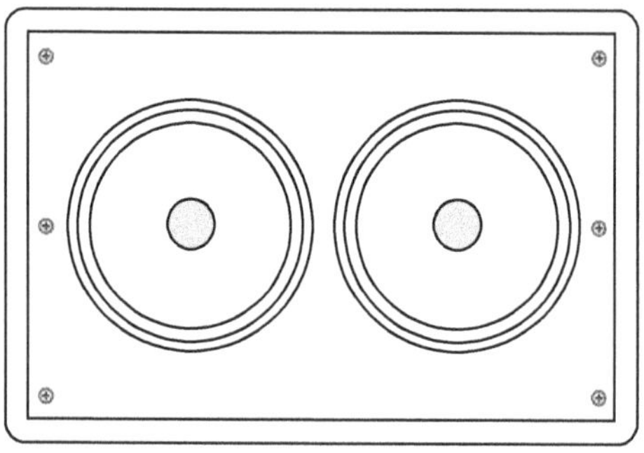

LEFT: A floating baffle is free to flex and vibrate in one dimension

RIGHT: A fixed or rigid baffle should not vibrate at all

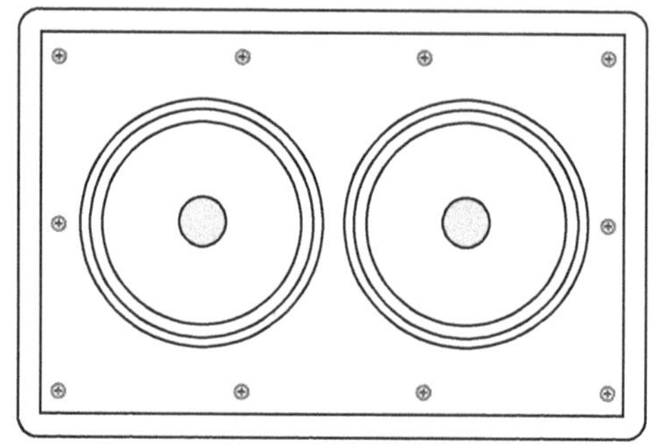

Mounting the speaker drivers from the rear

Mounting the drivers from rear is the most common way in musical instrument speakers; front mounting predominates in hi-fi speakers. Rear mounted drivers invariably use T-nuts. As a bolt (2) is tightened, the T-nut "bites" more and more, sinking deeper into the timber of the front baffle (4).

Using wood screws to directly attach drivers to the baffle is OK for smaller ones, but not strong enough for larger and heavier speakers.

Unlike front speaker mounting, rear mounting enables manufacturers to wrap the speaker cloth directly around the front baffle, no additional "stretcher frame" is required (next page).

The inverted horn speaker loading

As for the opening hole, normally we cut one in the front baffle, mount a driver on top of it and don't give it a second thought. However, the shape or profile of that opening alters the acoustic interface between the speaker's cone and the surrounding air and so has a profound impact on the tone of the cabinet!

One intriguing idea dating back to 1950s and 60s was to use an inverted horn front loading of a driver in a sealed enclosure. The inverted horn aperture results in increasing the velocity of air, which would equal that of a much large speaker, but without requiring an excessive movement of the actual diaphragm.

This minimizes distortion, especially of bass frequencies, but, more importantly, improves the low frequency performance so much, that an 8" driver would go as low in bass as a 12" speaker.

The high frequency improvement (a much higher upper frequency extension and wider dispersion) results from the acoustical diffraction that occurs at the sharp edge of the front loading opening. A small sound source has a better dispersion than a large one, and a small round front port behaves like a small virtual sound source.

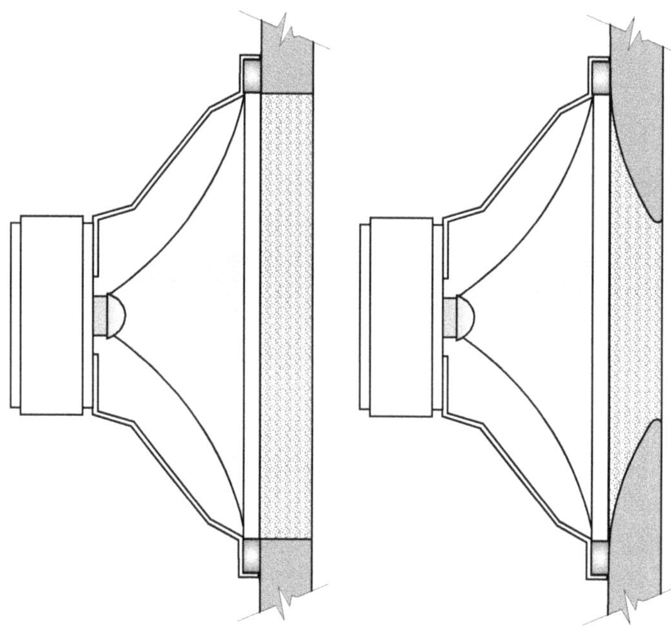

ABOVE: Normal (driver opening with square edges) baffle on the left versus inverted exponential horn baffle (tapered opening for the driver) on the right.

BELOW: How to use T-nuts to attach speakers to the baffle 1) T-nut pressed into the hole in the baffle 2) 3/16 or 1/4" bolt 3) Speaker frame (basket) 4) Baffle (speaker mounting board) 5) Washer 6) Foam gasket

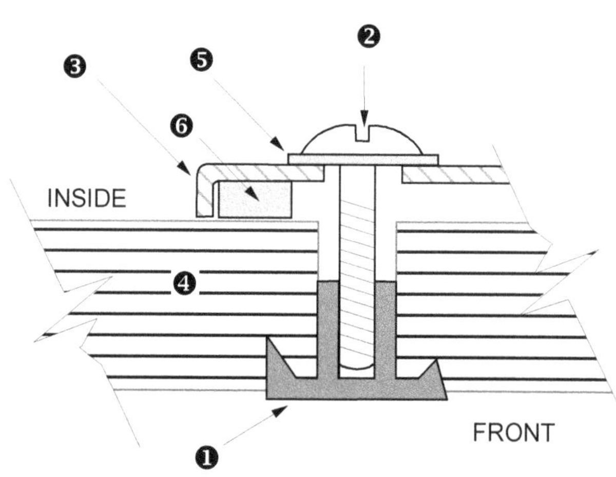

The front-mounting options

The usual frontal surface mounting method results in restricted air flow and rear reflections. This can be reduced by using flush mounting, but requires precisely routing the baffle. A better option is to use a chamfered internal edge.

With rear-mounted drivers, speaker cloth is simply wrapped around the front baffle. A removable rear cover is then required to gain access to the enclosure's interior and to the drivers.

With front mounted drivers, a stretcher frame is wrapped in grill cloth and attached to the fixed front baffle with Velcro or pins. If the rear panel is removable, the stretcher frame could also be screwed to the front baffle from the inside.

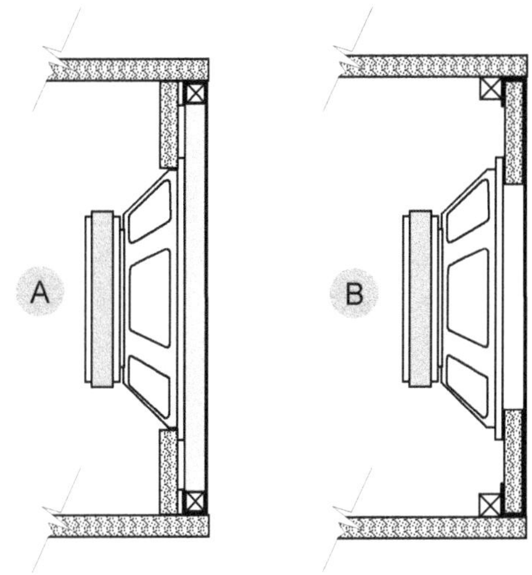

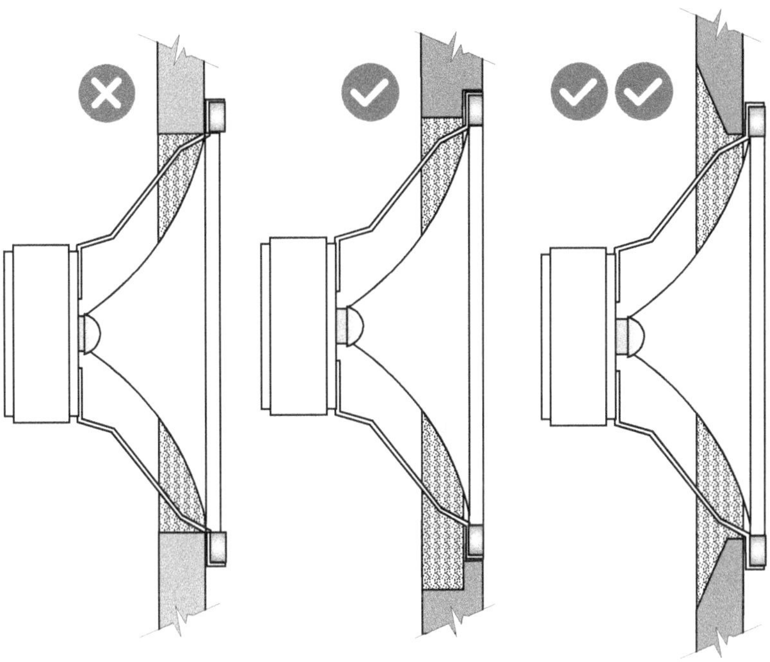

ABOVE: Front mounted driver (A), with grille cloth wrapped around a stretcher frame, which is then attached to the fixed front baffle with Velcro or pins, or permanently screwed into it from the rear (only if there is access from the back of the enclosure). In B) the driver is rear-mounted, and speaker cloth is wrapped around the removable front baffle.

FAR LEFT: The usual surface mounting method results in restricted air flow and rear reflections

MIDDLE: Flush mounting achieved by routing the baffle is better

LEFT: The best way is to chamfer the internal edge

The knock test on guitar speaker cabs

You may have been to a doctor who tapped your back or abdomen with their fingers and wondered what possible use such a seemingly primitive check could be. Known as percussion tapping, it's one of the "hands-on" methods used to discern the condition of internal organs such as the liver, lungs, and even muscles.

Each part of the body should "sound" a certain way, and if it doesn't, it indicates an abnormal condition that needs further investigation. For instance, a resonant sound should be produced by a hollow, air-filled cavity or structure such as lungs; a dull ("flat") sounding lung may indicate a presence of water or a solid mass (tumor). Abdomen tapping should produce a "tympanic" type of sound, relatively loud and higher-pitched.

An analogous test can be done on loudspeaker cabinets, guitars (esp. acoustic ones), and other musical instruments. In high fidelity systems, turntable plinths, hi-fi racks and acoustic panels can be evaluated that way.

The test itself is super easy, but interpreting its results is another matter. Speaker cab builders know that various material choices (solid timber such as pine, MDF, plywood, chipboard) sound different when tapped while still in the form of boards, planks, or panels; of course they will sound different once made into speaker enclosures.

When tapped, the sound produced by a loudspeaker will depend on its cabinet construction and damping (bracing) of its panels. With hi-fi and bass speakers, their cabinets (when knocked on) should resonate as little as possible; the flatter and duller the sound, the better.

In the case of resonant guitar speaker cabs, whose tonal balance makes up the instrument's total sonic signature, we don't try to minimize vibration and tame down enclosure resonances, but use them to achieve a certain tone.

Seasoned musicians don't need to try a dozen or so cabs in a music shop. They can make preliminary judgments and narrow the choices down simply by knocking on each of them in turn and comparing their acoustic responses.

THE QUAD - MODULAR CABS FOR EVERY CONCEIVABLE STAGE SITUATION

Having three or four identical cabs affords you maximal flexibility in almost all situations, from quiet practice, through playing in small clubs, to a large stage. Only a few basic options are illustrated here.

For best sound, the cabs should always be connected in parallel. This rules out 4Ω and 8Ω drivers - the speaker(s) in each cab must be of the 16Ω impedance! 50 watt power handling is assumed here, but that can vary as per your requirements, from 20 watts per driver (or cab), to 100 watts or more.

This discussion assumes your amp head has a 4-, 8- and 16-ohm outputs, or you can use heads of different power ratings, from low powered ones (5-15 watts), to 100 or even 200 watts.

BELOW: Using one or more of the modular enclosures with the same amp head. The individual cabs don't have to be identical, as illustrated, but the amp must have multiple impedance outputs, ideally 4, 8, and 16Ω

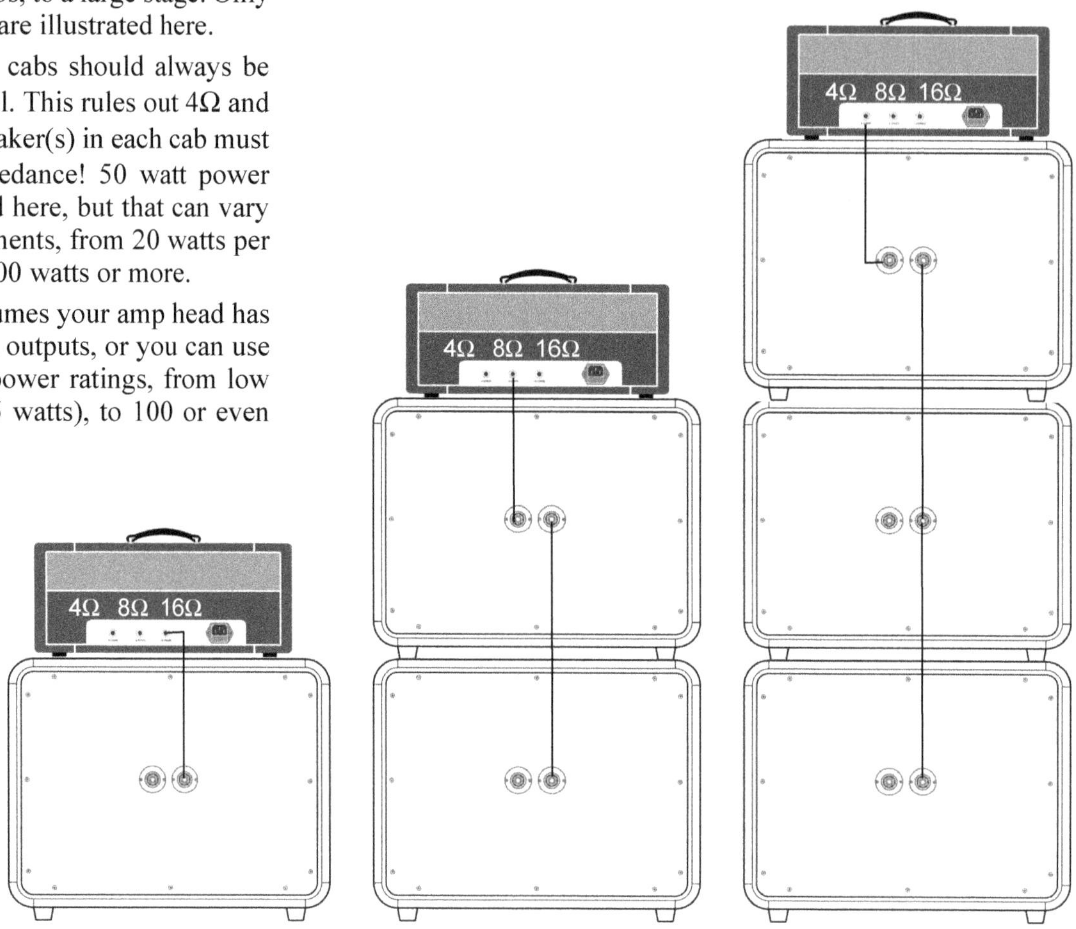

1x12", 16Ω, 50 watts 2x12", 8Ω, 100 watts 3x12", 5.3 Ω ,150 watts

You could use three cabs facing forward and one as a monitor, facing you (the band). Instead of stacking three (illustrated) or even four cabs vertically, you could space them horizontally, in-line, or in an arc, in a curved arrangement.

Instead of one large 4x12" cab that requires two people to handle it, these compact babies can be easily carried by one person, two-at-a-time, and all four can be placed on a back seat of a sedan, for instance.

On the debit side, making four cabs is more expensive and time-consuming than making one, and you don't get the higher sensitivity resulting from the speaker mutual coupling effect.

Two identical drivers in the same cab would increase the cab's sensitivity by 3dB, the sensitivity of three drivers would be 10log3 = 4.8 dB higher, while four identical drivers would be 10log4 = 6 dB louder!

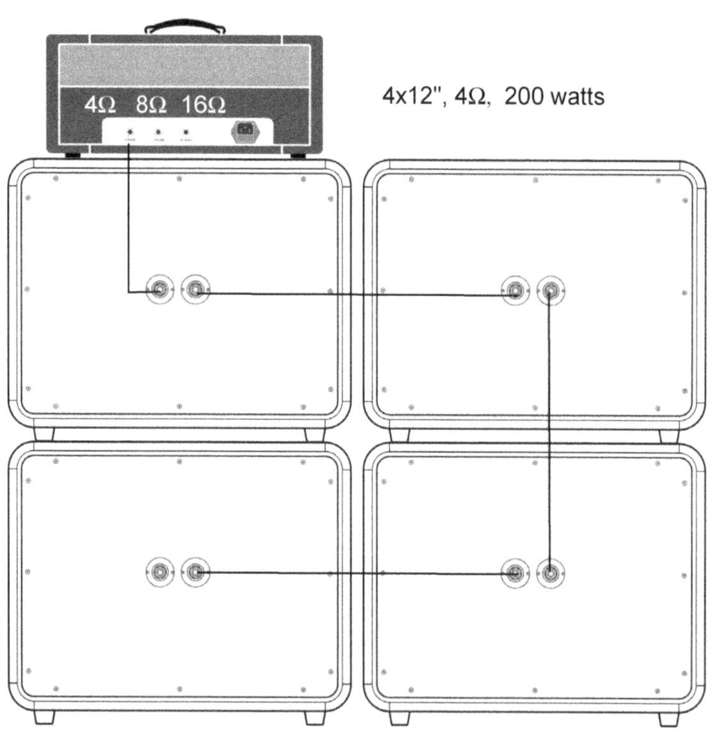

4x12", 4Ω, 200 watts

TIME & MONEYSAVING HACK: THE NEW YORKER CAB

Using bee boxes as a frame for a small speaker cab

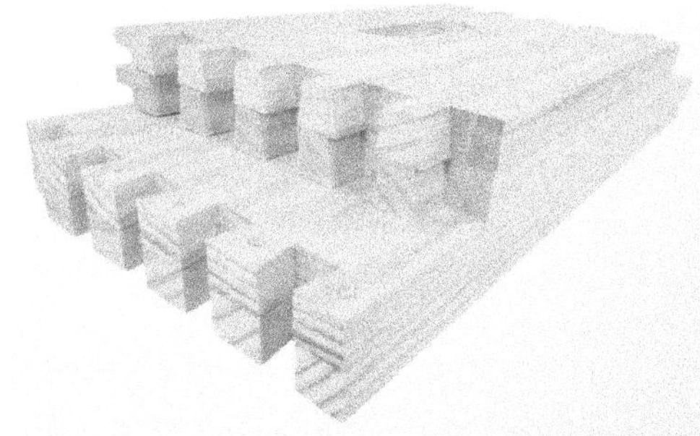

For most of us dovetailed joints are hard and time-consuming to make. Special router bits and jigs are required (unless you are a sadomasochist and want to do the manually, one-by-one), so the beauty of this hack is that is saves you lots of time and hassle.

Langstroth hive or "bee boxes" come in two sizes, "8 Frame" and "10 Frame". An "8 Frame" measures 506mm (L) x 355 mm (W), the "10 Frame" box is 506 mm (L) x 406 mm (W). The solid pine panels are either 20 or 22 mm thick.

Both boxes come in five different sizes: Full Depth 245 mm, WSP 193 mm, Manly 168 mm, Ideal 147 mm, and Half Depth 121 mm. The first two are suitable for our use as small speaker cab, the others are too shallow.

Some suppliers offer two different grades of timber, A-grade (with no knots) and B-grade (contains knots). The ones sold in Australia are made either from the local Hoop Pine or imported New Zealand pine.

A 10 Frame full depth A-grade flat pack sells for around AU$35 (US$21).

Once assembled, the four flat-packed 10 Frame panels make a box (frame) with outside dimensions 506 mm (L) x 406 mm (W). The internal dimensions are 18 1/4 inches (463.6 mm) by 14 1/4 inches (362.0 mm).

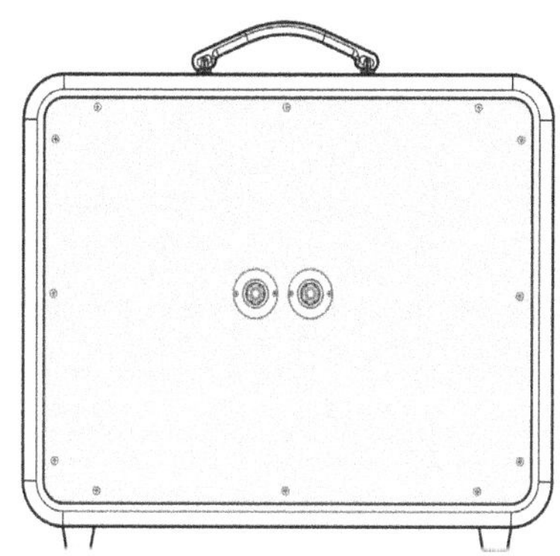

Assembling the 4-sided frame is easy. Simply glue the dovetailed surfaces and clamp the frames together while the glue dries overnight.

Alternatively, if you haven't got the clamps, use a couple of small nails or screws on each of the four joints. You can remove them afterwards or even leave them in place to add to the strength of the joints.

Open or closed back?

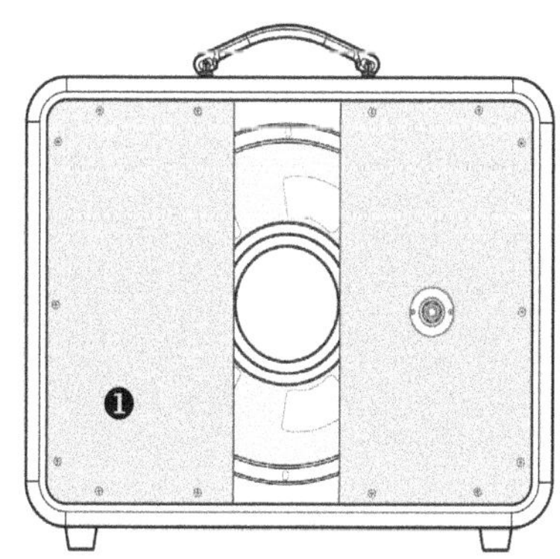

The sensitivity of your finished speaker cab will mostly depend on its type. Open back cabs will be more efficient (higher sensitivity), meaning they will produce a louder tone (for the same electrical signal from the amp) than sealed boxes.

However, perhaps more importantly, the tone and the dynamic behavior (the response) of the two types of cabinets will also differ.

I'd recommend making at least a couple of bee boxes using identical drivers, one with an open back, the other of the sealed variety and then comparing their sonics to see which one you prefer. You can always convert the other cab to the preferred type by changing its back cover!

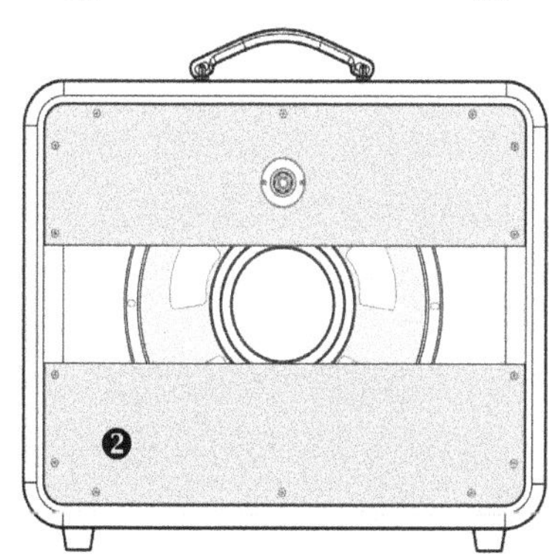

Depending on the size of your available plywood sheet, you can use either a vertical (1) or horizontal (2) arrangement of the back "cover".

Most combo amps use the later (usually for cable storage and the placement of a reverb tank) , but you have no such concerns in the case of a passive cab.

The New Yorker logo

New Yorker was the longest-running American car nameplate, in continuous production by Chrysler from 1940 until 1996, and one of the brand's most prestigious models. Since so many were made in that half-a-century, "New Yorker" metal emblems are plentiful and relatively cheap on eBay.

The smaller one (that we will use for these cabinets) measures approx. 112 x 30 mm, while the larger one is 175 x 50 mm. Two integral pins at the back are 20 mm long and 4 mm in diameter.

You don't have to use the pins (which require drilling a hole in the front baffle). Break them off, grind or sand the back of the logo to make it flat, and simply glue it to the front baffle.

A blast from the past: a few cool ideas for the New Yorker front baffle design

Gibson BR-6 was a small combo produced for a decade, from 1946 until 1956, when it was replaced by a modernized version, model GA-6. This particular design was the second version introduced in 1948.

Its 10" speaker was adequate for the 10-12 watts produced by its feeble tube amp. The cab itself was 14 3/4" (375 mm) high, 18 1/4" (464 mm.) wide and 7 1/2" (190 mm) deep.

Since we are using a 12" speaker, we will add two rectangular sections on top of the two wide horizontal slots.

Our "New Yorker" emblem is much wider than the simple "G" used by Gibson in BR-6, so the middle island in the center of the horizontal bar will have to be much wider, too.

The design of the speaker baffle, or, more precisely, the driver cutout, is only limited by your imagination. A few ideas from the past (modified by yours truly) follow below and on the next page.

Of course, the more "meat" (or rather "timber") you leave in front of the driver, the more the front-propagating sound waves will be obstructed. If you make two identical cabs with different cutout designs, you may be able to hear sonic differences between them.

To cut the shapes out use a router or a jigsaw. Afterwards, the required extent of sanding will depend on how steady your hand is and how well your eye-hand coordination works, in other words, how smooth or straight your cutouts are.

Finally, for our American and other "imperial" friends, keep in mind that all dimensions are in mm (millimeters)! CL means "Center Line" and "cc" means "center-to-center", referring to the distance between two holes.

BELOW and NEXT PAGE: a few cool front baffle cutout designs

#1: Horizontal bar with a central logo

#2: Letter "A"

#3: Palm trees

#4: Petals

#5: Rectangular vertical slots

#6: Two vertical bars

#7: Concentric circles

#8: Diagonal chevron shape

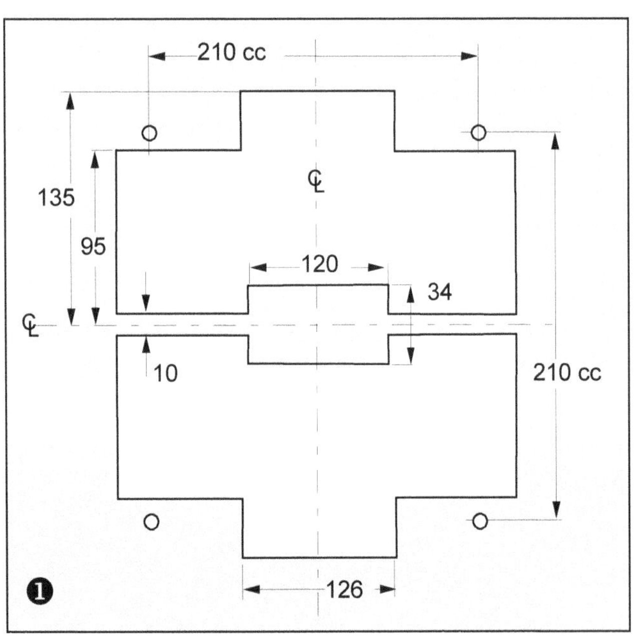

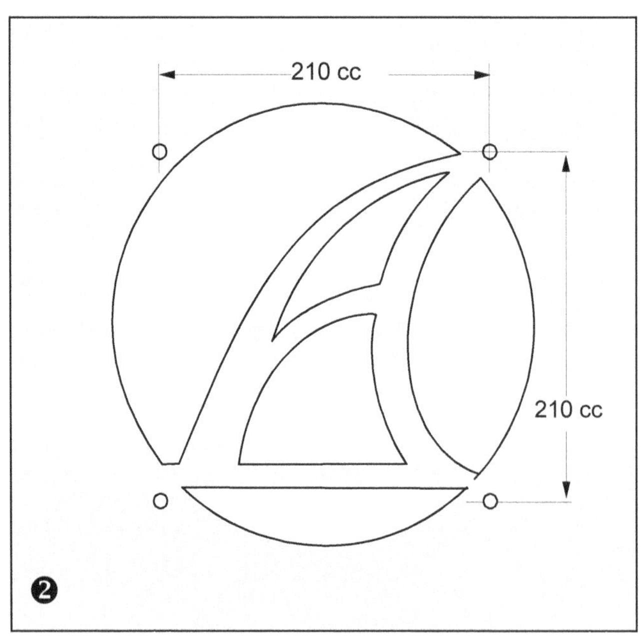

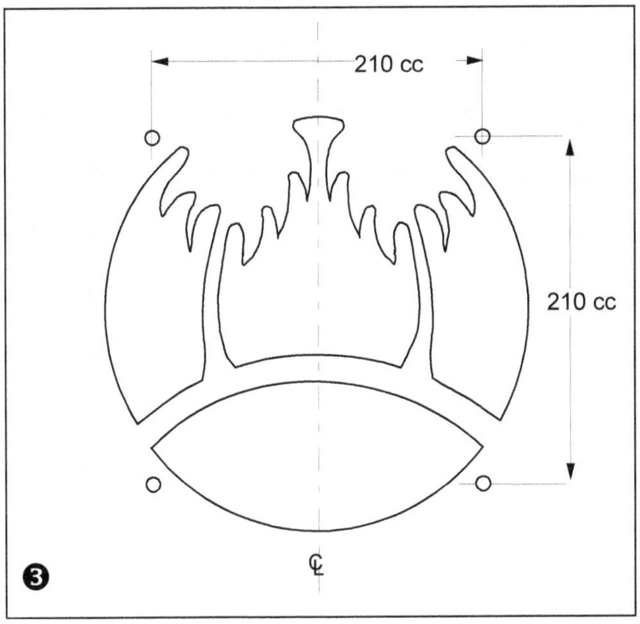
210 cc
210 cc
℄

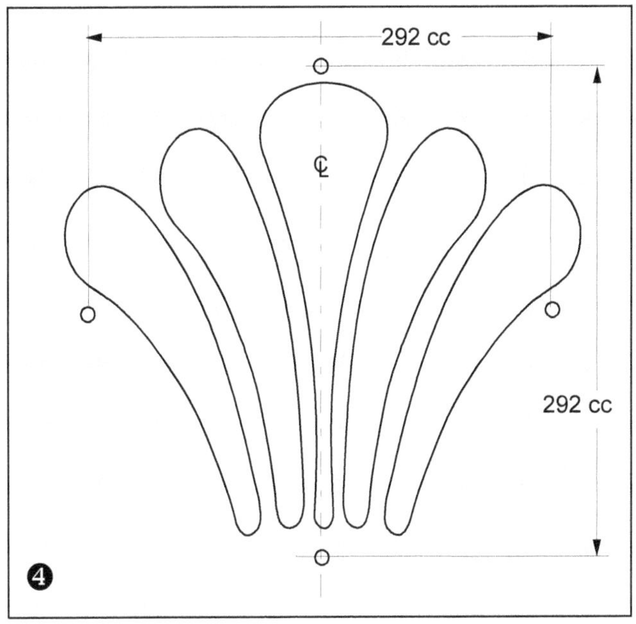
292 cc
℄
292 cc

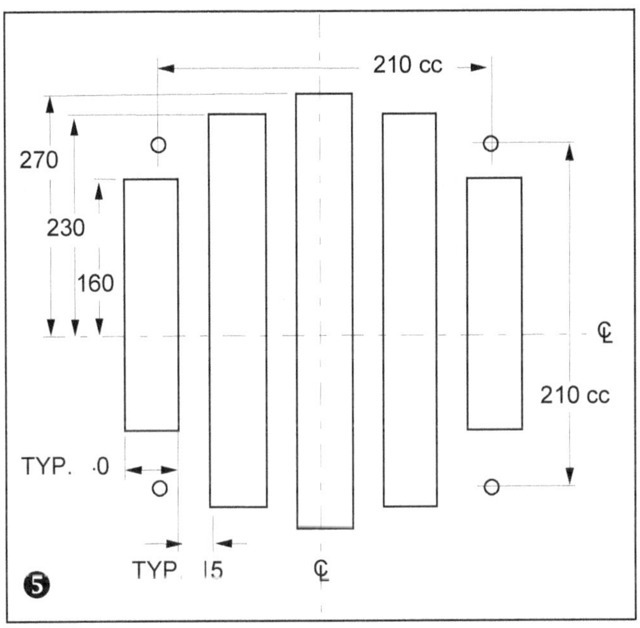
210 cc
270
230
160
℄
210 cc
TYP.
TYP. 15
℄

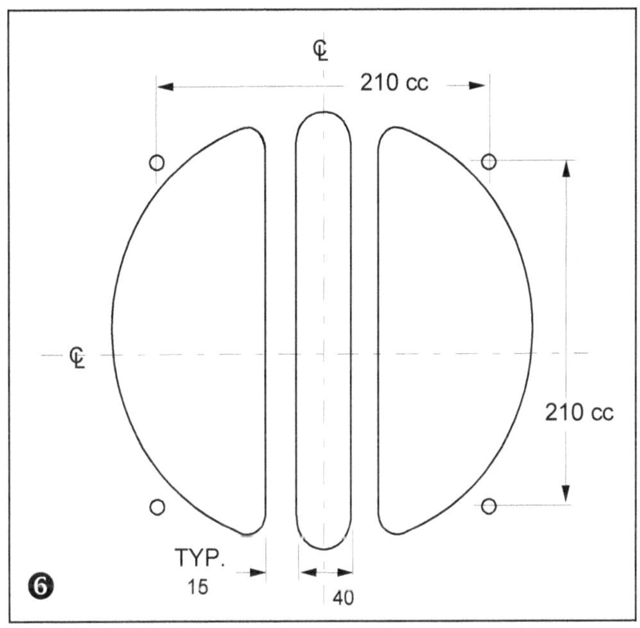
℄
210 cc
℄
210 cc
TYP.
15
40

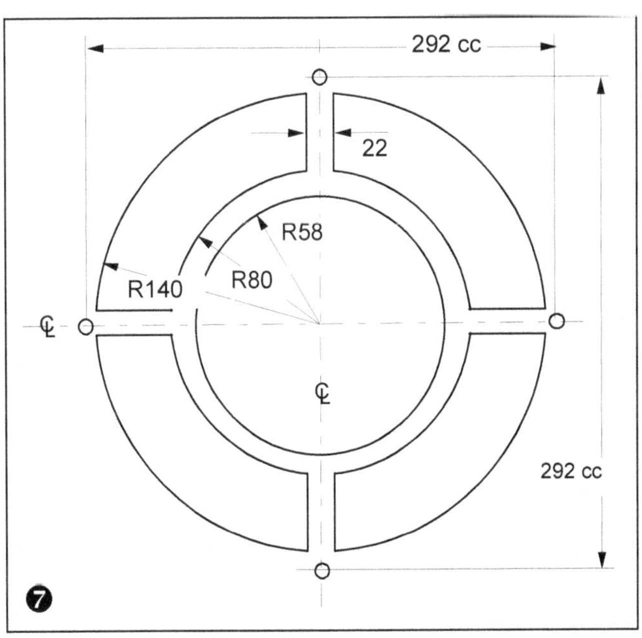
292 cc
22
R58
R140
R80
℄
℄
292 cc

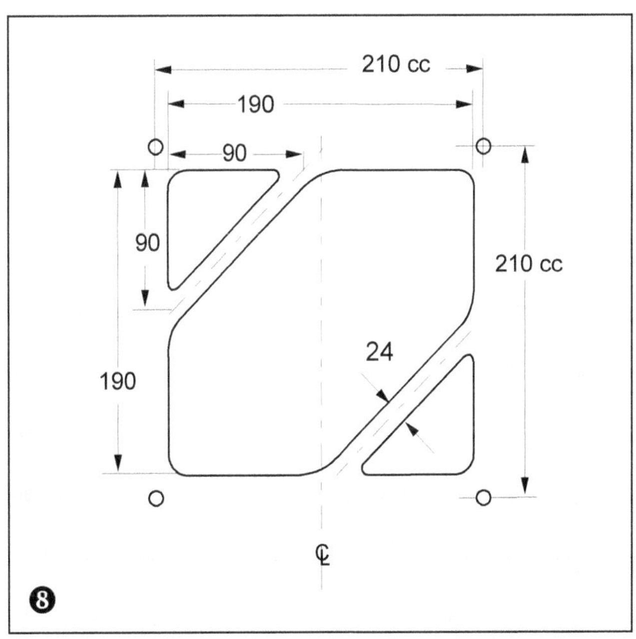
210 cc
190
90
90
210 cc
24
190
℄

ABOVE: The New Yorker cab (front baffle design #1) with a vintage 12" Australian-made Etone driver
BELOW: The New Yorker cab (front baffle design #5) with a 12" Jensen driver
Both cabs use 12mm plywood baffle and back cover, and solid machined pine legs, photographed before the cabs' internal surfaces were lined with acoustic padding (see BELOW)

LEFT: Both cabs are of the sealed box variety, so we lined the sides and the back cover with acoustic lining (recycled cotton) salvaged from a vintage Hammond organ

The finished cabs

We had quite a few 12" drivers lying around the workshop, so we made two more bee box cabs with a simple circular front opening. After cutting the back cover for one of them from a standard plywood sheet, we were left with a small piece, which was then cut and used as a partial back cover for the other cab. The piece wasn't wide enough for the customary horizontal arrangement (top and bottom strips), so we used a vertical design, strips on the sides (1).

ABOVE: The New Yorker cabs #3 and #4 from the front. The open-back cab was fitted with an extremely dull sounding driver (2) pulled out from a Jet City combo amp, so we added a vintage Japanese 3" Alnico tweeter (3), salvaged from a Sony tape recorder. The sound opened up and the missing sparkle appeared!

BELOW: The rear view of the two cabs

DIY PROJECT FOR THE INTREPID: THE ANNIHILATOR

The way it used to be: Gibson GA-79 RVT and Watkins Dominator

The 1961 Gibson GA-79 RVT was a 2 x 15 watt stereo combo guitar amp, housed in its instantly recognizable wedge-shaped cabinet. Apart from two horizontal slots and a Gibson logo, the front panel carried no speaker, which is somewhat perplexing. The two 10" Jensen P10Q speakers fire from the left and right angled baffles, thus creating a sonic dead spot in the front.

Across the Atlantic, Watkins did not use the flat front baffle section, opting for a simpler V-front cabinet, which first appeared in 1956 in an upgraded Westminster amp, followed by the Dominator in 1957. Just like GA-79, it had two 10" Elac drivers firing diagonally to its sides, but this is where the similarity ends. Electronically, the Dominator was an ordinary mono amp, with a pair of EL84s producing around 17 watts of push-pull power.

Compared to standard rectangular speaker cabs and combo amp, both of these designs achieved a wider sound dispersion and a broader soundstage, albeit at the cost of increasing production cost, which is probably why (to my knowledge), no other commercial guitar amp maker followed suit in all those decades after these early examples. Even Gibson reverted back to bland & boring; Dominator MkII, introduced in 1968 and produced until 1972, was housed in an ordinary-looking 1 x 12" cabinet.

FRONT VIEW

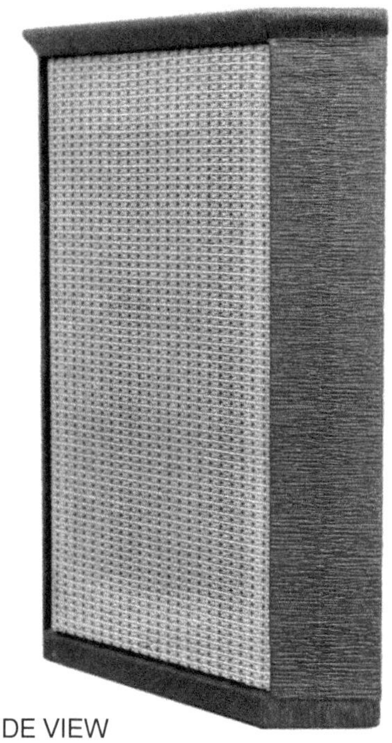

SIDE VIEW

ABOVE: Gibson GA-79 was 18.5" (47 cm) tall, 25 5/8" (65.1 cm) wide and only 10.5" (26.7 cm) deep.
BELOW: The ad for Watkins Dominator

Our design brief

Since the shape used by Gibson is more versatile, let's use it for The Annihilator.

To keep our cab relatively small and light, we have reduced the length of its sides (228 mm compared to Gibson's 255mm), which means we will not be able to fit 10" side drivers.

However, modern 8" drivers pack as much punch as the 10" drivers of the 1950's, so nothing is lost as far as the power is concerned.

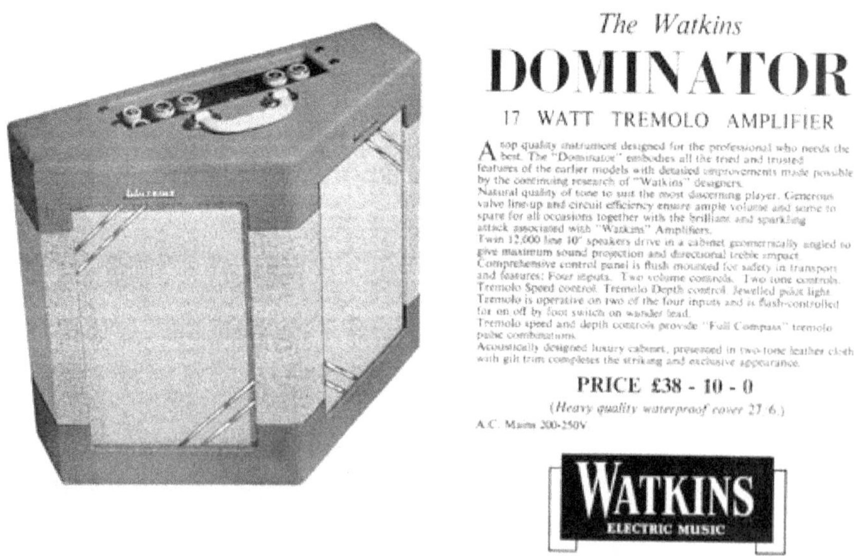

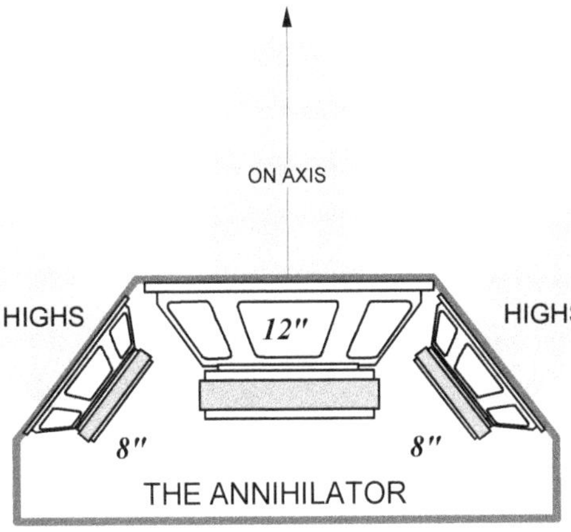

The high frequency sound dispersion patterns for Gibson GA-79 RVT cab (BELOW) and our Annihilator (ABOVE). Notice the acoustic "hole-in-the-middle" dip in the sound pressure level (loudness) for the Gibson-style enclosure.

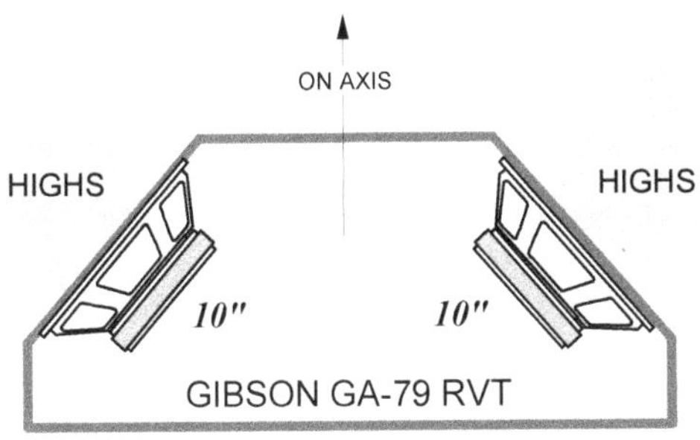

Recommended speaker configurations and connections

OPTION 1: Series-connected side drivers, then paralleled with the front driver. The overall impedance is equal to each of the side drivers. In our case Z=8Ω, Z_{TOTAL}=8Ω, the 12" driver is a 16Ω speaker.

With the same power ratings (side drivers P watts, front driver 2P watts), the total cab power rating is 4P. For instance, if the 8" side drivers are rated at 25 watts and the front driver (12") is rated at 50 watts, we get a 100 watt cab.

OPTION 2: All three drivers in parallel, but with side drivers of twice the impedance of the front driver and half of its power rating.

The two smaller drivers in parallel have a total impedance of 2Z||2Z = Z, which in parallel with Z of the large driver results in the total cab impedance of Z/2 Ω.

With an 8Ω front speaker and 16Ω side speakers, the cab's impedance is 4Ω. The smaller side drivers should be rated at half the power of the front driver, so the total power rating of the combination is 2P. 50 watt front and 25 watt side drivers make a 100 watt cab.

OPTION 3: The parallel connection with all three drivers of the same impedance and power rating. The total impedance is easy to calculate, Z/3 (2.7 Ω for 8 Ω drivers and 5.3 Ω for 16 Ω drivers). The cab power rating is 3P.

For example, with three P=30 W drivers, the cab power rating would be 3P or 3x30 = 90 watts.

Reducing the length of the sides enables us to widen the front panel, from Gibson's 282mm to 330mm, wide enough for even 12" drivers. As luck would have it, our sheet of the marine grade hardwood plywood is 610 mm wide (1220 mm long), the exact dimension used by Gibson all those years ago. This significantly minimizes the amount of cutting required and saves us lots of cutting and sanding time.

Our cab will be of the open back persuasion, so we can use any offcuts or remnants for top and bottom panels at the back, their width is not critical at all.

We reduced the cab's height from 470 to 440 mm overall. Since our marine ply is 12 mm thick, the front and the side baffles (and the fixed rectangular side panels) will be 440-2x12=416 mm tall.

While 12 mm may seem too thin, keep in mind that it is a very strong and rigid material, more than adequate for such relatively small cabinet.

This is a resonant enclosure, without bracing of any kind; its vibrations are part of its tonal signature.

Plus, cost-wise, a 1220 x 610 mm sheet cost us only AU$44; going up to 18mm would more than double the material cost to AU$92 per sheet.

OPTION 1:

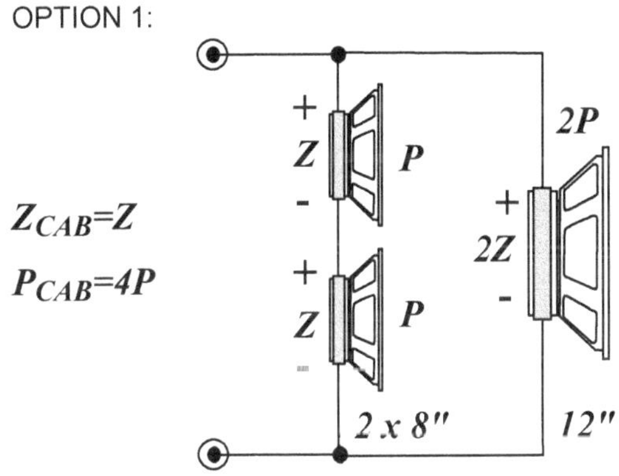

OPTION 2:

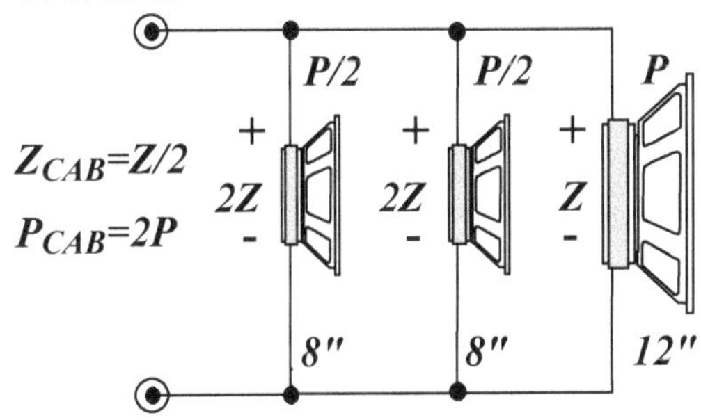

OPTION 3:

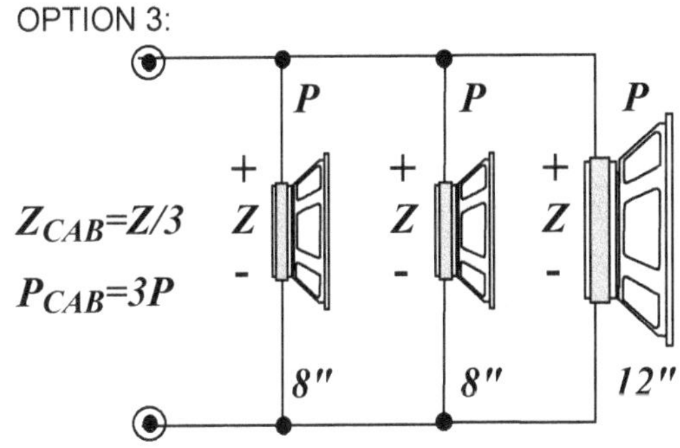

The shape of things to come

The dimensions of the frame and the front (central) baffle are shown on the drawing below.

Since this an open back cabinet (or, more accurately, a semi-open back cab), the shape and size of its back cover is not important.

Measurements

The impedance curve (next page) is relatively tame, the highest resonant peak of 24 ohms at 58 Hz is followed by a much lower peak, 11.5 ohms at 200 Hz.

The SPL curve has two corresponding innocuous peaks (1 & 3) at the resonant frequencies (58 and 200 Hz) but also peaks at 120 Hz (2), 300 Hz (4), 650 Hz (5), 900 Hz (6), 3,300 Hz (7) and 5,000 Hz (8).

These peaks are interspersed with dips, the most serious ones occurring at 140 Hz, 280 Hz, 800 Hz and 1600 Hz.

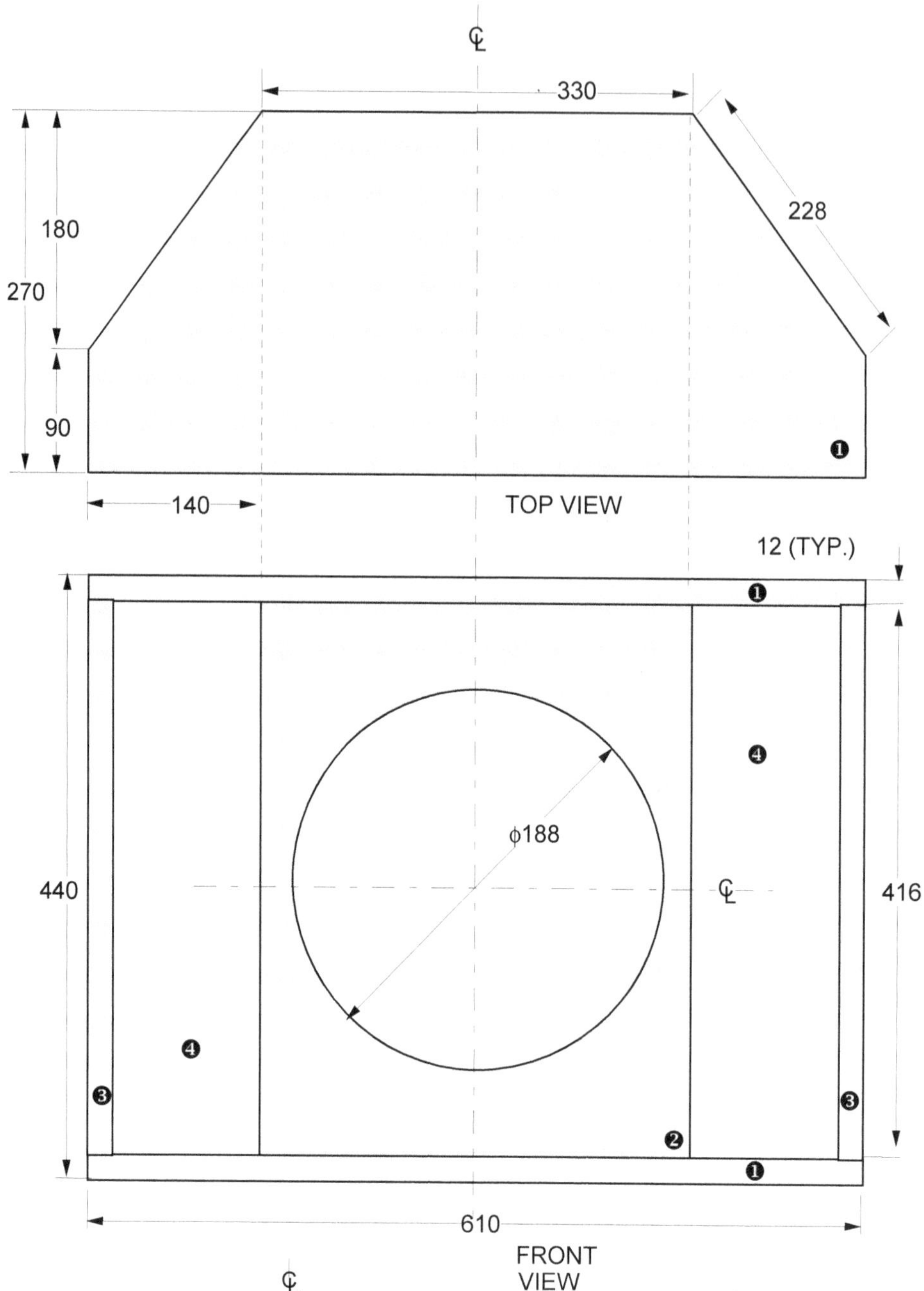

ABOVE RIGHT:

1) Top & bottom panels (2 of)
2) Front baffle (1 of)
3) Side panels (2 of)
4) Side baffles (2 of)
5) Rear panels (2 of)

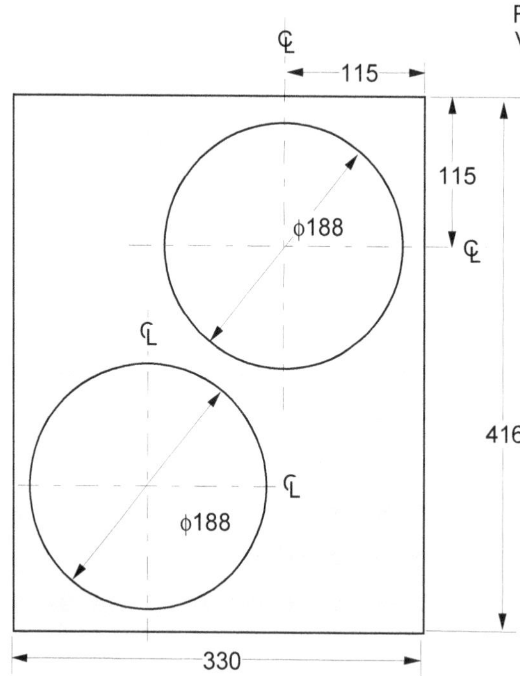

RIGHT: Front baffle for the 2 X 8" driver arrangement

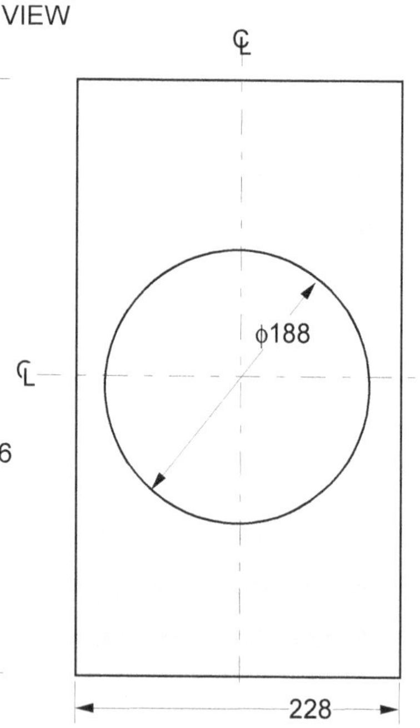

FAR RIGHT: Side baffles for 1 X 8" side speakers. For 6" speakers, reduce the diameter of the opening to ϕ136 mm.

ABOVE: the finished cab from the side/front and from the back. The second jack makes it possible to daisy chain The Annihilator with another speaker cab.

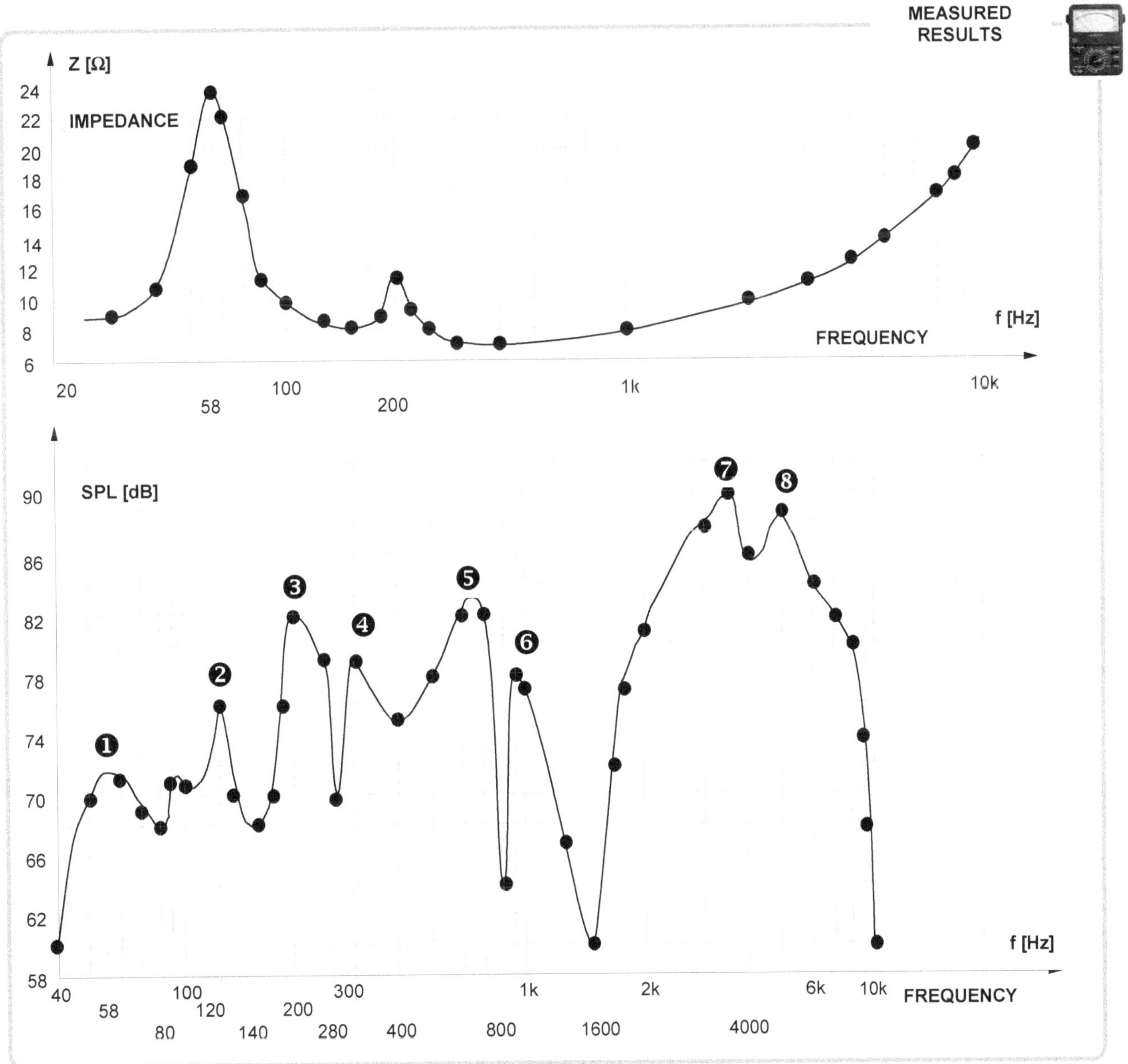

The Enveloper: An octet of 6"x9" drivers creates omnidirectional dispersion in the forward hemisphere

For this all out design you'd need eight identical 6"x9" drivers. The front pair and the two diagonally - facing pairs of drivers are placed horizontally, while the two 90° (sideways) facing drivers, one on each side, are vertically oriented.

A full 180° coverage is thus achieved, enveloping the whole stage or studio, truly deserving of its proposed name, The Enveloper.

Both the shape of the enclosure and the use of oval drivers will reduce internal standing waves and thus minimize peaks and dips in the cab's SPL curve.

Should you find it difficult to source eight vintage oval drivers, standard round speakers can also be used.

Using 4Ω drivers, connect each of the four pairs in parallel (2Ω), and then string the four pairs in series (4x2=8Ω), for an 8Ω cab.

Using 8Ω drivers, connect each pair in series (16Ω), and then parallel the four pairs (16/4=4Ω), for a 4Ω cab.

Using the same approach with eight identical drivers of the 16Ω persuasion, you would get an 32Ω/4 = 8Ω cab.

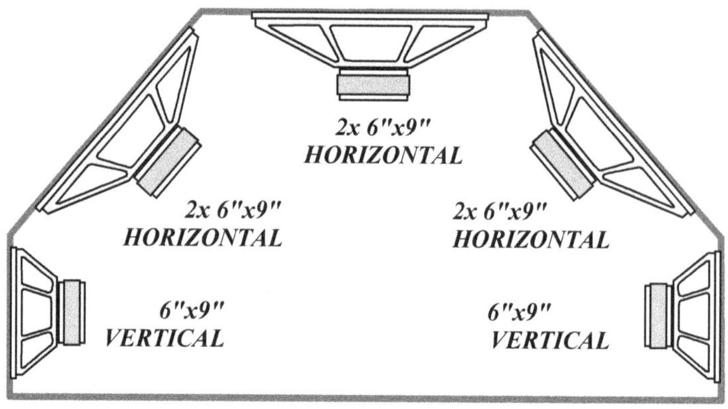

ABOVE: The Enveloper (top or bottom view)

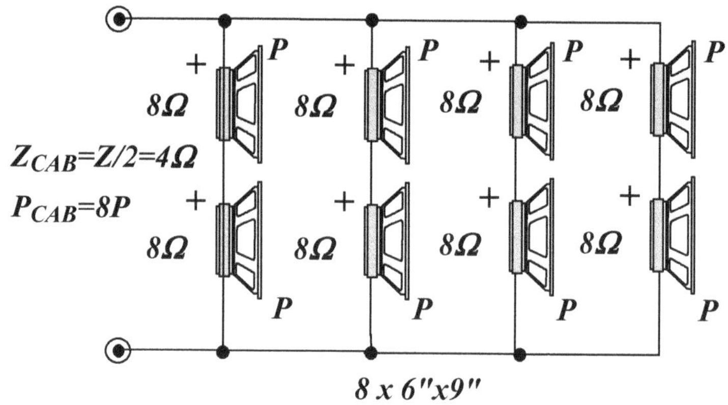

The Pulsator: The Annihilator with an up- or down-firing woofer

A 12" or 15" woofer can be placed upward or downward facing. The downward facing is the typical subwoofer firing direction, so the high wavelength (low frequency) sound waves bounce off the floor underneath, and is more practical of the two, enabling the guitarist to place his or her amp on top of the beast.

However, with the woofer firing upward, a space ball (almost 360° 3-D sphere) of pulsating bass is created, truly awesome.

For Option 1, the two side drivers should be chosen so that both their power rating (each) and impedance is half that of the larger driver. If the upward or downward firing driver is rated at 100 watts and 16Ω, the side-firing drivers should be rated at 50 watt and 8Ω.

For Option 2, the two side drivers should have double the impedance of the larger speaker, 16Ω & 8Ω respectively, for a total impedance of 4Ω.

OPTION 1:

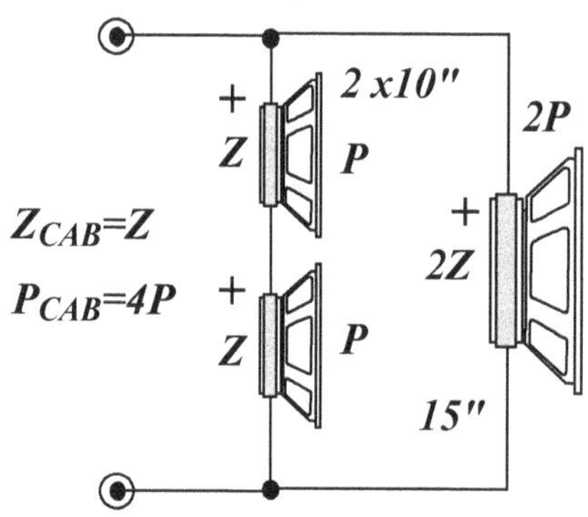

OPTION 2:

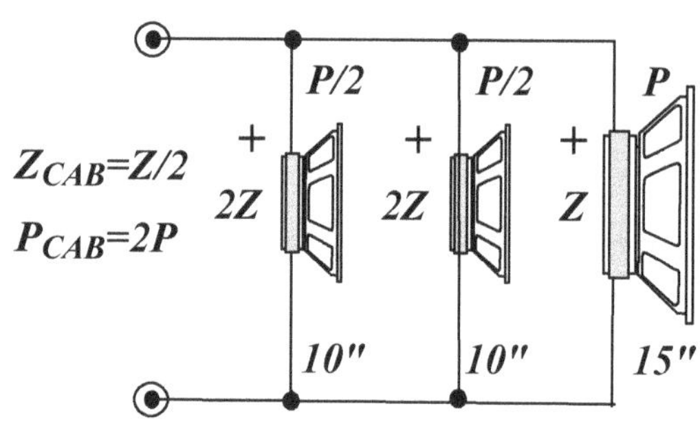

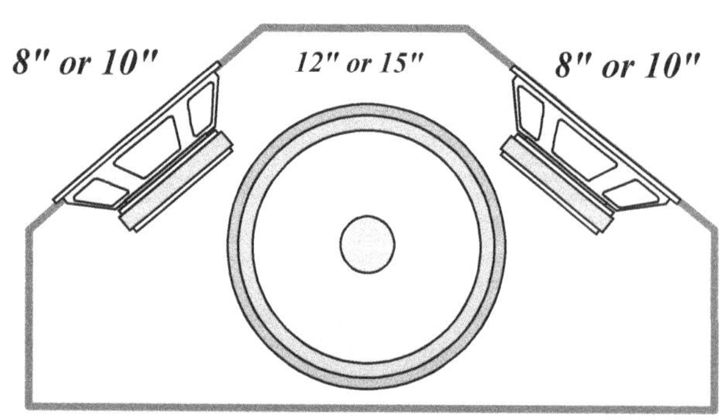

ABOVE: The Pulsator (top or bottom view)

DOUBLE TROUBLE - THE DIPOLE AND BIPOLAR ENCLOSURE

How to turn an ugly but sturdy bedside table into a respectable speaker box

These weird-looking bedside tables were drastically reduced so we got a pair for next-to-nothing. The construction is solid, thick veneered MDF panels and plinth (1).

As luck would have it, the pet-entry opening at the front door (2) was the perfect size for a 10" speaker!

The outside dimensions are 55cm (high) x 45cm (wide) x 40cm (deep). The interior dimensions are of more interest, as always, 43cm x 42cm x 38cm. The enclosure weighs 10.6 kg, always a good sign of rigidity and stability.

The front and back baffles

The back of the bedside table (3) was thin and flimsy and had to be discarded and replaced by a snug-fitting plywood baffle, which will become the front of the speaker cab, since it features a larger driver. We fitted one with a 15" driver, the other with a 12". The front of the bedside table with the existing cutout (2) became the back baffle of the cab (4), both cabs fitted with 10" speakers.

Actually, in this enclosure there is no "front" and "back". You can use it either way, the 15" driver firing to the front and the ten incher firing to the back, or the other way around; it all depends on what type of sound and projection you prefer in that particular room or venue!

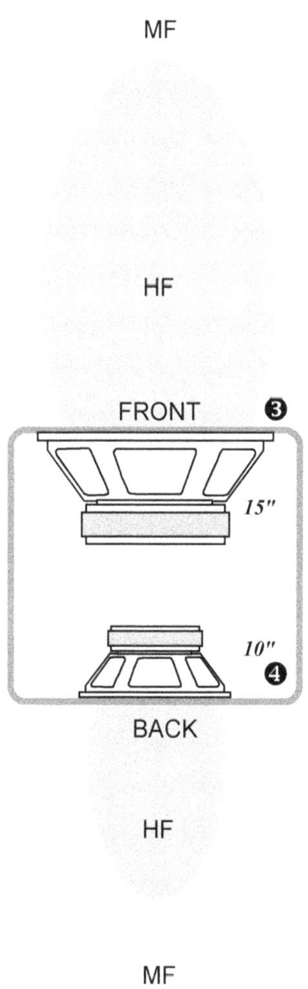

The electrics

One of the most often asked questions is "Can drivers of different diameters, power ratings, and impedances be used in the same cab?" The answer is yes they can!

However, it's a "qualified" yes, meaning there are some warnings and caveats. Let's look at the two drivers we used in one of the two cabs, and perform some basic "back of the napkin" calculations to see how the two would work together and what those warnings are in this particular case.

First, let's look at the two drivers, 16Ω, 30 watts and 8Ω, 50 watts. When paralleled, imagine we have a 100 watt head driving this cab, and we are cranking its master volume up ... which driver will reach its power (thermal) limitation first?

There are few possible ways to solve this problem, but since this is an educational book let's do both. Asking the right question is half of the problem solved, and in this case that question is "What is the maximum current allowed for each of these drivers?"

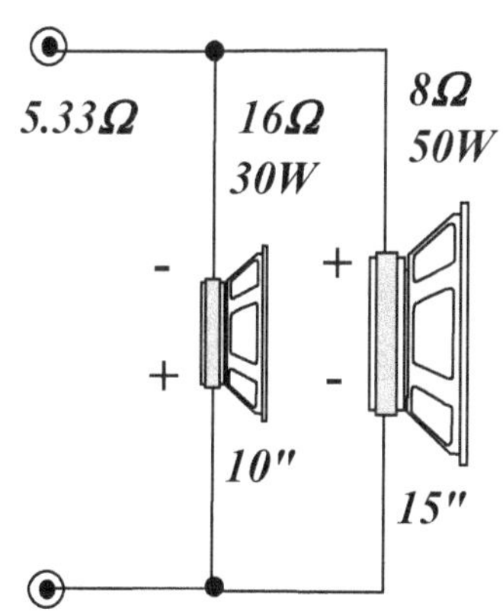

RIGHT: Double Trouble's electrical diagram using dissimilar speakers, 8Ω, 15" and 16Ω, 10".

The 5.33Ω combined nominal impedance can be used with both 4Ω and 8Ω amplifier outputs!

LEFT: The simplified depiction of the mid- and high-frequency (HF) dispersion of such a dipole cab.

Due to their long wavelengths, the LF (low frequencies) from both drivers merge into one 360° sphere.

The 15/10 (drivers' diameters in inches) version of Double Trouble (ABOVE LEFT), and the 12/10 version (ABOVE RIGHT)

The power dissipated on the impedance (resistance) R is $P=I^2R$, so $I=\sqrt{(P/R)}$. For 10" driver, $I = \sqrt{(30/16)} = \sqrt{1.875} = 1.37$ A (Amperes). For 15" driver, $I = \sqrt{(50/8)} = \sqrt{6.25} = 2.5$A.

Assuming the nominal (declared) impedances (8 and 16 ohms) are accurate, which current limit will be reached first? If the current through 10" driver is "I", the current through 15" driver will be twice as high or 2I, since it 8 ohm impedance is half that of driver A (16 ohms).

Because the same signal voltage is fed to both paralleled drivers, the current through them distributes (divides) according to their respective impedances.

So, assume the 10" driver reaching its maximum current I=1.37 A. The current through the 15" driver would then be twice as high, 2.74A, but its maximum current is only 2.5 A.

Thus, the limiting driver is not the lower-powered 10" driver but the higher powered 15" driver! Its impedance is twice as low as the impedance of driver A, but its power rating isn't twice as high: 50 is less than twice 30 watts, or 60 watts.

Finally, what is the maximum power of the two drivers in this cab? Since the 15" driver is the limitation, the maximum current through it is 2.5 A, meaning half of it (1.25A) will flow through the 10" driver.

The maximum cab power is $P = P_A + P_B = I_A^2R_A + I_B^2R_B = 1.25^2*16 + 2.5^2*8 = 25$ watts + 50 watts = 75 watts.

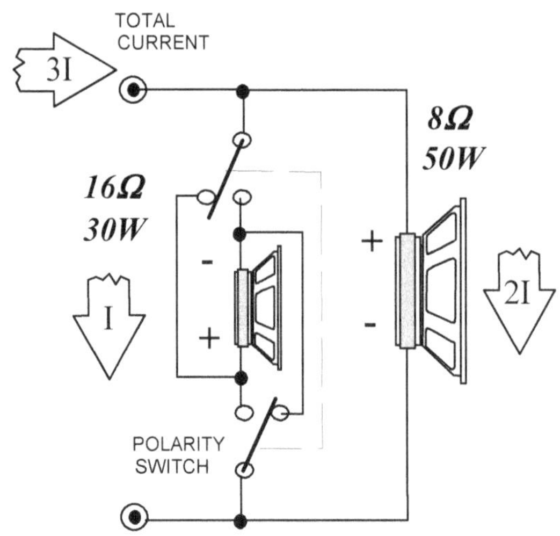

ABOVE: The current distribution in the speaker cabinet and the way to use a switch to make both configurations available to a guitar player.

The double-pole-double-throw (DPDT) switch is shown in the "IN PHASE" position, where the positives of the two drivers are connected, as are the negative terminals.

This is the bipolar arrangement; in the dipole arrangement the two drivers are out of phase.

The measurements: "Double trouble" 15/10 cab with drivers in- and out-of-phase

The two measurements with drivers in- and out-of-phase (electrically) were performed under the same conditions (the same room, the same distance between the SPL meter and the cab, the same signal level, etc.). We aren't concerned about the absolute values of the impedance or SPL (sound pressure levels), but the differences between the two configurations.

The impedance curves of the two cabs are almost identical above 200 Hz, the in-phase cab's impedance marginally higher (about 0.5 ohms on average). The in-phase arrangement shows only a single resonant peak, around 120 Hz, while the out-of-phase cab shows two peaks, just like bass-reflex enclosures, the lower one around 80 Hz and the higher peak at around 120 Hz.

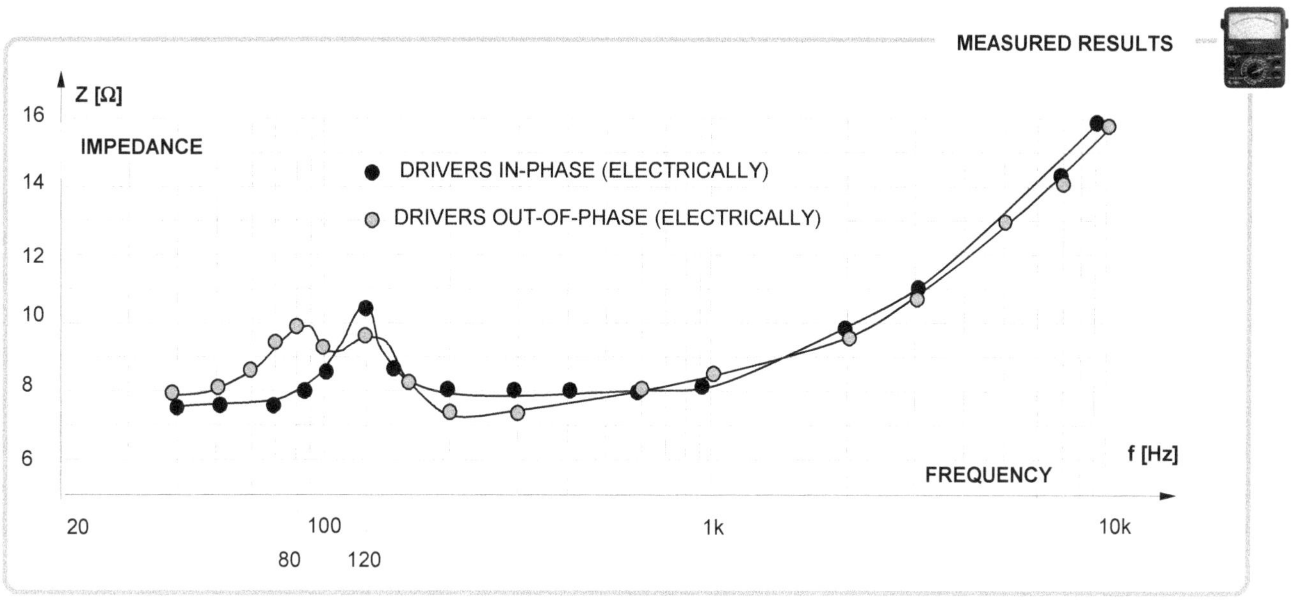

The SPL curves couldn't be more different. The in-phase version shows peaks around 175 (1), 350 Hz (2), 600 Hz (3), 1,050 Hz (4) and 3,000 Hz (5). The SPL drops rapidly above 7 kHz (6), so that at 10 kHz it is down 13 dB (from 75 to 62 dB.

The out-of-phase version goes much deeper; from a peak at 105 Hz, it drops to 60 dB at 40 Hz. The upper bass frequencies between 120 and 400 Hz are much more attenuated compared to the in-phase version, with a significant dip around 300 Hz (7). It also peaks around 300 Hz (3), but then it dips at 900 Hz (8), before sharing another peak with the in-phase option at 3 kHz (5).

Another sharp dip at around 6 kHz (9) is followed by a peak at 7 kHz. After that, there's a sharp drop, just as with the in-phase version, so at 10 kHz, the SPL is down a whopping 19 dB, a drop from 73 dB at 7kHz to 54 dB.

The two versions sound very different, the out-of-phase version with a more prominent lower but subdued upper bass, thus being more suitable for a bass guitar.

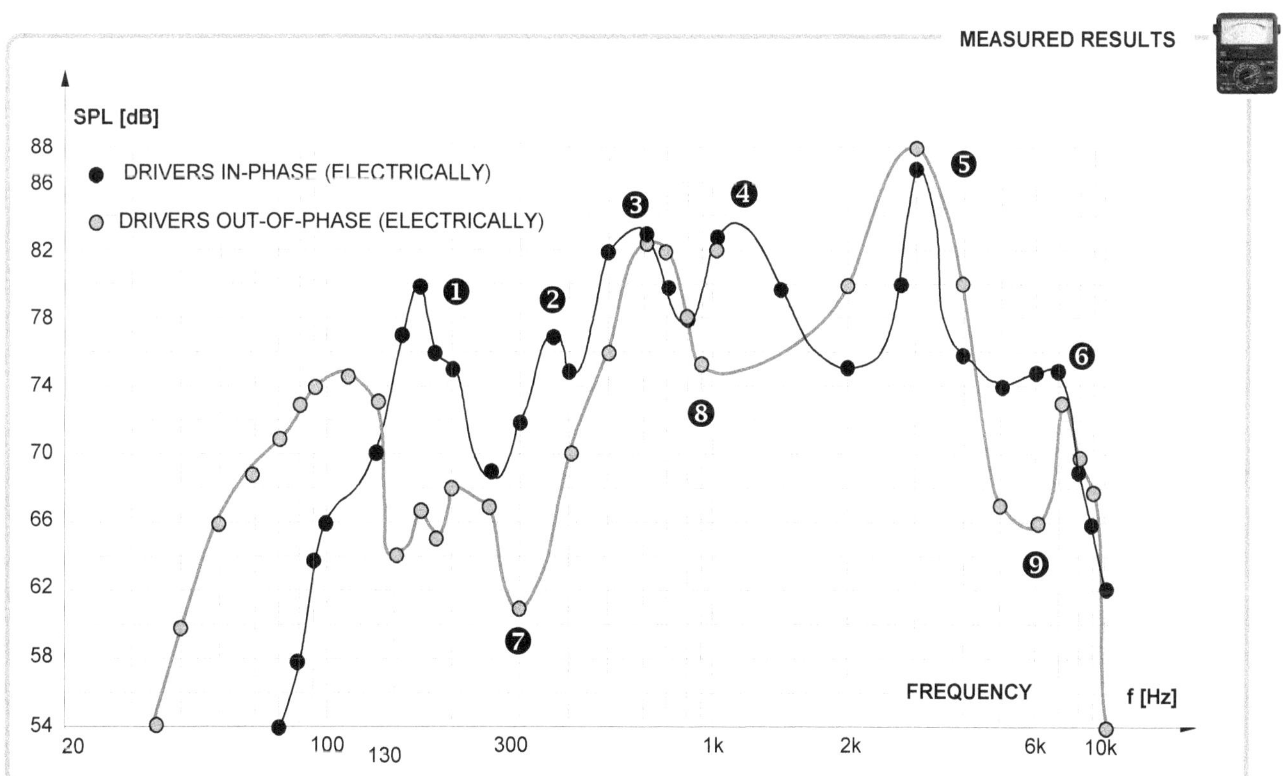

DISSIMILAR DRIVERS WITH A SLOT OPENING AND DISTRIBUTED PORT LOADING

The Ikea cab

These frames (used either as wall shelves or as one of their wardrobe models) from Ikea's Stuva range came in two sizes (depths), 60x64x30 cm and 60x64x50 cm. We bought two 30cm deep frames on a clearance sale a few years back and wondered what to do with them. Then, a crazy idea was hatched ...

We all know that MDF, solid pine and plywood enclosures make great guitar cabs, but how would a frame made of twin-walled, cellulose-filled Ikea panels sound? There was only one way to find out ...

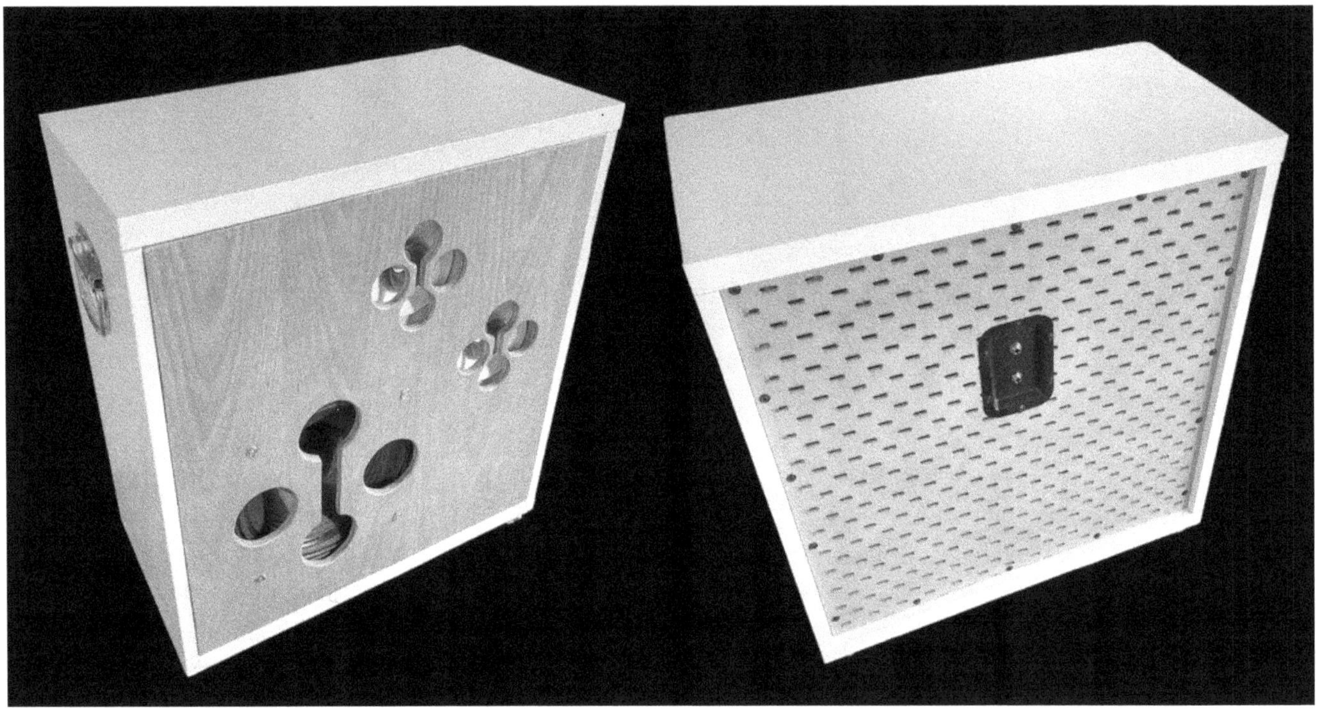

The frame was wobbly, held together by eight "twist & turn" metal pins used in modular furniture, but once the four joints were glued and the front baffle added (12mm plywood), it become quite rigid.

For the slotted back cover we used Ikea Skadis pegboard in "wood color" (also comes in white), made of fiberboard with clear acrylic lacquer finish. The idea was for the back to act as a "distributed port" baffle.

Its 56cm width was precisely the internal width of the Stuva frame, so only one cut was needed to shorten it from its original 76cm length to fit the internal height of the Stuva frame.

The drivers used

The frame was large enough to accommodate two 12" drivers, but we had a couple of 6"drivers salvaged from old Sanyo hi-fi speakers (8 ohms 15 watts each), so we paired them up (in series, to yield 16 ohms) with a vintage 12" Aussie-made Etone driver.

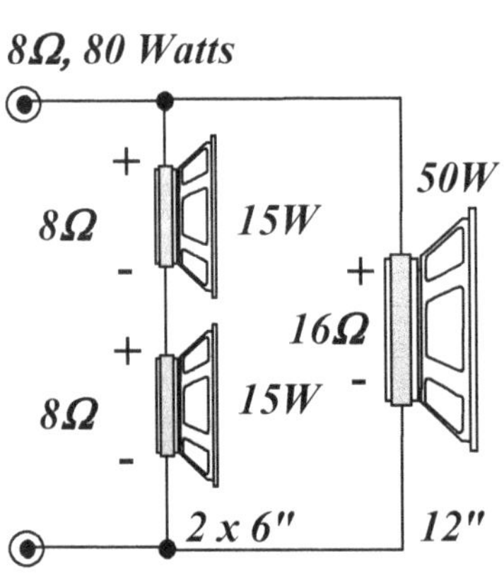

RIGHT: The internal view of the finished cab

LEFT: The electrical interconnection diagram

The front baffle

In this projects we used a cool slot diffuser idea from the mid-20th century, named after Kolster Brandes Ltd. and popularized by Gilbert A. Briggs in his books and magazine articles on loudspeakers.

Using a slot in front of a driver cone was shown to greatly reduce high frequency beaming and thus improve speaker cabinet's horizontal dispersion.

The diffuser lowered the on-axis sound pressure levels at higher frequencies and thus greatly reduced the difference between on- and off-axis response, making them more uniform and equally distributed.

As an additional benefit, the diffuser tamed the bright, treble-dominating character of some cabs, which was especially objectionable when standing right in front of them (on-axis).

The modified version, with four additional symmetrically - placed circular cutouts, avoids resonance effects and the restriction of the output at bass frequencies by allowing more air (otherwise trapped) at those long wavelengths through. For smaller diameter driver, such as our 6" speakers, slots should be around 3/4" wide (about 2 cm). For use with 8", 10" and 12" drivers increase the width to around 1" (2.5 cm).

The diameter of the four circular cutouts should be around 3-3.5"(76-84 mm) for a 12" driver; for smaller drivers the diameter should be proportionally decreased.

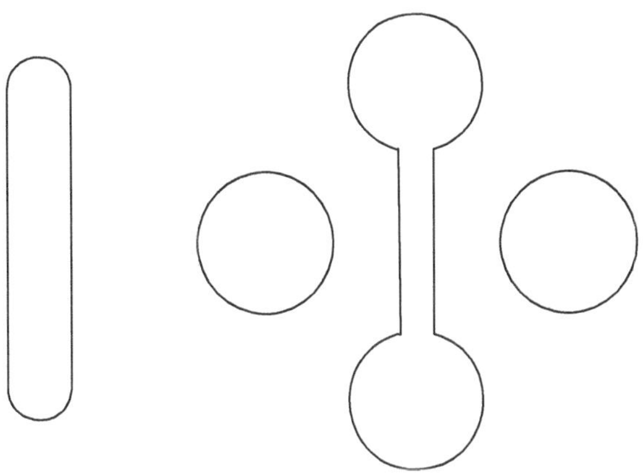

ABOVE LEFT: The KB (Kolster Brandes) slot diffuser and its modified version that avoids resonance and the restriction of the output at bass frequencies by allowing more air at those long wavelengths through. The large circle is there for reference only, that is the usual cutout, so you can measure it and scale it to suit a driver of any size (diameter)

ABOVE RIGHT: The KB diffuser implemented on the two smaller 6" drivers in the Ikea cab

RIGHT: The JBL EON700 series of powered PA speakers uses a similar slot in front of their 10", 12" and 15" woofers.

DRIVER SIZE		A	B	C
8"	(20.3cm)	18	12	7.2
10"	(25.4cm)	23	15.2	9.2
12"	(30.5cm)	28	18.5	11.2
15"	(38.1cm)	35	23	14

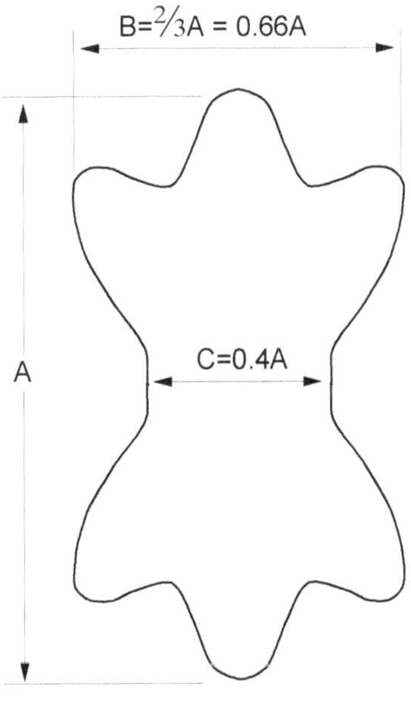

The measurements

The impedance curve, while conventional in the treble region (the usual rise in impedance due to drivers' inductance), is unusually flat in the bass region. There are no resonant peaks at all, except a small rise around 115 Hz (1). The overall impedance is higher than the nominal 8 ohms, the lowest it dips is 7.5Ω at 300 Hz (2). I would declare this cab to be a 9 or 10Ω cab.

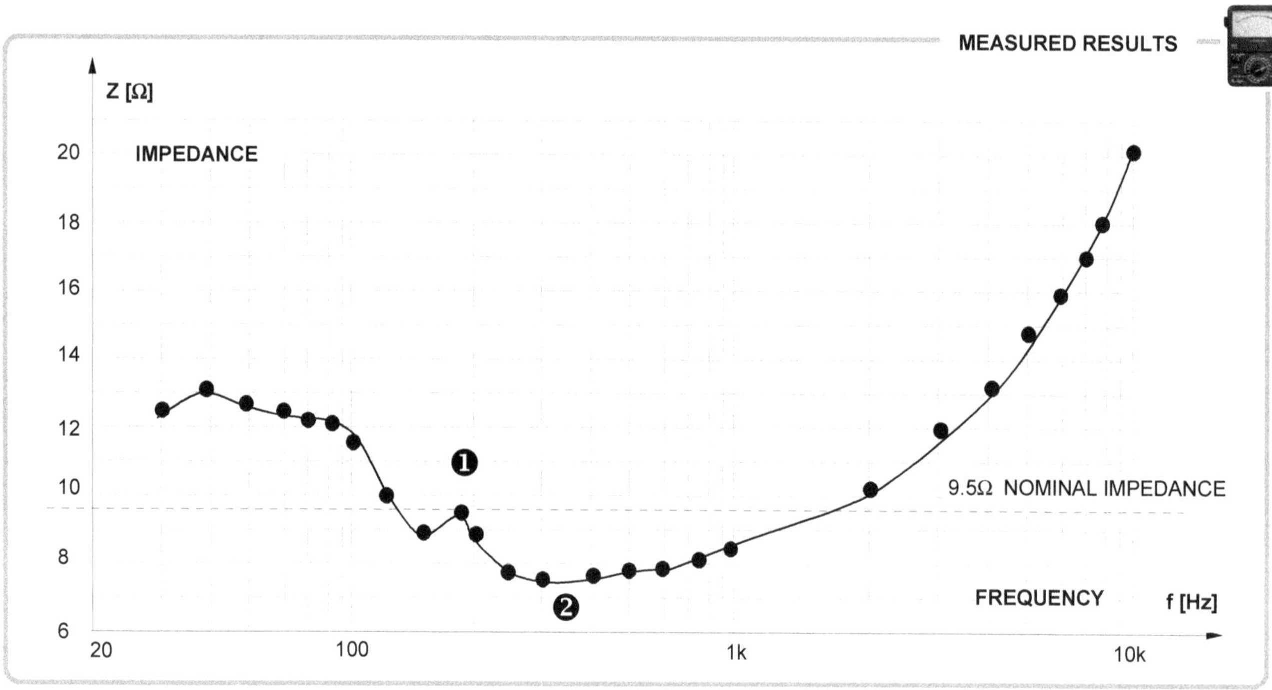

If we choose a referent 0dB in the usual fashion (so the areas above and below it are approximately equal), the 76 dB level seems as good or better than any. In that case, the -3dB frequencies are at 80 Hz and 15 kHz, making this design a clear winner in terms of the high frequency extension. Interestingly, the steep HF drop does not continue past 16 kHz, the SPL level rises again for a minor peak at 18 kHz and then drops to 59 dB at 20 kHz.

After the internal padding was added, a few changes were noticeable. Two adjacent peaks at 120 and 240 Hz (3) and the minor one at 145 Hz (4) were flattened, as were the two peaks at 5.5 and 8 kHz (5). Two major dips (6) and (7) were eliminated, but a new one appeared at around 2.5 kHz.

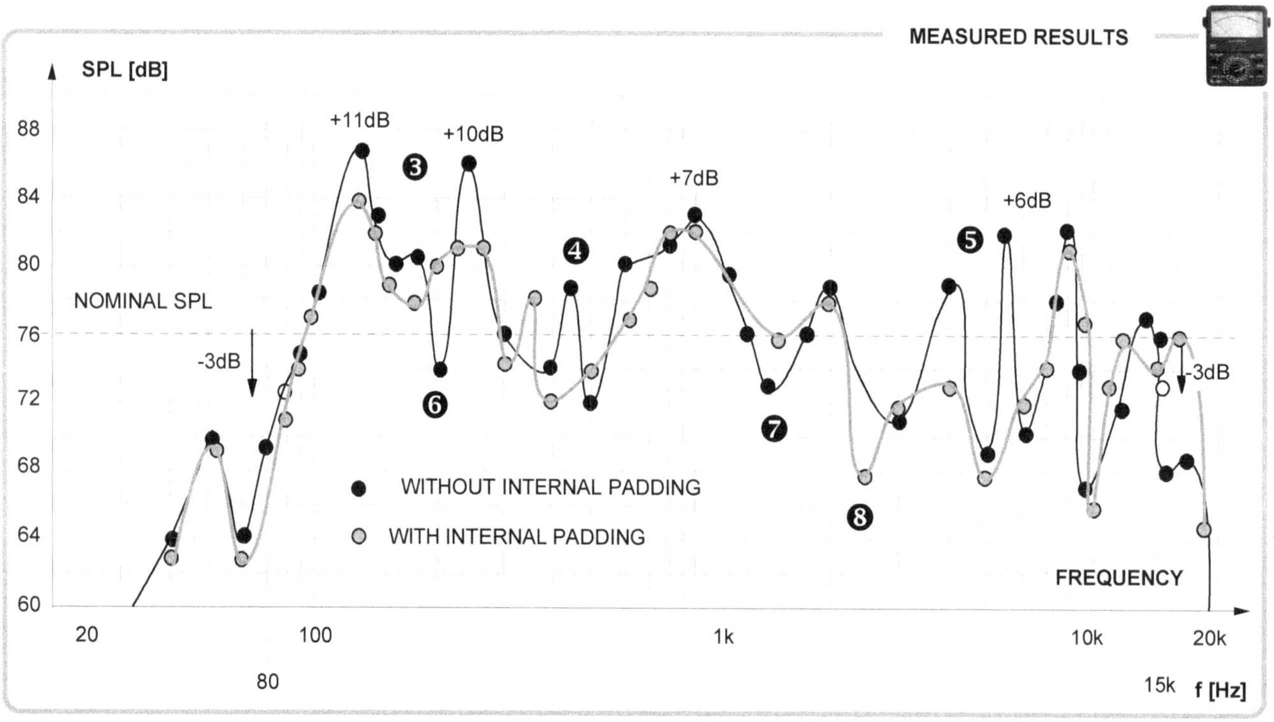

THE WOODWORKING CORNER: CABINET CONSTRUCTION TIPS & TRICKS

9

Not happy with relatively cheap, no-personality black boxes? Or, refuse to pay obscene prices for "boutique" speaker cabs, which, let's be honest, may look good, but sonically may not be any better than the B&B (Bland & Boring) ones?

Get your trusty jigsaw out, make friends with the lovelies at your local hardware and timber supply store, and build your speak cab!

I've been a father, a husband, an engineer, a manager, a martial artist, a seminar presenter, a book writer, an audio expert, an amplifier designer & builder, and many other things in my relatively long life, but a cabinet maker or woodworker isn't on that list.

But, if I could do it, so can you, which means you have no excuses ... just do it!

- TYPES OF TIMBER USED IN SPEAKER CABINETS
- JOINING THE SIDES OF THE CABINET'S FRAME
- THE FINISHING TOUCHES: HOW TO STAIN OR VARNISH A CAB
- HOW TO COVER A SPEAKER CAB WITH TOLEX
- CABINET BUILDING TOOLS
- WIRING YOUR SPEAKER CABS

TYPES OF TIMBER USED IN SPEAKER CABINETS

Plywood

Plywood sheets are made by slicing wood from a tree trunk in thin layers or "plies" and gluing them together in alternating directions. Because wood layers are glued in cross directions, plywood is more stable than solid wood and rarely warps or buckles. The number of plies or layers varies from three to 19, but is always an odd number to inhibit warping.

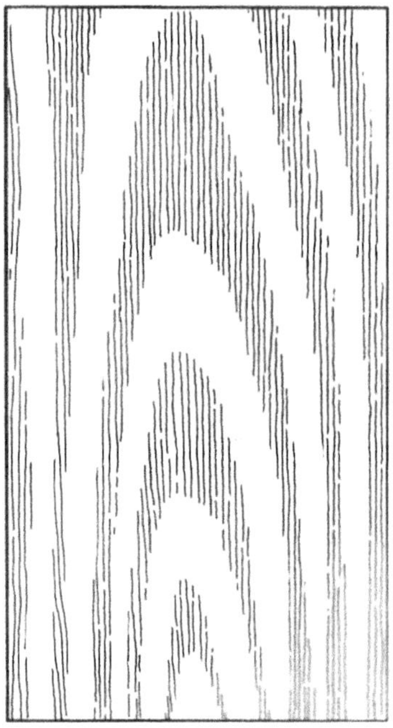

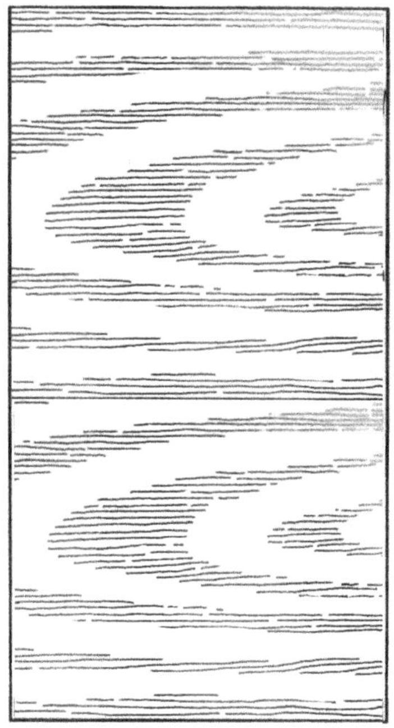

ABOVE LEFT: Standard plywood sheet (wood grain along the sheet's longer dimension)

ABOVE RIGHT: Counterfront sheet (wood grain across the sheet)

ABOVE: A close-up view of 9-ply plywood sheets stacked on top of each other. The color of the alternate layers differs since their grain runs in perpendicular directions.

Two common sheet sizes are 610 x 1,220 mm (0.61 x 1.22 meters) and 1,220 x 2,440 mm (4x8 feet), with thickness in the 6-30 mm range. 9, 12, 15, 18, and 24 mm thick sheets are most commonly used in smaller speaker cabs.

Letters A, B, C or D indicate the grade. A two-letter system is used, the first for the face, the second for the back. A-grade is the highest quality, flawless and smoothest, B has minor flaws and is less smooth, but generally fine for speaker cabinets that will not be covered in Tolex.

Grade C has more visible flaws and knots up to 1.5 inches in diameter, and grade D can have large flaws and knots up to 2.5 inches in diameter. The cost obviously increases from Grade D to Grade A, with AA grade the most expensive.

Marine-grade plywood is lightweight and made with high-strength waterproof glue. Since it is generally much more expensive; unless you can buy it cheaply on sale, it isn't necessary for indoor speaker cabs.

Particle Board

Particle Board (a.k.a. particleboard, chipboard, and low-density fiberboard) is a low density board made of large wood particles in the core and finer particles on sanded surfaces, which are combined with a synthetic resin or other suitable binder and pressed.

Standard particleboard is cheaper, denser and more uniform than plywood. Moisture resistant particleboard sheets are also available.

In a single layer particle board chips of the same size are used throughout, giving it a consistent density. A three-layer particle board has higher density outer layers and a lower density interior. A density of a graded density particle board varies gradually throughout its thickness.

Timber veneered 18, 25 or 33 mm thick particle board panels are another option. American Cherry, American White Oak, European Beech, Jarrah (Australian dark hardwood), Tasmanian Oak, Radiata Pine, Tasmanian Blackwood, and Rock Maple are just some of the veneers commonly available in Australia.

Prices vary significantly depending on the veneer. For instance, 18 mm sheets range from AU$125 for Radiata pine, to AU$300 for a jarrah-veneered 2400 x 1200 mm sheet.

As for the thickness, Tasmanian Oak sheets (2400 x 1200 mm), for example, range in price from AU$130 (18mm) and AU$185 (25mm), to AU$215 for 33 mm thick sheets.

In comparison, a standard particle board sheet (no veneer) of the same size is priced at AU$48. An MDF board of the same size is slightly more expensive, AU$52.

Medium Density Fiberboard (MDF) and High Density Fiberboard (HDF)

Medium Density Fiberboard (MDF) is an engineered wood product with a smooth surface and uniform density core. MDF is made by breaking down hardwood or softwood residuals into wood fibers, combining it with wax and a resin binder and forming panels by applying high temperature and pressure.

MDF can be painted, cut, routered and cleanly drilled without splintering or chipping. MDF sheets can be easily cut with a saw, jigsaw or router, making them ideal for speaker enclosure frames (and front & rear baffles too).

HDF fibers are much smaller than medium-density fiberboard (MDF), making it less prone to warping, cracking, and chipping. HDF is more water-resistant than MDF, and its density does also make it stronger. However, it is much more expensive and only comes as 3 - 8 mm thick sheets. You may have seen HDF boards since they are often used as backing panels for furniture.

Some HDF boards use resins that contain formaldehyde and other volatile organic compounds, so the risk of harmful out-gassing is present.

While an 8mm frame isn't viable, you may use HDF sheets for the back cover, and in resonant enclosures even for their front baffles. Both MDF and HDF are denser and more stable than particleboard.

Solid dressed timber

Dressed timber is solid timber that has been planed and smoothed to a consistent finish. It is used for flooring, paneling and furniture- or cabinet-making. The three classes are D1S (Dressed 1 Side), D1S, D1E (Dressed 1 Side and 1 Edge), and DAR (Dressed All Round).

Due to its low cost (compared to more exotic timbers and hardwoods) the most commonly used soft timber in guitar speaker cabs is pine. It is also lightweight, so pine cabs aren't prohibitively heavy.

Solid wood is a natural product so the density of such boards will vary, meaning even identical speaker cabs will sound different. Apart from this consistency issue, solid timber is hygroscopic (it attracts moisture), so the sonics of such cabs will vary with humidity and temperature.

Even if you decide to use solid timber for the four sides of your speaker cab, the front baffle and the rear cover should still be made of plywood, MDF or particle board. Using wide planks of solid timber such as pine is the easiest and fastest way to build cabinets. As a soft wood, pine is easy to work with, but it warps and splits, and the wider the board the more likely they are to happen.

The use of standard 235 mm or 285 mm wide timber boards for your cab's frame will minimize cutting - only four 235 or 285 mm long cuts are needed to make a frame.

235 mm wide x 19 mm thick x 2.4 m long premium grade dressed Radiata pine boards cost AU$44.50 (AU$78.90/m^2), while 285 mm wide boards of the same thickness and length retail for AU$55.65. Shorter boards (235 mm x 19 mm x 1.8 m) cost AU$33.20, while 285 mm wide boards of the same thickness and length cost AU$43.90 (AU$85.68/m^2).

Glued strip boards instead of solid timber boards

18mm "Finger Jointed Beech Panels" (1) are generally used for indoor applications such as shelving and bench tops. Three sizes are available: 1200x405mm sell for AU$49, 1200x600 cost AU$59, and 1800x405 are priced at AU$89. However, at AU$122/m^2 they are around 50% more expensive than the solid dressed pine planks.

The strength of the glued-up board is in no way compromised; in fact, glued joints are stronger than the timber itself. Breaking up one solid board into five to ten narrow ones could also be beneficial acoustically. The unwanted vibration of the cabinet propagate through the solid plank unimpeded.

Numerous interfaces filled with glue, although of negligible thickness, still break up the propagation path and impede such vibrations.

Another interesting product is coreboard or blockboard. The core is made by edge jointing and gluing together strips of softwood (2) such as poplar, spruce and pine.

The heart sides are alternated to minimize twist or warp. The core is then dressed and cross-banded. After drying, face veneers are glued on (3), with the grain in the same direction as the core strips.

Although it is stiffer and lighter than plywood, blockboard does not split or splinter while cutting because of the presence of softwood in its core. Sheets are available in a range of different thicknesses, from 12 to 50 mm. The standard size of blockboard is 2440 x 1220 x 30 mm.

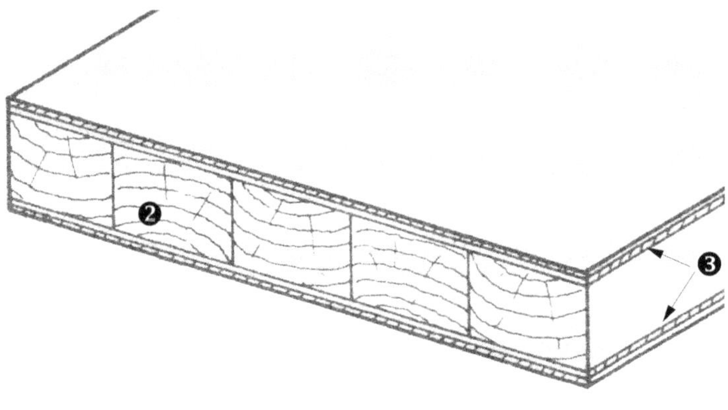

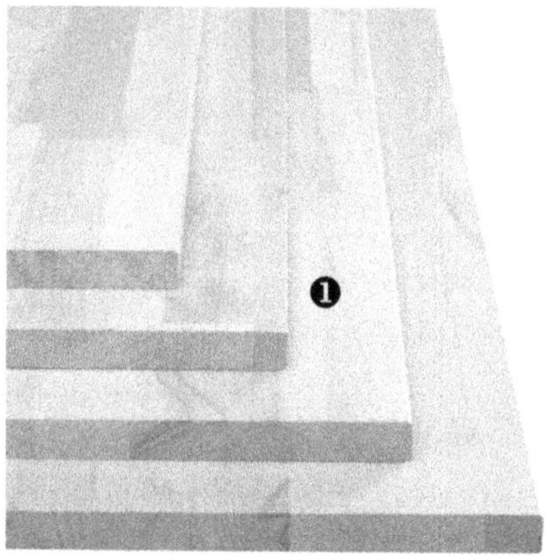

JOINING THE SIDES OF THE CABINET'S FRAME

The furniture joint

Cheap modern "furniture" is usually made a low density chipboard, covered in vinyl wrap. Better quality furniture has an MDF or plywood frame, which is then veneered using real wood veneer and glued to a solid timber edge (a "round off").

Biscuits and/or rebated slots (as illustrated) may be used to add further strength.

Due to its complexity (professional woodworking skills and tools are required), cost, and the fact that a thin veneer layer can easily be damaged in transport, this method of construction is never used in musical instrument cabs.

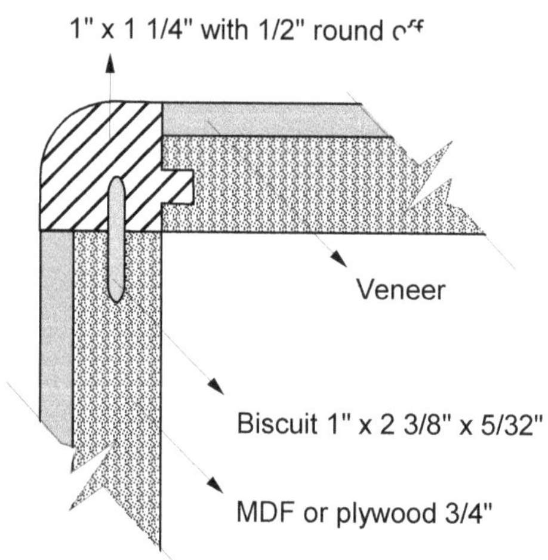

The simple joints even you and me can make

The easiest way to join frame sides is a butt joint. Glue the required surfaces and clamp them together until the glue dries. Alternatively, for an even stronger frame, use glue blocks (battens), and/or the screws.

Method C) isn't suitable for plywood, since there's a great chance of the screw splitting the board (5). Method D) is superior for another reason, the screw heads cannot be seen from the outside, and the screws are not in the way, so edges can be beveled or mitered with a router.

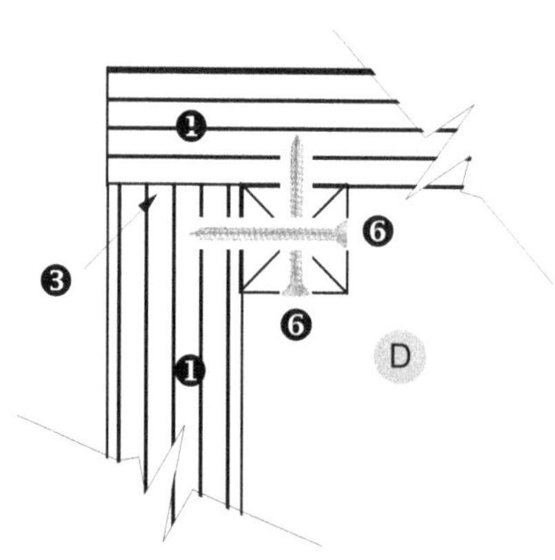

A) Rabbet (or "rebate") joint method of making a plywood speaker box (frame)

B) Speaker plywood box with beveled (miter) joints

C) Butt-joined MDF or particle board box with external screws

D) Butt-joined MDF, particle board or plywood box with internal screws (previous page)

1) 19 mm plywood sheets 2) Glue block 3) Glued surfaces 4) Caulking (flexible sealing compound) 5) MDF or particle board 6) Wood screw

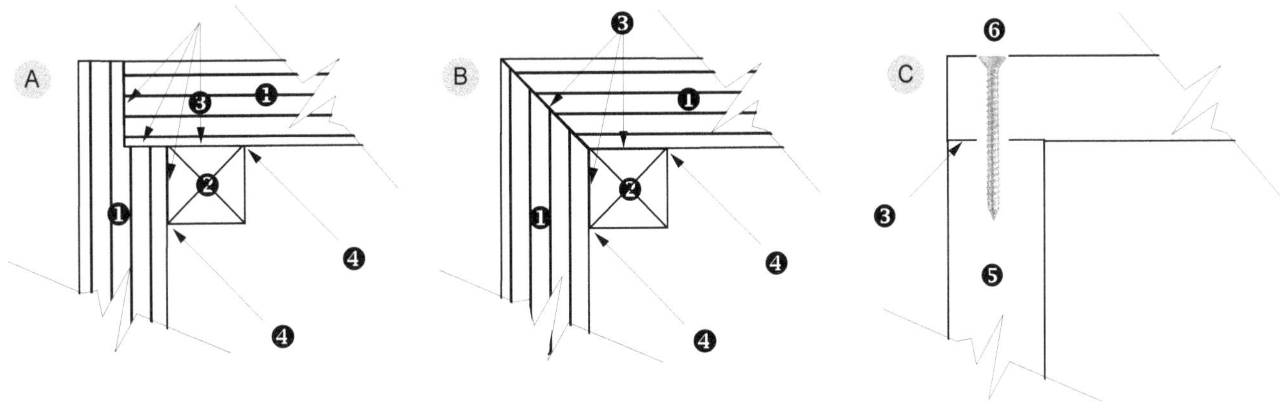

More complex joints for solid timber box frames (sides)

Both mitered and rebated joints require the use of a router, but rebated joints are faster and easier to make.

RIGHT: Two common types of mitered joints, box joint (left) and a dovetail joint (right). Both require the use of a router,

BELOW: Cabinet frame made using rebated joints. Gluing them together is usually strong enough, but nails or screws can be used in addition to glue, especially if the enclosure is to be covered in Tolex.

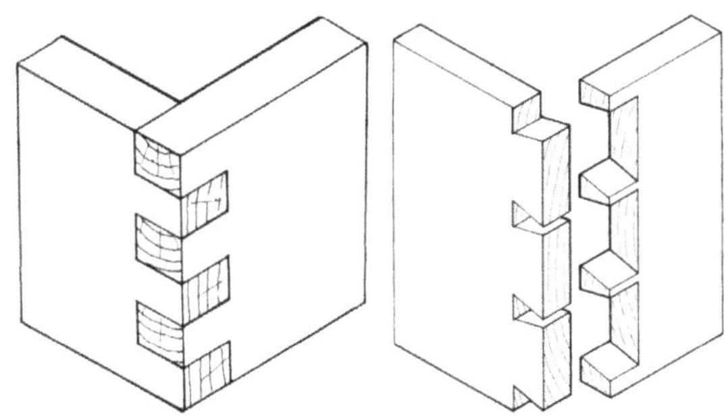

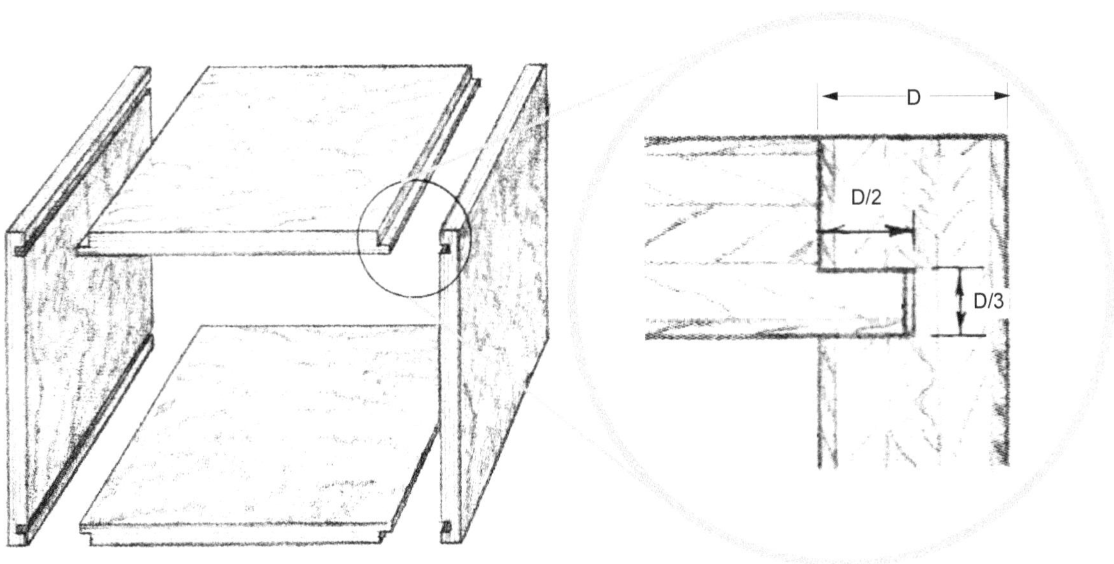

Wood screws

Wood screws are of the self-tappered variety. The spiral thread bites cuts and bites into timber as the screw is turned. Screws are specified by nominal diameter or "gauge", ranging from 0 to 2; the larger the number, the bigger the screw (dimension "D"). For instance, the diameter of a #5 screw is just over 3mm (1/8"), while #14 screw is double that, 6mm or 1/4".

The length is measured from where the material surface is meant to be to the end of the fastener. That means countersunk screws are measured over the whole length (L); all other types are measured from under the head (L-H).

Standard lengths are 6-9-12-16-18-22 mm (under 1 inch), the 1" and longer lengths (25-32-38-44 mm) are more commonly used in making enclosure frames.

TPI is the number of threads per inch, used only with American fasteners. Metric fasteners use TP (Thread Pitch), the distance between threads expressed in millimeters

Self-countersinking screws (6) have tooth-like cutting nibs under the head that help them sink flush with the surface without a need for a countersink bit, saving you time and effort.

In other words, they cut their own recess, even in hardwoods, and thus reduce the possibility of timber splitting.

Round or domed heads are usually chosen for decorative purposes, using either cross recess or slotted drive.

Mushroom (1), button (2) and wafer head screws are slightly domed, larger diameter and lower profile than round heads, usually with slot or cross slot.

Flat heads (3, 4, and 5) are countersunk (cone shaped under the head) with a flat top.

The most commonly used "drives" are the slotted drive, the Phillips drive ad the square drive.

Square drive or Robertson screws have a square slot, greatly reducing the risk of slippage.

There are even combination square and Phillips drive screws, giving the choice of the positive hold of a square drive bit or the convenience and availability of a Phillips bit.

Frearson drive is also x-shaped like the Phillips drive, but the two are not compatible.

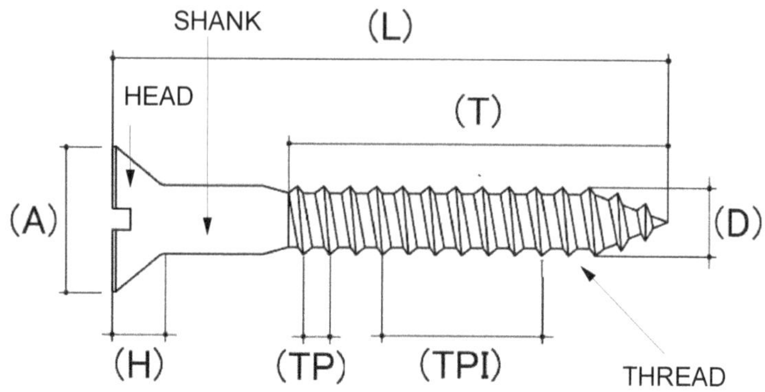

ABOVE: Partially threaded ("T" smaller than "L") flat top wood screw with a slot drive L= length, T= thread length, A=head diameter, D=diamater, H=head height or depth, TPI=threads per inch

BELOW: A few common types of wood screws

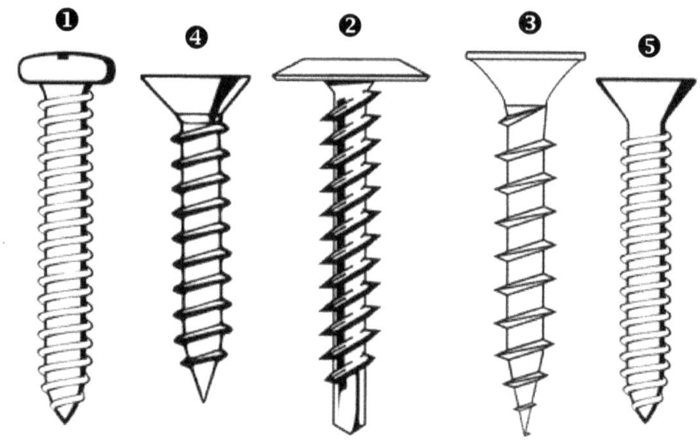

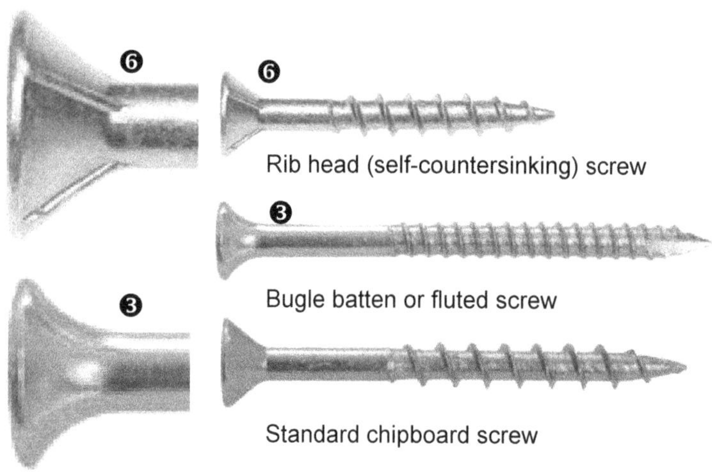

ABOVE: Chipboard screws can be recognized by a flat head, a long shank (unthreaded section, and a large thread pitch

Wood glue

The cheapest and thus most commonly used wood glue is the PVA (PolyVinyl Acetate) type. This pH neutral, inexpensive, white milky liquid dries clear and isn't toxic.

Polyurethane glue is another contender, with only 20-30 minutes of clamp time, and is fully cured in 24 hours.

FURTHER READING

We cannot possibly cover woodworking techniques to any meaningful depth in this loudspeaker manual. Also, as that prominent contemporary philosopher, Inspector Harry Callahan (a.k.a. Dirty Harry) said in one of his best movie scenes ("Magnum Force", 1973), "A man's got to know his limitations!"

To go beyond the relatively narrow limitations of my woodworking skills, I'd recommend two general woodworking books "Woodwork: The Complete Step-by-step Manual" and "Collins Complete Woodworker's Manual".

Both are in full color, with simple and practical explanations, and, best of all, older editions are available at a much lower price.

Routers are very powerful and versatile tools, but a specialist topic we couldn't possibly cover here in any meaningful depth. The best starting points are "Woodworking with the Router" by Hylton, "Router Book: A Complete Guide to the Router and Its Accessories" by Warner, and "Taunton's Complete Illustrated Guide to Routers" by Bird.

THE FINISHING TOUCHES: HOW TO STAIN OR VARNISH A CAB

Sanding paper

The cheapest sanding sheets used crushed glass, which is fairly soft and wears out rapidly. They can be recognized by sand-like color. Stay away from those.

Garnet sheets are far superior, although they may be the same sand color, just to confuse the situation. Others are darker, dark yellow, orange, even red. As a natural mineral, garnet grains fracture during sanding, presenting fresh cutting edges , so you could say that such sanding sheets are "self-sharpening".

However, these broken-off tiny shards of garnet, if left behind, can get stuck into the wood grain and ruin sanding with finer grits.

Another option are sheets using aluminium oxide grain with zinc stearate coating, which don't get clogged so quickly and are thus often marketed as "no clog" sheets.

The hardest and most expensive are sanding sheets with silicon carbide. Usually dark gray or black, the "wet & dry" variety has waterproof paper backing and water is used a lubricant during sanding. there's also a self-lubricating option for finishes that would be harmed by water.

Sanding sheets come in a variety of grades, from very coarse to ultra fine. 40 is a very coarse grit, 60 = course, 80 = medium, 120 & 180 = fine, 240 = very fine, 400 = extra fine, 600 = super fine, and 1200 = ultra fine.

An open coat sandpaper has larger spaces between the abrasive grains, making it more resistant to dust clogging and thus more suitable to resinous softwoods. Closed coat varieties are densely packed with abrasive grains and provide a larger sanding area and faster sanding.

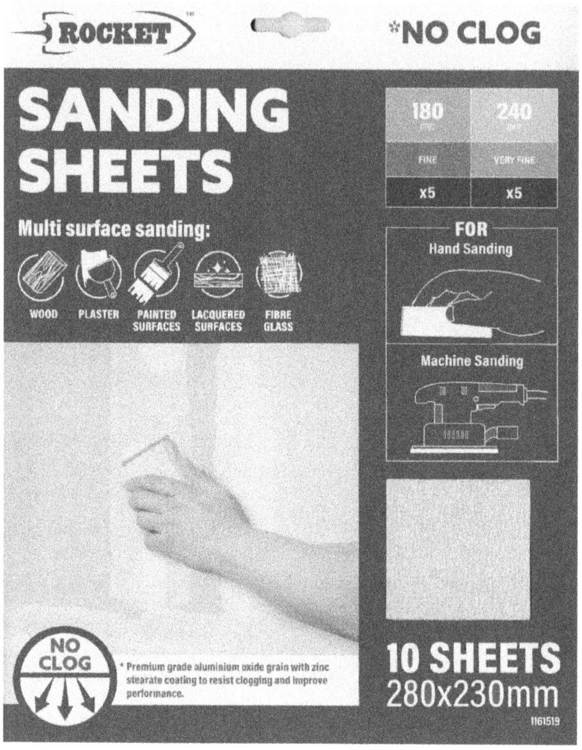

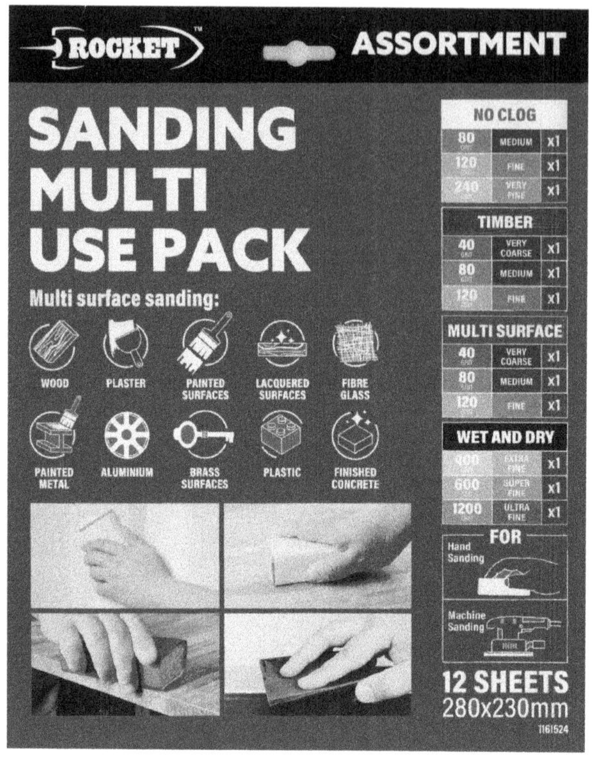

Staining and varnishing your cabinet

Varnish is a clear coating applied to wood to protect it from damage and wear. It also comes as "stain & varnish" coating with various colors added to resemble a darker timber.

A pine cabinet can be stained and varnished in one of a dozen or more finishes such as teak, jarrah (Australian hardwood), cedar, maple, walnut, mahogany, merbau, teak, and others. Varnish is available in gloss, satin and matte (flat) finishes.

Varnish is usually applied with a brush, but a short-pile roller may be used as well. For a smooth and even finish it should be applied with the grain (along the grain).

Oil-based varnish is made from a mixture of oil and resin, which gives it its durability and water-resistant properties. It's thinner and requires more coats than the polyurethane varnish, with very long drying time between coats.

Water-based or acrylic varnish dries quickly and does not yellow like oil-based varnishes. However, it is not as durable or water-resistant, and does not penetrate into the wood or spread as uniformly as other varnishes, lacking the self-leveling qualities of solvent-based varnishes.

A big plus is that you clean the brushes and your hands with water, unlike oil-based ones where mineral turps (turpentine) are used.

Polyurethane varnish also dries quickly and resists yellowing, but is more expensive than the other two. It is tougher, harder wearing and more durable, providing a more rigid finish. Since the coat is thicker, fewer coats are required. Polyurethane varnish can be water or oil-based.

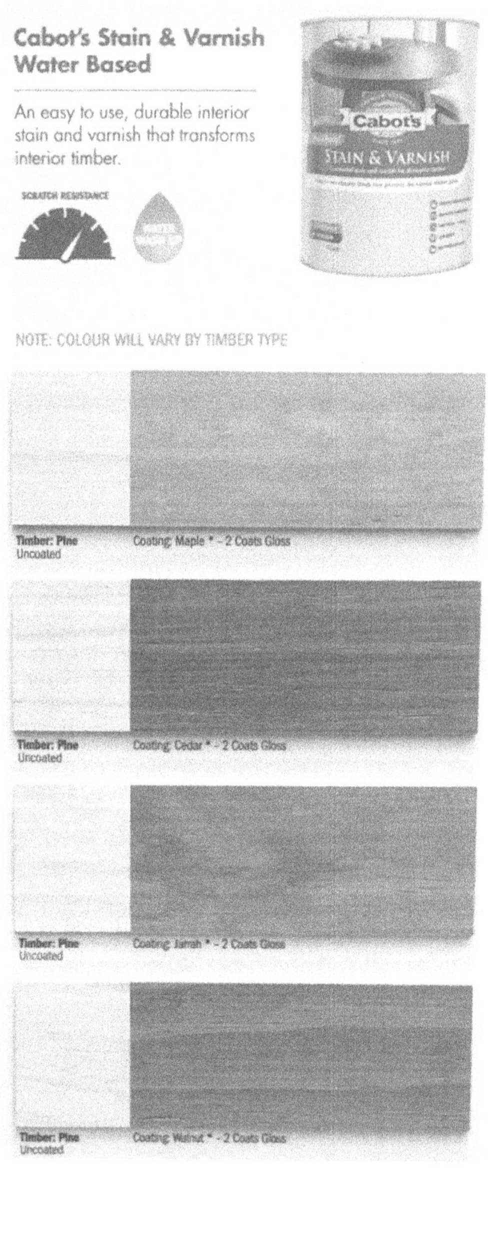

RIGHT: Just some of the dozens of real wood samples on display in our local hardware warehouse. Pine was used as the stained & varnished timber base so a fairly accurate idea of the stained cabinet's finish could be had.

HOW TO COVER A SPEAKER CAB WITH TOLEX

How much Tolex do you need?

Retailers and upholstering shops buy Tolex-type materials in 1.28 meter wide rolls, but sell it to you per lineal meter. Thus, you first task is to go back to your primary school maths class and solve the "How many lineal meters would little Johnny need to cover his frame that is ... wide, tall, deep, thick ..." type of problem.

Assuming you have a finished cab that needs covering, the first step is to add its width and twice its height (W=2H). If the total comes to less than 1,280 mm, you will be able to fit it sideways (as illustrated below). If not, you'd need to orientate your cutout lengthwise ...

Add about 2" (around 50mm) an all sides, that is the material that will cover the thickness of the boards used to make your cab's frame (usually 19 or 24 mm) and the overlap.

The usual way is to cover three sides with one piece (1) and then the bottom (2) with a separate, smaller piece. That way you will have two symmetrical Tolex joints (3) at the bottom, usually positioned just under the cabinet's feet.

As for the backboards, for a closed cab, you will also need to allow enough material to cover the back panel (4). For an open-back cab, the two smaller horizontal or vertical sections (if any), can usually be covered by the leftover pieces of Tolex.

Instead of sideways positioning, as illustrated (across the width of the Tolex roll), if the width W plus twice the height H of your cab (W+2H) exceeds 1,280 mm (the width of the Tolex sheet), you will need to position your pattern lengthwise. That will inevitably lead to more wastage, but if you want the finished cab to have no visible (top and sides) joints, there is no alternative.

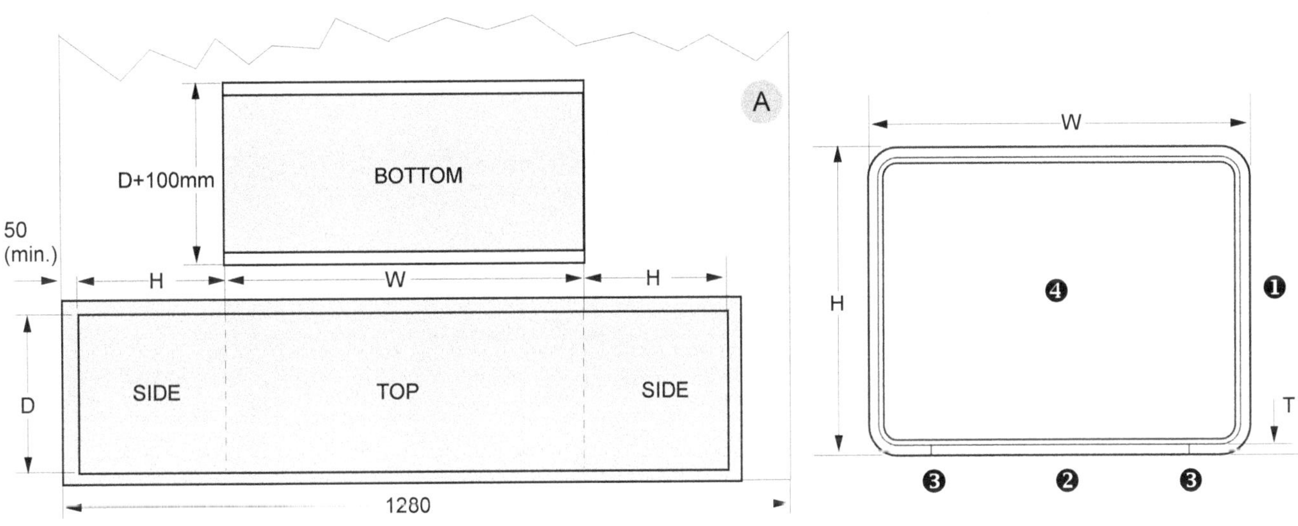

Preparation

Find a clean flat surface, a large table, a timber floor, etc. and lay your Tolex out flat. If creased, leave it in the sun for 10-20 minutes or go over it with a hairdryer (not too close to damage it).

Cut the two pieces as we've just discussed, a large piece 100mm (or more) wider than your cab for the two sides and the top, and a smaller piece for the bottom. Leave a minimum of 50mm (2") on the side and place the bottom edge of your cab on the larger sheet and lay it on its side.

Make sure your cab is sanded smooth and that its edges are rounded. Check that the metal corner protectors will sit flush against the cabinet sides. Wipe all the dust and sawdust off, then wipe it off again, using methylated spirits if necessary. If you are planning to pint the cab internally, now is the time to do so, not after it's covered in Tolex.

You'll need a sharp trimming knife. Box cutters have an 18 mm blade with snap off points. The problem is in their plastic body and flimsy support for the blade.

Stanley knife has a die-cast zinc body and much firmer support for the blade so it won't wobble or break as a easily as box cutter blades do.

Kwikgrip (or similar glues) come in 100mL tubes (AU&10.70), 500mL (AU$23.-), 1L (AU$31.60), and 2L (AU$56) metal cans. Not cheap, so figure out beforehand how much glue you will need.

If you prefer to spray the adhesive on, larger 350g cans are AU$21.90. You'd need quite a few of those for a speaker cab, so I'd forget about that option.

Gluing

Stir the glue until completely mixed. Take a suitably sized brush and evenly apply a thin layer of glue on the larger piece Tolex, but only within the outlines of the cab. Do not (yet) apply the glue to the 50mm+ edges.

Apply glue to the top and sides of the cabinet and to the smaller (bottom) piece of Tolex.

Wait a few minutes until the cab and the two pieces material are touch dry and tacky.

Place one side of the cab down onto the larger sheet, on the marked lines, as per step A (previous page). Roll the cabinet so that the top is now facing down onto Tolex, and finally tumble it over again to the other side.

Flip it over once more so it is now sitting on the bottom. Go over Tolex on the top and sides, rubbing it and removing any bubbles of trapped air.

Flip the cab so its facing bottom up and apply glue to the bottom side of the cab. Once it's touch dry fold the ends of the larger piece from each side over to the bottom of the cabinet.

Place the bottom piece on top, over the folded sides of the main piece.

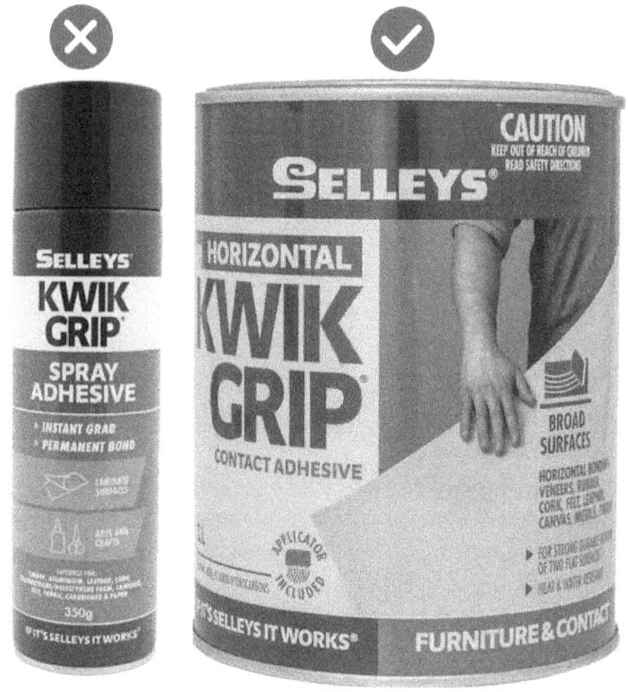

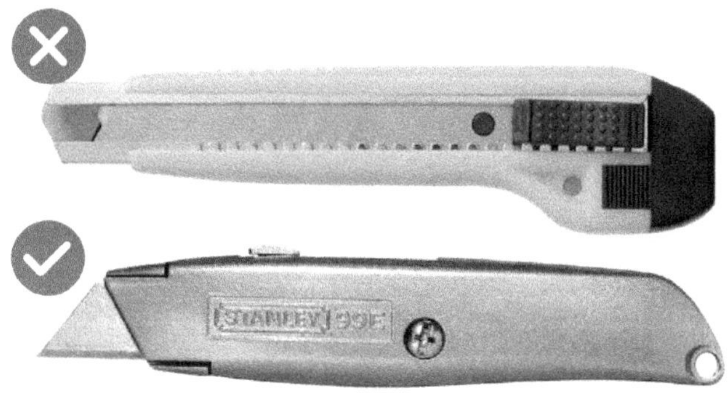

Trimming

Using a straight edge cut across through both Tolex layers (B).

Peel off (C) and discard the strips cut from the top layer (on both sides of the cabinet).

Finally, peel back the top Tolex layer and peel off and discard the strip cut from the bottom layer (that was overlapping under the top layer). Repeat on the other side.

Push the edges of the trimmed bottom piece right against the overlapping edges of the larger piece. Providing your cuts were clean, sharp and straight, it is possible to join them so that the joint is almost undetectable,

Now apply the glue to all the edges, fold them over and carefully cut and trim the corners.

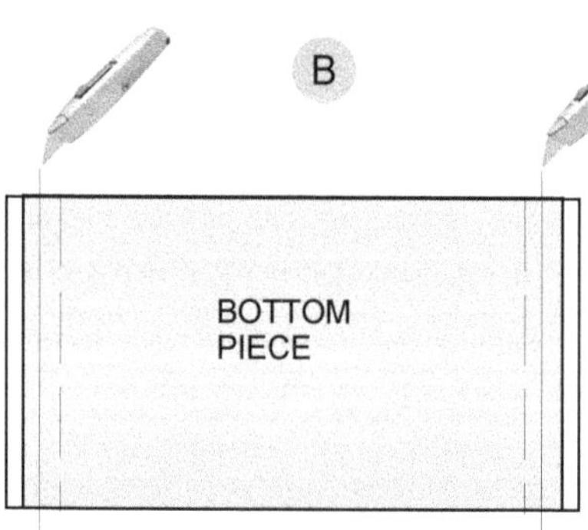

Glue the bottom piece in place. Using a long ruler or a straight edge, cut across through both layers of Tolex.

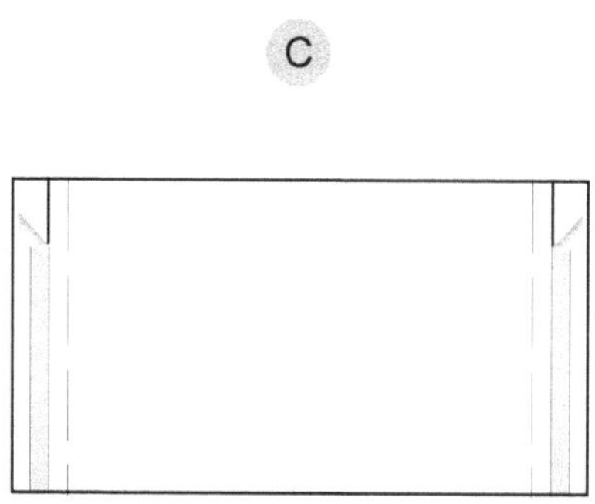

Peel off and discard the strips cut from the top layer (on both sides of the cabinet).

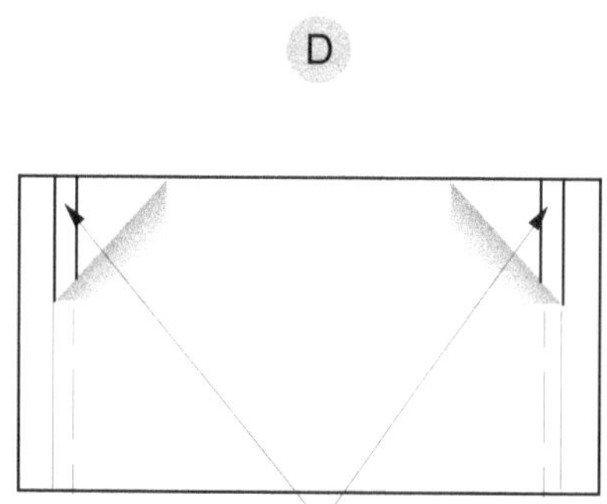

Peel back the top Tolex layer and peel off and discard the strip cut from the bottom layer (that was overlapping under the top layer). Repeat on the other side.

CABINET BUILDING TOOLS

Must have tools

Most musicians and DIY amp and cabinet builders don't have professional woodworking tools. You can make a speaker cab with no other power tools but a jigsaw. However, for multiple cabinets or more intricate designs, such as those with beveled edges, or enclosure frames other than a simple rectangular ones, you should consider investing in more advanced power tools such as a band saw, mitre saw and a router.

While a jigsaw (1) is a must have, a router (2) and routing bits (3) are optional. While all sanding can be done by hand, an electric sander (4) with various attachments (5) will save you a lots of time and effort. Good quality sanding sheets (6) are required in either case.

The two electric tools everyone should have around the house, indispensable in this day and age, are a cordless drill and a cordless screwdriver (7), together with a set of drill bits (8). make sure you buy them as a set of two, or at least from the same manufacturers, so they use the same batteries. Our Makita set included the two tools, a battery charger and three 18V batteries. One died just after the warranty expired, but seven years later the other two batteries are still going strong.

You will also need a bunch of smaller tools and accessories (9), such as a measuring tape, a ruler, scissors, a marking pen, a stapler and a utility knife.

Nice-to-have tools: compound mitre saws and circular saws

Basic 1600 Watt corded compound mitre saws start at AU$99 (about US$65). The blade diameter is 210 mm (8¼"). A larger (blade diameter 305mm or 12") and more powerful (2000 Watts) dual bevel siding compound mitre saw are much more expensive, AU$399 or more.

Typical capabilities are 45° left - 45° right for bevel cuts and 50° left - 50° right for mitre cuts.

Corded circular saws for cutting large panels such as those used in cabinet construction range from AU$69 for a local Ozito brand, AU$119 for AEG, and AU$169 for Makita. They are typically rated at 1200-1300 watts, with 185mm (7-1/4") diameter blades.

Cordless varieties start at AU$349 for skin-only, meaning without any batteries or charges, which can cast you that much again, so going cordless (and why would you if you make your speaker cabs at home?) is a much costlier proposition.

For wood cutting, blades come in 24, 30, 32, 40, 45, 50, 60, 64 and 80 tooth varieties. The motors' power ratings vary from 1200 watts up, typically 1400/1450, 1650/1700, up to 2000 watts.

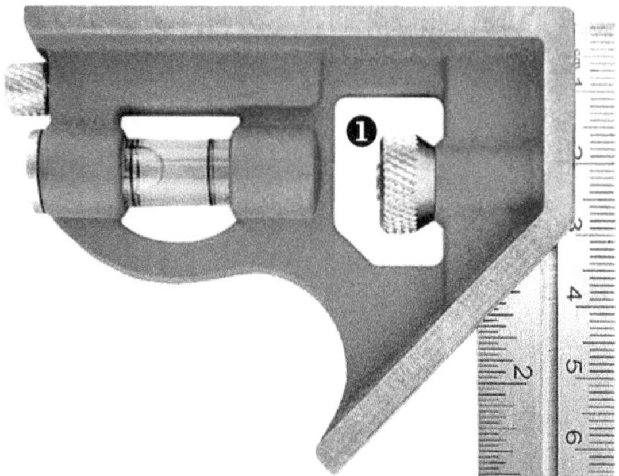

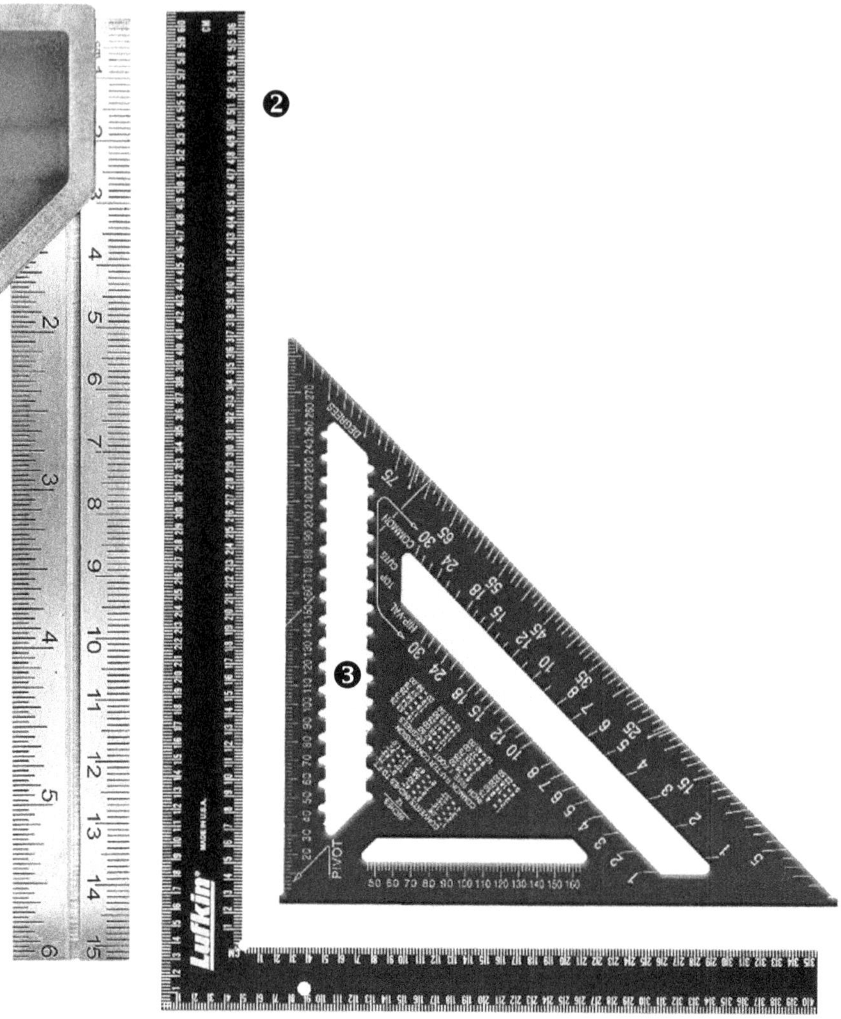

1) 150mm/6" sliding combination square with metric and imperial markings on the stainless steel ruler. It can mark 90° and 45° with an adjustable depth for repetition marking and scribing.

A built-in level bubble is handy for approximate marking, as is a metal scriber contained in the handle.

2) Two-sided 200mm x 300mm metric & imperial (12" x 8") carpenter's square

3) 300 mm metric square is a versatile layout tool that can be used as a rafter square, try square, protractor or power saw guide. Smaller 180mm squares are also available, as are imperial (in inches) squares and dual-marking squares.

Drills, jigsaws, compound mitre saws, circular saws, routers and other power tools are serious hazards and must be treated with respect and always given your full attention. Never operate such tools when tired, intoxicated, ill, absent minded. Remove dangling jewelry, lose clothing or anything that could be caught in the moving parts. Make sure all electric tools are regularly safety tested and tagged.

When replacing blades, router bits and other parts, don't rely on the on- off switch; always unplug the tool form the power socket or remove its battery!

Routers

A router is a power tool that uses a fast rotating collet (a segmented band or sleeve) to hold sharp, interchangeable bits (cutters) of various profiles that extend through the router's base aperture. Tightening the collet nut locks the bit in place.

Router bits come in ¼" and ½" diameters. to suit the collets of the same size. Some routers have exchangeable collets of both sizes, most only use a single size bits.

A motor housing has a pair of handgrips mounted on each side. Pressing the grips causes the housing to slide ("plunge") down towards the timber to be routed, the cutter moves through the opening on the router's bas plate and into the wood.

As the downward pressure is released, two inbuilt springs, sliding on a pair of stainless steel columns, push the motor housing back up.

TYPICAL SPECS:

- Base diameter: 160x148mm
- Plunge capacity or depth: 0-60mm
- Collet capacity: 12.7mm (1/2")
- Continuous rating input: 1,650W
- Net Weight: 5.5kg
- No Load Speed: 22,000 RPM (fixed speed router)
- No Load Speed: 8,000 - 23,000 RPM (variable speed router)
- Weight: 6.3 kg

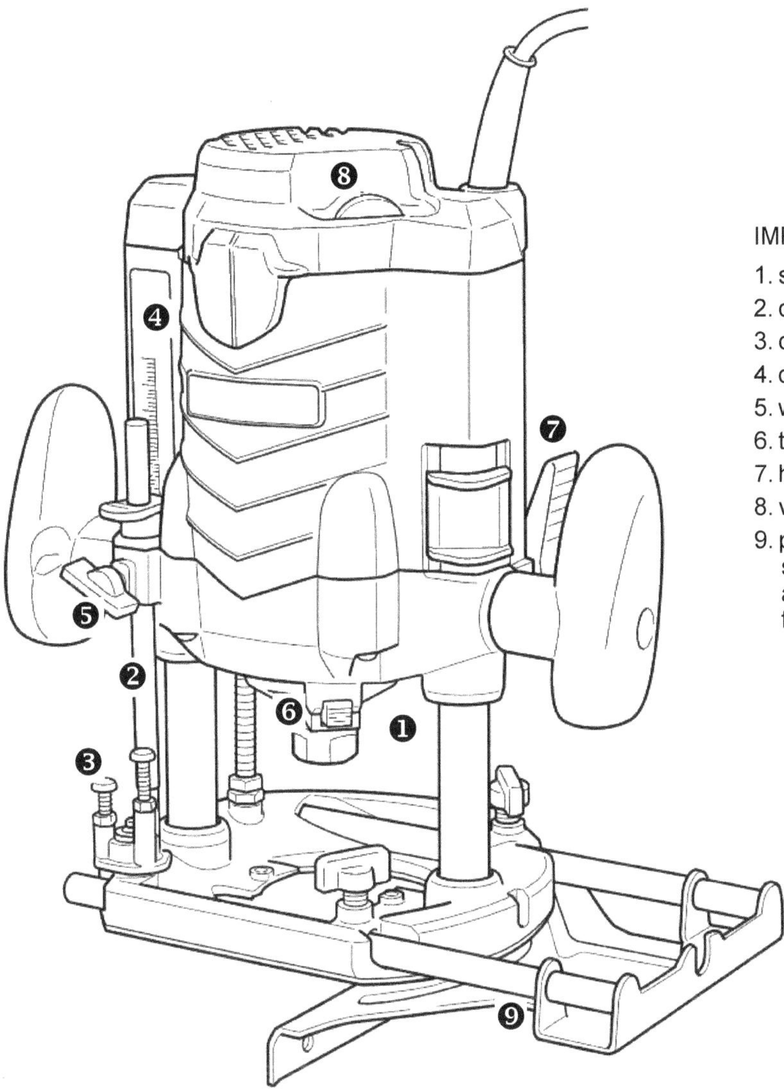

IMPORTANT FEATURES:

1. spindle lock and collet
2. depth rod
3. depth stop turret (stopper block)
4. depth indicator
5. wing nut - plunge lock
6. threaded column for depth fixing
7. height (depth) lock lever
8. variable speed dial control knob
9. parallel guide (guides the router parallel to a straight edge and allows rotation to cut circles and arcs, very useful for speaker openings in front baffles).

Keeping the router pressed down is tiring (when choosing between two routers choose the lighter one!) so most plunge routers have a plunge lock mechanism, either as a threaded column with nuts (on older units) or inbuilt into one of the hand grips on more modern designs. That enables a plunge router to be used as a fixed router.

Most full power plunge routers are mains powered, but some cordless models are also available. "Trim" or fixed routers are smaller, and most are now cordless.

A standard 3-stop (5-stop on some models) adjustable turret stop is used to control the bit-depth range, and on some models additional micro adjustments can be made to the bit depth.

A variable speed control is useful if working with materials of varying hardness (density), but also when using cutters (bits) of different sizes. The variable speed settings typically span from 9,000 RPM for 2.5-3.5" bits, to 23,000 RPM (or even higher) for a full scope of routing with 1" bits. The lower speed indicates the biggest bit you can use and the fastest speed determines the smallest size.

A slow speed isn't needed if working with soft timbers such as pine. Slowing the speed down is useful for harder wood or other dense boards. 18,000 RPM is the slowest acceptable speed for producing a cut so smooth that no sanding is required. If choosing between routers of the same power rating, buy one with the higher speed - it will produce a smoother cut.

Some brands of routers feature a soft start feature in addition to their variable speed functionality, where the rotational speed is gradually increased from a low speed to the set (desired) speed.

This avoids the common problem of starting delicate cuts with a huge burst of power, as well as minimizing the likelihood of injury due to the sudden jump of the router caused by the initial jolt of power.

Plunge depth capacity is not an issue in cab making, since all routers have at least 50 mm (2") capability, and no speaker cab cabinet that I know of uses a thicker material.

WIRING YOUR SPEAKER CABS

Small tools

Just like other hand tools illustrated, flat nose or long nose pliers come in various sizes, 5" (130 mm) or 6" (160 mm) are perfect for our purposes. Most have a rubberized grip and are made from chrome vanadium.

Long nose pliers have serrated jaws for stronger grip, and, although they do incorporate a cutting section, invest in good quality side cutters, they are much more versatile and easier to use.

The curved nose pliers are not necessary but often come as part of a set of three, just like the one on the photo. Sometimes a whole set on special (on sale) is cheaper than just one tool sold separately.

Electronic parts shops aren't the only place to buy hand tools. They are often cheaper and of better quality in craft shops, automotive parts shops and even fishing shops. Yes, they all sell pliers, cutters and even soldering irons.

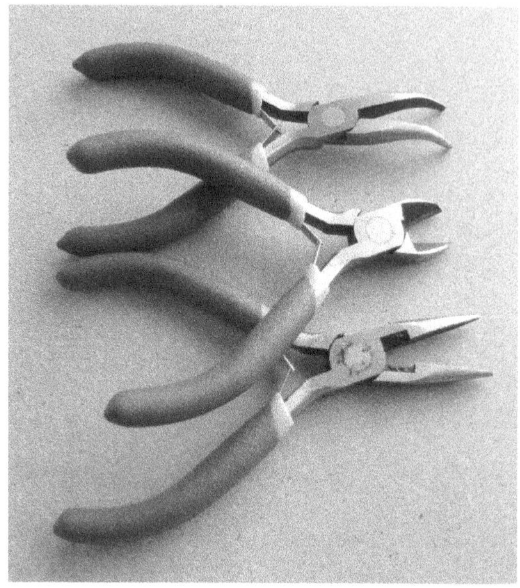

1) On-off switch
2) Tip temperadure display
3) Temperature regulator knob
4) Hand grip
5) Solder roll
6) Tip cleaning steel wool

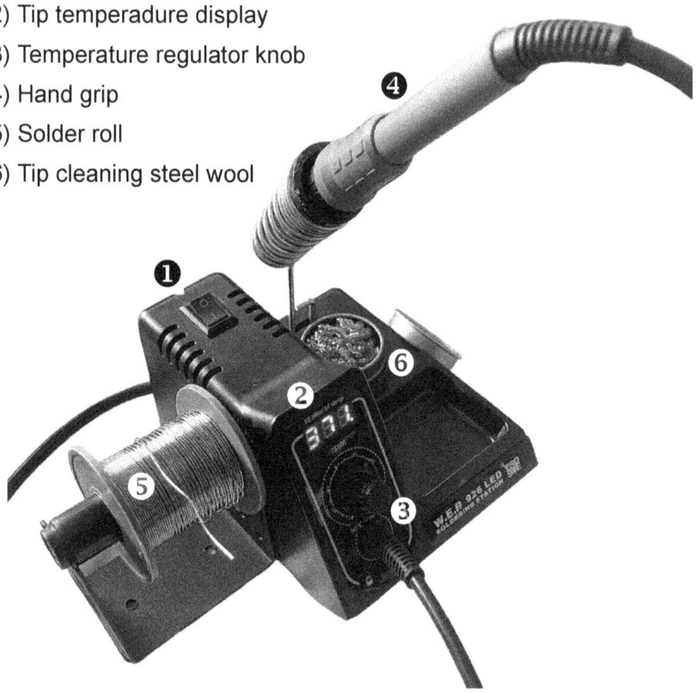

ABOVE: Temperature-regulated 60 watts soldering iron
BELOW: 60/40 and 63/37 are typical tin/lead ratios in solder

Soldering irons

Unless you are building amps, there's no need for an expensive Weller soldering station. Even the most basic fixed temperature soldering iron is fine for a few soldering joints you need to make in a speaker cab.

However, China-made temperature-regulated 40-60 watts soldering irons are so cheap, no point buying a fixed temperature one.

WEP 926 LED is a typical model in its class, selling for under AU$50 (US$35) online. It works well and is quite well-made and reliable.

The job of the solder is to bind components, terminal strips, printed circuit boards, tube socket lugs, and wires together.

As the name suggests, Flux-core solders have a core of flux, whose job is to "reduce" metal oxides on the surface of components to be soldered and improve the mechanical contact between the solder and the metal surface.

To "reduce" means to "return oxidized metals to their metallic state." It also helps solder flow (spread out). The alloys most commonly used in electronics are 60/40 tin/lead (Sn/Pb) and 63/37 Sn/Pb solder.

A brand new soldering tip must be tinned first, covered with a thin coat of solder. During use, the tip of the soldering iron will get dirty, covered in gunk. Gently clean it by rubbing it against a wetted sponge.

Proper soldering

Press the tip onto the wires/contacts to be soldered together for a few seconds. Then touch the hot joint with the solder. The solder should melt upon contact. If it doesn't, the connection hasn't been heated enough.

You should never place solder onto the tip of the soldering iron, only on the joint to be soldered! The more component leads-wires and the larger they are, the longer it will take for the joint to heat up enough to take on the solder and melt it. Knowing how long to heat the joint comes with practice.

The solder should flow smoothly and freely and should cover the whole joint relatively quickly. It should not 'blob" and fall off. If that happens, the surfaces are not clean enough; they are dirty, greasy, or oxidized and must be cleaned before attempting to solder the joint again. The tip should be touching the component lead or wire and the terminal lug or pin simultaneously.

Health hazards

In many countries lead solder was made illegal a few decades ago. Lead-free solders don't flow as well and are a cumbersome to work with. As of this writing, there are no federal laws in the USA prohibiting lead in solder.

Lead is toxic, affecting the brain and the nerve system, and the human body cannot eliminate it. Luckily, it does not get absorbed through the skin, so touching lead solder is not a hazard, and there is no lead in the solder fumes. The most significant risk is the oral transmission of particles from your hands through the mouth, so washing your hands after touching anything electronic (new components, solder, old amplifiers, etc.) is mandatory.

Also, just because there is no lead in solder fumes does not mean they are harmless. Antimony, bismuth, germanium, indium, and other elements are present in solder and its fumes. The primary source of fumes is the flux and not the solder.

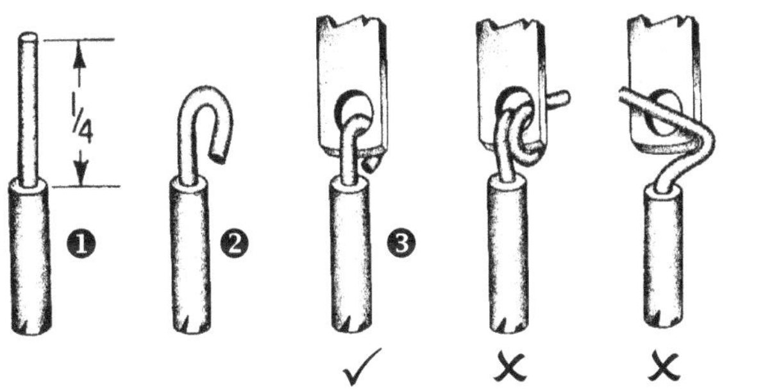

ABOVE: The three steps involved in preparing a wire for soldering onto a terminal. The copper lead should only be looped once through the hole, so future unsoldering is fast and easy. The two rightmost examples are wrong, one being a double loop, the other an improper one (wire just tacked on, not going through the hole).

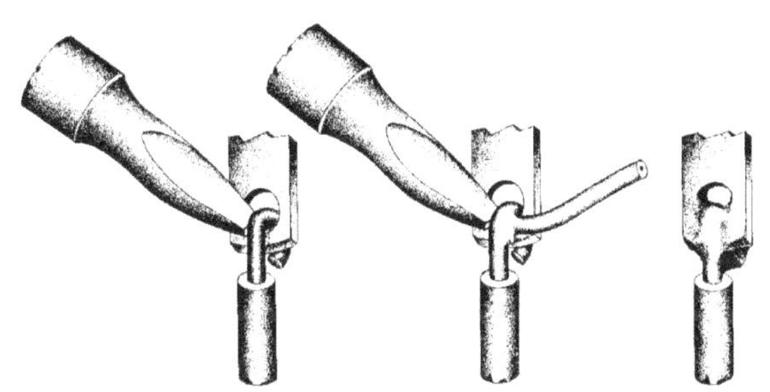

ABOVE: Once the copper lead is looped through the hole, the tip of the soldering iron should be touching both the lead and the lug, thus heating both together. The solder is then brought against the lead and the lug and melts to cover both without even touching the soldering iron's tip.

Hookup wire

The most commonly used hookup wire is AWG18/19, with a current rating of 7.5 A (amperes). As for its construction, you will see stranded wires specified by two numbers, for instance 24x0.2, meaning 24 conductor strands of 0.2 mm diameter each.

Although AWG (American Wire Gauge) tables are for a single, solid and round conductors, stranded wires can also be specified that way.

For instance, "22 AWG 7/30 stranded wire" is a 22 AWG wire made from seven 30 AWG strands. The overall AWG of a stranded wire (equivalent to a single solid wire) is simply the sum of the diameters of its individual strands.

Thus, because of the internal spaces between strands, a stranded wire is always thicker than the solid wire of the same AWG size.

The higher the AWG number, the thinner the wire, and the lower its current rating.

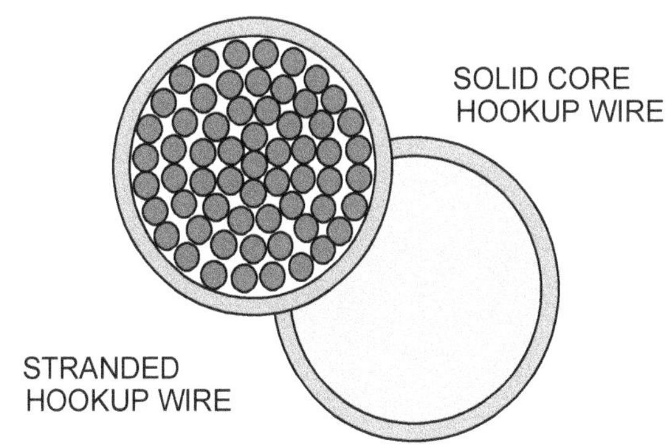

DIY PROJECT: Rebuilding a Hohner Contessa CA-300 combo amp's damaged cabinet

When this extremely rare combo guitar amp was listed on eBay USA, we just had to have it. The amp's large case fell apart in transport, not helped by the fact that the seller wrapped it in cellophane. No bubble wrap, no Styrofoam, no cardboard box, nothing! Luckily, the chassis and the tubes were OK, as was its awesome 15" Alnico speaker. We got all of our money back from eBay.

All components were original, and all looked distinctly American; no European parts were used at all. Hohner did not make this amp, after all, but a company called Sano, based in Irvington, New Jersey. Its baby sister, model CA200 (with two EL84 in the output stage) was released with the Sano badge shortly after.

Providing its timber is structurally sound, a broken-in-transport speaker cab or a combo amp could be rejoined. However, this amp also suffered water damage at its bottom; the particleboard was all puffed up, the bottom joints simply crumbled apart, making repair impossible.

The quotes from local Perth cabinet makers ranged from AU$400 to AU$800! And that was just to make a pine frame (the four sides), without Tolex covering. Never the ones to accept such blatant blackmail, we bought A-grade 19 mm solid pine boards for $50 and made the frame ourselves in a couple of hours.

Since the cab was to be covered in Tolex anyway, no complicated and time-consuming joints were used - the four sides were joined, glued & screwed together using butt-joints, strengthened by thick internal aluminum L-profiles for bracing.

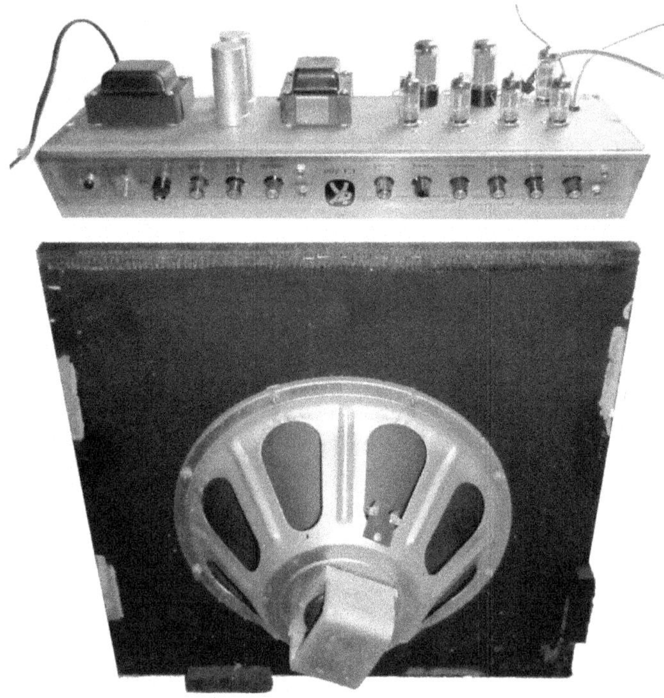

ABOVE: By some miracle, the rear cover, the front baffle, the speaker and the chassis were not damaged. Even all the tubes survived!

ABOVE: The totally awesome Hohner Contessa CA-300 tube guitar amp in a rebuilt cabinet. The front baffle and the rear panel were reused. A true sleeper of an amp, better built and waaaay better sounding than Fender Twin Reverb. Fully made in the USA by Sano for Hohner, it had point-to-point wiring, two channels, tremolo, reverb and a 15" Alnico speaker.

MODIFYING & CONVERTING VARIOUS TYPES OF SPEAKER CABS FOR GUITAR USE

10

If woodwork and cabinet building isn't one of your stronger points, don't despair.

The cheap and easy options described in this chapter will still get you a unique looking and awesome sounding speaker cabinet.

All that is required is an open mind and some lateral thinking. Keep an eye on your local items for sale and sooner or later you will score a bargain, just as we did with these examples!

- PROJECTOR ENCLOSURES, RECORD PLAYERS AND OTHER CABINET DONORS
- VINTAGE AKAI SS-SERIES PORTABLE SPEAKERS
- THE NEW YORKER: A RETRO HI-FI SPEAKER BECOMES A GUITAR CAB
- A SOLID STATE MARSHALL COMBO AMP TRANSFORMED INTO A QUAZI-HORN CAB
- A RARE BASS COMBO AMP THAT SOUNDS BETTER AS A GUITAR CAB
- THE BUSHMAN - CONVERTING AN OLD CONSOLE RADIO/TURNTABLE INTO A GUITAR CAB
- A VINTAGE PHILIPS RADIOGRAM BECOMES A HIGH POWER GUITAR CAB
- CONVERTING A SEALED PA SPEAKER INTO A GUITAR CAB

PROJECTOR ENCLOSURES, RECORD PLAYERS AND OTHER CABINET DONORS

Bell & Howell Filmosound 179 speaker cab

One sunny summer day, this Bell & Howell Filmosound 179 speaker cab was listed on our local Perth Facebook Marketplace. Made in the USA between 1945 and 1949, this model featured a 16Ω, 25 watt, 12" Alnico speaker. It was used with Model 179 B&H 16mm film projector, which had a 6V6 push-pull amplifier at its bottom.

This cab was being sold without a speaker, but was in a relatively good condition. We had a 12" electrodynamic Alnico Philips speaker (1) from a console radio of the same vintage (late 1940s), a perfect pairing. There is plenty of space for a variable DC field-coil power supply on the side (2), or inside the bottom compartment (3).

The input jack (4) was added at the rear.

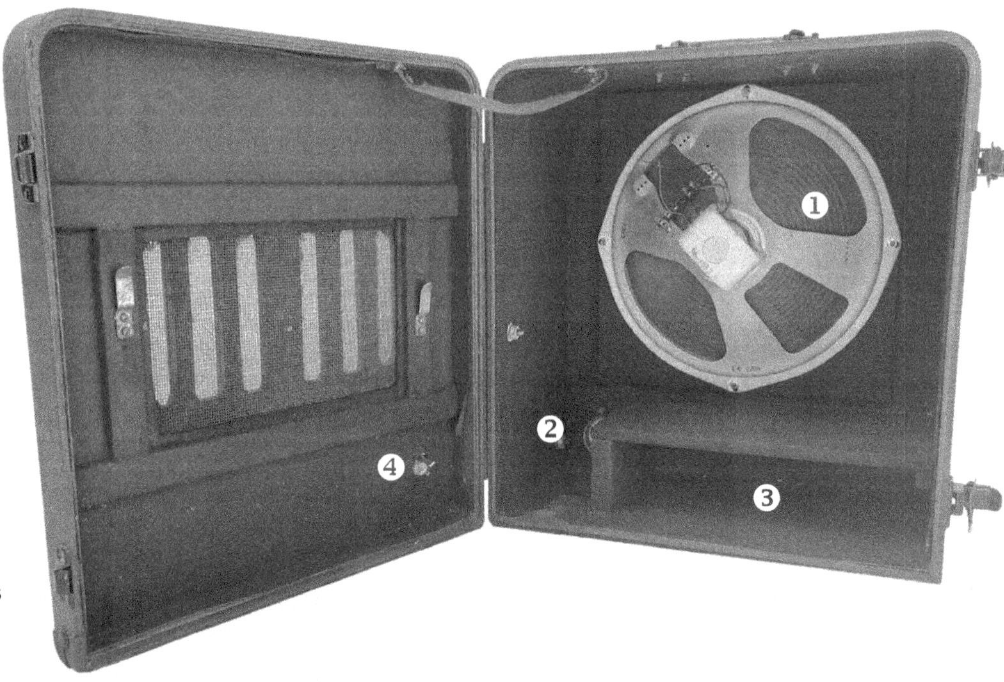

RIGHT: The 12" electrodynamic (field coil) Philips speaker installed in the B&H cabinet.

The photo was taken before the installation of the power supply at (3) and the associated mains and field coil wiring.

The nameless projector speaker cab

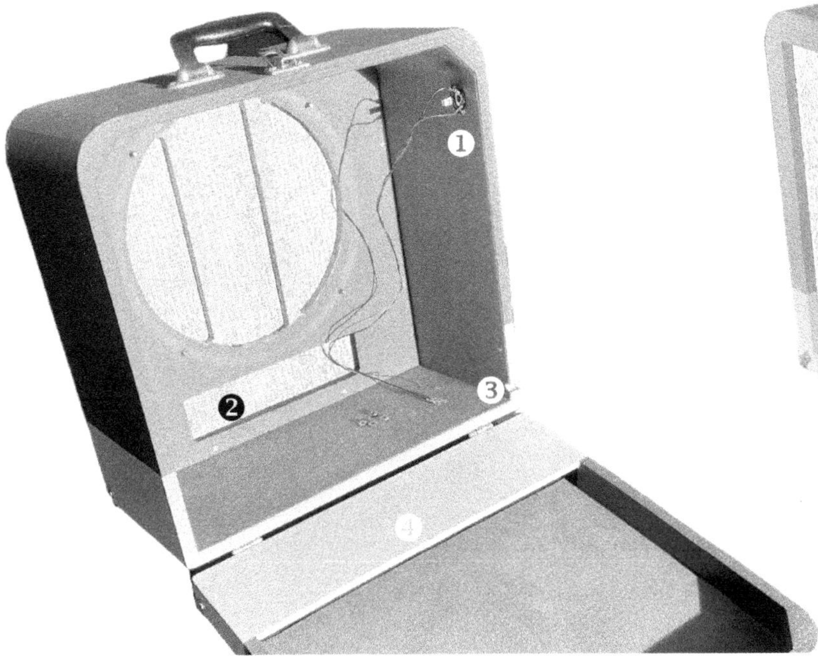

This beautiful two-tone (gray and cream) vintage speaker cab had no logo or markings of any kind. It was most likely used with a 35mm 1960s projector. Cosmetically, it was 10/10, totally unmarked, it seemed unused.

The phone jack (1) and speaker wiring were added on, originally the speaker cable exited the cabinet through a slot on its side (3). When not in use, the cable was coiled up and stored in its compartment (4). All interior surfaces were lined with a green felt, while the two-tone exterior was gray and cream.

Notice the rectangular bass reflex port (2) below the 12" speaker cutout.

Impedance and SPL measurements

After loading the cab with Celestion G12T-100 driver, we performed our standard measurements with the back panel closed, then repeating the test with the open back.

The open back impedance curve shows a single peak (31Ω) at 80 Hz. In comparison, as expected, the impedance curve for the closed back shows much lower peaks, 13Ω at 52 Hz, 16Ω at 80 Hz and 11Ω at 160 Hz. However, unexpectedly, there were three of them.

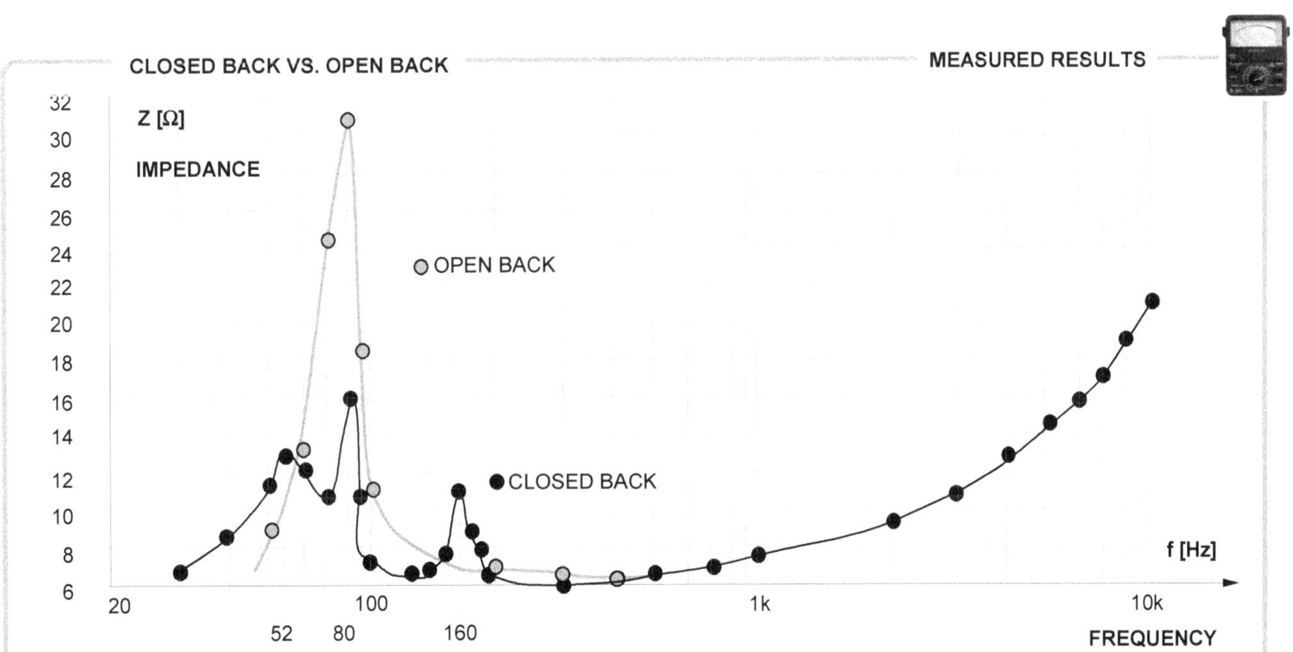

ABOVE: The impedance curve of the ex-projector cab with the open and closed back

CLOSED BACK VS. OPEN BACK — **MEASURED RESULTS**

BELOW: The SPL (sound pressure level) curve of the ex-projector cab with the open and closed back

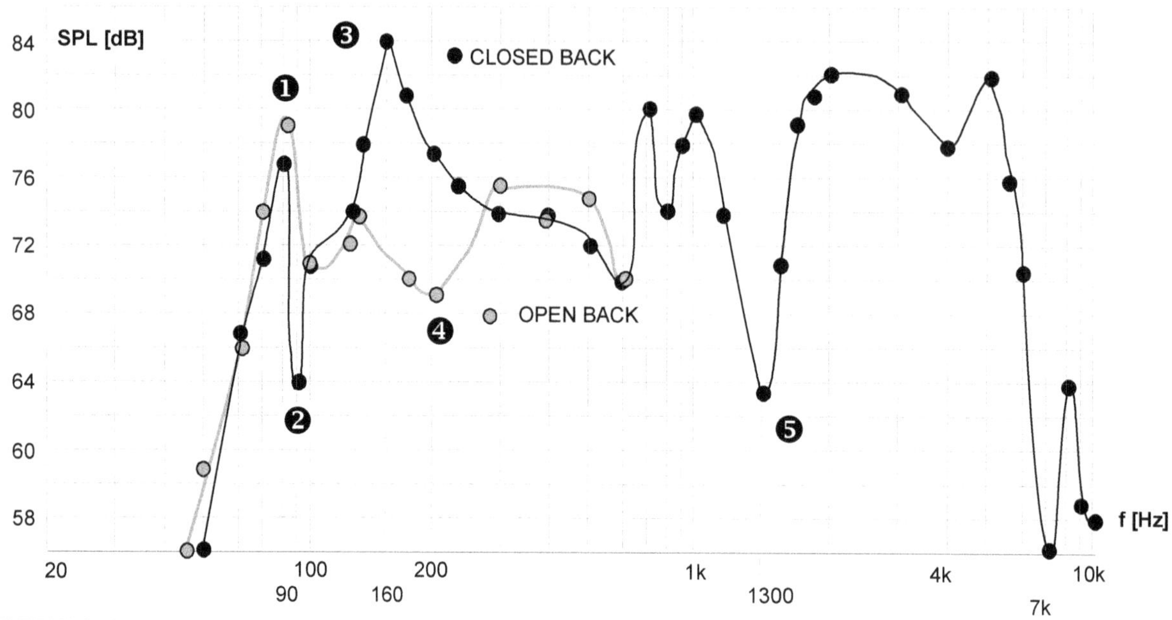

After 300 or so Hz, the impedance curves are identical, showing the typical smooth rise to about 21Ω at 10 kHz.

As expected, the mid- and high-frequency SPL curves for the open and closed back are almost identical above 600 Hz. Even below 80 Hz the difference is negligible. Both peak around 80 Hz (1), the closed back then immediately dips around dB at 90 Hz (2), and peaks again at 160 Hz (3). The open back cab has a minor peak at around 120 Hz and a moderate dip at 200 Hz (4).

After a serious dip 17 1,300 Hz (5), the characteristic is more-or-less flat up to 5 kHz, when it steeply drops a whopping 26 dB (82 to 56 dB).

Portable vintage record players make awesome micro speaker enclosures

Measuring 380 x 180 x 315 mm (15" x 7.1" x 12.4"), this 1957 portable turntable (Philips Disc Jockey Major) had an inbuilt single-ended tube amplifier. Ours had the original Mullard EL84 tube, worth more (used) than what we paid for the whole record player (US$25).

Remove the plinth (1) and the tonearm (2), add a 6.35mm phone jack and use it as a tiny tube guitar amp with its existing 5.5" (14 cm) speaker (3).

The ON-OFF switch and volume control knob (4) is obscured on the photo by the headshell. The tone control knob is on the same side, in the back corner (5).

Assuming their timber cases are of solid construction (and most are), such portable vintage record players can easily be converted into 1x8", 1x10", or 2 x 6"x9" cabs.

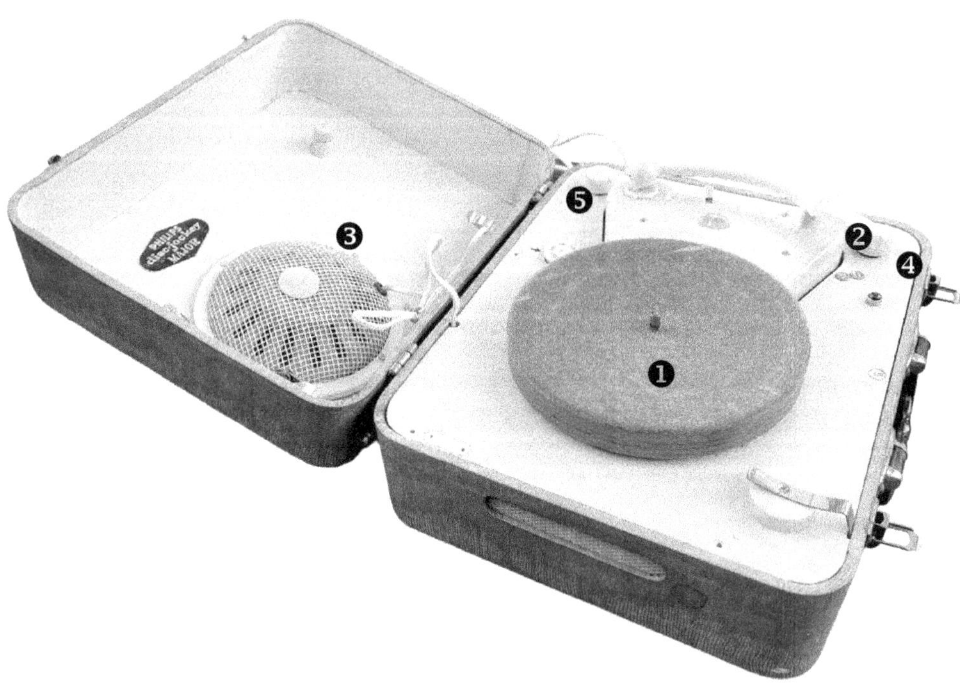

You may keep the existing tiny speaker, but, since it will be used as a tweeter, must feed it through a capacitor, which will form a 1st order high pass crossover filter with the speaker's resistance, and attenuate the lower frequencies.

That way most of the power will be dissipated on the added larger driver and the existing low-power speaker will not burn out if you drive the cab with a 15-50 watt amp.

ABOVE: How cute can one get? Add a larger speaker to the other side of this vintage portable record player's case and you have a Munchkin-sized studio recording and room practice guitar cab

AKAI SS-SERIES PORTABLE SPEAKERS

A brief overview of the Akai SS lineup

The unfortunately named SS speakers (an acronym for "Suitcase Speakers" perhaps?) by Akai hail from 1965 or thereabouts. They came in a whole series of models, ranging from the smallest/simplest/cheapest to the top of the range SS-110: SS-30, SS-40, SS-50, SS-55, SS-70, SS-88A, SS-90, SS-100, SS-110.

A pair of SS-30 weighted only 3 kg. Designed for the Tape Recorder 1700 and had a 12x18cm oval driver . The frequency range was specified as 100 Hz - 12 kHz, 8ohms, 5 Watts power rating (each).

SS-55 is the smallest of the 2-ways cabs, a 20 cm woofer and a 6 cm horn type tweeter. The frequency range was specified as 60 Hz - 15 kHz, same power and impedance as SS-30 (8 ohms, 5 Watts).

Due to their bulkier size (428mm wide, 532mm high and 260 mm deep), the weight of a pair shot up to a whopping 11 kg.

SS-88A was quite similar, only with a higher upper frequency limit (17 kHz).

SS-90, rated at 30 watts, had four 16cm woofers and one 6cm HF driver.

ABOVE: Akai SS-50 speakers positioned side-by-side but front and up facing, just like stage monitors

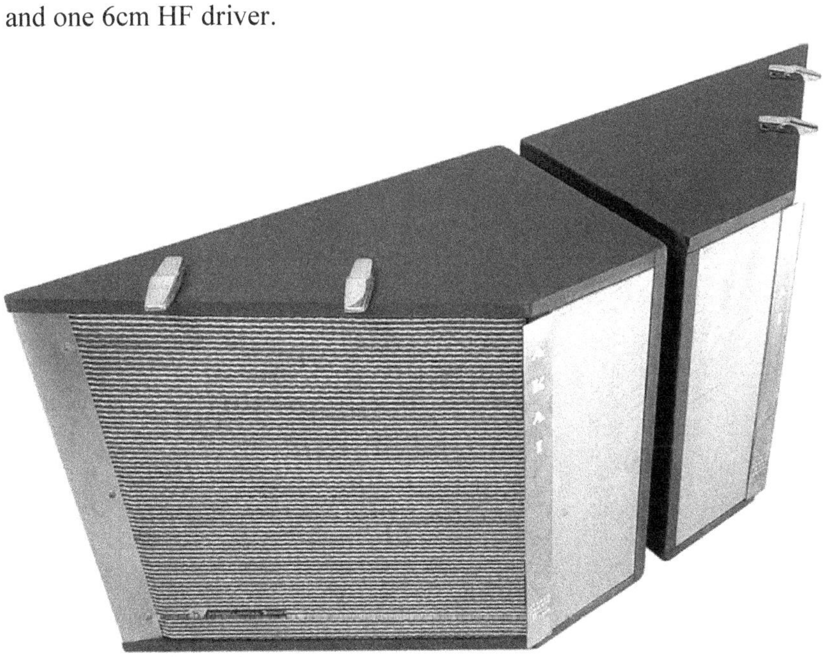

LEFT: Akai SS-50 speakers positioned side-by-side, left and right facing

Finally, the SS-100 had two Alnico drivers, 10" woofers and 2.5" horn tweeter, while SS-110 had a 10" coaxial Alnico driver in a bass reflex enclosure, both rated at only 10 watts. A pair of SS-110 weighed just over 16 kg.

Akai SS-50 portable tape recorder speakers

Our pair of SS-50 were sold with the M8 reel-to-reel tape decks. Rated at 8 Watts, 8 ohms each, they were supposed to have 8" "mechanical 2-way" full-range Alnico drivers made by Pioneer.

Unfortunately, at some stage, someone replaced them with cheap and nasty ceramic magnet Radio Shack (Tandy) drivers from the late 1970s.

The original mechanical 2-way, full range drivers are quite sought-after now and fetch hundreds of dollars. Lesson learned? Ask the seller if you can open the speakers up to check the drivers.

Anyway, we only paid AU$50 or so for the pair, so we cannot complain too strongly. Plus, we could take the photos for this case study.

To use the two cabs together, a 6.35mm socket must be added to one of them, so they could be daisy-chained. The sockets can be wired in series or parallel fashion, yielding a total impedance of either 16 ohms (series connection) or 4 ohms (parallel hookup). See the wiring diagrams on page 127.

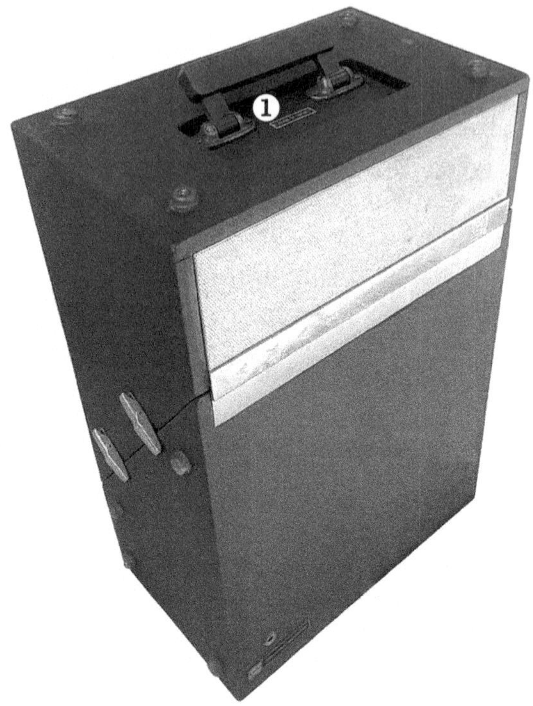

ABOVE RIGHT: The two boxes clip onto each other making them easy to carry around as a single cab. The cleverly designed handle (1) on one of the triangular boxes has it's own recess, so when that box is used in the upright mode it doesn't protrude.

THE NEW YORKER: A RETRO HI-FI SPEAKER BECOMES A GUITAR CAB

The speaker box

A pair of these teak hi-fi speaker boxes in flawless condition were spotted on a verge during council pickup days and ended up in our stash, where they sat pretty for a few years.

While looking for projects to include in my "Tube Guitar Amplifiers" book, an idea was born. Why don't we convert it into a combo guitar amp by installing a vintage tube amp with its controls at the existing cutout for a bass-reflex port (2)?

Now, a few years later, let's use the second enclosure to add another speaker and connect them in a push-pull arrangement. There is plenty of room inside the enclosure, so even a 12" speaker can be installed.

One end of the front baffle had a funky strip of embossed aluminium as a decoration (5). The cab could be turned with strip on top or at the bottom.

Teak finish and dark cloth make the vintage speaker box look classy. A vintage- looking currently produced leather handle was added on top. The etched aluminum panel (5) is for esthetic purposes only, simply glued to the front baffle.

The drivers

The speaker inside (3) was a 6.5" Australian-made ROLA, so the enclosures were probably locally made and sold with one of the many radio-turntable consoles popular at the time. The stamp on the speaker says 65/8, most likely the 8th week of 1965.

Judging by the size of its magnet, I would rate it at ten watts, possibly 15. Using a digital LCR meter on "R" (AC resistance) function, the readout was 11.4Ω (4), so we could call it a 12 ohm driver.

The added driver was a 25 watt 8 ohm speaker from a Hammond organ. Connecting them in parallel will work perfectly, at least as far as the overall impedance and the distribution of power is concerned.

The added, more powerful speaker will draw a higher current from the amp due to its lower impedance (8Ω compared to 12Ω), with the original 15 Watt speaker drawing proportionally lower current. The overall power rating of the cab with the added driver will not be the full 40 watts (15 watts + 25 watts), but will depend on the distribution of the impedances (and thus currents) between the two paralleled drivers.

The wiring diagram

In this case, the maximum allowable current through the original driver is $I_{Amax} = \sqrt{(15/12)} = \sqrt{1.25} = 1.11$ A (Amperes). For the added driver, the max. current is $I_{Bmax} = \sqrt{(25/8)} = \sqrt{3.125} = 1.77$ A (Amperes).

The original speaker's current limit will be reached when the voltage reaches the $V = I_{Amax}{}^2 * R = 11.1*12 = 13.3$ Volts. Since the same signal voltage is fed to both paralleled drivers by an amp driving the cab, the current through the added driver will then be $I_B = V/8 = 13.3/8 = 1.66$ A, marginally lower than its rated current of 1.77 A.

So, the new power rating of this enclosure can be estimated as $P = I_{Amax}{}^2 * R_A + I_B{}^2 * R_B = 1.11^2 * 12 + 1.66^2 * 8 = 14.8 + 22 = 36.8$ watts.

In this instance that is pretty close to a figure we would get by simply adding the 15 and 25 watt driver ratings (40 watts), but in many other instances that will not be so.

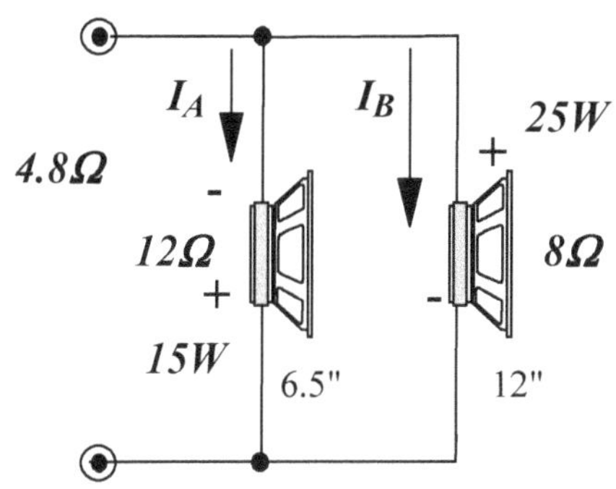

ABOVE: The interconnection diagram showing the driver sizes, impedances and estimated power ratings

As you have just seen, it all depends on the power ratings of the two drivers and on their impedance ratio, which will determine the current distribution through the pair and which driver's dissipation limit will be reached first.

The enclosure's new 37 watts rating is perfect for smaller (15-20 watts) amps with a pair of EL84 tubes, medium-sized push-pull amps with a quad of EL84s (up to 30 watts of power) or even lower powered 6L6 amps working in Class A, producing 20-25 Watts.

A SOLID STATE MARSHALL COMBO AMP TRANSFORMED INTO A QUASI-HORN CAB

The fault (destroyed output transistors) in this 100 Watt solid-state Marshall MG100FX combo was uneconomical to repair. Due to high labor costs in Australia, fixing it would cost much more than what the amp was worth in the used condition - about AU$150 at best.

So, we removed the amp chassis from its top compartment (6) and kept using it as a test speaker cab in our amp repair workshop.

Referring now to the photos and side cross-section on the next page, the amp chassis sat on the MDF shelf (1), which, together with the back crossbar (2) and the back cover (not shown) made the bottom part of the cabinet a sealed chamber for the two nondescript 12"drivers.

The fact that the crossbar can easily be removed gave us an idea of a versatile 3-in-1 cab.

Three possible configurations in one cab

Screw the rear cover back in (without any modification) and you have a sealed (acoustic suspension) cab, just as originally intended.

You may wish to add thick acoustic padding to at least two of the four inner surfaces and the back cover.

Alternatively, remove the back cover, as in this photo from the rear (right), and get an open back cab.

To convert the cab into a vented enclosure, carefully detach the crossbar (2), move it up and attach it firmly to the top panel (3) of the cab. You will need it there so the new back cover can be attached to it.

Then, cut a new, larger back cover, ensuring a snug, airtight fit, so it completely seals the back of the whole cab (not just the bottom speaker chamber as before).

You may need to add two short vertical battens (support pieces) on both sides (4), effectively extending the bottom ones (5).

Now you have a vented enclosure, with the former front opening for the amp chassis (6) acting as a rectangular port.

The rear view of the original enclosure (ABOVE) and internal details (BELOW)

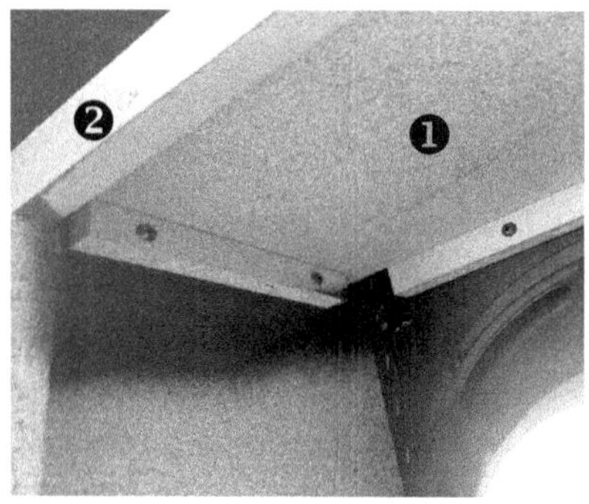

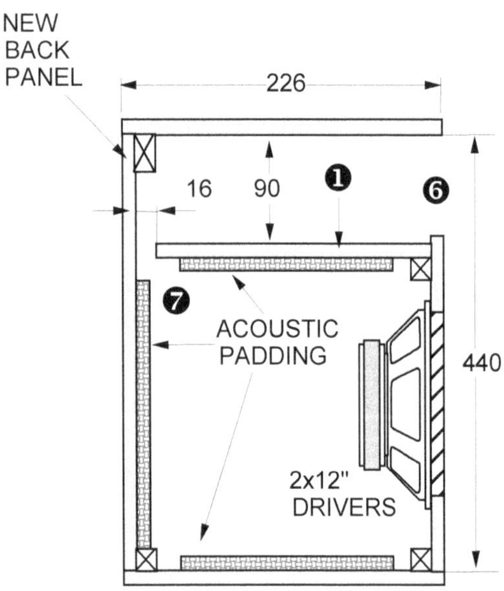

ABOVE: The side view of the modified Marshall enclosure (all dimensions in mm)

BOTTOM CHAMBER HEIGHT: 334 mm
UPPER CHAMBER HEIGHT: 90 mm
CAB WIDTH: 636 mm
CAB DEPTH: 226 mm

The rear sound waves produced by speakers' cones will now rush through the narrow (16 mm) slot where the crossbar used to be (7), enter the wider upper chamber (90 mm tall) and exit through the rectangular port (6).

I called this narrow chamber-wide chamber progression a quasi-horn, but it's performance cannot be compared to a real, properly designed and tapered exponential horn, of course.

Nevertheless, this last, vented option, sounded better and louder than the first two, especially in the lower registers, even with such budget drivers.

The first rule-of-thumb of driver assessment is: "If the speaker isn't good enough for its manufacturer to proudly claim it as theirs by slapping on their sticker or logo, it is most likely of the cheapest possible (most economic to make) kind."

These drivers had no brand name stickers, only the "300RFL10-2"stamping, which turns out to be a made in China speaker model by Chunil Corp.

The detailed specs published online indicate a resonant frequency of 81 Hz, a 1.6"(40 mm) voice coil diameter and a smallish 17.5 oz (500 g) magnet, resulting in the RMS power rating of 65 watts.

A RARE BASS COMBO AMP THAT SOUNDS BETTER AS A GUITAR CAB

Creative Sound Zero is a rare vintage Japanese-made solid-state bass combo amplifier with a quasi-horn-loaded 15" woofer.

Why "quasi" horn? Because simply slanting the two front baffles (6) does not result in horn action!

The nameplate at rear specifies a power output of 60 watts. The speaker cable (7) is terminated with a standard 6.35mm jack; it can be unplugged from the amplifier and used with a different amp.

Let's perform our standard simple measurements, the impedance-vs-frequency and SPL-vs-frequency to see if its impressive looks and quality construction are matched by its performance on the test bench.

ABOVE: Creative Sound Zero EB-601EF bass amp next to Marshall 4x12" cab for size reference

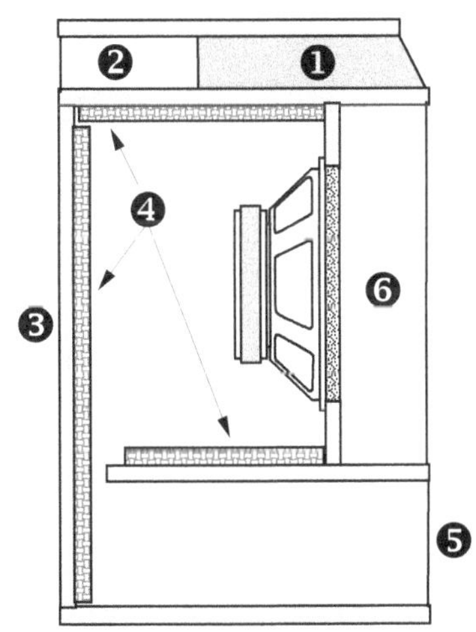

ABOVE: Side-view cross-section of the Creative Sound Zero EB-601EF bass amp's cabinet

1) Amp

2) Rear compartment for storing the power cable

3) Back panel

4) Acoustic padding

5) Front grille

6) Front "horn" flare

RIGHT: The view of internal construction details after the removal of the back panel

Creative Sound Zero EB-601EF 1x15" cab impedance and SPL measurements

The singular resonance peak of the impedance curve at 100 Hz makes it obvious that, despite its appearances, this is not a bass-reflex enclosure. It is not tuned at all, otherwise the curve would have the typical two-humped shape (two lower resonant peaks).

We didn't have any published specs of this obscure bass amp, but the SPL curve leaves a lot to be desired. The two prominent peaks, at the resonant frequency of 100 Hz (1) and then an almost equally loud one at 2 kHz (2), the midrange of this cab will sound severely "scooped out" due to that wide dip between 300 Hz and 1.2 kHz (4).

The only reasonably flat part of the curve is from 120 Hz to 300 Hz (3).

If we (arbitrarily) proclaim the average SPL level to be 79 dB, the -3dB frequency range is very narrow, from about 95 Hz to only 4 kHz. Lowering the average or nominal SPL level to 73 dB gives us two humps (3) and (2) and the -3dB frequency range of about 65 Hz to 7 kHz.

For a bass cab, despite its large size (15") speaker, EB-601EF didn't go deep enough; it sounded better as a guitar cab!

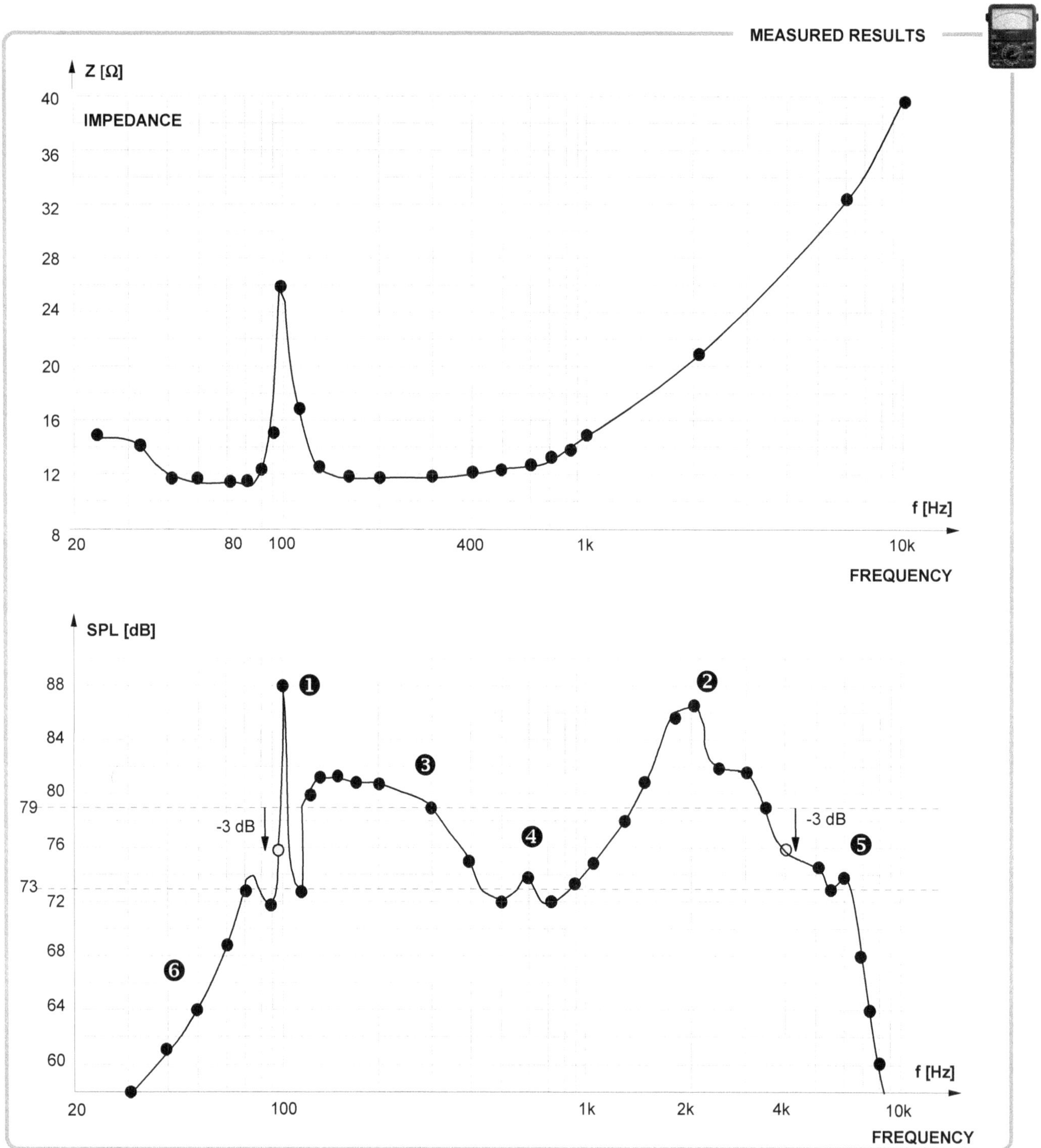

THE BUSHMAN - CONVERTING A RADIO/TURNTABLE CONSOLE INTO A GUITAR CAB

How to get a vintage console for free and even make a tidy profit on the resale of its parts

This 1952 UK-made Bush radio-turntable console had a SW-MW-FM radio and a Garrard turntable, all in mono, of course (stereo was still a decade away). The Celestion speaker was in an incredible condition for its age. We cleaned it up and sold it for AU$200, a profit of AU$150, after the $50 cost of the radio console was deducted.

Remember, such truly ancient drivers may be fine for a while, but their adhesives have dried up after 70+ years and it's just a matter time before the paper voice coil buckles and detaches from the cone. Also, the driver was of a 4Ω persuasion, unsuitable for a parallel connection with another one (who wants a 2Ω cabinet?!?), so we replaced it with a modern 8Ω, 12" driver.

Since the turntables in most of these vintage consoles were made by Garrard, the most sought-after UK brand of turntables among hi-fi fanatics, you may get lucky and make a further profit on its sale, enough to buy yourself a boutique speaker for this project, resulting in net zero cost.

The radio and single-ended tube amplifier also had valuable components. Both the power transformer and the output transformer are perfect for a small single-ended guitar amp with 6V6 or EL84. In this case EL84 (6BQ5) output tube was used, made by Mullard in UK, another $50 value.

ABOVE: 1952 10" Celestion Alcomax (Alnico) driver, model Z10W0

The conversion

After removing the turntable and the tuner/amplifier, we spent a whole day (about 8 hours) cleaning, repairing, sanding and re-varnishing the cabinet. The turntable's cutout was rectangular, so we enlarged it and made a new baffle for the internal speaker (7).

The front plywood baffle and its circular opening (6) were large enough for a 12" driver, savings us heaps of time and effort in enlarging the opening or making a new baffle.

The rear wave of the front speaker combines with the front wave of the internal speaker, firing downwards, but also with its rear radiation, which is channeled through the four small round holes and a large rectangular opening in the horizontal shelf (9).

Lifting the lid causes the rear radiation of the horizontal driver to reflect off it and radiate forward and to the sides of the cab, just like the lid on a grand piano (page 11).

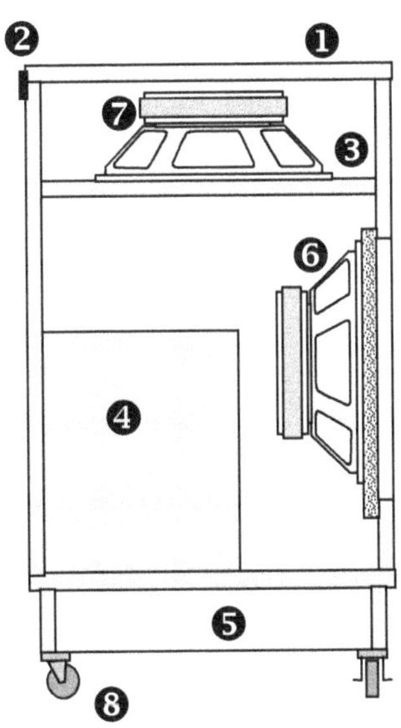

LEFT & RIGHT: The side view cross-section of the Bush console and the view of the finished Bushman cab after the conversion

1) Hinged top lid
2) Hinge
3) Upper shelf and compartment (where turntable used to be)
4) Side box with a door
5) Plinth
6) 10" speaker replaced by a 12" driver
7) Added 12" speaker
8) Added lockable castor wheels
9) One large rectangular and four small circular openings where the tuner scale and knobs (controls) used to be

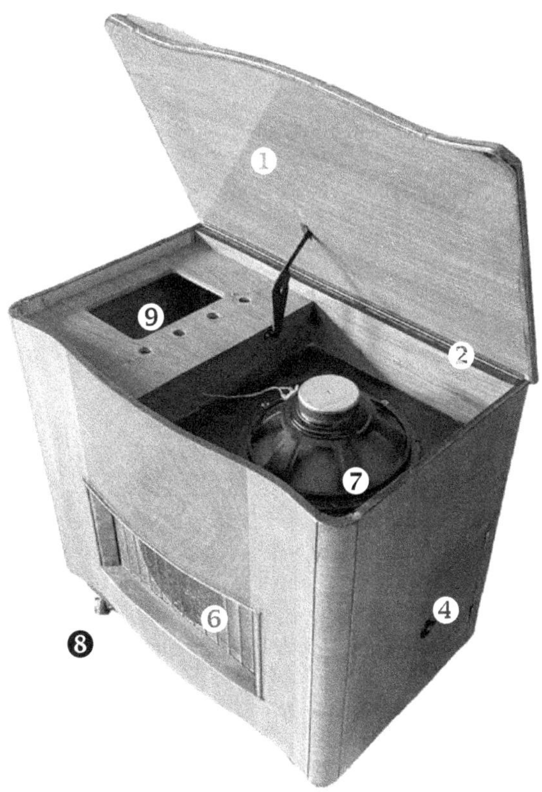

ABOVE: The finished Bushman in all its vintage beauty

ABOVE RIGHT: The felt-lined internal compartment is quite large (large enough for LP records), perfect for storing guitar, speaker and other cables, even microphones and other smaller pieces of audio gear. Alternatively, remove the thin plywood box and enlarge the internal volume of the cab.

RIGHT: The internal view of the original console after the back cover was removed: 6) tube tuner and amplifier 7) the original Celestion speaker 8) internal storage compartment

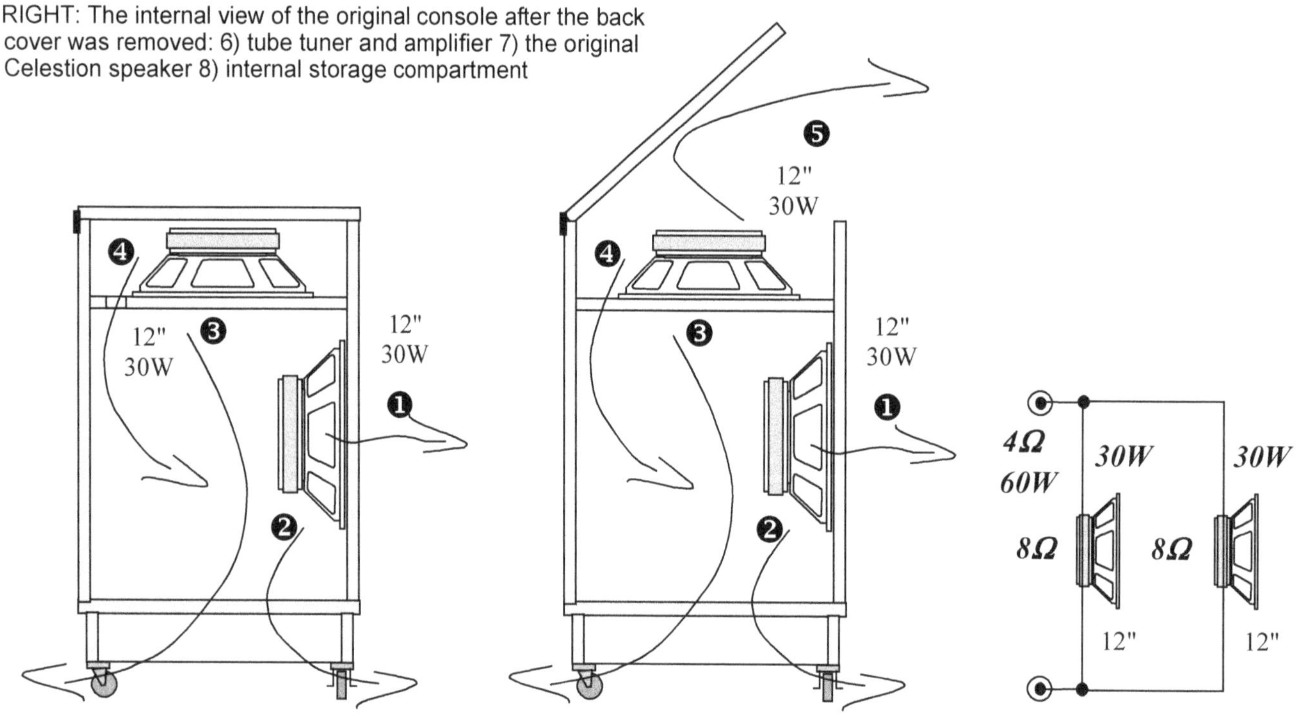

ABOVE: The side view of The Bushman cab showing main sound propagation channels of its speakers. Each was a 12", 4Ω, 30 watt driver salvaged from a Kawai organ

1) Forward radiation of the front speaker

2) Rear radiation of the front speaker, channeled through the bottom gap between the plinth and the floor

3) Front radiation of the internal (horizontal) speaker into the bottom chamber, and then channeled through the bottom gap between the plinth and the floor

4) Rear radiation of the internal (horizontal) speaker with the top lid closed, channeled through the holes in the horizontal shelf, into the main chamber, and then through bottom gap between the plinth and the floor

5) Rear radiation of the internal (horizontal) speaker with the top lid open, reflected of the lid and radiated in the semi-spherical forward plane

The measurements

"What a difference a day makes" was a huge1959 hit for the late and great dame Dinah Washington. In this case, I've renamed this project's theme song to "What a difference a lid makes".

The impedance curve with the top lid open shows a simple single resonant peak curve around 75 Hz. Closing the lid removes that peak totally and shifts it down to 60 Hz, but flattens it from 7Ω (a very good result compared to most other cabs that shoot up to 50, 100 or more ohms!) down to only 4.2 ohms!

The SPL levels of the open lid cab go up and down like a yo-yo: after the first peak at 75Hz (1), sever dips at 80Hz (2), 150Hz (3), 259Hz (4), 400Hz (6), and 600Hz (8) follow, interspersed with peaks at 180-Hz, 300Hz (5), and 500Hz (7).

Above 700Hz the responses of the cab open and closed lid are almost identical, certainly within the error margin of our relatively crude measurements. The only difference is in the 4-6 kHz region (9), where the response of the open lid cab (thicker gray line) dips 10dB (86-to-76dB) .

The closed lid cab sounded warmer and fuller, but I'm sure some guitarists would like the tonal idiosyncrasy of the open lid cab.

CLOSED VS. OPEN TOP LID — **MEASURED RESULTS**

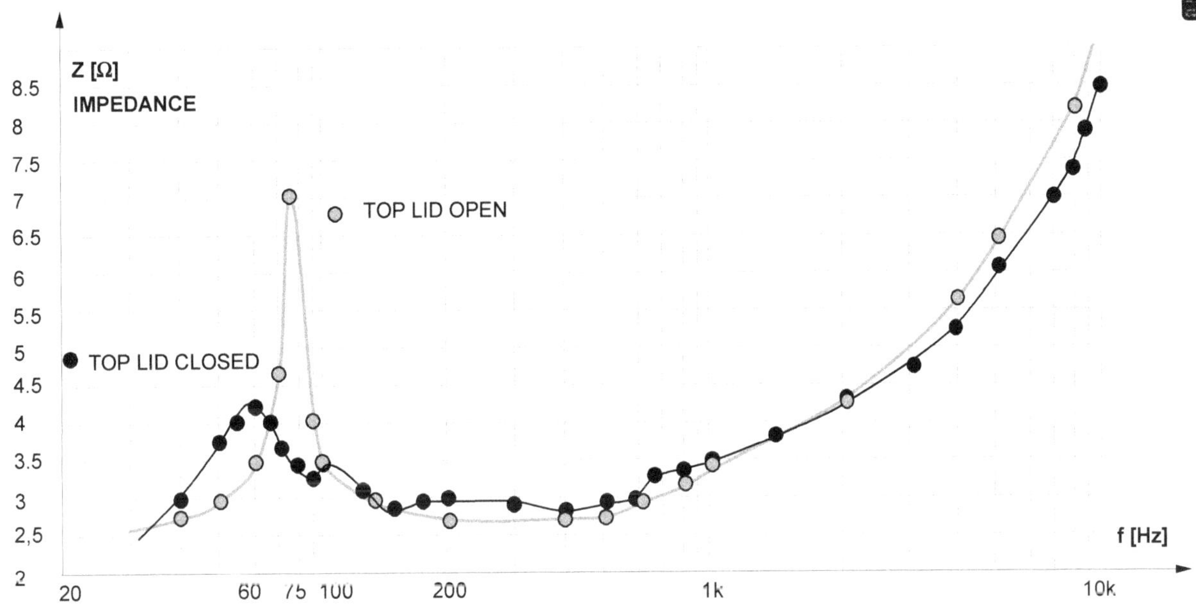

ABOVE: The impedance curve of the Bushman cab with the open and closed top lid
BELOW: The sound pressure levels of the Bushman cab with the open and closed top lid

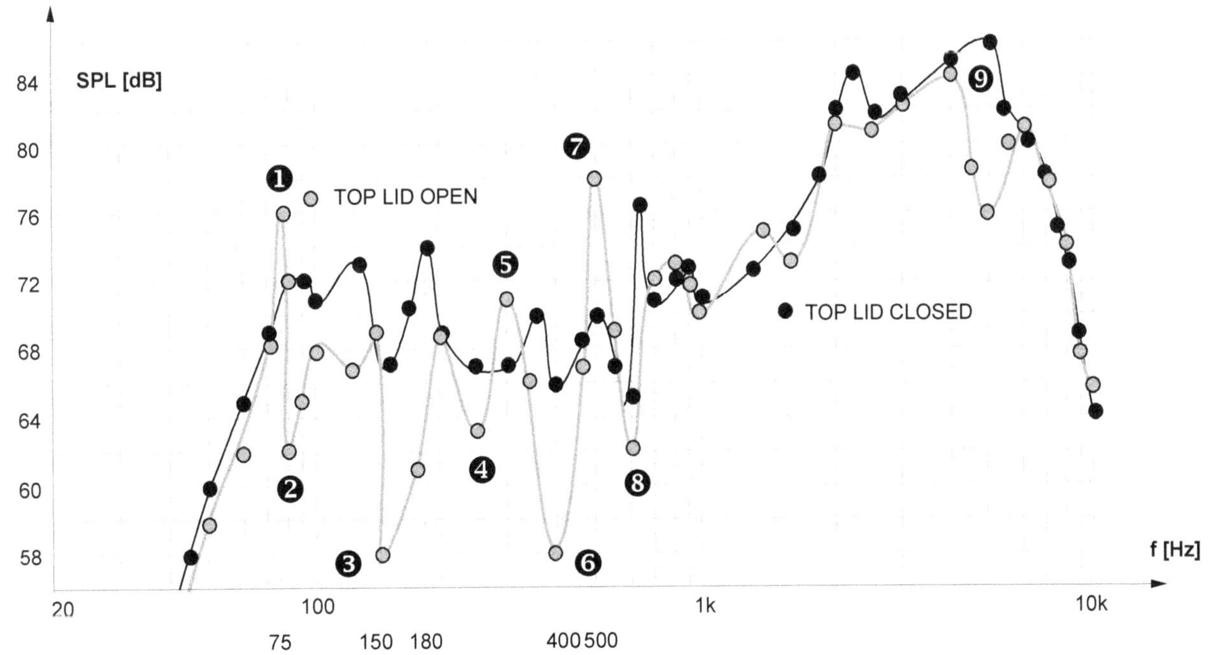

A VINTAGE PHILIPS RADIOGRAM BECOMES A HIGH POWER GUITAR CAB

The cabinet donor: Philips console radio

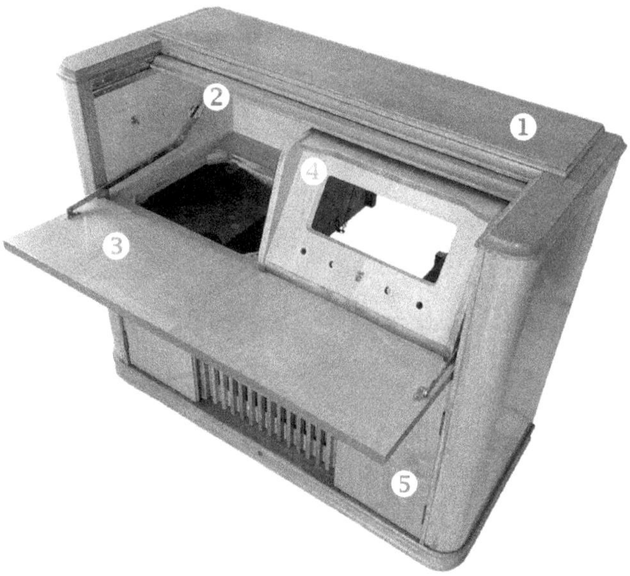

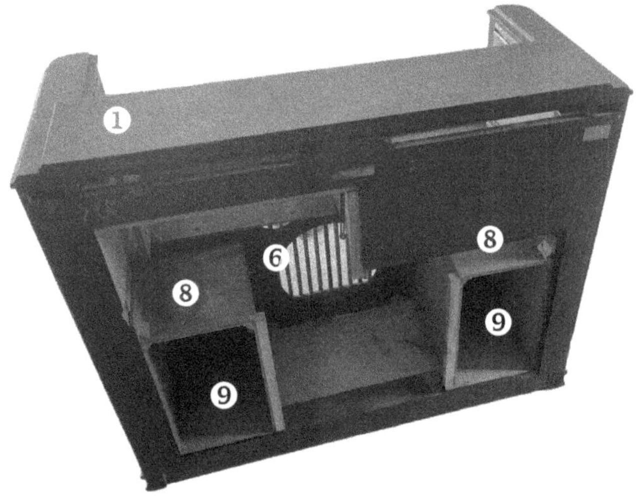

ABOVE: The Philips radiogram after the removal of all electronics and general cleanup

This Australian-made Philips console was slightly newer (late 1950s), bigger and much sturdier than the Bush radiogram. It was in a better condition, too (no sanding or varnishing needed, the lacquered finish was fine), but it had no speaker, so we got it for free.

The rear top cover (1) was an integral part of the frame, but the front top (2) and top front section (3) were interconnected by a cleverly-devised lever mechanism, so they opened together.

While the gramophone sat on its own horizontal shelf, the radio/amplifier was housed in a separate plywood box (4), which could be retained or easily removed.

The bottom section below the shelf was comprised of two identical side compartments (for LP records), each with its own front-opening door (5).

The middle compartment housed a 12" speaker and couple of smaller tweeters on a single baffle (6), protected by timber slats (7) and a steel mesh.

Possible speaker placements

Due to its design, apart from the obvious front-firing 12" driver, this cab presented many more opportunities and additional driver mounting options.

The two side compartments are small enclosures in their own right. They could be removed or used to direct the sound from two oval 6'x9" or round 10" drivers. One option is to mount them on the top shelves of the two compartments (8), firing downward into the boxes and then channeled to the front (with the two small doors open). The other is to mount them inside the compartments, on added small vertical baffles (9), firing directly forward.

The photo of the cab's rear (above right) was taken after the removal of the solid plywood back cover. If sealed properly and made stiffer by additional internal bracing, that back plate would completely seal the back of the side compartments (9), as well as the main cavity. However, unless we install a new solid shelf on top of the existing partial top shelf (on the left side, where the turntable was), this cab would never be a sealed box, since there is a significant leak around the front top (2) and top front sections (3).

The pair of smaller (10") drivers on the cross-sectional view (next page) seem to fire into the back of the front 12" driver, but in reality they fire beside it, through the two side compartments, exiting through the open doors (5).

Configuration "A" is identical to many projects already studied; ideally, each of the two smaller drivers in series should have half of the 12" driver's impedance and half of its power rating.

This also applies to the arrangement "B", but the impedance of the one or two top-firing drivers will depend on how they are connected and if one or two are used.

With 3 x 12" drivers in parallel, all should be 16Ω and the same power rating. Three in parallel would yield 16/3= 5.33Ω, so the two smaller drivers can be 2x4Ω or 2x8Ω. Then, due to their higher combined impedance (8 or 16Ω), their power rating (each) can be lower than half of the 12" driver's nominal power.

With two of the three 12" in series (Y and Z), as illustrated in "B", their power rating should be half of the driver X power and their impedance should be half of driver X impedance.

Or, if their power rating (each) is equal to that of driver X's, their impedance should be 1/4 of driver X impedance, which means 16 ohms for driver X and 4 ohms for each of drivers Z and Y.

Z and Y together would then have 8Ω impedance and draw twice as much current as the 16Ω driver X.

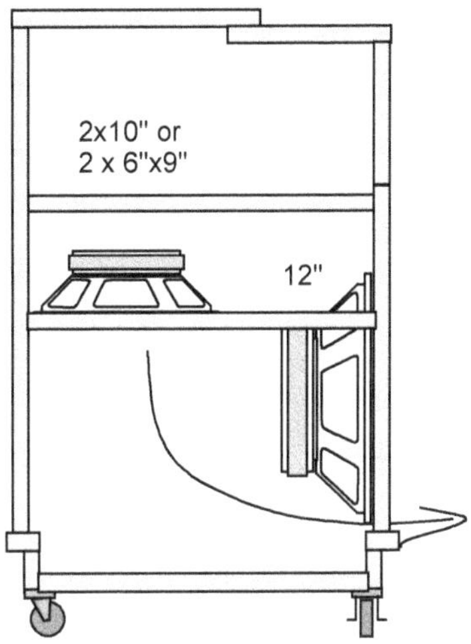

ABOVE: A lower-powered 3-driver option, upper compartment unused (side view)

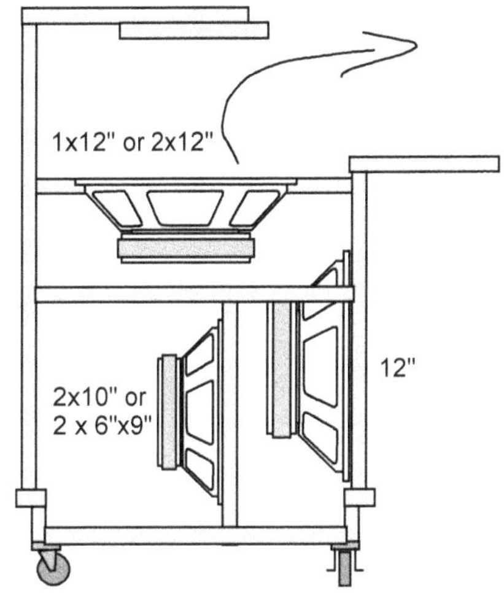

ABOVE: A high power 4- or 5-driver option, utilizing the upper compartment (side view)

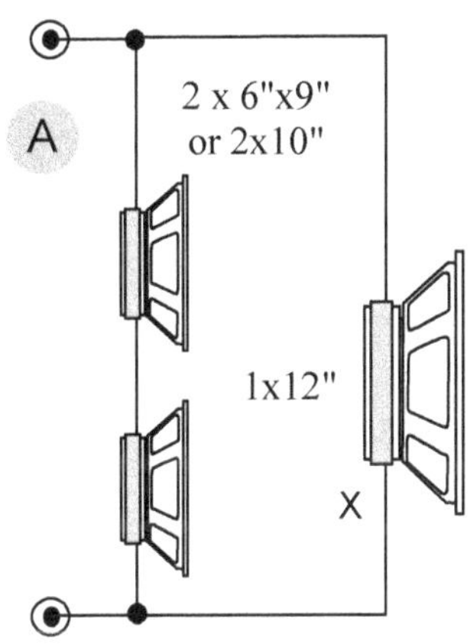

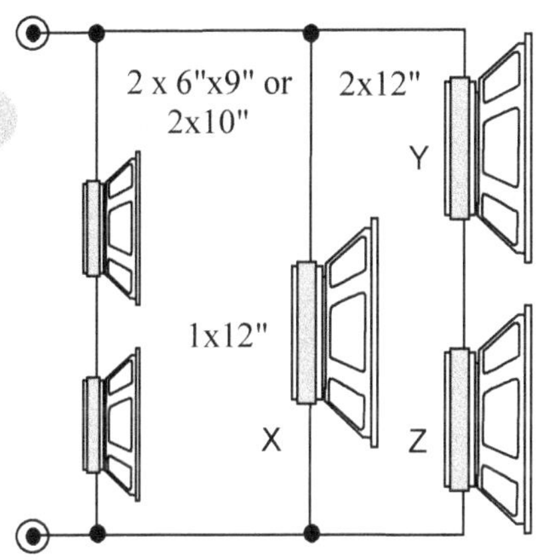

The measurements

We only had available a 12"16Ω Etone driver (50 watt rating) and a 15"8Ω driver rated at 150 watts. The 15 incher was mounted on the new horizontal plywood shelf (photo on the next page), firing downwards.

The impedance curve with the lid closed shows a 12Ω resonant peak around 40 Hz (1) with a tiny secondary peak (obviously of the smaller 12" driver) around 60 Hz (2). These are excellent results, both resonant frequencies were below the lowest tone a guitar could produce.

Opening the lid eliminated the minor peak, increased the major peak to just under 14Ω (3) and shifted it even lower, to around 37 Hz.

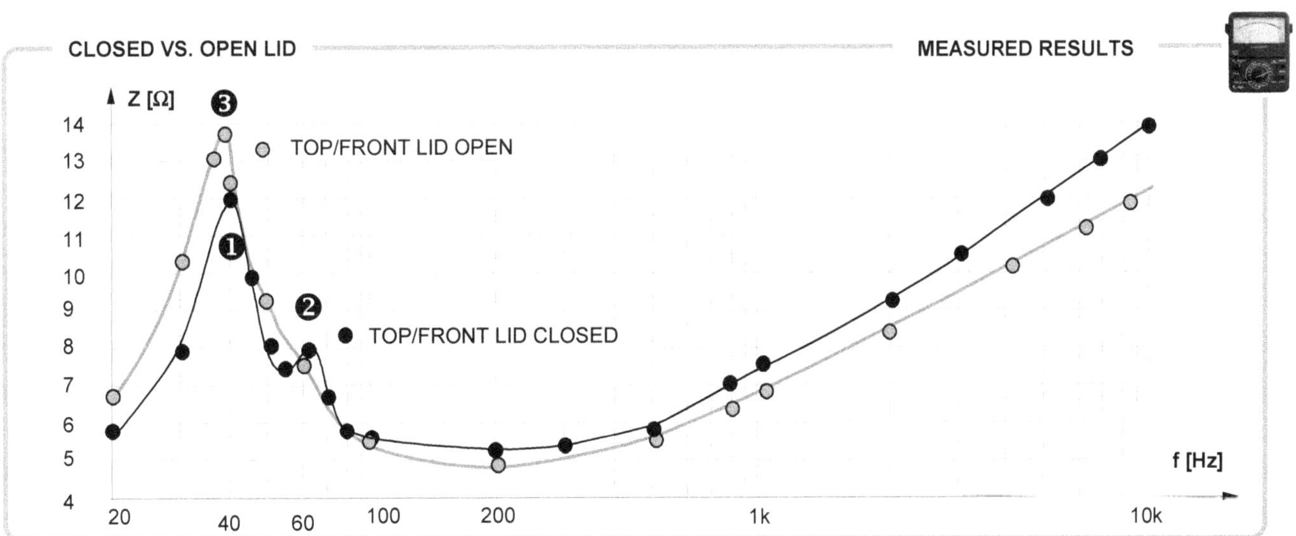

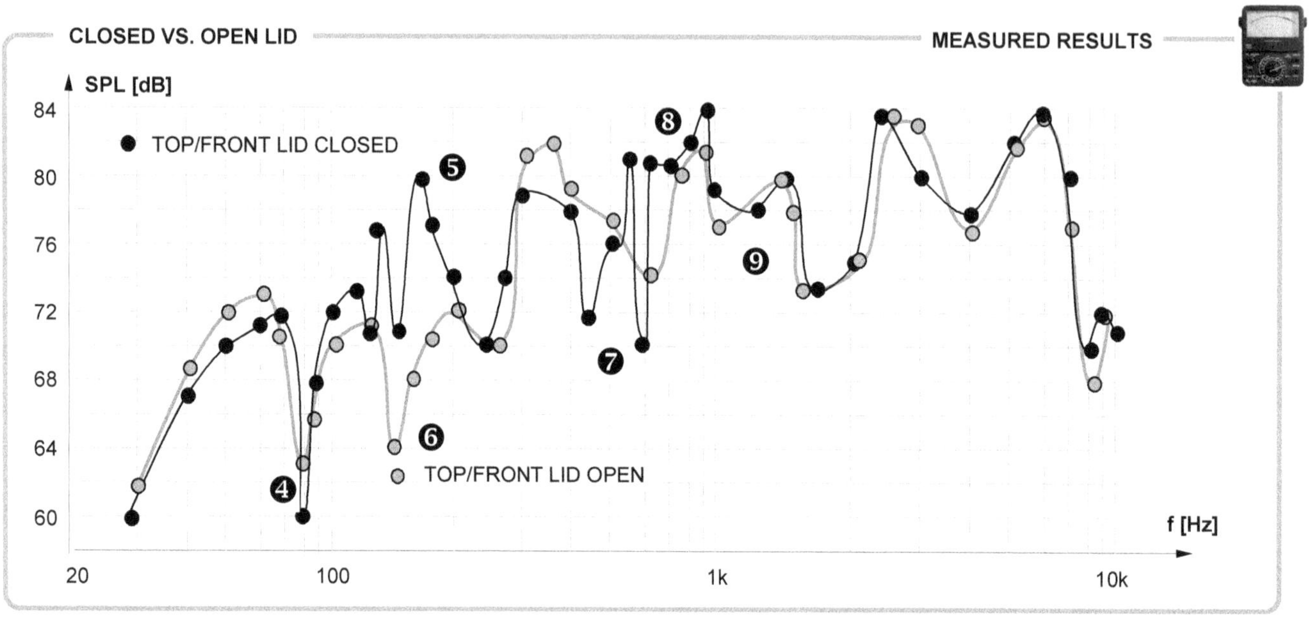

With a closed lid the enclosure suffered from significant peaks and dips, first dip at 80Hz (4), a couple of peaks between 100 and 200 Hz (5), two dips between 400 and 600Hz (7), and a relatively constant shelf between 600 and 1,000 Hz (8).

Opening the lid made most of these fluctuations milder, but added a dip at 150Hz (6), thus weakening the 100-200Hz bass. Above 1kHz there was no discernible difference between the open and closed lid (9).

RIGHT: The rear view of the internals before the drivers were wired up and back cover affixed. 1) 15" driver, 2) 12" driver 3) the added plywood shelf onto which the 15" driver was mounted

CONVERTING A SEALED PA SPEAKER INTO A GUITAR CAB

The box

This solidly-built (3/4" MDF all round) Australian-made PA speaker was missing its tweeter and two midrange drivers. The 15" woofer was in a great condition, as was the cabinet (cosmetically), so we grabbed it for AU$20 (US$13). It came with a round metal grille (woofer protector), but we'd removed it before the photo was taken.

Buying the MDF panels, Tolex, handles, feet, and corner protectors would cost way above AU$100, not to mention the functioning 15" driver (AU$199-150). Plus, it would take almost a day to build this cabinet from scratch.

There wasn't enough real estate at the top for two round 6" drivers, and the existing holes would make it impossible to seal the front baffle. Too much effort was needed to patch the holes up and achieve a tight fit of the new drivers, so we decided to dispose of the front baffle and not to reuse the original 15 incher.

Instead, we would put our trusted jigsaw to work, cutting out the permanently fixed front baffle around its perimeter.

This would leave an edge 1/2" or so wide, that will act as a furring or a frame, onto which a new baffle will be screwed in from the front.

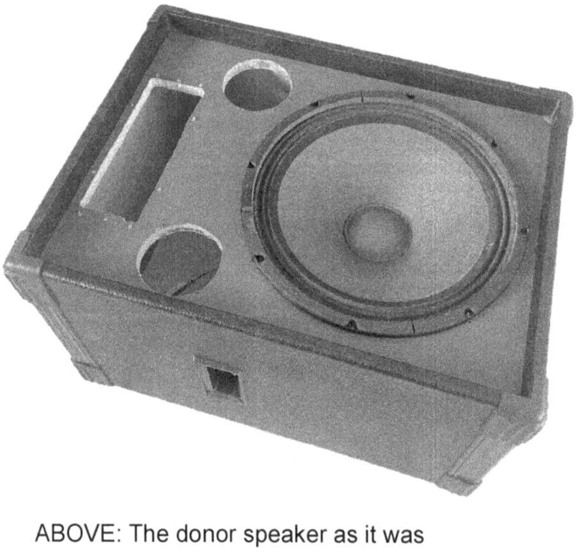

ABOVE: The donor speaker as it was

Why from the front? There was simply no other option, the back cover was also permanently glued to the cab's frame, so there was no access into the cavity from the back. We would still bolt the new drivers from the back of the baffle, and then just drop it into the old cab.

The driver configuration

The new driver configuration will be 1x10"at the bottom + 2 x 6"x9" oval speakers on top.

We got a pair of Stromberg Carlson Alnico beauties on eBay from the USA, just what the doctor ordered for this conversion project.

Stromberg Carlson was a reputable company that made many types of hi-fi and PA amps in the 1950s and 60s, but, as far as I know, they never made loudspeakers, so one of the American speaker makers supplied it to them.

Each driver measured 7.1 ohms of DC resistance, so they would most likely be declared 10Ω speakers. We will connect them in series, and then in parallel with Celestion G10R-30 ten incher, rated at 30 watts, with DCR=5.9Ω.

We don't know the power rating of the two oval speakers, so let's start with what we do know. At its maximum power, the current through the Celestion 10" driver will be $I_1 = \sqrt{(30 \text{ watts}/5.9 \text{ ohms})} = \sqrt{1.875} = 2.255$ A!

We also know (Kirchoff Laws) that the currents through the two branches will distribute themselves in accordance with the impedances of each branch: $I_1/I_2 = R_2/R_1$.

Notice the inverse relationship: the currents will be inversely proportional to the impedances of each branch. The branch with higher impedance will pass lower current, and the branch with the lower impedance will pass proportionally higher current.

BELOW: Drivers mounted onto the front baffle and wired up

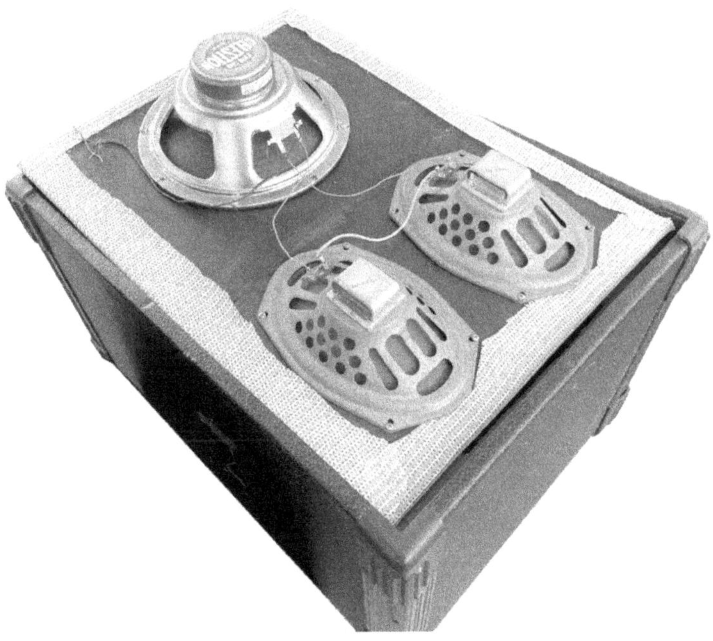

4.2Ω, 43 Watts

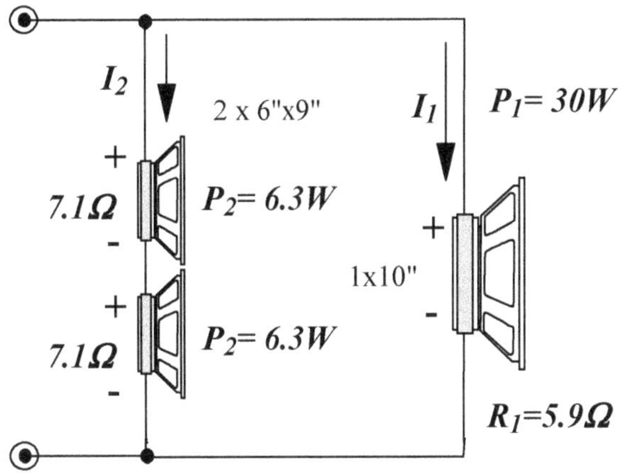

ABOVE: The electrical interconnection diagram

BELOW: The only way we could think of to transfer the driver's oval shape to a piece of paper was to use the shading technique (a pencil required), pressing against the seal, and then cutting the shape with scissors.

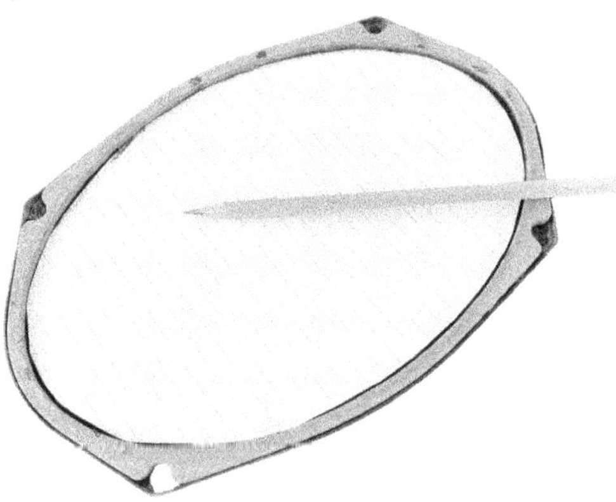

We know that $I_1/I_2 = R_2/R_1 = 14.2/5.9 = 2.4$, so the current through the two oval speakers will be 2.4 times lower than the current through the Celestion ten incher.

Now we can calculate $I_2 = I_1/2.4 = 2.255/2.4 = 0.94$ A, and the power dissipation on each oval speaker at the point of the maximum power dissipation on the 10" speaker:

$P_2 = I_2^2 R_2 = 0.942 * 7.1 = 6.3$ watts.

Therefore, the total power rating of this cab will be 30+2x6.3= 42.6 watts.

By our estimation, each oval speaker can easily dissipate 6 or so watts, I would rate them at 10 watts (each) at least.

So, perhaps counter-intuitively, in this case the power limitation is the bigger driver, the 30 watt Celestion speaker.

The overall DC resistance is 5.9||14.2Ω = 4.17Ω (we measured 4.3Ω) or around 5-6 Ω AC impedance. Thus, this cab could be safely used with both 4Ω and 8Ω amplifier outputs.

The measurements

The manufacturer-declared resonant frequency of Celestion G10R-30 is f_S=94 Hz, and, although we haven't measured the f_S of the two oval speakers, we see a double impedance peak when all are together in this cabinet.

The first peak at 110Hz is the 94Hz peak of the Celestion driver, raised by the sealed cabinet by 110/94 = 17%. Likewise, the second peak at 145Hz is that of the two oval drivers, which means, assuming an equal percentage rise, that their un-mounted resonant frequency was 145/1.17 = 124 Hz.

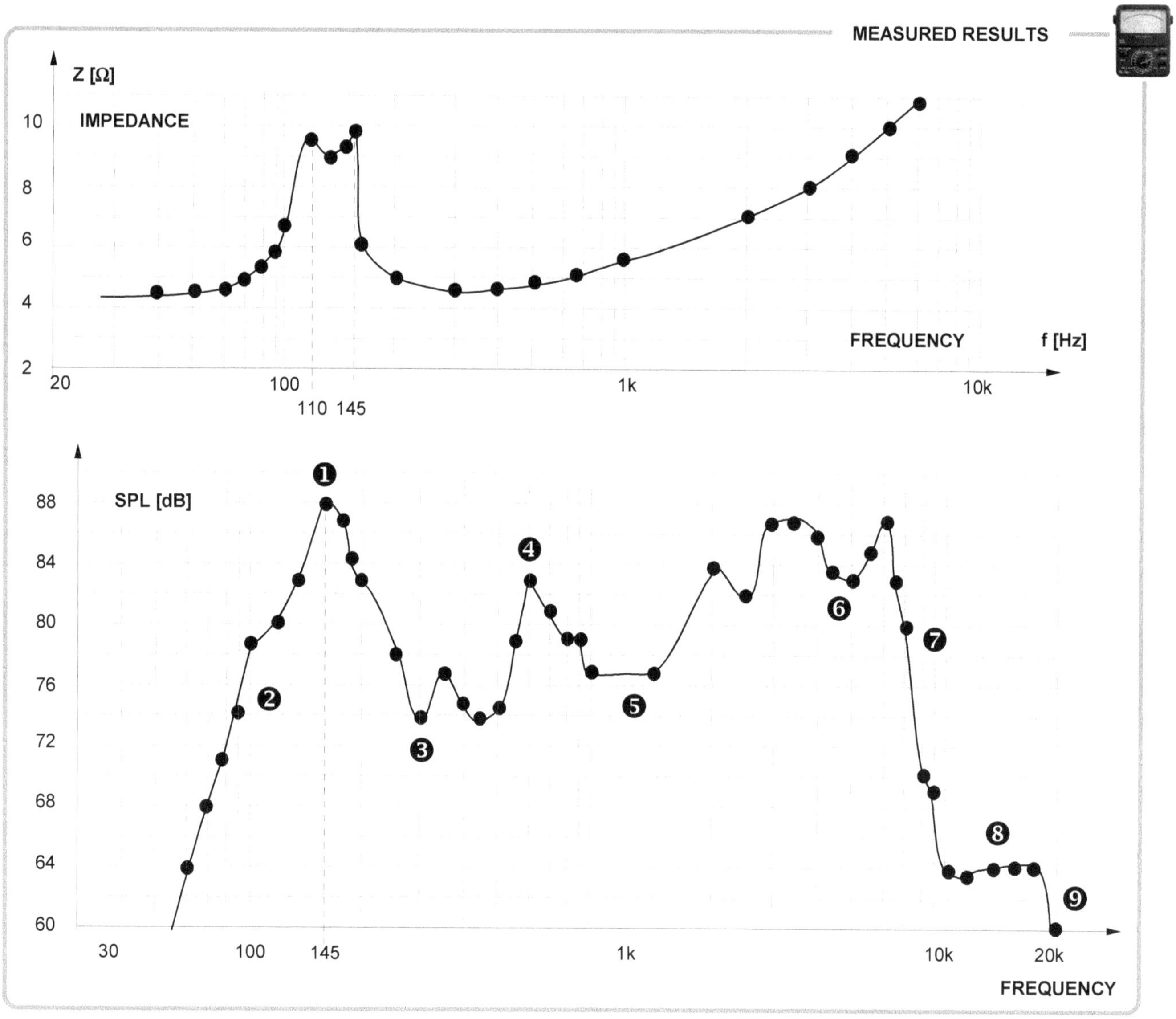

A peak at the upper resonant frequency of 145 Hz was prominent (1), below which the response dropped abruptly (2) in a straight fashion, as it did at higher frequencies (3), only on this side the fall settled around 300Hz and remained constant in the 300-600Hz region, only to rise again and peak at 700Hz (4).

The rapid fall between 700 and 1,000Hz was followed by a 1-1.5 kHz shelf (5). Another, higher shelf is prominent between 2 and 6kHz, with a minor dip around 5kHz (6). Above 7kHz the fall was rapid (7), obviously the two oval speakers could not mitigate it due to their low power ratings, but they did stabilize the response above 10khz. Between 10 and 18kHz this horizontal "shelf"(8) is most likely due to the two oval drivers continuing to produce sound long after the 10 incher gave up. Finally, at 20kHz, the response dropped to 60dB (9).

The cab sounded very dynamic, sparkly and exciting. The steep drop below 145Hz meant there was no muddy and annoying "one-note" bass to obscure the higher registers. The scooped-up upper bass-lower mids (3) were barely noticeable, but could be easily compensated for by the tube amp's tone controls.

The drop after 7kHz attenuated the guitar's upper harmonics just enough, and although the final shelf (8) was on average 82-64=18dB below the previous one (6), some of that sparkle and shine remained. AU$89 for the Celestion 10-30, AU$80 for the two vintage ovals shipped from the USA, and AU$20 for the cab, a total of AU$189, an incredible value-for-money for one of the best sounding cabs in this book!

LOUDSPEAKER FAULTS & FAILURES: QUICK CHECKS, SIMPLE FIXES, AND EASY MODS

11

Guitar and bass speakers and speaker cabs are incredibly tough and resilient, but they do fail occasionally.

So, it's important to understand how and why they fail, and how to minimize such failures.

When it happens, musicians should be able to perform a few quick & easy checks & tests to identify the nature of the fault, and, in some cases, to fix the simple problems themselves.

- HOW AND WHY SPEAKERS FAIL
- QUICK SPEAKER CHECKS
- HOW TO MEASURE SPEAKERS' SENSITIVITY, IMPEDANCE, AND FREQUENCY RESPONSE CURVE
- QUICK & SIMPLE SPEAKER DRIVER MODIFICATIONS

HOW AND WHY SPEAKERS FAIL

Considering how hard they work (both mechanically/structurally and electrically) and how relatively primitive they are (in terms of their design and the materials used), it is a small miracle guitar speakers (and speakers in general) work as well and for as long as they do.

Thermal compression and overheating in loudspeakers

The two main types of loudspeaker compression are the voice-coil thermal compression and the displacement or speaker cone excursion compression.

When the driving signal current flows through a speaker's voice coil, it heats it up. The resistance of a voice coil may be very low (a few ohms), but the currents flowing through it are pretty high, so the power dissipated into heat on such a physically small coil of wound wire becomes significant.

As a result, the temperature of the coil increases substantially. That's why many tweeters use ferrofluid cooling. They are much smaller than woofers (the built-up heat cannot quickly dissipate) and can easily burn out due to overheating.

The resistance of a hotter copper voice coil at a temperature T will rise to the value $R_T=R_0[1+\alpha(T-T_0)]$, where R_0 is the resistance at a reference temperature T_0 (usually given in tables for 20°C) and α (alpha) is the temperature coefficient of resistance for the particular metal.

For copper $\alpha=4.041*10^{-3}$ [1/°C] or 0.4 percent, so if the DC resistance of the voice coil is 5.6 Ω at 20°C, at 70°C will rise to $R_{70} = 5.6[1+0.004041(70-20)] = 6.73\ \Omega$, an increase of 20.2%!

If you thought that 70°C was hot, keep in mind that copper or aluminium wire in the voice coils of powerful speakers working close to their rated power levels can reach a hundred degrees more, yes, 170°C!

For practice, repeat this calculation for such an extreme temperature. The answer: 9.0 Ω, an increase of a whopping 60%! At 250°C (a speaker working beyond its limits) the resistance of the voice coil will be double its room temperature figure.

So, if you ever noticed that your speaker cab sounds different at a start of a practice session or live performance and an hour or so later, you are not imagining things. As a consequence, its loudness (the SPL level produced) will reduce once is speakers' voice coils heat up, a reason why guitarists often keep cranking their amps' master volumes up as the session progresses!

This is a very dangerous practice, since it may take the speaker beyond its power limits, causing overheating and a very likely failure.

BELOW: As an experiment, we took this already damaged Australian-made vintage FluxTone driver and connected it to a DC power supply. After about 10 seconds, the smoke appeared, but it took almost a full minute for the voice coil to burn out. The paper former was blackened (1) but kept its shape and structural integrity. The enameled copper wire also got charred (2), the change of its color cannot be easily seen on an B&W photo but is easily discernible in real life or on a color photo.

Dynamic compression and extreme cone travel beyond its maximum excursion limits

The displacement compression happens when the driving signal pushes the speaker cone beyond its maximum excursion limits. Dynamic compression causes a change in tonal balance. If a woofer is overdriven, its contribution relative to the midrange driver and tweeter (if used) is reduced. The speaker will sound brighter or even shrill, with more prominent midrange and treble frequencies.

Below we see a VC at rest (A) and at a point of the maximal permissible excursion (d_{MAX}). At that point the magnetic gap is still fully filled with turns of the voice coil, the same active number of turns. Beyond that point, with a yet larger audio signal, the VC will more even further forward, so now a smaller number of its windings (turns) is inside the gap. The magnetic and thus the mechanical force on the VC and the attached cone (not shown) is not lower and is not linearly proportional to the amplitude of the music signal! A dynamic "compression" has happened.

Of course, the same effect happens with a large negative signal, where the VC moves backwards (to the left).

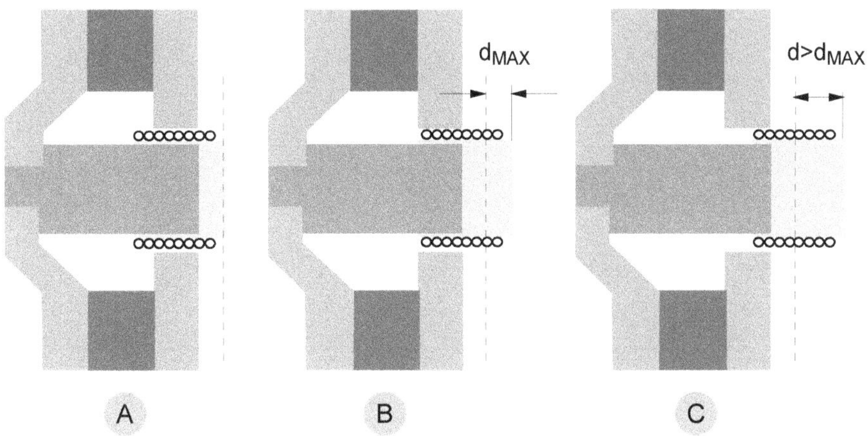

LEFT: How loudspeaker's voice coil travels during compression and overload

A) No audio signal - speaker's voice coil (VC) at rest (fully retracted)

B) The largest allowable excursion d_{MAX} of the voice coil

C) Audio signal too large, the VC travels beyond its allowable maximum limit (d larger than d_{MAX})

NOTE: The spider (voice coil and cone support), the cone and the dust cap are not shown to preserve clarity

Amplifier and loudspeaker "clipping" creates new, unpleasant sounding harmonics

Dynamic compression in speaker drivers is similar to "clipping" in amplifiers. The term describes that the peaks of large signals (a sinewave signal illustrated below, but this applies to all "peaky" signals) are clipped, the amplifier or the speaker simply cannot reproduce such peaks due to electrical or mechanical compression.

A voltage input versus output is illustrated in the case of an amplifier, and a VC+cone displacement "d" is the variable illustrated in the case of a loudspeaker. Such dynamic compression creates two major problems.

In the case of a speaker, the produced SPL is not as high as it should be, so the loudness of the speaker isn't proportional to the amplitude of the driving signal produced by the guitar amp.

An even bigger problem is the change of tone. We will use a simple case of a pure sinusoidal tone (like the one produced by a function or tone generator for testing purposes). Once its peaks are clipped, the distorted signal resembles more a square wave than the original sine wave.

In time domain, a distorted signal (thicker line) does not look that bad. but in the frequency domain, where frequency components of the signal are visually presented, we get a better picture.

A spectrum of a square wave (next page) shows a fundamental tone (the same frequency as the original sine wave f_1) and an indefinite series of higher harmonics.

Only odd harmonics are present, third, fifth, seventh, and so on. remember, a third harmonic is a sine wave with three times higher frequency than the fundamental ($3f_1$).

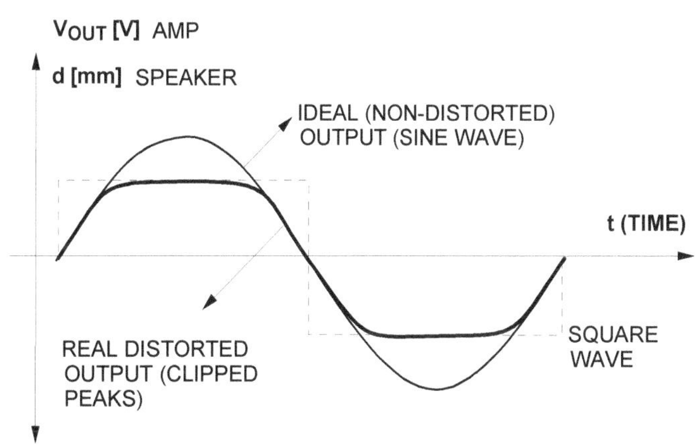

ABOVE: In time domain, a distorted signal (thicker line) does not look that bad. "Sure, it's peaks are clipped or rounded, but it will still sound good?" you may ask. No, it will sound awful. On the next page you will see why ...

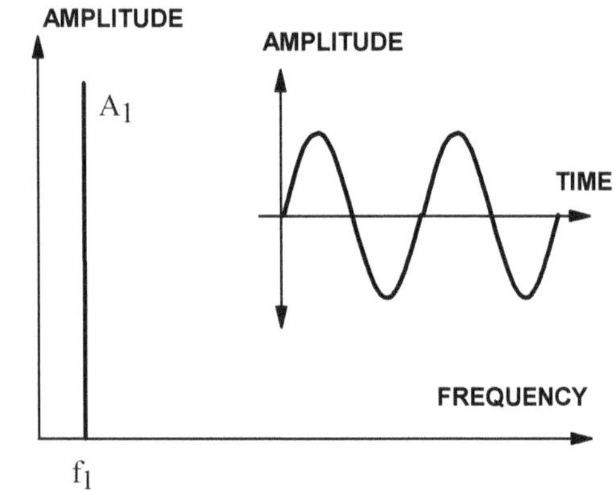

ABOVE: The spectrum of a pure (undistorted) sine wave shows a single frequency component (a "fundamental" tone)

RIGHT: The spectrum of a square wave of the same frequency as the sine wave shows the fundamental plus odd harmonics (3rd, 5th, etc.) and their amplitudes relative to the amplitude of the fundamental A_1

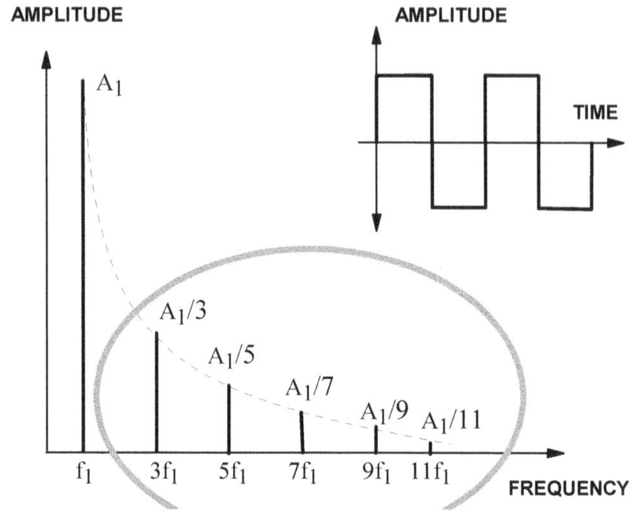

DISTORTION TONES (HARMONICS) ADDED BY THE SPEAKER OR AMPLIFIER

This means that clipping and dynamic compression produces new tones which were not present in the signal produced by your guitar or your amp. Since odd harmonics are not harmonically-related to the original, they sound shrill and unpleasant.

Interestingly, the asymmetrical distortion, where only the positive or negative peak is flattened or clipped, results in the creation of even harmonics (the 4th, the 6th, etc.), which generally sound harmonious and pleasant, add "body", "warmth" and "character" to the sound.

This is especially noticeable in hi-fi. Single-ended amps clip asymmetrically and such clipping often goes unnoticed since it sounds warm and "musical". In push-pull tube amps the even harmonics (distortion components) generated in the output stage get canceled in the output transformer, which leaves the odd harmonics "exposed" or "unmasked". That is why most audiophiles and many guitar players prefer the rich, "textured" and "nuanced" sound of single-ended tube amps.

Failures caused by decentered cone

A de-centered cone doesn't result in an immediate demise of a speaker but in "coil rubbing", the scraping of VC windings and/or the VC former against the magnet structure. The result is unpleasant distortion and gradual stripping of the coil wire insulation, which can then lead to short-circuited turns and further cascading effects.

Many primary failures can cause VC decentering and coil rubbing, a deformed or loose magnetic structure (A), due to a mechanical shock, such as when a speaker is dropped on the floor, a spider that has lost its elasticity and gets deformed (B) and a failure (such as drying out) of the glue that bonds the voice coil to the spider and cone.

The overheating of the voice coil

The overheating of the VC assembly, a result of overdriving the speaker electrically, can cause an electrical failure (secondary fault) or a mechanical failure.

In the former case the magnet wire's insulation will melt and evaporate, leading to shorten turns and an even higher current through the voice coil, a cumulative process that will speed up its destruction.

Sometimes the voice coil former softens up and buckles under heat (C) before the magnet wire or its insulation melts. In that case such deformity will again cause identical symptoms as (A) and (B) .

Failures caused by the overexcursion of the cone

You may recall X_{MAX}, one of the parameters specified by speaker manufacturers. For the 8Ω Jensen P12R driver, X_{MAX} is +/-1 mm, while for its 16Ω twin, the X_{MAX} is only +/-0.75 mm! Not much, isn't it?

Repeated over-excursions of the cone can lead to the spider losing its elasticity and becoming deformed, leading to the same secondary fault as in (B). In extreme cases, the VC can hit the back plate ("bottom out") and become deformed (E), resulting a warped voice coil and coil rubbing, or, on the other side of the excursion, the whole voice coil can jump out of the magnetic structure, twist to one side and get stuck. In that case the VC cannot return to its normal resting position (F).

Foreign objects stuck in the air gap

Foreign objects, both magnetic and nonmagnetic, can get stuck between the voice coil windings and the magnetic structure, as in (D), or inside, between the VC former and the center slug (CS). Sometimes the stuck piece will completely prevent the coil from moving, in which case, with time, assuming the signa keeps driving the speaker, the VC will overheat, since there is no cooling effect of the VC and cone movement.

If the VC is still able to move, the symptom will be similar to coil scraping - reduced acoustical output, distortion and mechanical scraping noise.

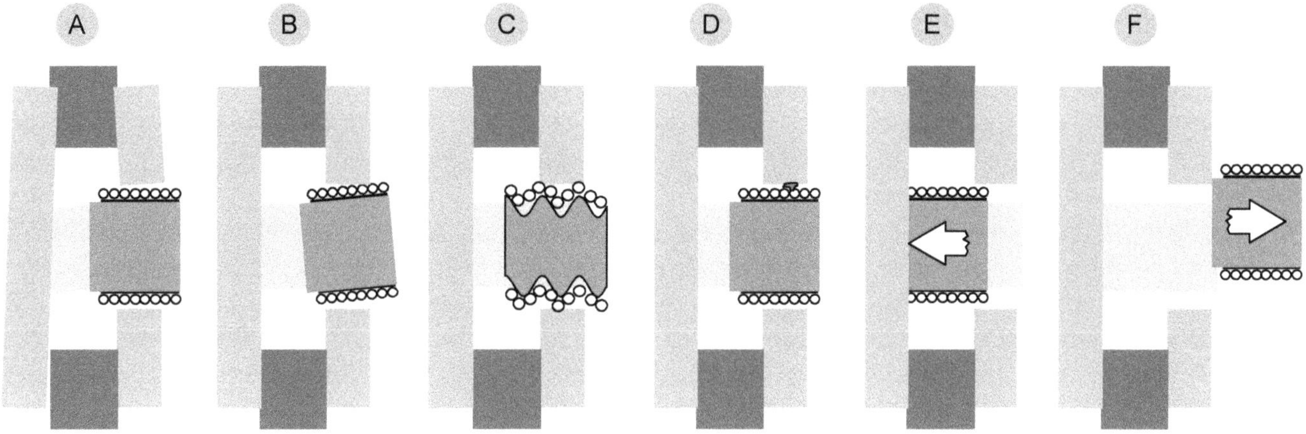

ABOVE: The most common mechanical failure modes. Some are primary faults, others secondary, caused by a primary fault, for instance when overheating of the VC causes its deformation, which then causes coil rubbing and distortion

Wire and wire joints failures

The solid wire used to wind the voice coil is not soldered directly top the drivers terminals. The constant flexing would cause fractures and an eventual breaks. The VC wire is brought out and glued along the surface of the cone to the termination point where it is soldered to a flexible "tinsel" wire that connects it to the driver terminals.

Made from superfine copper or silver strands, the flexibility is achieved by braiding the multiple strands together and adding interwoven cotton to prevent chafing and fractures.

Failures are still possible, of course. The soldered joints fatigue, crack and fail, resulting in either a total loss of sound or an intermittent crackling. resoldering the connections is an easy fix.

However, the solid VC wire may still fracture in the vicinity of its attachment point to the voice coil, in which case repair may or may not be possible.

QUICK SPEAKER CHECKS

The visual inspection and the push check

There are two quick and simple ways to check loudspeaker drivers' mechanical and structural health. The first is the visual inspection. There should be no holes, tears, or rips in the paper cone.

If there are, think carefully about buying such a speaker. Reconing can be done by experts (if a speaker is rare and valuable and hard-to-replace) but is costly.

The second test involves gently pushing the speaker cone by hand, both inward (from the front of the cone) and outward, pushing the cone from the speaker's back, where the metal frame (basket) and the magnet are.

The cone should move freely and return to its original position.

ABOVE: Due to high stresses around its perimeter, rips most often happen around the edge of a cone

If any sudden resistance or rubbing can be felt or if any "scraping" sound can be heard, that means the voice coil is rubbing against its surroundings. The "spider" or speaker's suspension has been damaged and/or misaligned, or there is foreign matter stuck in the air gap. Such speakers may work to some extent (produce sound) but will sound distorted and will eventually fail.

Sometimes a speaker passes the push test but its VC windings are loose (due to overheating and the melting of the adhesive used to glue the voice coil to its former) and will rub against the magnet in operation.

How to remove the dust cap and inspect & clean the air gap

The voice coil of this 12"' speaker made a scratching sound when the cone was gently pushed inward. In most cases that is a death sentence for a speaker. If the voice coil former is deformed and rubbing against the magnetic structure around it, there is nothing you can do to fix it.

If you are lucky and the rubbing is cased by a speck of foreign matter stuck in the air gap between the VC former and the magnet, it may be possible to dislodge and remove it. In this case we managed to do that and, after gluing the dust cap back to the cone, return the speaker into service.

In this case the dust cap was a piece of cloth, impregnated with a stiffening agent and pressed into a spherical shape. To remove it, take a thin sharp blade and gently separate and pry open a small section around its circumference.

Then, hold the dust cap with two fingers and, moving around its edge, gently separate it from the cone. The strong glue used is much stronger than the paper from which the cone is made, and will cause small pieces of the paper cone to remain attached to the fabric of the dust cap, leaving a somewhat rough edge (1). This will mostly be covered up by the dust cap once it is glued back onto the cone, so don't worry too much about it.

Now you'll be able to see the center piece (2) around which the voice coil (3) moves, and the two wires entering from the outside of the cone (4) and connecting to the voice coil.

Don't pry or stick anything into the gap trying to clear it of debris. Blow air into it while gently pushing the coil in from both sides of the cone - this may be sufficient to dislodge small nonmagnetic particles stuck in the gap.

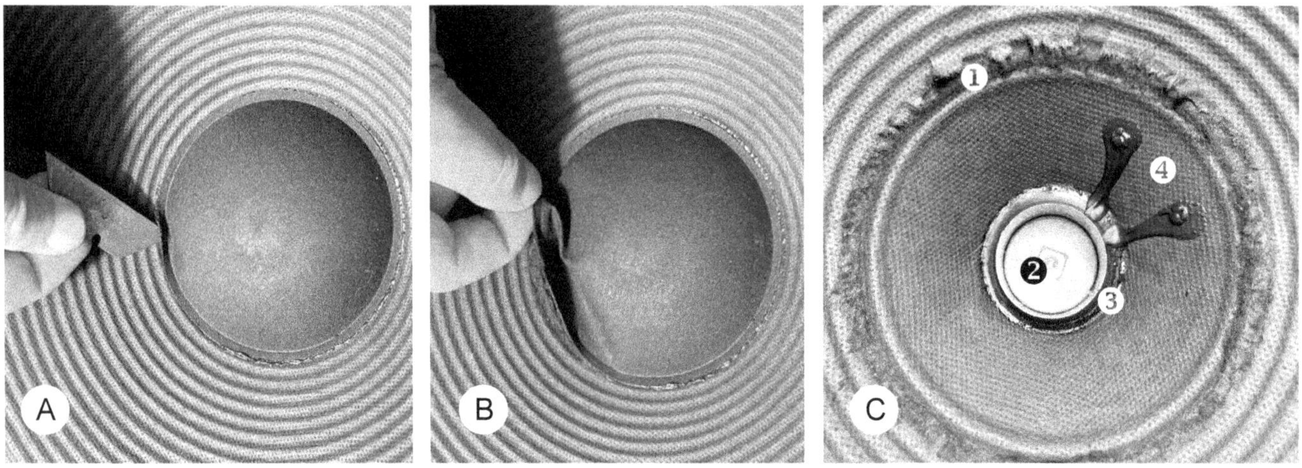

The battery test

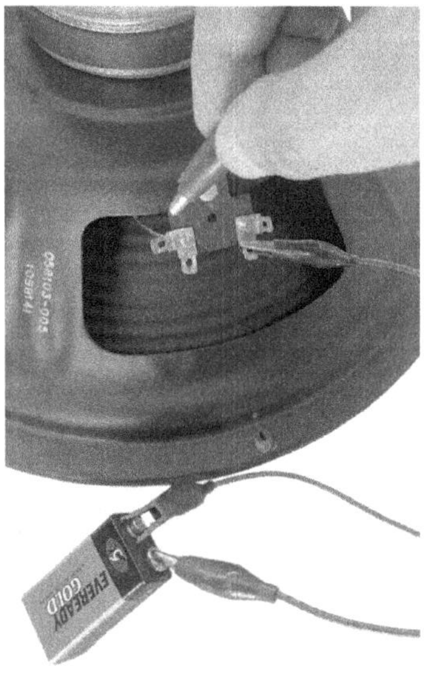

Take two leads terminated with crocodile clips and connect a 1.5V or 9V battery across speaker terminals.

With battery's + pole on speaker's + terminal, speaker's cone should quickly move forward (outward) and stay there! With the battery's + pole on the speaker's - terminal, the speaker cone should promptly move back (inward) and stay there.

If the speaker cone moves in the opposite directions of those just described, the battery's polarity markings are the opposite of the markings on the speaker terminals.

For some weird reason, a few speaker manufacturers of the last century used to mark their speakers' terminals the opposite way of what is customary today. Some, like the vintage speaker pictured, had no markings at all!

Multimeter DC resistance check

Once a speaker passes these two quick checks, it is ready for electrical tests. Set your multimeter on a low resistance range and measure the resistance across the speaker's terminals. Since this is a DC (direct current) resistance (DCR), it will be lower than the speaker's nominal impedance.

For instance, an 8 ohm nominal impedance speaker will usually show a DCR of between 5.5 and 6.7 ohms. That is fine and confirms that the speaker's voice call has continuity, i.e., that it hasn't burned out. The speaker in the photo had a DC resistance of exactly 6.0 ohms.

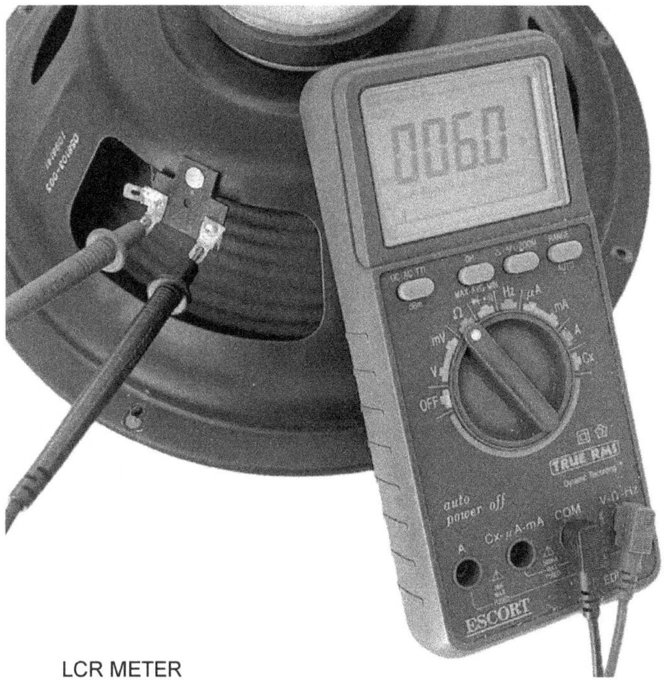

LCR meter "beep" test

Finally, the best and most conclusive test is by connecting a digital LCR (on "L" or inductance range) meter across speaker terminals.

The LCR meter will send current pulses through the voice coil and show the voice coil inductance on its LCD (display) at the selected test frequency. These pulses should be audible; the speaker should beep about once a second.

Most LCR meters can test at 60 Hz or 1 kHz, although some can also test at 10 and even 100 kHz. Use the 1 kHz setting, the most meaningful frequency. Most guitar speakers cannot reproduce a 10 kHz signal at any significant level.

An added bonus of this check is that the loudness of these beeps can be used as a rough (subjective) indication of the speaker's sensitivity. The higher the speaker's sensitivity or SPL level (dB/W), the louder the beeps will be.

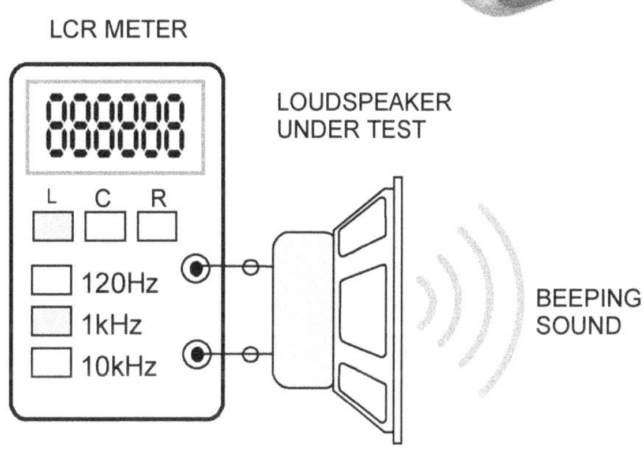

ABOVE: With an LCR meter set on "L" or inductance range, a pulsating tone should be heard in the speaker. The higher the speaker's sensitivity or SPL level (dB/W), the louder that tone will be.

HOW TO MEASURE SPEAKERS' SENSITIVITY, IMPEDANCE, AND FREQUENCY RESPONSE CURVE

Most bass and guitar players have no idea about their amp's or speakers' technical specifications and performance parameters or even an inclination to find out. If you want to know more, performing some basic tests isn't difficult, although it requires an access to a few basic instruments.

The good news is that quality multimeters, LCR meters, and sound pressure meters have become very affordable, and used instruments can be found at bargain prices. Even function generators and handheld LCD (liquid crystal display) oscilloscopes are getting cheaper by the day.

Why is it useful to know the sensitivity of your cab's speaker(s)?

If you like how your amp-cab combination (or your combo amp) sounds, but would like to make it louder, the cheapest and often the only way is to substitute the existing speaker(s) with a more sensitive ones. Remember, even if a replacement speaker is only 3 dB more sensitive, the increase in loudness will be equivalent to doubling the power of your amp (with your existing speaker)!

So, (let's say you are a VOX AC15 dude), before you rush out and fork out your hard-earned cash on AC30, investigate how the current speaker in your cab compares to others, and which replacement speakers would give you that +3dB (double power) boost.

Likewise, just in case VOX amps don't light your fire and you've missed my point, that would be the difference of selling your 50 watt Marshall or Fender head and buying a 100 watt one!

How to measure the sensitivity of speaker drivers and speaker boxes

You can measure the sensitivity at one particular frequency or repeat this test at various frequencies to get a further insight into how its sensitivity changes. The key is constant 2.83 V_{RMS} reading on the voltmeter. If the voltage drops or rises, adjust either the volume control on the amp (1) or the amplitude control on the function generator (2) to bring it back to the required constant level.

Better quality, lab-grade function generators usually have the capability to provide the higher currents required to drive the speakers directly, and a constant voltage output. If yours doesn't, use an amplifier that will ensure the sufficient current capability and a constant output voltage.

Hi-fi amps are generally superior to guitar amps in those respects, as well as having a much wider frequency range and a flatter frequency characteristic.

Since most guitar speakers and speaker cabs are relatively efficient and sensitive, especially compared to hi-fi speakers, a 2.83 Volts test signal (to produce 1 watt input on the speaker's 8 ohm nominal impedance) will be very loud, so take a quick reading of your SPL meter's indication and turn the function generator off quickly.

Speakers of other nominal impedances will require test voltages of different RMS values. Since power $P = V^2/Z$, voltage $V = \sqrt{(PZ)}$. In this case the power level is P=1 watt, so the formula simplifies into $V=\sqrt{Z}$.

For 8 ohms, as we know, $V = \sqrt{8} = 2.83$ V, for a 2-ohm cab $V = \sqrt{2} = 1.41$ V, for 4 ohm speakers and cabs $V = \sqrt{4} =$ 2 V, 2.45 V for 6 ohm speakers, and 4 V for 16 ohm speakers/cabs.

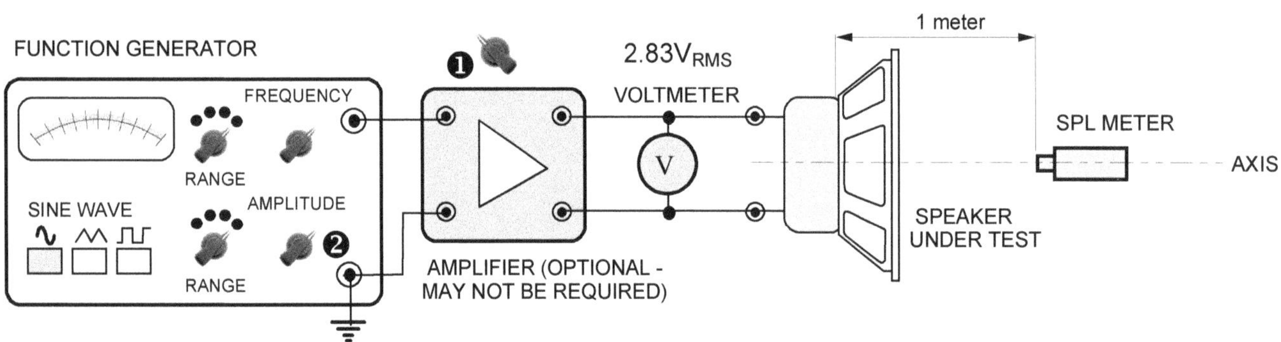

ABOVE: The test setup for measuring the sensitivity (dB/Watt) of loudspeakers . Test tone(s) from a digital or analog source can be used instead of the function generator.

How to plot the loudspeaker impedance and SPL curves using the current source method

You can use this simple setup to measure speaker drivers by themselves or whole speaker boxes. Speaker manufacturers perform this test in special anechoic test chambers, so your results will be different from theirs. Your test room will include sound pressure from both direct sound waves and the reflected ones; it will also include the background noise level, which can be significant.

No matter how quiet your room may be, there will always be a residual background noise level. The aim is to make it constant, by avoiding sudden noises such as planes flying over or cars driving by, dogs barking, etc.

Thus, it pays to do this in a room well-insulated from outside noises and make that room as acoustically "dead" as possible. Minimize hard, reflective surfaces and increase the surface area of soft furnishings, carpets, blinds, and curtains. Consider placing a thick rug at (3) to reduce reflections off the hard floor.

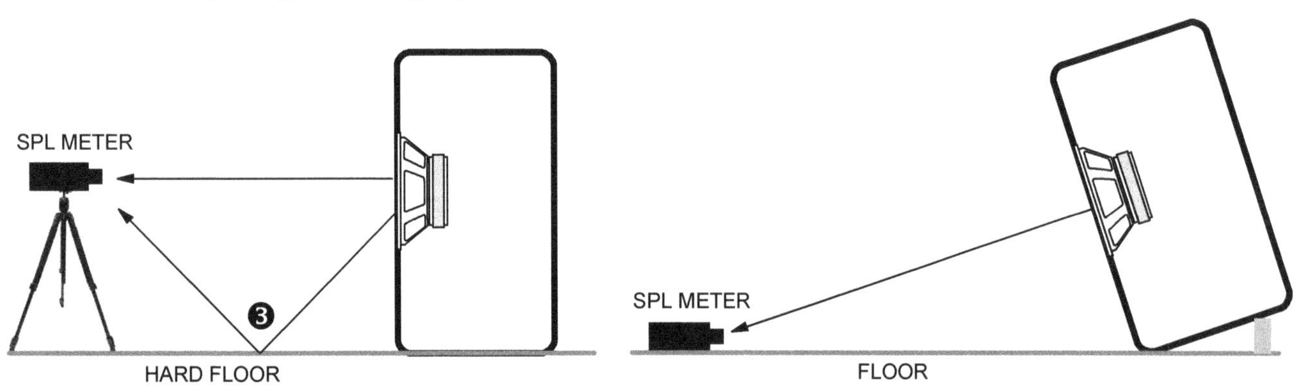

ABOVE: Two ways to measure SPL levels. Choose one or the other and stick to it for all your measurements.

The placement of the cab under test in the room will make a huge difference at lower (bass) frequencies. Avoid placing the cab close to a rear wall, and avoid corners by placing the cab in the middle of the room.

You'll have to stand behind the SPL meter to take readings, which may impact the sound field around the meter and the speakers and slightly alter the measured results, so make sure you always stand in the same spot and don't move around while taking the reading.

Apart from the variable frequency of the test signal (produced by a function generator), all other variables must remain constant. Both the function generator and the amplifier (if used, as in the last test setup on the previous page) must have a flat gain-vs.-frequency characteristic over the entire measurement frequency range.

Most importantly, since usually we are not really interested in the absolute values of SPL but in relative comparisons between cabs, speakers and our modifications (before and after), make sure you to all tests at the same SPL meter distance and in the same cab's position in the room.

Had the speaker's impedance been constant with changing signal frequency, the ratio of the two measured voltages V1 and V2 would remain constant. However, since the speaker's impedance is highly frequency-dependent, the ratio of the measured voltages will also change with frequency.

The voltmeters (multimeters on "AC Volts" range) must be capable of operation up to at least 20 kHz, so cheap multimeters are out of the question.

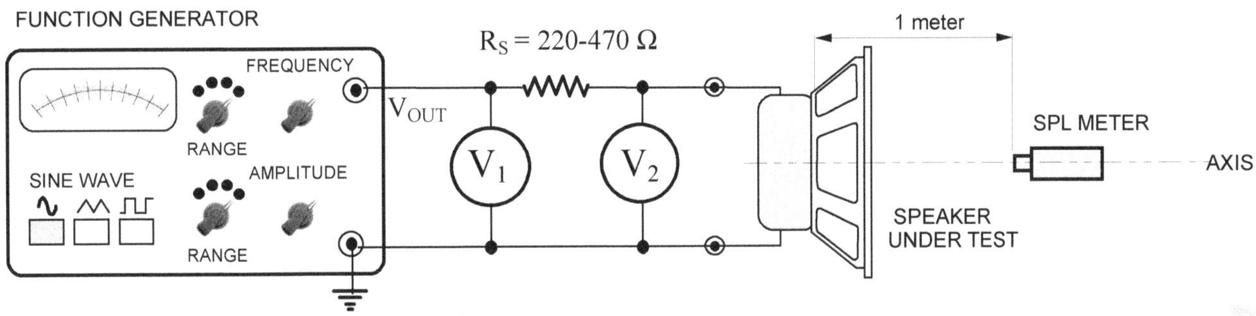

ABOVE: Test setup for measuring the sensitivity (dB/Watt or dB/2.83V) of speakers and speaker cabs, and for manual (point-by-point) plotting of a loudspeaker's impedance - vs - frequency curve

SPEAKER IMPEDANCE BY TWO VOLTMETER METHOD

$$Z(f) = R_S V_2(f)/[(V_1(f)-V_2(f)]$$

$V_1(f)$ and $V_2(f)$ are measured results at various frequencies, Z(f) is the calculated impedance modulus at each test frequency "f".

Impedance curve for Celestion Super 8 speaker — **MEASURED RESULTS**

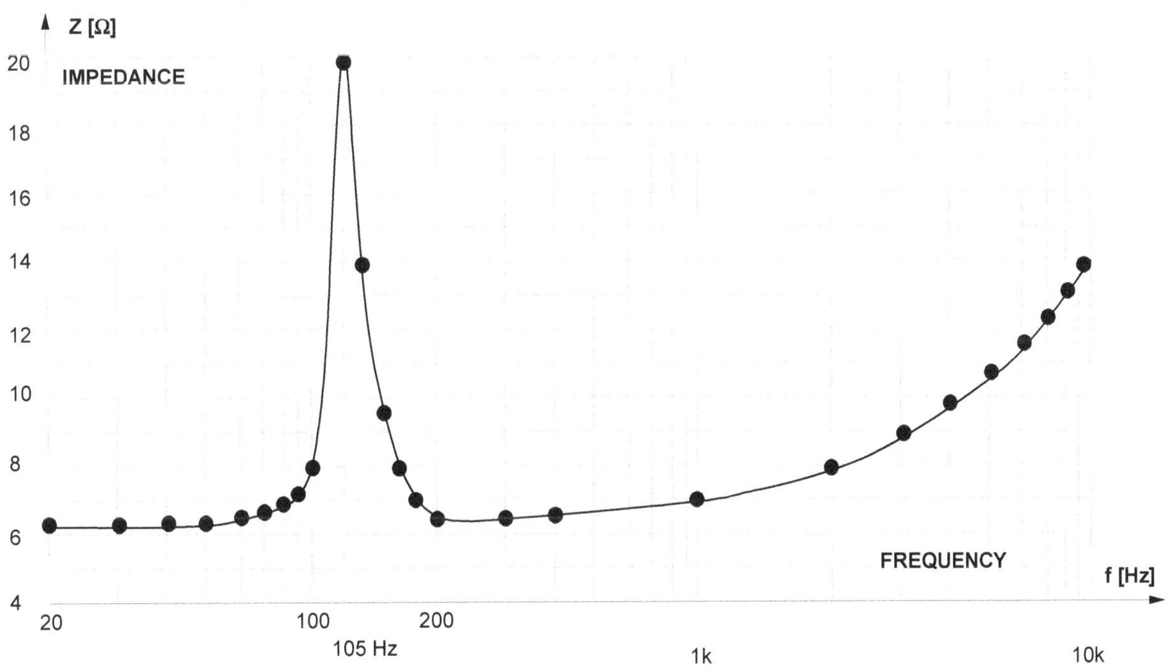

Impedance modulus versus frequency test results for Celestion Super 8 loudspeaker.

In the experiment described here, from 20 to 100 Hz, we took measurements every 10 Hz, then from 100 to 200 Hz every 20 Hz, from 200 to 1,000 Hz every 100 Hz, and then every 1 kHz. After punching in all the figures into a spreadsheet, we got the recognizable dynamic speaker impedance curve. The speaker was suspended in the air, without any enclosure.

SPL (sound pressure level) meters

The operational principle of SPL meters (the three pictured below) is fairly simple. An electret condenser microphone (1) converts the sound pressure into voltage, which is then amplified, filtered and displayed on an analog (2) or digital display (5). The accuracy of these instruments is in the order of ±2 dB.

A bit of simple arithmetic is required on the older, analog models, where he range switch (3) indicates the dB figure that must be added to or subtracted from the meter's reading. The analog SPL meter indicates around -1dB (2) on the 60 dB range (3), meaning the reading is 60-1=59 dB.

The newer models display the overall figure directly (5), although the range switch or button still must be adjusted up or down.

SPL meters offer at least two weighting options (6), usually "A" and "C", and "Fast" and "Slow" display response (7). A-weighting has an amplitude-frequency characteristics that attenuates low and high frequencies and limits the meter's response to midrange frequencies where the human ear is the most sensitive. Thus, it is used mainly for noise checks in health & safety applications.

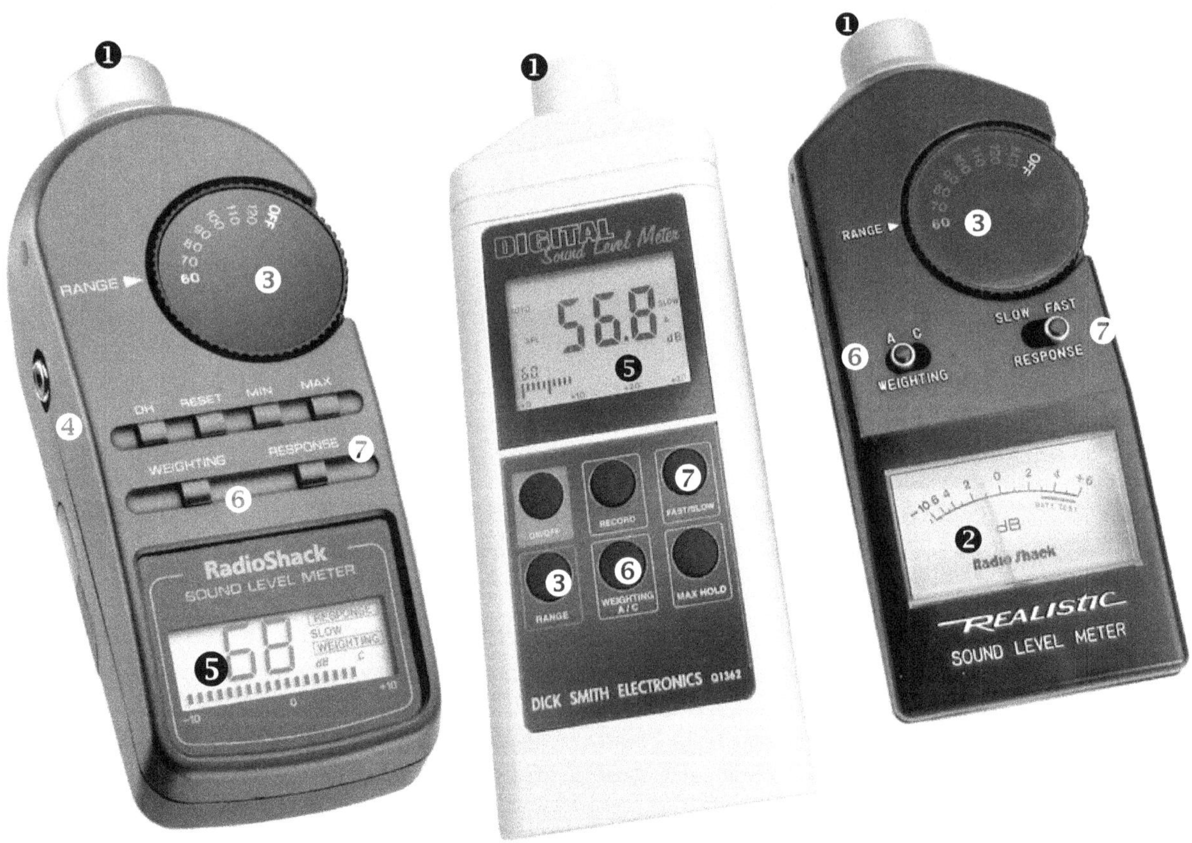

C-weighting has a much flatter frequency characteristic and is more suitable for audio measurements.

There is also a signal output of 1 V_{PP} (open circuit, full scale at 1 kHz) with less than 2% distortion (4).

For the new kids on the block, there is even a NIOSH Sound Level Meter app for iPhones, which can be used with both the inbuilt and external microphones.

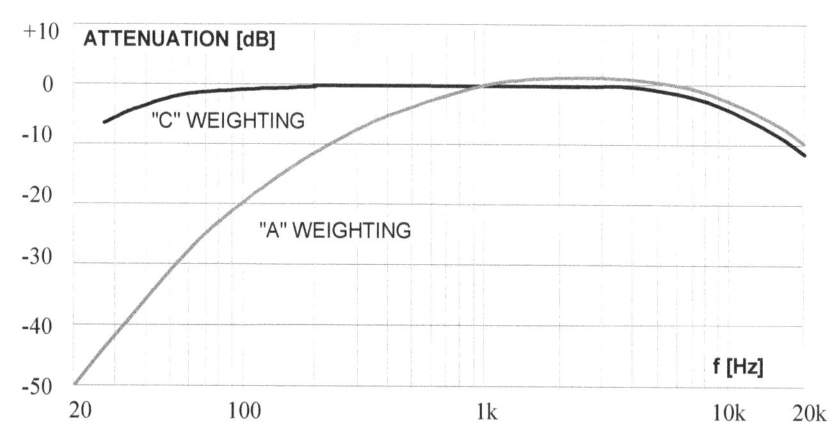

QUICK & SIMPLE SPEAKER DRIVER MODIFICATIONS

The way it used to be: The Trifecta

As with most other examples of innovative loudspeaker engineering, PACO (Precision Apparatus Company) Model L4 (early 1960s) was a high fidelity speaker. Nevertheless, it illustrates a few important principles and improvement possibilities for guitar speakers.

A top-mounted tweeter (1) and a bass-reflex port (2) at the bottom of its front baffle are immediately noticeable, as are its three identical woofers/midrange drivers (3). Mounted in close proximity to each other, they act as a single mass in moving air, but without the inertia (sluggishness) of larger speakers.

Due to their smaller and lighter cones and voice coils, their upper frequency limit is significantly extended. This phenomenon is due to the fact that the drivers in such close physical proximity are acoustically-coupled.

The total area of three 6" drivers (3 x 176 cm^2) is 528 cm^2, larger than the area of a typical 10" driver (452 cm^2).

Two simple yet effective measures were taken to improve the performance of Model L4. A lightweight diffuser of the table tennis (ping-pong) ball size was bonded to the voice-coil former. It dispersed the mid-high frequencies and widened the normally narrow beam of the driver.

Each driver also had three struts glued to its paper cone and to the ball diffuser, eliminating the edge distortion and increasing the rigidity of the cone, thus extending the piston-operation of the cone up to 7-8 kHz.

This trick also lowered the resonant frequency of each driver down to 40 Hz, something only much larger and more expensive speakers could hope to achieve.

So there you have it. By using the couple of hacks just mentioned, you can improve the upper frequency extension of your dull-sounding speaker cab, adding sparkle and chime and widen its high frequency dispersion.

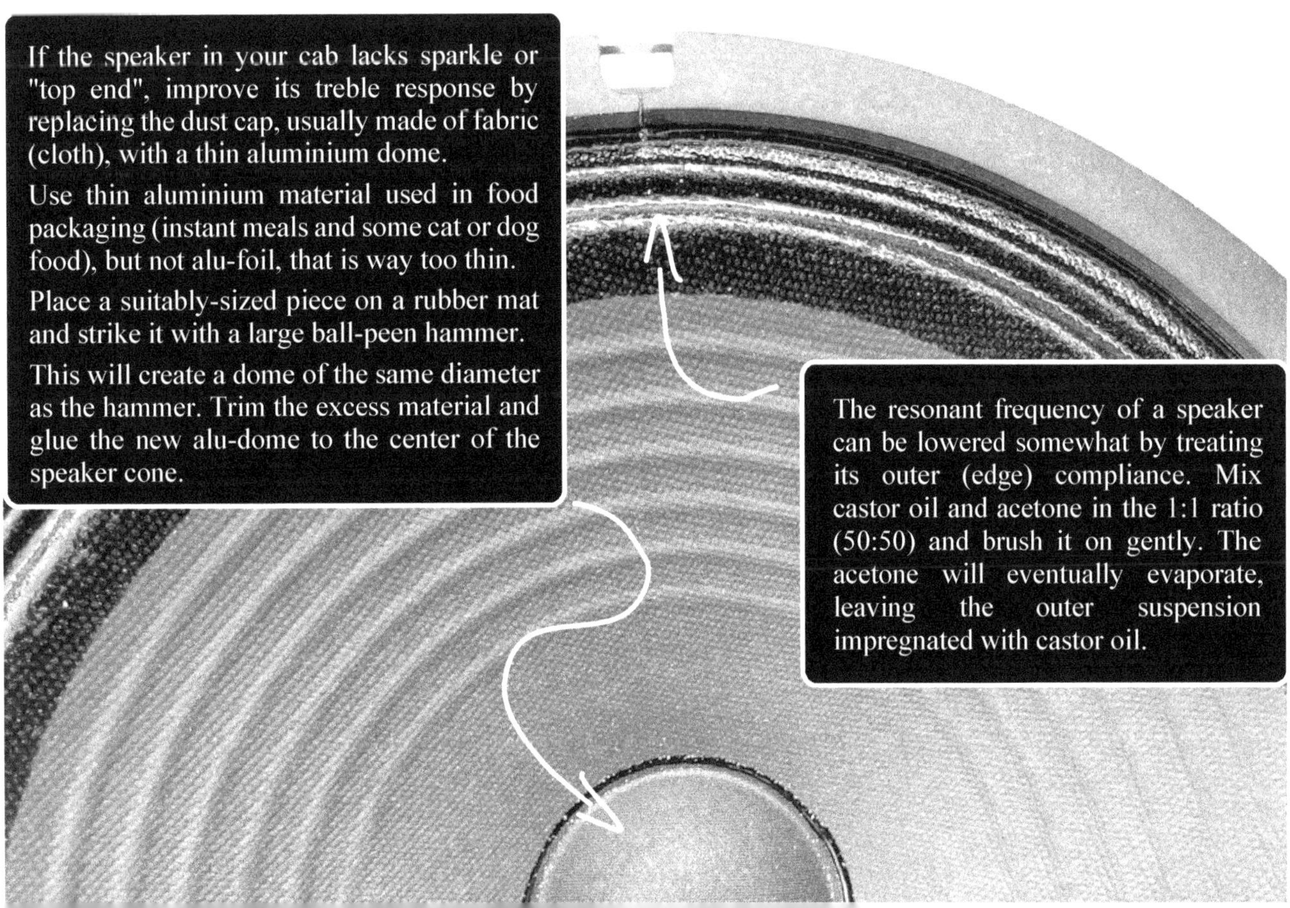

MINIMIZE CONE BREAKUP EFFECTS **DIY PROJECT**

Impregnating a speaker's cone with varnish or any other type of sealant will increase both its stiffness (desirable) and mass (undesirable) and thus change its physical parameters, some for better, others for worse. However, such a drastic step is irreversible - if you don't like the sonics of the newly-treated driver, there is no way of removing the coating and reversing the modification. This hack is not only faster and cheaper than the cone-coating option, but is fully reversible, too.

Buy a PU (polyurethane) foam self-adhesive seal strip for doors and windows from your local hardware or home-depot store. Cut four strips of a suitable length (which will depend on the size of your speaker) and stick them onto the cone at an angle, as illustrated below.

Four strips are usually sufficient, but you may experiment by using six or even eight. For this photo we used a white sealing strip so it's in contrast to the dark gray cone, but usually a black seal (as pictured in its packaging) is less obtrusive and looks better.

You can check the results by measurements, comparing the speaker's SPL-vs-frequency curves before and after, especially above 2 kHz, in the cone breakup region, the land of peaks and valleys.

Alternatively, if you don't have the required test equipment or cannot be bothered, simply play your guitar through your amp and the modified cab and see if you can hear any tonal difference.

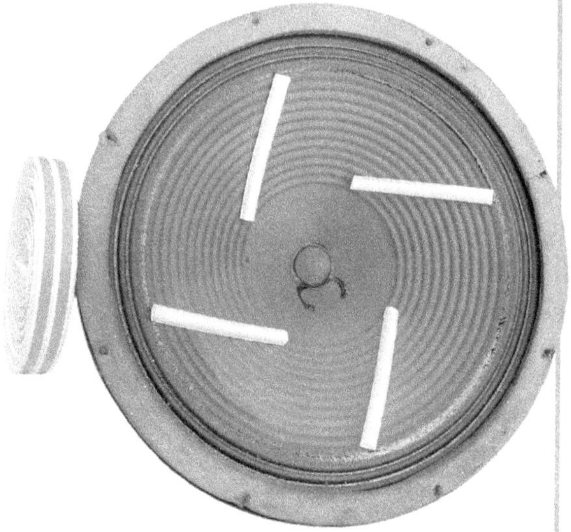

NOTES

THE ELECTRONIC CORNER: CROSSOVERS AND FREQUENCY COMPENSATION NETWORKS

12

Crossovers aren't often used in guitar cabs, but are important in bass, keyboard and PA speakers, not to mention their hi-fi brethren.

Adding a tweeter and blending its output with the existing speaker(s) in your cab to improve its tone and add sparkle to its top is not difficult. It could be a far better option than selling your "dull sounding" cab and buying a more expensive one ...

Likewise, just like tone controls on your guitar amp, frequency compensation networks can be used to tame down the bass resonance and boominess, or to smooth out the shrill treble of cabs that are too bright and brittle.

- CIRCUITS 101: RESISTIVE, CAPACITIVE AND INDUCTIVE IMPEDANCE
- HOW CROSSOVERS WORK
- PRACTICAL CROSSOVERS
- COMPENSATION NETWORKS

Unless you've had some formal training in electronics or an extensive practical experience in amplifiers and other electronic circuits, you will not be able to understand crossovers in any meaningful depth without the knowledge of passive electronic components and filters. In essence, that is exactly what a crossover is, a passive network of interconnected passive electronic components (resistors, inductors and capacitors) that is as a filter between an amp and various speakers (drivers) inside a loudspeaker box.

The following few pages are a quick course to get you up to speed with concepts such as impedance, phase (angle), resonance and the like.

CIRCUITS 101: RESISTIVE, CAPACITIVE AND INDUCTIVE IMPEDANCE

The equivalent of resistance in DC circuits is *impedance* (symbol Z) in AC circuits. While resistance is always a real number, a number of ohms, impedance is a complex number, comprised of its real part (resistance) and its imaginary part or *reactance*: $Z(j\omega) = R(j\Omega) + jX(j\omega)$.

The three basic passive components used in electronics, represent three fundamental physical properties, resistance, capacitance and inductance. As always, to make our journey smooth and gradual, we will start with ideal resistor, ideal capacitor and ideal inductor.

Ideal resistors

An ideal resistor has a constant resistance, meaning it does not change with temperature, and, more importantly, does not change with the frequency of the AC signal or with the magnitude of the current flowing through it.

The ideal resistor has no inductance and no capacitance. Its reactance is zero. This means that it behaves in the same way in both DC and AC conditions.

The vector diagram shows the current and voltage phasors "in phase" or parallel to one another.

If the AC current flowing through an ideal resistor is a sine wave of frequency f, the voltage drop that current will produce on the resistor will also be a sine wave, of the same frequency and the same phase.

The amplitude of the voltage drop can be determined by Ohms' Law, just as for DC conditions:

$V_R = I_R * R$

An ideal resistor is a linear device, there is no amplitude distortion of either current or voltage, there is no phase shift between them, so no phase distortion either.

The higher the current through an ideal resistor, the higher the voltage drop across it, so the V-I graph is a perfectly straight line.

In any operating point Q, the slope of this curve or the tangent of the angle between that line and X-axis is constant and equals the ratio of the voltage and the current in that point or the resistance R.

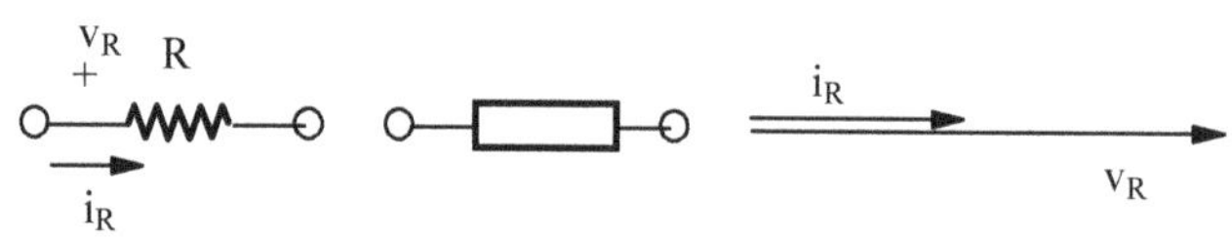

ABOVE: Ideal resistor, its common symbols and the vector diagram

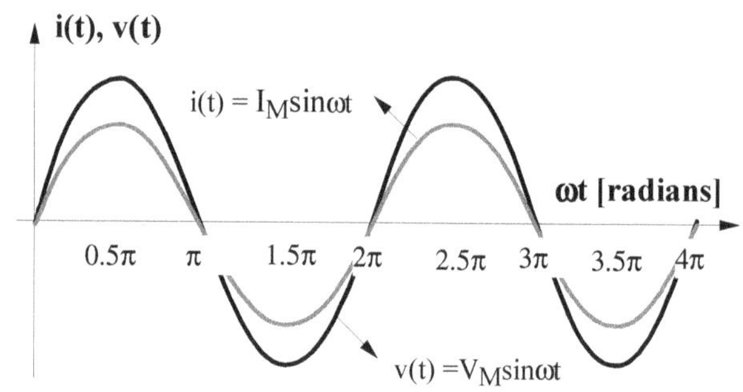

ABOVE: Ideal resistor: voltage and current are in phase (no phase shift)

BELOW: The impedance of an ideal resistor is pure resistance, independent of frequency.The V-I relationship is linear, with constant slope (tan α), equal to resistance R.

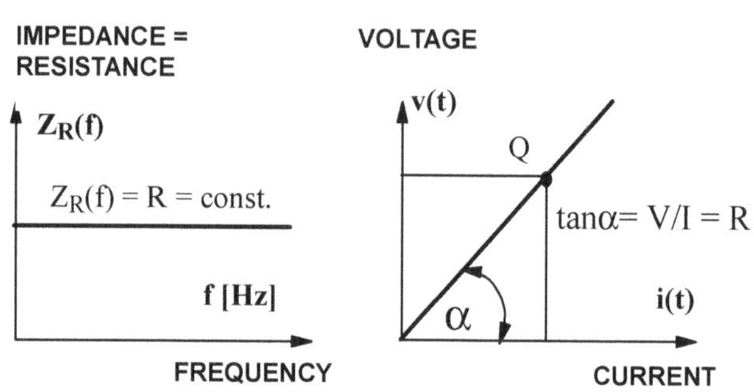

Ideal capacitors

An ideal capacitor would have an infinite resistance for DC current, so it would not allow any DC current to pass. The AC current through an ideal capacitor precedes the voltage by 90 degrees. If you cannot remember this fact, you can deduct it logically.

Just remember that a discharged capacitor (no voltage across it and no charge held inside it) needs a current to start flowing first, in order to charge it. So, the current "leads", while the voltage across it develops slowly (raises), or "lags" behind the charging current.

The impedance (reactance) of an ideal capacitor is inversely proportional to frequency f and its capacitance C: $Z_C = 1/(2\pi fC)$. The higher the frequency of the signal, the lower the capacitor's impedance.

This fact will have important repercussions in amplifier circuits, where unwanted or parasitic capacitances will form low pass filters with adjacent resistances and divert high frequency signals to ground, thereby limiting amplifiers' frequency range.

A 220μF electrolytic capacitor in a tube amp's power supply will have an impedance (reactance) of $Z_C = 1/(220*10^{-6}*2*3.14*100) = 7.2\Omega$ at 100 Hz, but only 0.072Ω at 10kHz. A 100 times higher frequency means 100 times lower impedance!

In audio, capacitors are used for the following purposes:

1. To block the flow of DC currents (in audio and guitar amplifiers).
2. To bypass AC signals around resistors (in audio and guitar amplifiers)
3. In various filters, such as speaker crossovers and compensation networks, tone controls in amplifiers and phono preamplifiers' RIAA filters.
4. As smoothing capacitors in amplifiers' power supplies, to reduce ripple (AC component on a DC power line) and to provide energy storage.

Ideal inductors (chokes)

An ideal inductor (choke) has no resistance for DC current, so there is no energy (heat) loss. The voltage on the choke precedes current, or you can say that the current lags voltage by 90 degrees, the opposite of a capacitor.

An inductor is named after a physical phenomenon called "electromagnetic induction", while a "choke" is its popular name, due to its effect on alternating currents, which it tries to suppress or "choke".

In other words, an inductor or choke resists any change in the current that flows through it by inducing an opposite voltage which then tries to push the current in the opposite direction so as to neutralize the current that caused the induction in the first place. That phenomenon is called *self-induction*.

A real choke has a low impedance for DC currents (so there is some energy loss) and a high reactance to AC currents (increasing in a linear fashion with rising frequency).

This desirable combination of properties makes chokes useful in power supplies, as ripple filters, and in speaker crossover networks.

The reactance of a choke is $X_L = \omega L = 2\pi fL$. Thus, the higher the frequency of an AC signal, the higher the reactance of a choke and the higher the choke's attenuation of the signal.

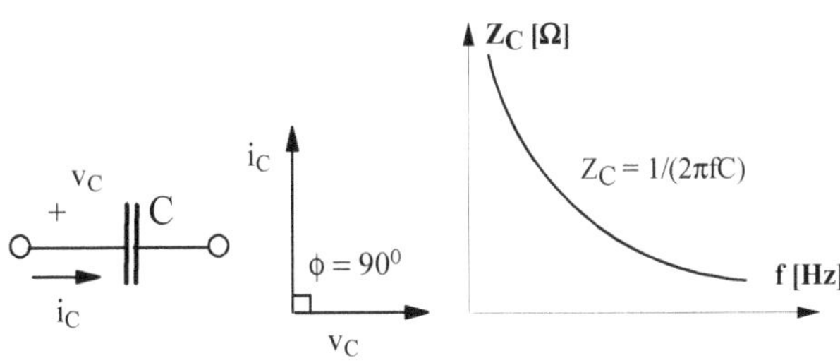

ABOVE: Ideal capacitor and its vector diagram: current leads voltage by π/2 radians or 90°. The impedance of an ideal capacitor is inversely proportional to frequency

BELOW: The phase relationship between the AC voltage across and current through an ideal capacitor

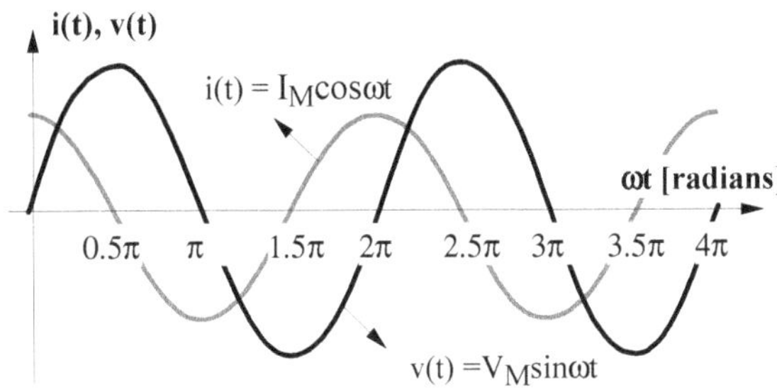

IMPEDANCE OF AN IDEAL CAPACITOR

$Z_C = 1/\omega C = 1/(2\pi fC)$

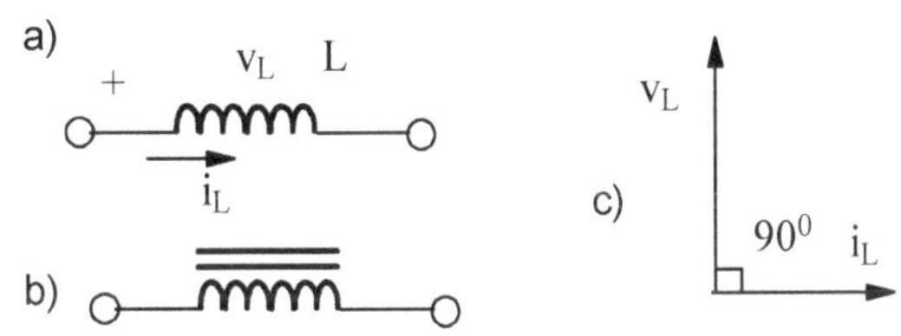

a) Ideal air inductor (no magnetic core)
b) Ideal inductor with magnetic core
c) Vector diagram shows voltage leading current by π/2 radians or 90°

IMPEDANCE OF AN IDEAL INDUCTOR

$Z_L = \omega L = 2\pi fL$

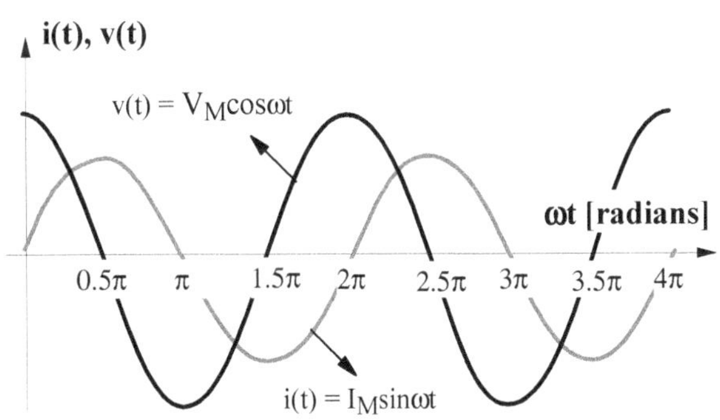

ABOVE: The phase relationship between the AC voltage across and current through an ideal inductor

RCL circuits

If a series RCL circuit is connected to an AC source, the same current flows through all three components. The voltage on the resistor is in phase with the current, but the voltages across the capacitor and the inductor are shifted in phase. One is lagging the current by 90 degrees (capacitor), while the other (inductor) is leading the current.

Since the voltages V_L and V_C are opposing each other, at a certain frequency f_R they will become equal and cancel each other out. Such a frequency is called a *resonant frequency* and the phenomenon is called *series resonance.*

$X_L=X_C$ or $\omega L=1/\omega C$ Solving the equation for ω, $\omega_R=\sqrt{(1/LC)}$ or $f_R=\sqrt{(1/LC)}/2\pi$.

At the resonant frequency f_R the overall impedance of the circuit is at its minimum, and equals the resistance R. The current through the circuit (assuming constant input voltage) is then at its peak. At all other frequencies the resultant impedance is larger than R. Its amplitude is $Z=\sqrt{[R^2+(X_L-X_C)^2]}$ and its phase angle is $\phi=\arctan[(X_L-X_C)/R]$.

In the parallel RCL circuit the resistor, capacitor and inductor are connected in parallel, so the same voltage is across each of them, and the currents through them vary with frequency. A similar situation (resonance) occurs there, the only difference being that at the parallel resonant frequency f_R the equivalent impedance is at its maximum (R), and the impedance at all other frequencies is lower than R, just as with loudspeakers and their impedance curves.

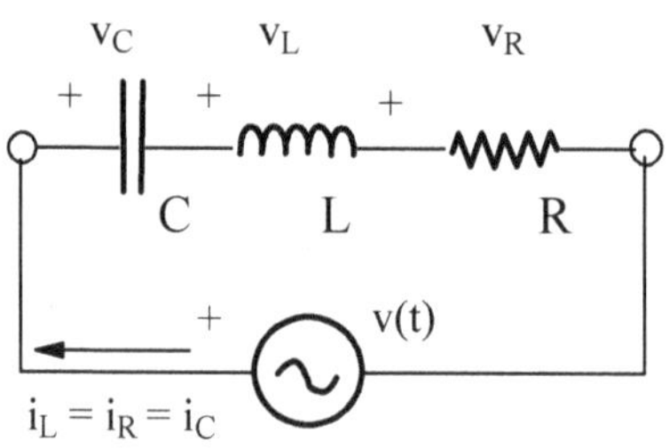

ABOVE: Series RCL or resonant circuit

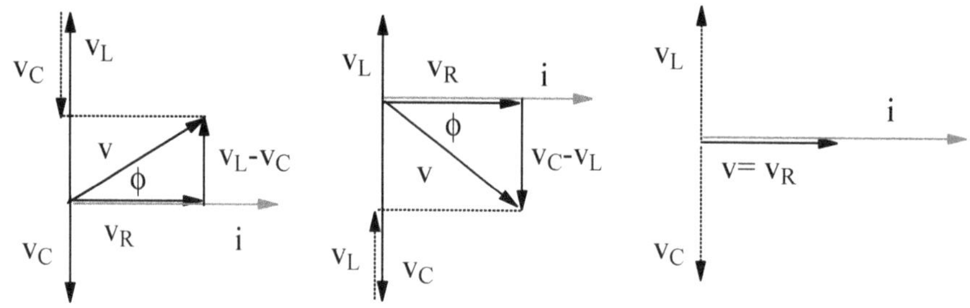

ABOVE: Vector diagrams of the series resonant circuit
a) predominantly inductive (voltage leading current)
b) predominantly capacitive (voltage lagging current)
c) At the resonant frequency the inductive reactance cancels the capacitive reactance. The equivalent circuit is a pure resistance.

RESONANT FREQUENCY

Angular frequency: $\omega_R=\sqrt{(1/LC)}$
$f_R=\sqrt{(1/LC)}/2\pi$.

HOW CROSSOVERS WORK

What are crossovers?

Crossovers are extremely important in high-fidelity loudspeakers, much less so in guitar amps. However, the three other types of musical instrument cabs, PA cabs, bass guitar cabs, and keyboard cabs make extensive use of crossovers, since they typically use 2-way designs (a woofer plus a tweeter) or even 3-way designs (a woofer, a midrange driver, and a tweeter).

In all these cases each driver must be fed only the frequencies that fall into its usable frequency band. Speaker crossovers filter out the range of frequencies the particular driver cannot or should not reproduce due to unacceptably high distortion or irregular SPL-vs.-frequency response, with lots of peaks and dips.

Electronic circuits that perform such a frequency-selective attenuation are called filters. The filters used in loudspeakers are called crossovers since they have to cross over to each other in a smooth and seamless way.

Crossovers can be passive (usually placed inside speaker enclosures) and active.

Passive filters in speakers' crossovers use passive components, resistors (R), capacitors (C), and inductors or chokes (L). They are called passive because they do not amplify the signal and thus require no external power to operate.

However, they filter higher voltage and current signals, so the L, C, and R components used in passive filters must be rated for such high power levels. They are physically larger and more expensive than those used in active crossovers. As such, they introduce power losses which reduce speaker sensitivity and SPL levels of the whole amp-speaker system.

Active crossovers are actually frequency-selective or frequency-shaping preamplifiers. They can be realized with vacuum tubes, discrete bipolar transistors or FETs, or, most commonly, with ICs (integrated circuits). In the signal chain they are placed between the signal source and the power amplifier.

A two-way crossover has two filters, a low-pass filter, whose output feeds the low frequency or woofer amplifier, and a high-pass filter, whose output is connected to the input of a HF or tweeter amplifier.

These amplifiers can be but don't have to be identical, and that is one of the benefits of this kind of bi-amping. Tweeters require significantly less power than woofers, so can be driven by (very) low power amps.

A LF amplifier needs to drive the power hungry voice-coil of a bass driver, or "woofer" in popular jargon, and thus requires significant power (current) capability.

Active crossovers do not introduce power losses, so speaker efficiency is increased significantly. Amplifiers, especially low powered SET ones, will work at reduced power levels and produce lower distortion, have more dynamic headroom and sound better, with improved microdynamics and faster and cleaner transients.

Many passive crossovers use bipolar electrolytic capacitors and iron-cored inductors, which are nonlinear and cause significant distortion. This problem is eliminated in active crossover-based systems, which don't use inductors or elcos, resulting in further lowering of distortion.

Since the bandwidth of signals to each driver (woofer, midrange, if any, and tweeter) is strictly controlled and limited, the cross-modulation is if not eliminated then significantly reduced.

For instance, a distorting amplifier will pass a 2kHz tone and a 2.5kHz tone together, but will also produce a 500Hz intermodulation distortion tone and a 5.5kHz tone.

In a three-way active crossover the two original tones will pass through the midrange crossover. The 500Hz and 5.5kHz distortion products will be significantly attenuated by such filter.

The 500Hz tone will not propagate through the LF fitter and the 5.5kHz tone will not even appear at the input of the HF filter since the filters are independent. The end result will be a significant reduction in IM (Intermodulation) distortion.

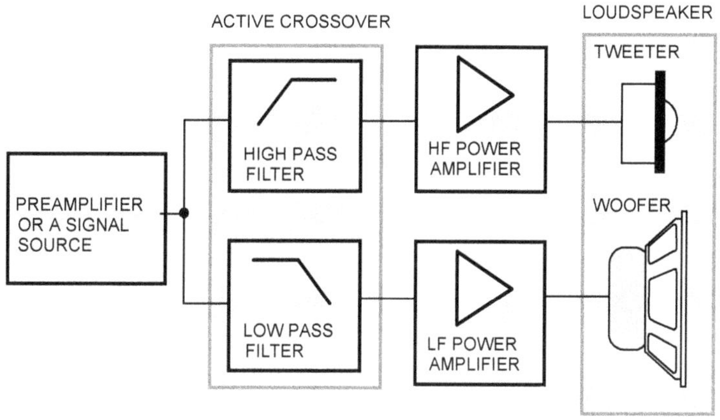

ABOVE: Block diagram of a 2-way active crossover setup (mono or single channel)

BELOW: Block diagram of a 3-way active crossover setup (mono or single channel)

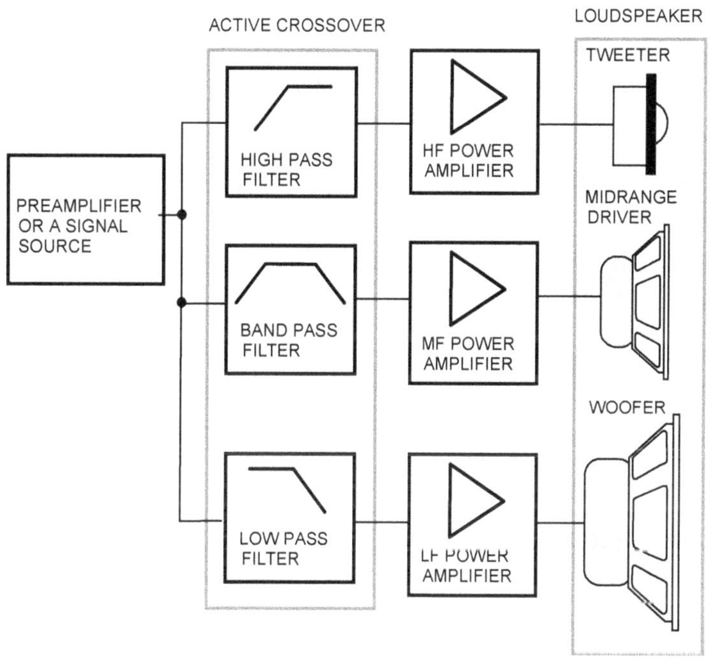

Assuming they feature volume or "level" controls for each frequency band, which are very easy to implement, active crossovers bring us the flexibility to adjust power levels of the high and low frequency bands. With inbuilt, passive crossovers that decision was made by the speaker manufacturer and no user adjustment is usually possible.

Active crossovers do not interact with the frequency-dependent speaker impedances as passive ones do, so a more precise and stable spectral division is achieved. The damping factor is also improved.

For a two-way active crossover system, two power amps are required (per channel!), three amps for a 3-way system, and so on. This adds complexity and cost, not to mention the introduction of new, system-specific problems.

The building blocks of crossovers - Filters 101

Electronics students study filters in great detail, but all we can do in a book of this kind is to briefly look at some simple yet often used and thus important filter types.

Crossovers range from the very simple 1st order types, with only one reactive element (capacitor or inductor) per driver, through 2nd order designs, with two LC elements per driver, to third, fourth, and higher-order filters.

Again, since they are by far most common in musical instrument speaker cabs (unlike in hi-fi speakers where complex higher order filters abound), we will only analyze 1st order filters, but will provide formulas for calculating the component values for a few higher order filters.

1st-order low pass filter

The capacitive version of this filter is a resistive-capacitive voltage divider. For low frequencies the reactance of the capacitor shunting the load is high and all input signal passes through the series resistor R and reaches the load.

As the signal frequency increases, the capacitor's reactance $X_C=1/\omega C$ gets smaller and smaller, meaning more and more of the input signal is shunted (diverted) to ground (COM), and less and less of the signal reaches the speaker.. The expression $X_C(f)$ indicates that the reactance of a capacitor is a function of signal's frequency.

At a certain frequency f_0 the capacitor's reactance (impedance) is equal to the resistor's resistance $X_C=R$ or $1/\omega C=R$. That -3dB angular frequency (where the output drops -3dB or to 0.71 times the low frequency value A_0) is $\omega_0=1/RC$, and since $\omega_0=2\pi f_0$, we get $f_0=\omega_0/2\pi = 1/(2\pi RC)$! This frequency is called a "corner", "break", "turnover" or "half-power" frequency.

To draw the Bode diagram and its asymptotes, first we assume that $f \ll f_0$. In that case A=1, or in dB A=20log1= 20*0 = 0 dB. This represents a horizontal line ❶.

If $f \gg f_0$, $A = f_0/f$ or in dB $A = 20\log(f_0/f) = 20\log f_0 - 20\log f$. For $f=f_0$, A= 0 dB, represented by a downward sloping asymptote, with a slope of -6 dB per octave or -20dB per decade ❷.

The inductive version of this filter is almost exclusively used in speaker crossovers. It is an inductive-resistive voltage divider. For low frequencies the reactance of the series inductor ($X_L= \omega L = 2\pi fL$) is low that all of the input signal passes through it without attenuation and reaches the load (the speaker).

As the signal frequency increases, the inductors reactance (impedance) increases and impedes the signal flow (attenuates the signal), so less and less of the signal reaches the speaker, just as with the capacitive variant.

For the LR filter, at the corner frequency the inductor's impedance magnitude is equal to the resistor's resistance, $X_L=R$, so $2\pi fL=R$, thus $f_0=R/2\pi L$.

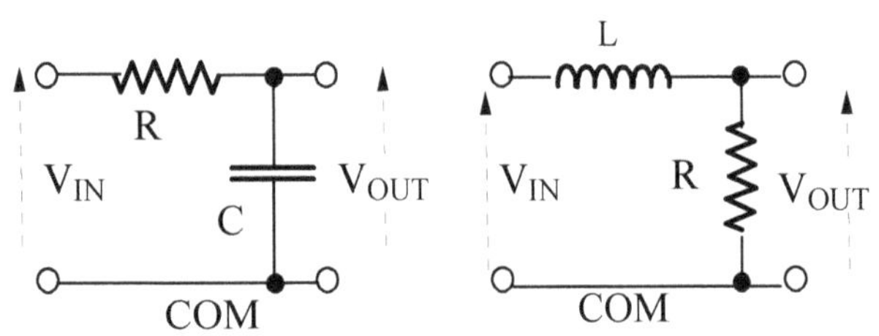

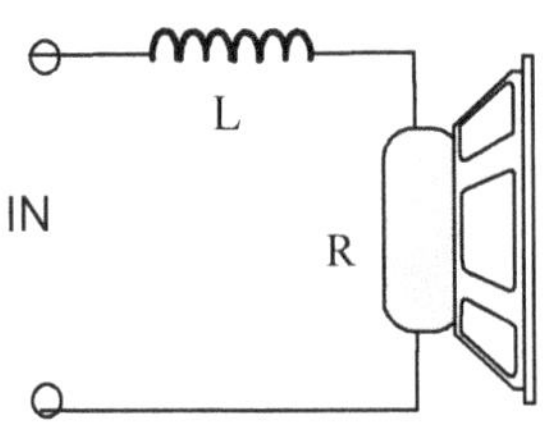

1st ORDER LOW-PASS FILTERS

$f_0 = \omega_0/2\pi = 1/(2\pi RC)$

$f_0 = R/2\pi L$

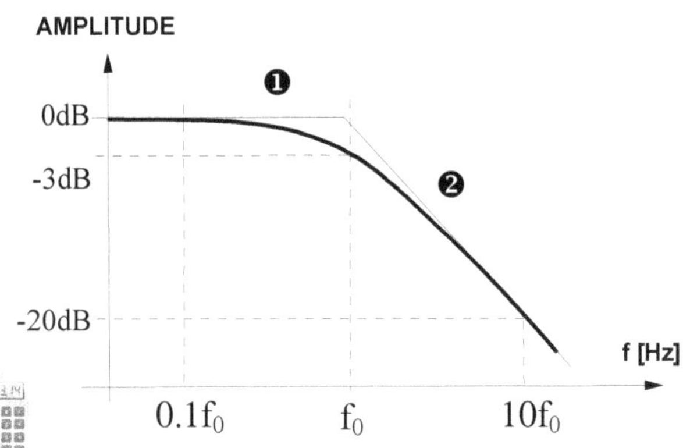

ABOVE: The Bode diagram (signal amplitude-vs.-frequency) for the 1st-order low pass filter

ABOVE LEFT: The simple practical crossover for the low frequency driver (woofer) consists of an inductor in series with the speaker. We simplify things by neglecting the speaker's inductance and capacitance and consider only its DC resistance (for instance 6.6 ohms for an 8-ohm speaker) or its nominal impedance (say 8 ohms).

1st-order high pass filters

The capacitive version of this filter is a capacitive-resistive voltage divider. For low frequencies the reactance of the series capacitor is very high (infinite for DC) so very little signal reaches the load (speaker). As the frequency increases, the capacitor's impedance reduces, so at high audio frequencies it is practically negligible, and all of the signal reaches the high frequency speaker (the tweeter).

The voltage attenuation of this passive circuit is $A_V(\omega) = R/[R+X_C(\omega)] = R/(R+1/\omega C)$.

Again, the -3dB frequency f_0 (where the output drops -3dB or to 0.71 times the high frequency value A_0) is $\omega_0=1/RC$, and since $\omega_0=2\pi f_0$, we get $f_0 = \omega_0/2\pi = 1/2\pi RC$!

BELOW: The simplest practical crossover for the high frequency driver (tweeter) consists of a capacitor in series with the speaker (R).

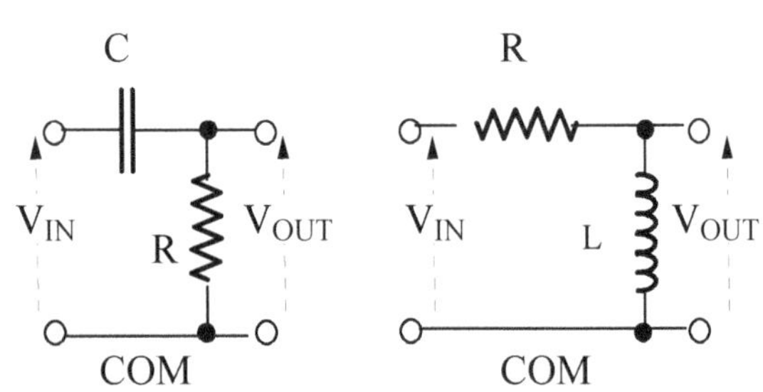

Notice that it is identical to the f_0 for the 1st order low pass filter. We draw the transfer function and its asymptotes in the same way as for the low pass filter.

If we now superimpose the two attenuation curves we get their resultant curve which isn't a curve at all, but a straight line, as it should be (RIGHT). The "crossover" at f_0 is smooth and seamless, at least in theory.

Before we conclude this section with the practical circuits, let's have a quick look at higher order filters and crossovers.

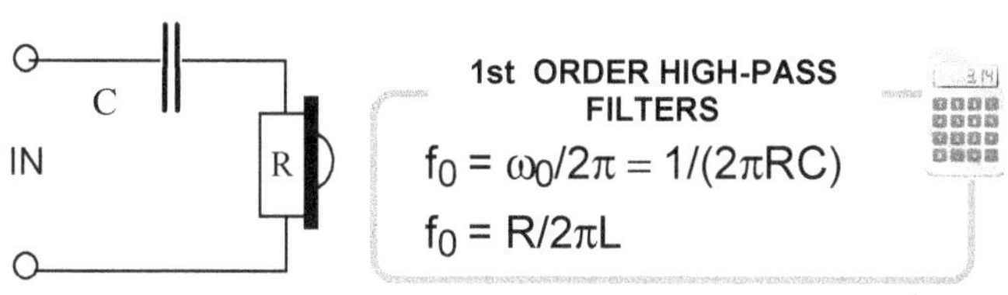

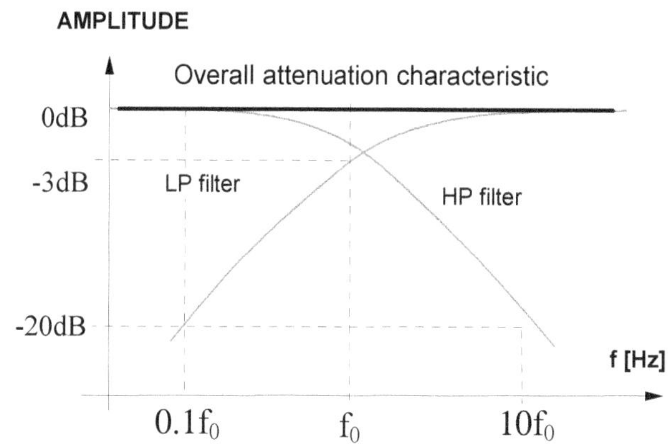

RIGHT: Ideally, the overall frequency curves of the LF and HF filter should add up to a straight line

Higher order filters

Butterworth-type filters are most common since they have the smoothest response and the sharpest possible roll-off without peaking.

Linkwitz-Riley, Chebyshev, and Bessel filters are occasionally seen in some speaker designs. The filter types were named by the mathematicians or engineers who first described them.

In all these crossovers, $C = 1/(2\pi f_0 R)$ and $L = R/2\pi f_0$, where π=3.14, f_0 is the crossover frequency (-3dB), and R is the coil resistance of the driver (woofer and tweeter).

Notice that filters start attenuating signals way before the crossover frequency f_0 (1).

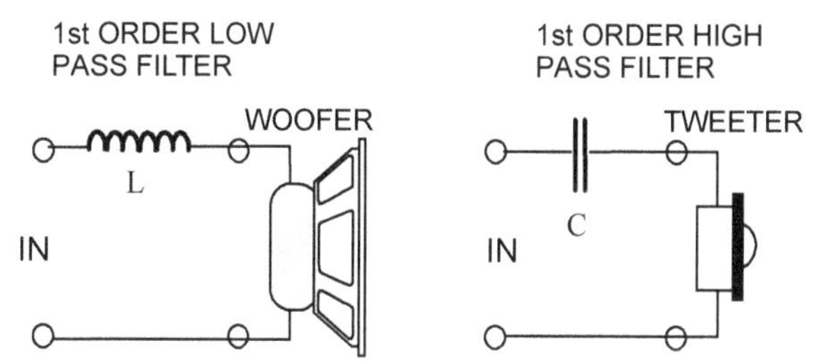

The 1st order (ABOVE) and 2nd order (BELOW) Butterworth filters (low-pass and high pass) for 2-way loudspeakers

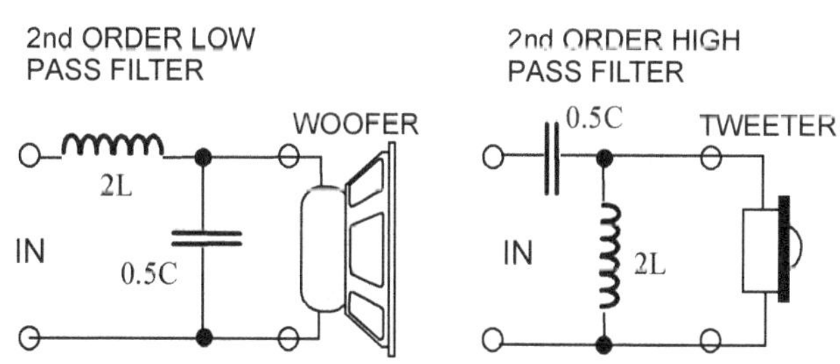

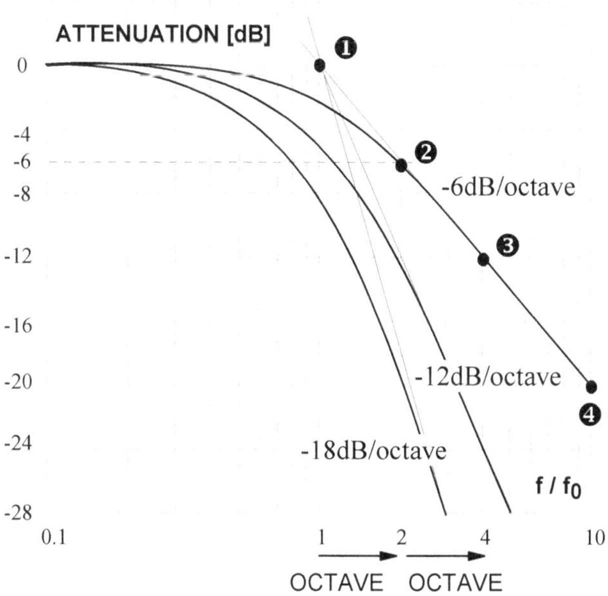

ABOVE: Attenuation curves and asymptote slopes for the 1st, 2nd and 3rd order low pass filters. High pass filter curves are a vertical mirror image ...

1st order filters attenuate signals at a rate (slope) of 6 dB per octave (2:1 ratio of frequencies), so at a frequency twice f_0 ($f/f_0 = 2$), the curve is down -6 dB. Another octave up (f/f_0=4) the curve is down -12 dB (3).

This attenuation is 20 dB per decade (a decade is a 10:1 ratio of frequencies), so in point (4), where the frequency is $10f_0$, the attenuation is -20 dB!

2nd order filters double that attenuation to 12 dB per octave, and 3rd order crossovers have attenuation curves with 18 dB/octave slopes.

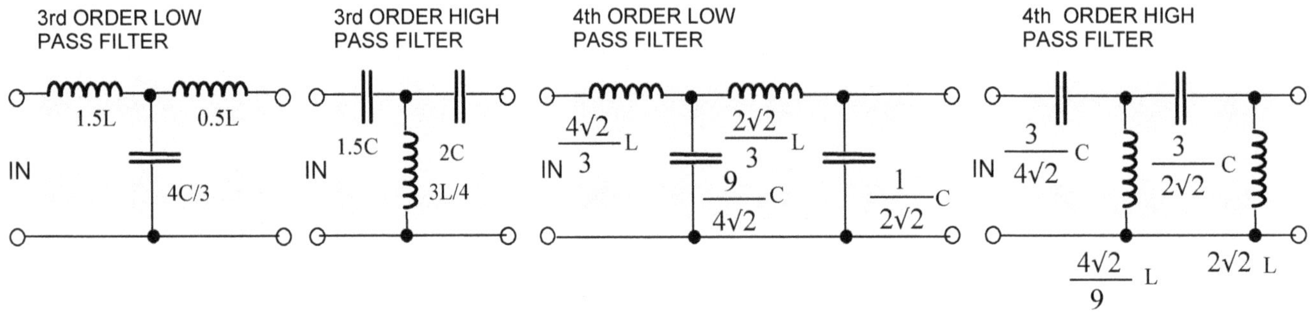

ABOVE: The 3rd and 4th order Butterworth filters (low-pass and high pass) for 2-way loudspeakers

PRACTICAL CROSSOVERS

How to calculate the required inductance and capacitance values

In hi-fi speakers, the woofer's output has to be attenuated above a certain crossover frequency; designers don't want the woofer to reproduce mid and higher frequencies, where its SPL curve fluctuates significantly due to cone breakup effects.

In guitar cabs that isn't usually an issue (except perhaps in bass guitar cabs), so most guitar speaker designs omit the series inductor.

However, the power ratings of tweeters are almost always much lower than the power rating of the woofers in our case the existing speaker in a guitar cab), so at higher power levels the tweeter will burn out.

Even the series capacitor will not attenuate the high frequency fast enough to protect the tweeter.

Therefore, we need to add an attenuator, either a single series resistors R_1, or a pair of resistors, a resistive voltage divider R_1-R_2.

Instead of a fixed attenuation by the two resistors, a more flexible option is to insert a variable attenuator, basically a dual rheostat (high power potentiometer) with two dissimilar values (typ. 8Ω and 30Ω). That way we can precisely blend the tweeter's output with that of the woofer.

Another reason for adding R_1 or R_1&R_2 is to match the SPL levels of the two drivers.

Tweeters are much more sensitive than woofers, so just connecting a tweeter in parallel with the existing guitar speaker will result in a shrill, treble-heavy tone, and no guitar player wants that.

There's a huge difference between adding a shimmery, sparkly top to your dull-sounding speaker box, and getting a screechy, irritating, rodent repelling and dog annoying ultrasonic generator!

Let's do a sample calculation of the required values for the 1st and 2nd order crossover components.

Let's assume an 8Ω woofer and an 8Ω tweeter, and a crossover frequency f_0=2kHz.

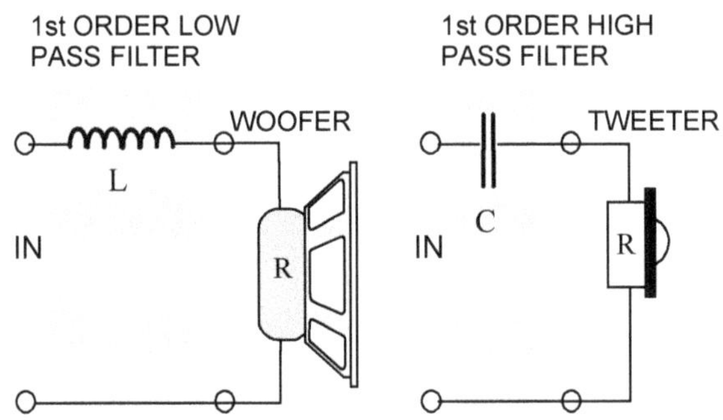

ABOVE: While a necessity in hi-fi speakers (to prevent the woofer from reproducing higher frequencies), the series inductor is almost always omitted in 2-way guitar or bass guitar speaker cabs (BELOW). The tweeter's HF output simply adds to that of the woofer

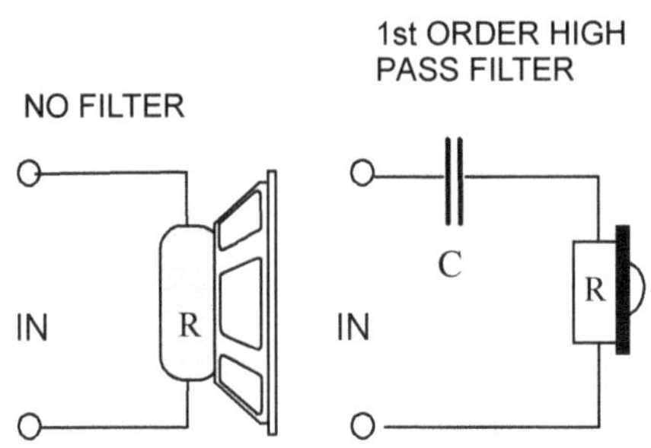

RIGHT & BELOW: To reduce the signal feeding the tweeter, either a fixed resistive voltage divider or a variable (adjustable) attenuator can be used.

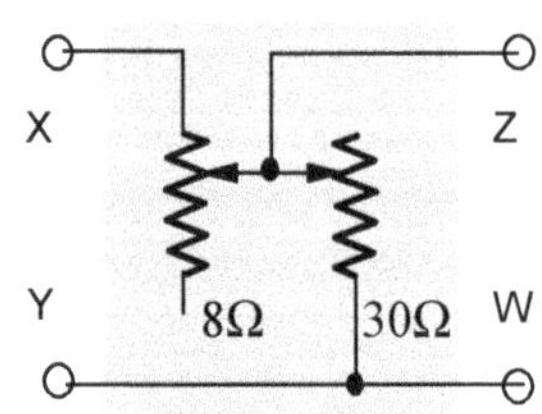

L-PAD ATTENUATOR

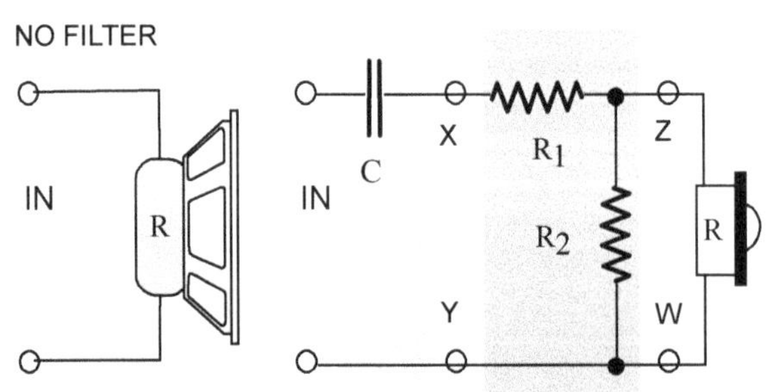

FIXED RESISTIVE VOLTAGE DIVIDER

For the 1st order crossover, $C = 1/(2\pi f_0 R) = 1/(2*3.14*2{,}000*8) = 9.947\ \mu F$.

For the 2nd order filter we need half of that value, or 5 μF. $L = R/2\pi f_0 = 8/(2*3.14*2{,}000) = 0.636$ mH for the 1st order crossover, and again, for the 2nd order filter we need double that value, or 1.27 mH.

What components to use?

Adding a crossover to your speaker cabinet can be done in a few ways. If you are using a simple 1st order XO you don't need any boards or terminal strips, simply solder an inductor (if used) in series with the woofer, and a capacitor in series with the tweeter/midrange driver.

Even 2nd order crossovers can be wired that way, but a much neater solution is to buy them pre-made. Many are sold online and their prices aren't any more expensive than buying individual components.

For instance, 2nd order (12 dB/octave) Linkwitz-Riley crossovers by Dayton Audio, with the listed price of US$$20.99, can be found on sale for only US$11.98 each. So a pair of filter boards (low and high pass) for one cab will cost you only US$24.-

A polypropylene film capacitor (4) and an air-core inductor (5) are used.

They are available in a range of common crossover frequencies, which aren't critical for musical instrument cabs anyway.

The only issue is that a pair of power resistors to attenuate the tweeter signal aren't included, although the PCBs are sufficiently large (100 mm x 60 mm) to accommodate them.

If you opt to include a continuously adjustable tweeter attenuator such resistors don't need to be added, and the wiring can be kept neat and tidy.

Let's see how to do that.

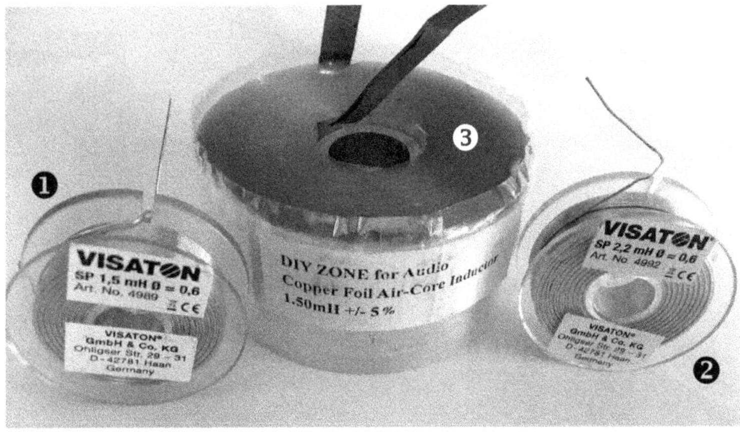

ABOVE: Air core inductors (1) and (2) are simple spools of round copper wire on a plastic bobbin. Copper foil conductors (3) are rare and expensive. For the same inductance value, they are much larger and cost many times more, but are highly regarded in hi-fi circles.

BELOW: A 3kHz high-pass filter for an 8 ohm driver by Dayton Audio. The complementary low-pass filter looks identical, uses the same value components (0.85 mH inductor and 3.3 μF capacitor), but the PCB tracks on the other side are different.

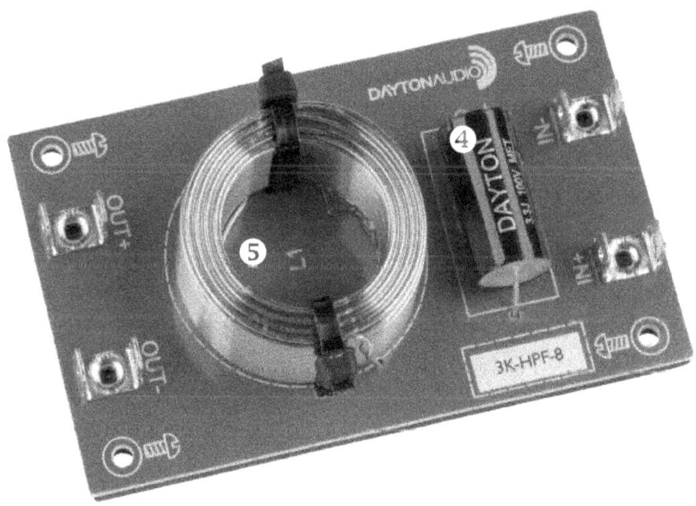

Adding an L-pad attenuator for tweeter volume control

If you are adding a tweeter to your dull-sounding guitar cab, a more elegant and flexible solution is to use an L-pad audio attenuator, so you can precisely control the balance between the tweeter and the woofer/midrange driver.

Rated at 50 Watts, Dayton Audio speaker L-pad attenuator was designed to be used with 8Ω drivers. The faceplate, volume control knob, and mounting hardware are included. The Parts Express part number is 260-255. There is also a 100 Watt version

At its maximum (fully CW), the wiper of the 8Ω section is at the input terminal (3), and the wiper of the 30Ω section is at the output terminal (2). The 8Ω resistor is out of the circuit, and the 30Ω resistor is connected across the load.

As the control knob is moved counterclockwise towards the MIN position, the larger and larger upper section of the 8 Ω resistor is connected between the IN and OUT terminals, thus acting as a voltage divider and reducing the voltage across the tweeter.

At the same time, a smaller and smaller section of the 30 Ω resistor is connected between the OUT and COM terminals, in parallel with the tweeter.

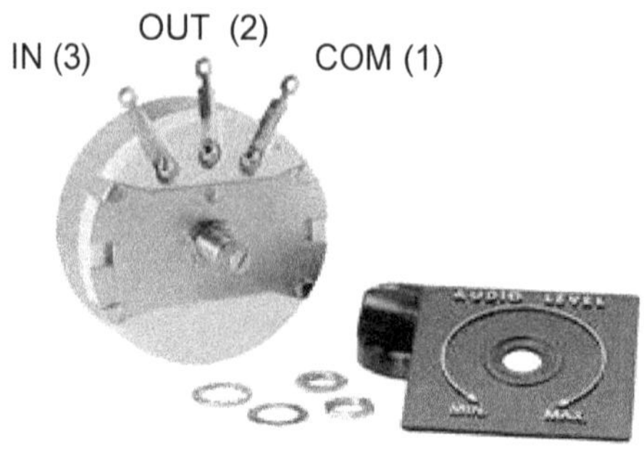

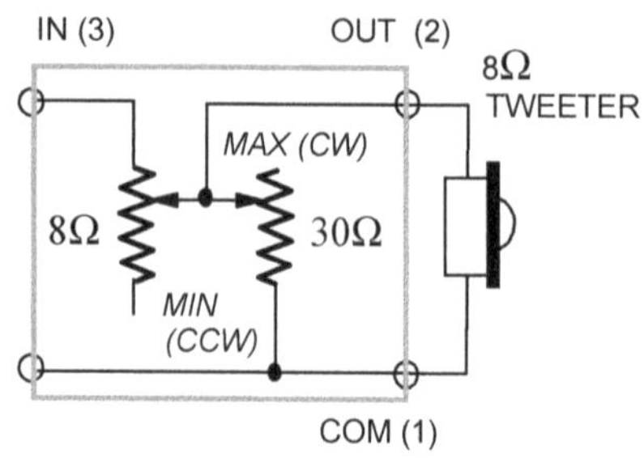

The 8Ω attenuator consists of two sections of different resistance, 8Ω and 30Ω.

Due to such reducing resistance, more and more of the load current is diverted away from the load and shunted to COM. At the minimum position (fully CCW), the 30 Ω resistor is out of the circuit, and the OUT terminal is shorted to GND, completely bypassing the load. The 8 Ω resistor is now connected between the IN and OUT terminals, the speaker terminals "OUT" and "COM" are short-circuited, so there'll be no sound from the tweeter at all.

The way it used to be: HF (tweeter) controls in vintage Jensen cabinets

Although designed for use in high fidelity speaker cabinets, the 15" Jensen Concert Type H coaxial driver was used in the Gibson GA-Cb guitar amp. Gibson designers included a Jensen four-position range control switch (ST-784). The four position control provided the tweeter attenuation so the user could blend in high frequencies at different levels. Gibson specs claimed a "50 to 15,000 cycle" frequency range.

Obviously, as with all coaxial drivers, its tweeter was mounted inside the woofer driver.

COAXIAL CONTROLS

LEVEL CONTROL, ST-783. A 500-ohm L-pad for use in the input of 15-inch Co-axial speakers. Complete with satin finish brass cup escutcheon, bar knob and mounting screws **$3.00**

RANGE CONTROL SWITCH, ST-784. A 4-position selector switch, operating in the h-f range, to adjust speaker performance to any type of program material. Complete with satin finish brass cup escutcheon, bar knob and mounting screws **$2.80**

COMPENSATION NETWORKS

Capacitive compensation of inductive loads (speakers)

The reactance of a dynamic (moving coil) speaker increases in a linear fashion with rising frequency ($Z_L=2\pi fL$), so this simple type of RC compensation (called a Zobel network) aims to compensate for this rising impedance and make the amplifier load independent (or less dependent) on frequency.

To put it somewhat crudely, the inductance and capacitance are electrically opposite and are often used to compensate for or cancel/neutralize each other!

If L_S/R_S is made equal to the time constant C_ZR_Z ($L_S/R_S = C_ZR_Z$), then the load as seen by the amplifier's output is purely resistive and thus independent of frequency.

Usually R is made equal to R_S, so then $R_Z= R_S = \sqrt{(L_S/C_Z)}$! From here we can find the required value of the compensating capacitor as $C_Z = L_S/R_S^2$.

Place (solder) the Zobel network directly across the driver's terminals. Compare the sonics of your speaker cab before and after.

If you don't like the sonics of the compensated cab, the change is reversible. Simply remove the Zobel resistor and capacitor.

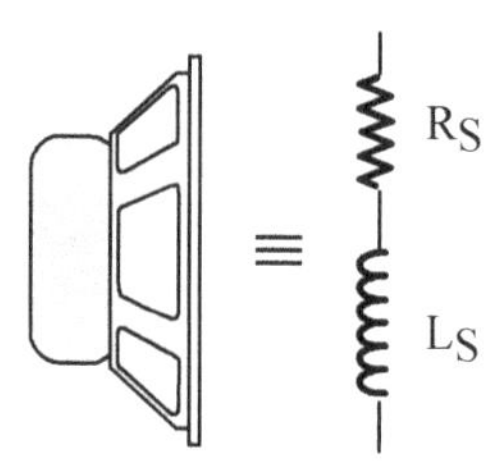

ABOVE: A simple model of a dynamic speaker driver

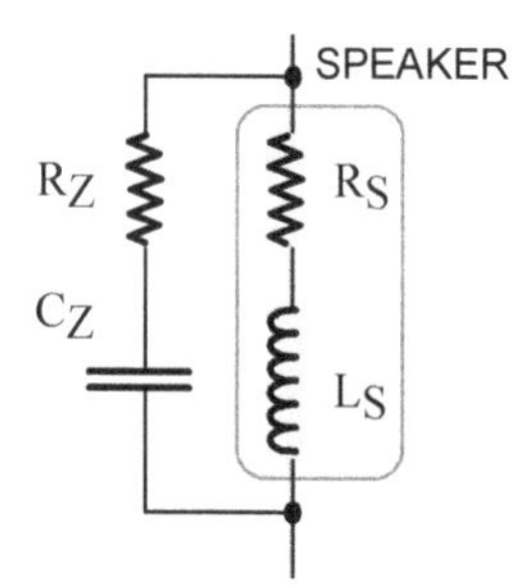

ABOVE: RC compensation of an RL (inductive) load

Choosing the required components

ZOBEL NETWORK VALUES

$R_Z=R_S$

$C_Z = L_S/R_S^2$

The Zobel resistor value is equal to the nominal impedance of the driver: $R_Z=R_S$. For a driver of an 8Ω nominal (declared by the manufacturer) impedance, choose an 8Ω resistor, for a 4Ω driver a 4Ω resistor, and so on - you get the picture.

Calculate the value of the capacitor from the nominal impedance and the driver's L_S value: $C_Z=L_S/R_S^2$.

As an example, let's say you speaker has an 8 ohm nominal impedance and 1.2 mH (0.00012 H) voice coil inductance. Thus, $R_Z = R_S = 8\Omega$ and $C_Z = L_S/R_S^2 = 0.0012/8^2 = 0.0012/64 = 0.00001875$ F = 18.75 µF.

Remember, the 8Ω nominal value of the driver's DC resistance may or more often may not be accurate, so don't lose sleep if you cannot find a capacitor of that precise value. Unless you buy a bag of 100 capacitors of the closest standard value, in this case either 15 or 22µF (18µF film caps exist, but are not common), and measure their capacitance one-by-one, you will not be able to find it, since it is a nonstandard value. So, a close value will do.

Due to manufacturing tolerances the capacitance values will range +/-20%, especially with electrolytic caps!

Choosing the right capacitor for the Zobel network

The component values are just the first step. What about the voltage- and power-ratings of the required capacitor and resistor?

What kind of AC voltage will that Zobel capacitor experience across its terminals? Let's say your speaker is 8 ohms, 100 watts. Power is voltage squared divided by the impedance, or $P = V^2/R$. expressing the voltage from that equation we get $V = \sqrt{(P*R)} = \sqrt{(100*8)} = \sqrt{800} =$ 28.3 Volts.

BELOW: Bipolar electrolytic capacitors are marked "BP" and have the "notches" (2) on both ends of their cans (casings)

Bipolar electrolytic capacitors used in speaker crossovers usually have a rating of 50 or 100 Volts, so they would be suitable. You can recognize them by their "BP" markings (1) and also by the shape of their cans, they have a "notch" on both sides (2). Ordinary electrolytic caps only have a notch on one side. They are not suitable for AC applications such as this one.

The 105°C is the cap's maximum temperature rating, of no concern in this application unless you play your guitar and use your speaker cab in Hell.

A better choice would be to use a film capacitor (a different technology). Their voltage ratings go up to 600-1,000 Volts, even higher, so any voltage rating over say 50 V is fine for this application. Film caps are not polarized, so one less issue to worry about.

A code imprinted on the case or body of a capacitor usually declares what the dielectric is. The codes are: KP - polypropylene film & foil, KS - Polystyrene film & foil, KT - Polyester film & foil, MKC- metallized polycarbonate, MKP - metallized polypropylene, MKT - metallized polyester, MKY - metallized low-loss polypropylene, and MKL or MKU - metallized lacquer (cellulose acetate).

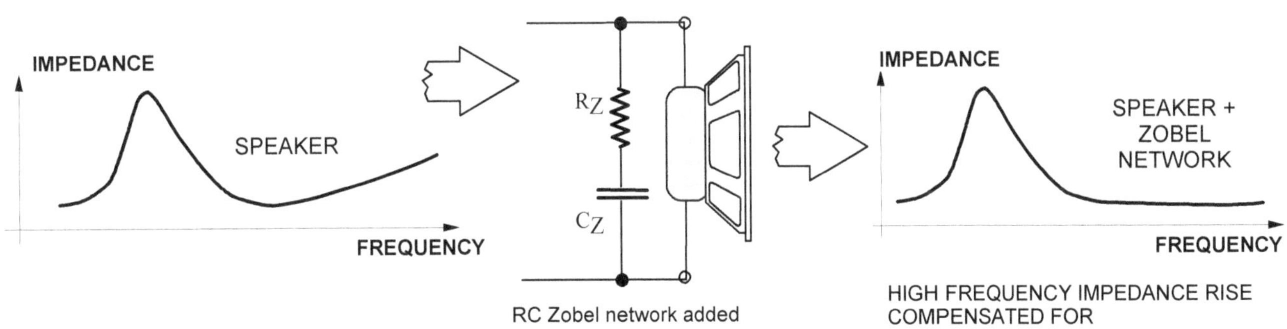

ABOVE: The impedance curve of a typical dynamic speaker before and after adding a Zobel network

Resistors have a power rating, which depends on the maximum power (heat) they can dissipate. Remember, all resistors are mini heaters, just like the ones in your toaster, kitchen hot plate or electric room heater. What power rating is required for the Zobel resistor R_Z?

Let's choose one audio frequency towards the upper range of a guitar speaker, say f=5,000 Hz, and calculate impedances. That will gives an idea about the distribution of currents between the two branches (the speaker and the Zobel network) and the required power rating of the Zobel resistor.

Using the same values as in the last example, the impedance of the 18 μF capacitor at f=5,000 Hz is $Z_C = 1/(2\pi fC) = 1.77\ \Omega$. The impedance (reactance) of the speaker's voice coil is $Z_S=2\pi fL = 37.7\ \Omega$.

Since the voltage drops across capacitors and inductors are not in phase with the voltage drops on the resistors in series with them, we cannot simply add them up or subtract them, vector algebra must me used.

However, this is just a quick estimate, so for this purpose the error made is negligible.

So, very roughly speaking, we have 10Ω, the impedance of the Zobel branch (8+1.77=9.77 Ω), in parallel with 38+8 = 46 Ω (let's call it 50Ω), the impedance of the speaker. The ratio is 1:5, so five times more current will flow through the Zobel resistor than through the speaker.

The worst case is approached as the capacitor's reactance becomes smaller and smaller, approaching zero, so the full signal voltage is across R_Z.

Some amplifiers suffer from high frequency oscillations, which will increase the current flowing through R_Z.

Thus, its power rating must thus be the same as the speaker's power rating, which means a large wirewound resistor must be used, such as the one pictured (RIGHT).

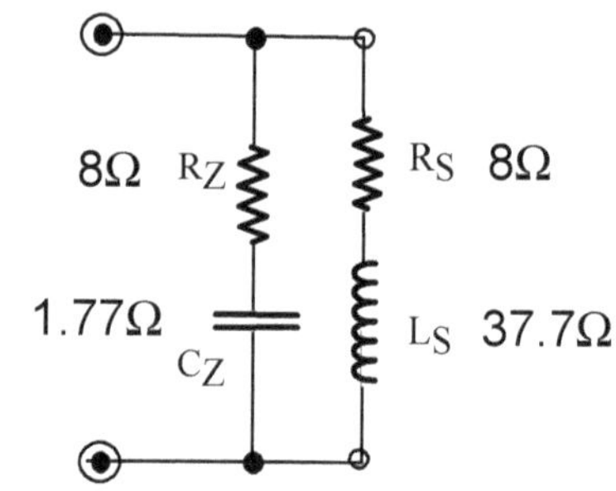

ABOVE: The impedances at 5kHz signal frequency

BELOW: 8Ω 100 watt Made-in-China wirewound resistor

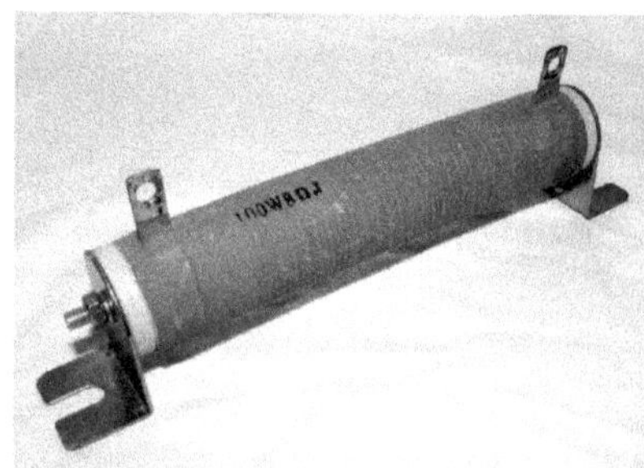

Flattening the resonant peak

The resonant peak (3) of the speaker can also be compensated for by installing a series RLC resonant circuit (4) in parallel with the speaker. Such a circuit has its own resonant frequency at which its impedance drops to a low level determined by the resistor R.

By calculating the R, C and L values so its resonant frequency equals that of the speaker, its impedance curve (5) will be a mirror image of the speaker's impedance curve, so the two effectively cancel or linearize each other in the vicinity of the speaker's resonant peak.

At resonance, the capacitive and inductive reactances are equal: $X_C=X_L$ or $1/(2\pi fC)= 2\pi fL$. If we express the frequency f from that equation we get $f=1/[2\pi \sqrt{(LC)}]$ or approx. $f=\sqrt{(0.025/LC)}$! We have one equation with two variables, L and C, so we can choose one and calculate the value of the other.

For instance you may have an inductor (inductance L) you want to use and many capacitors to choose from, the capacitance you need is $C=0.025/(L*f^2)$. Likewise, if you choose the capacitor (capacitance C) and need to calculate the value of an inductor you need to buy, use $L=0.025/(C*f^2)$.

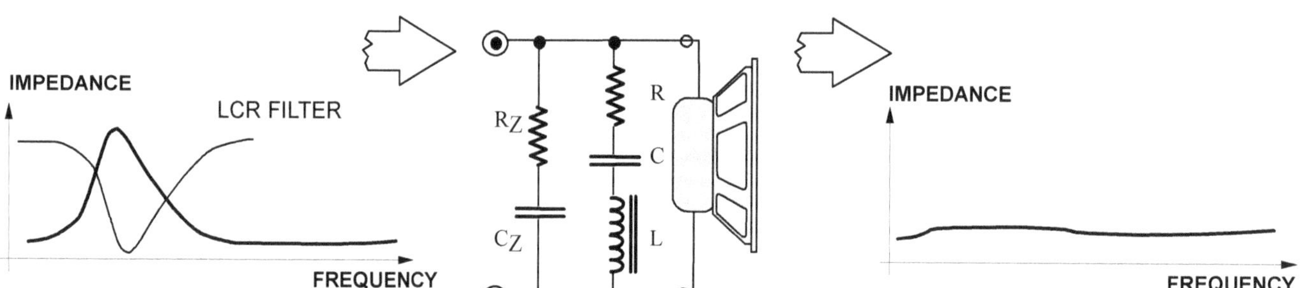

RCL series resonant circuit (notch filter) added

ABOVE: Two simple passive networks added, a Zobel RC circuit to compensate for the rise of speaker's impedance at higher frequencies, and a series resonant RCL filter to flatten or even completely remove the speaker's resonant peak

LESLIE AND OTHER ROTATING-TYPE SPEAKER CABINETS

13

Although originally designed to add that swirling magic to electronic organs' tone, Leslie and Leslie-type speaker cabs (more accurately "combo amps", since they contain an amplifier as well) can be used as guitar amps as well.

Apart from getting old and thus requiring major overhauls (most are now aged 50+), most vintage Leslies are bulky, heavy and difficult to transport beasts (not to mention "ugly").

"The Rotator", our super cool DIY project, shows you how to salvage a Leslie "engine", a rotating assembly from a cheap organ, and reuse it with any speaker of your choice in your own super-duper cab. Do not skip this section!

- HOW LESLIE AND LESLIE-TYPE CABINETS WORK
- THE LOW-END: LESLIE 130
- THE HI-END: LESLIE 910 AND ITS RELATIVES, MODELS 900 AND 925
- KORY - AN AUSTRALIAN TAKE ON THE AMERICAN IDEA
- THE ROTATOR: BUILD YOUR OWN LESLEY CAB FOR A GUITAR AMP

HOW LESLIE AND LESLIE-TYPE CABINETS WORK

The way it used to be: Don Leslie's legacy

Despite their popularity, even Hammond's electronic tube organs sounded two-dimensional compared to the three-dimensional richness of pipe organs in churches and cathedrals or theater organs of the silent movie era.

Donald Leslie achieved a similar 3D effect of a sound swirling around a large reverberant space using simple mechanical means, designing his organ cabinet (which also included a tube amplifier) using two stationary speakers and two separate rotating drums.

In 1941, when Leslie made his first cabinet, The Hammond Corporation wasn't interested in it. In fact, it is claimed that Laurens Hammond loathed it. The dislike was mutual.

Don Leslie commented "I hate those damn things!" and started selling his invention as an amplifier/speaker cab for all electronic organs, including Hammond's competitors, Wurlitzer, Baldwin, Conn and Thomas being the most prominent of the era. Kawai, Farfisa, Technics, Korg, Roland and others followed later.

As an interesting twist of fate, Leslie sold his Electro-Music business to CBS in 1965, the same year CBS acquired another iconic brand, Fender.

In 1980, CBS sold the Leslie brand to Hammond, which went out of business soon after, in 1985.

In 1991, Suzuki Musical Instrument Corporation became the owner of both the Hammond and Leslie brands.

ABOVE: Leslie 910 was one of the biggest and most complex of all Leslie cabinets. It featured two separate enclosures for easier transportation. Together, they weigh more than 75 kg!

The mechanics behind the Leslie cabinet

There are a few different types of Leslie cabinets. The most common "budget" units have a "fixed" woofer (firing upward, downward or forward) and a rotating tweeter. To be more precise, the tweeter is stationery but it fires into a rotating horn. Then there are cabs with a single speaker and a rotating drum (no tweeter), and others with a fixed tweeter and a woofer that fires into a rotating drum.

Finally, the best models have two rotators. Facing downward, the woofer fires into its own drum, while the high frequency (HF) speaker, placed above the woofer, is coupled to a dual horn, which spins in the opposite direction from the woofer's drum.

One horn is a dummy, used simply to mechanically balance the active horn. A diffuser attaches to the horn and helps to diffuse (spread around) the treble frequencies.

The two speeds are labeled "Chorale" (slow speed) and "Tremolo" (faster swirl). On both speed settings, the HF horn rotates slightly faster than the LF drum, 50 revolutions per minute (RPM) for "Chorale" versus drum's 40 RPM, and 400 RPM vs. 340 RPM for "Tremolo" setting.

Although technically accurate, the fast and slow options (controlled by a musician through a manual or foot switch) are somewhat misleading - the Leslie effect is much more complex than either of these two effects. In fact, it's a great example of acoustic synergy, of various effects and physical (acoustical) phenomena working together.

The Leslie sonic magic demystified

Physically and acoustically, how is that Leslie magic created? What does that seemingly and deceptively simple rotating drum do to the sound?

As the drum rotates, its opening redirects the sound that has bounced internally off the enclosure's walls for while, through three louvres, one on the front baffle and two on the cabinet's sides.

The cabinet is sealed at the back, so these three openings are the only ways for the sound to exit the enclosure.

When the drum's "mouth" moves towards the listener, the sound gets louder and then its loudness drops as the drum rotates away in the opposite direction.

This is the tremolo effect or amplitude modulation, "amplitude" referring to the peak of the acoustic wave, which we perceive as loudness. This aspect of the Leslie's sonics gets more pronounced as we move closer to the cab.

Likewise, as the drum's mouth (in effect, the rotating sound source) moves towards the listener, the perceived pitch (the signal's frequency) seems to increase, and as it then moves (rotates) away, the pitch falls. You may remember this phenomenon as the Doppler effect from you physics class.

In the musical instruments world, this effect is called "vibrato". However, note that Fender erroneously named the tremolo effect in their amps as "vibrato" - it is nothing of a kind, so don't confuse the two. Very few guitar amps have ever incorporated a true pitch-shifting vibrato, Magnatone being a notable exception.

Last (but by no means least), the rotating drum "throws" and "swirls" the sound waves at least 180 degrees (half a circle) through the three cab openings, and these sound waves bounce of hard surfaces in the room thus undergoing multiple phase shifts before they reach the listener.

In contrast to the tremolo, this phasing *schwebung* (as our German friends call it), just like the electronic "phaser" effect, becomes more pronounced as you move away from the cabinet.

Now we understand that the Leslie magic is created by a complex interplay of three physical phenomena, whose "proportions" or dominance of one over the others changes depending on the placement of the Leslie cab in a room and on the listener's position in relation to the cabinet.

THE LOW-END: LESLIE 130

Leslie 130 looks like it belongs to the early 1970s, and it does. It was one of the simplest Leslie models. A12" woofer fired downward into a rotating drum, while a stationary 3" tweeter reproduced frequencies above 4 kHz. Both drivers had a nominal impedance of 8 ohms.

A fixed tweeter made the sound brighter than that of other Leslie cabs, and the Leslie effect itself was different, some would say less pronounced and even compromised.

Leslie's standard design solid state amplifier with an interstage transformer driver/phase splitter was used, with a pair of 2N3055 output transistors producing (claimed) 50 watts RMS.

Just as many other Leslie models, everything except mains power goes through that silly 11-pin cable (a 9-pin cable is used on some other models), making it a nightmare for a modern user.

Many Leslie units are now pulled out of attics and basements and sold without such cables, which got lost or damaged over the 50 or so years.

Thus, modifications are necessary to be able to use guitar jacks, and that is exactly what we did with this unit.

We added two jacks (4), the photo on the next page, one for the input signal, the other a stereo jack for a footswitch to control the off-slow -fast speeds of the rotating drum.

The 11-pin connection to an organ was retained and kept functional.

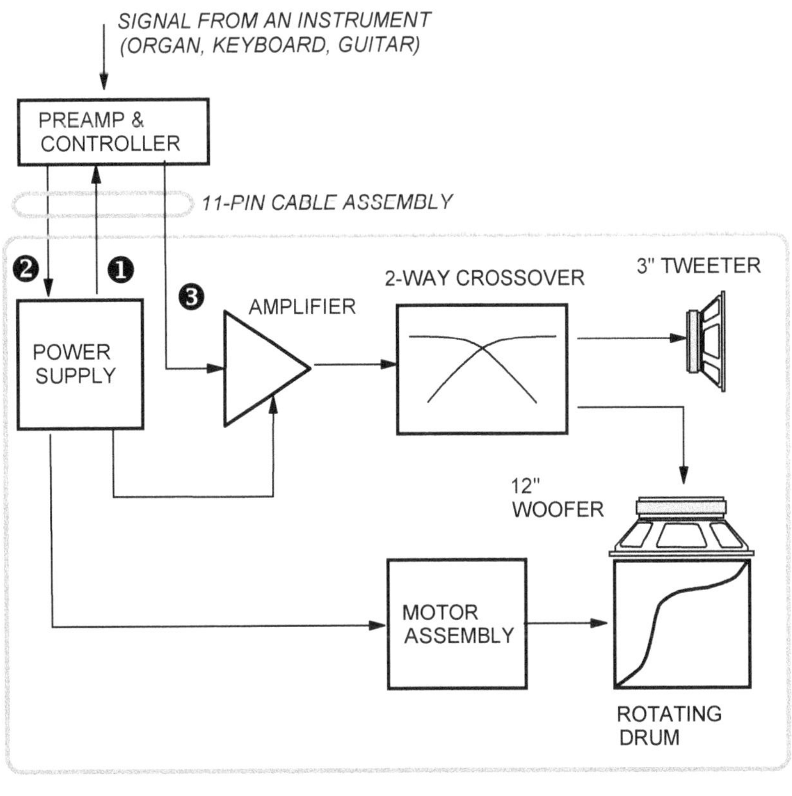

ABOVE: The topology of Leslie Model 130 was quite simple

ABOVE: Leslie 130 had no feet of any kind (?!?) so we added lockable castor wheels (5). To avoid the use of impossible to buy 11-pin connectors and cables, two jacks were added (4), one for the input audio signal, the other a stereo jack for a footswitch to control the off - slow - fast speeds of the rotating drum.

2 minute Leslie 130 modification for guitar use

MODIFICATION

As we have just seen, Model 130 differed from most Leslie and Leslie-type cabs, where the low frequency drivers are stationery and the tweeter horns rotate. For that reason its sound is brighter, with a less pronounced Leslie effect.

It is very easy to modify the cab's wiring to bring its sound closer to other mainstream Leslie cabs.

All you need to do is to unplug the two wires (orange & black) from the woofer's terminals (1), unplug the two input (2) wires (also orange & black) from the crossover terminal strip (3) and plug them directly onto the lugs on the woofer,

This effectively leaves the whole crossover and the tweeter (4) disconnected and inactive.

After that mod, all frequencies will be reproduced into the rotating drum, but the treble range (frequencies above 4 kHz) will also be reduced due to the woofer's inability to reproduce them as efficiently as the dedicated tweeter could, thus making the sound less bright.

This simple yet better sounding approach (especially if the cab is used for a guitar, where a tweeter is seldom necessary) is used in Leslie models 820 and 825,as well as model 125.

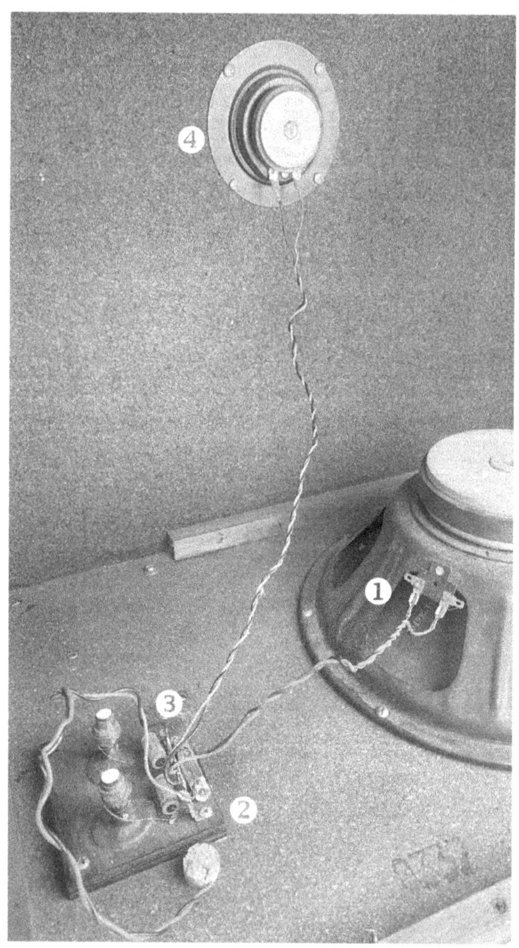

THE HI-END: LESLIE 910 AND ITS RELATIVES, MODELS 900 AND 925

As we have seen on its page 226 photo, Leslie 910 is a fridge-sized beast. While the upper cab is of a manageable size, the bottom cab is unnecessarily large. It could had been made half the volume or even less - the lone 15" woofer simply doesn't need such a big enclosure.

Many other strange design choices were made. The reverb assembly is not a tank, but a strange contraption where delay coils hang like drapes.

That would be one of the first mods I would make, replacing that large reverb delay line with a modern reverb tank, much less susceptible to transport damage and external vibrations.

For some strange reason, the treble horn can rotate at both fast and slow speeds, but the bass drum can only rotate at a fast rate; in other words, either both turn at a fast rate, or the treble horn turns slowly while the bass drum does not turn at all.

The removal of the upper back cover (photo on the right) reveals the already-mentioned large cavity with only reverb springs (photo on the next page) and a 15" woofer (1), firing down into a rotating plywood drum (2).

The lower compartment houses the power supply (3), the preamplifier (4), and three topologically identical amplifiers (5), each on its own printed circuit board, feeding three pairs of output 2N3055 transistors, one for each of its three channels (6, 7, & 8).

The lower back is covered by a louvered metal cover that carries the output transistors. The cover serves as their heatsink, a clever cost-saving measure.

In its original state (before any modifications and improvement), the only way for this and most other Leslie amps/cabs to function is through an "octal-style" Amphenol connector, in this case, a 9-pin version.

"Octal style" means that its size, shape and dimensions are the same as the octal tube socket, but with 9, 10 or 11 pins instead of 8 pins as in the octal socket. Such plugs and sockets are very rare, hard to get, and when available, very expensive.

Pin 1: Audio ground (common), Pin 2: Flute/Tibia audio in, Pin 3: unused, Pin 4: Main audio in (hot), Pin 5: 32 V_{DC} output to preamp/controller, Pin 6: Fast Motor Control, Pin 7: Slow Motor Control, Pin 8: AC power (neutral), Pin 9: AC power (phase or "live").

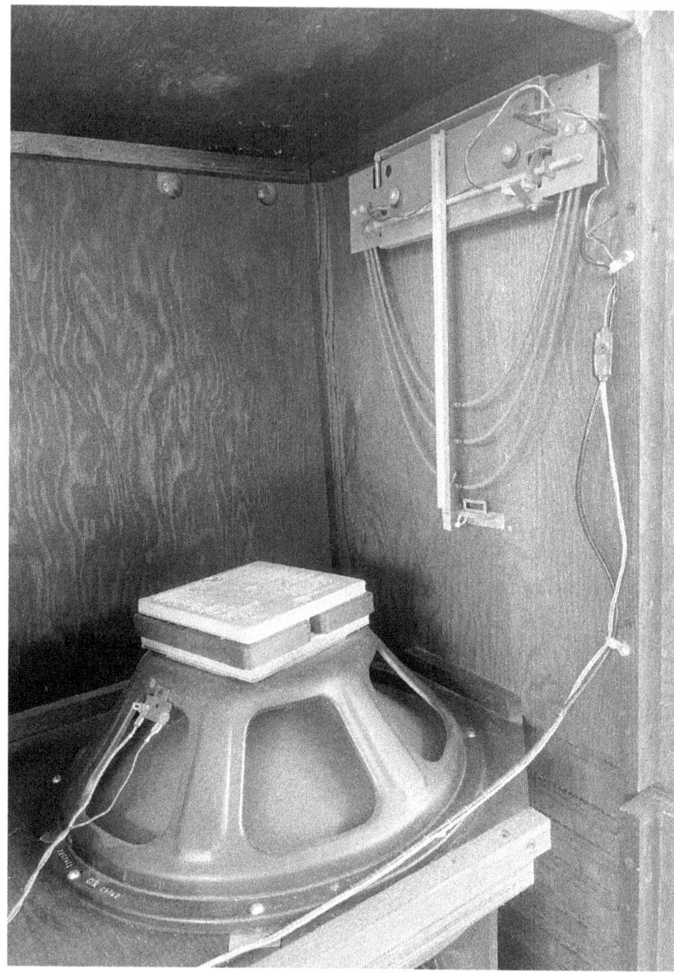

ABOVE: The internal rear view of Leslie 910 top cabinet

A. High frequency compression driver, firing upwards into the rotating horn assembly

B. The rotating horn

C. The motor driving the rotating horn

D. Two e cable that connects the upper and the lower enclosure

E. A pair of fixed oval 6" x 9" drivers, firing sideways to the left and right, dedicated exclusively to the production of the reverb effect, driven by the reverb amplifier

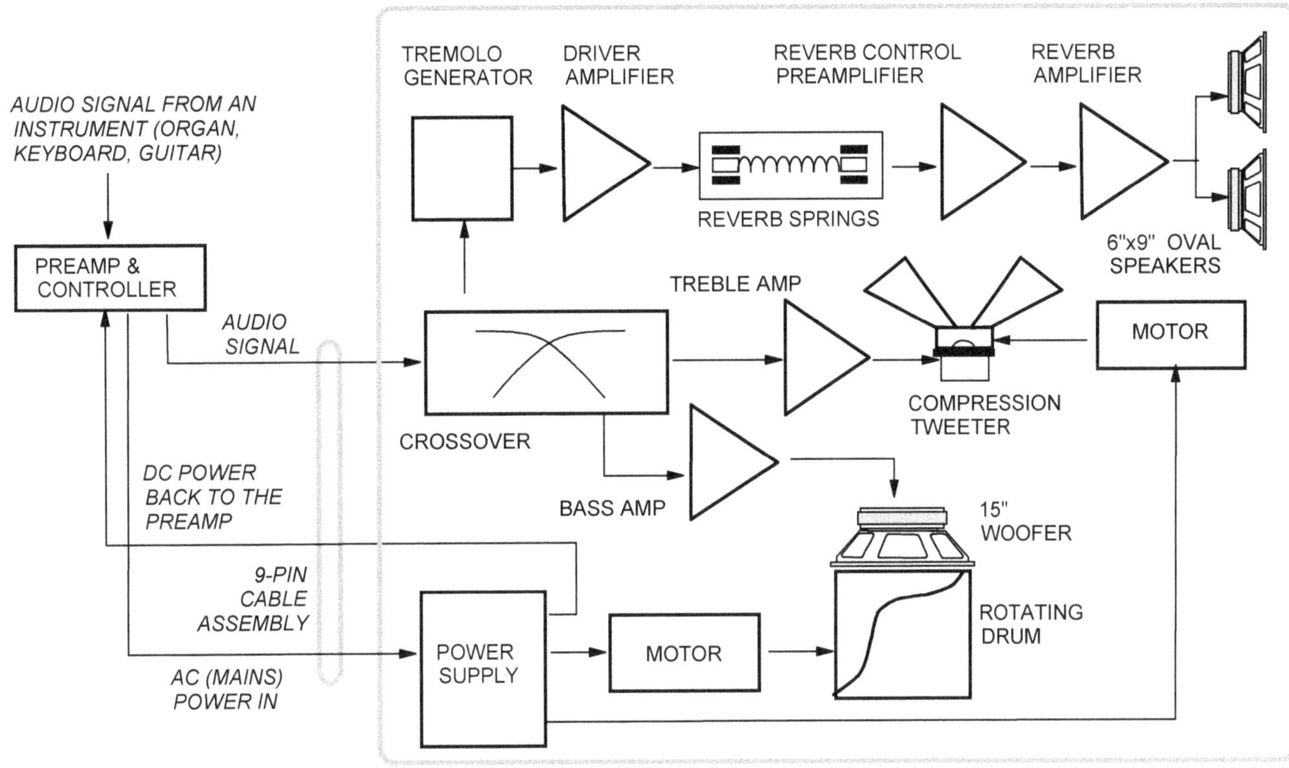

ABOVE: A case study in overcomplication - the topology of Leslie 910 and its relatives
BELOW: The four volume control trimmer potentiometers can be accessed from the outside, without opening the amp up.
1) The required 9-core cable plugs into the 9-pin male socket and to the dedicated Leslie preamp on the other side
2) Two large (thick & long) and cumbersome 9-pin 9-core umbilical cords connect the two parts of the cab

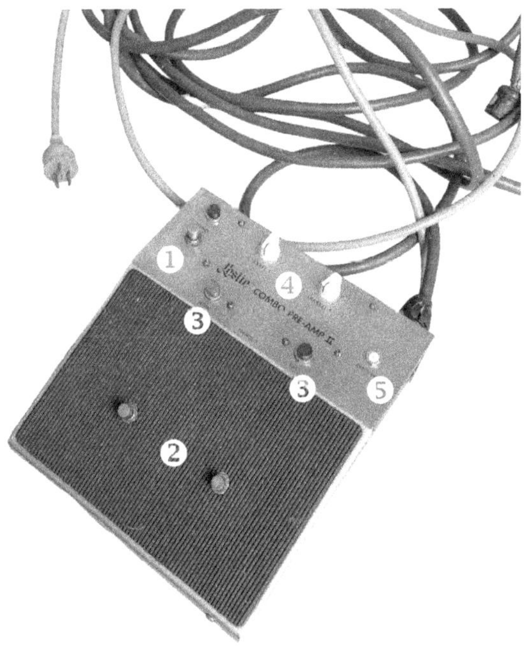

Leslie preamplifiers

The Leslie "Combo Pre-Amp II" is a simple 2-channel transistor preamp contained in a large foot stomp box style housing, with an on-off switch (1), slow and fast foot switches (2) and associated indicator lights (3) and two volume controls (4). The knobs are not original, and neither is the circuit breaker, which failed and had to be replaced (5)

Each channel has a standard 1/4" phone jack input at the back and its own volume control on top. Leslie preamps allow you to use their cabinets/amps with any source, guitar, organ, or keyboard with a simple 1/4" guitar cable.

On the output side, a proprietary 9-core cable using octal-style plugs interfaces with compatible Leslie cabinets.

Leslie Combo Preamp De Luxe is a more complex iteration (also using a 9-pin plugs), compatible with Leslie models 7420, 900, 910, 925, and 950.

When we bought Leslie 130 and Leslie 910 locally in Perth, we got Leslie Combo Preamp II as part of a bundle with two Hammond organs, but it is not suitable for either speaker/amp. It works with Model 910 but cannot control all its functions. The later models, Combo Preamp III and IV, used 11-core interface cable.

KORY - AN AUSTRALIAN TAKE ON THE AMERICAN IDEA

Esthetically, this Australian-made rotating speaker combo reflects its 1970s origins, but the conceptual design is very sound (no pun intended).

The enclosure is a combination of plywood and cheap chipboard, with the furniture grade veneer all round.

A 15" woofer fires upward (1) into the upper chamber (2). The woofer baffle is fixed, giving the whole cab rigidity. The back of the woofer fires downward into the bottom chamber (not shown on the photo), which has a large round port at the bottom, so the rear sound waves would bounce off the floor and back into the room on all four sides of the cabinet.

A horizontal shelf carries a 2-speed motor (4), which drives the single rotating tweeter horn (6) through a pulley and rubber belt system (5). The horn is balanced by an adjustable counterweight (7).

Both the HF and LF sound exit the upper chamber on all four sides (9), a great idea; most other rotating cabs have openings on three sides, some even on only two.

The compression tweeter isn't on the photo, but all the connections between the upper chambers motor and tweeter and the power amplifier housed in the bottom chamber are through a terminal strip (8).

This makes troubleshooting easier since all the voltages as well as continuity can be tested on the terminal strip connections.

The amplifier is a much more modern design than the Leslie amps of the era, which were late 1950s and early 1960's early transistor topologies, with interstage transformers among other, since then long abandoned practices.

The amp's connectors and controls are easy to decode: the power switch and pilot light (1), IEC power socket (2), 3-pin XLR socket for motor speed controls and switching (3), two instrument inputs, "Low" and "High" level (4), Volume and Tone controls (5), and a bank of three fuses (6), one for AC mains supply, and two DC fuses, one on each positive and negative DC rail.

The amp is on a steel chassis that bolts onto the side of the enclosure, making it easily removable for servicing.

THE ROTATOR: BUILD YOUR OWN LESLEY CAB FOR A GUITAR AMP

Just as no self-respecting guitar player would play a digital guitar amp on stage, digital effects are one thing (although some are OK), and real effects are something entirely different. And one of the coolest sounding real acoustical effects is the magic of a Leslie rotating speaker cabinet.

Although mostly used with Hammond and other vintage organs, the Leslie cab also adds its distinctive magic and three-dimensional swirl to guitars, other keyboards, and even vocals. So, if you've ever been jealous of keyboard players and their magical tones, you should be jealous no more! Dr. Iggy will show you how to make your own Leslie-style cabinet for next to nothing! Of course, we will not call it "Leslie-style cabinet " from now on, we'll show an incredible leap of imagination and name it - wait for it - "The Rotator"!

The Leslie module donor

A single rotating polystyrene drum in front of a stationary speaker was a simplified type of Leslie effect built into many Japanese-made Hammond organs of the 1970s and 80s.

We got a partially working Hammond Composer 144 organ for free, one of their mid-level models from 1982.

Apart from its Japanese-made speaker drivers, a USA-made Accutronics reverb tank and this Leslie module, nothing else was of any real value. Sadly, unlike the earlier Hammond models, these organs used integrated circuits, not tubes.

The "Leslie Tremolo Unit" was self-contained, mounted on its own plywood sheet, which was then simply screwed into the organ's wooden pedestal.

Four wires from the two motors' coils were terminated in a plug. Another connector with two wires supplied the signal to the speaker's voice coil.

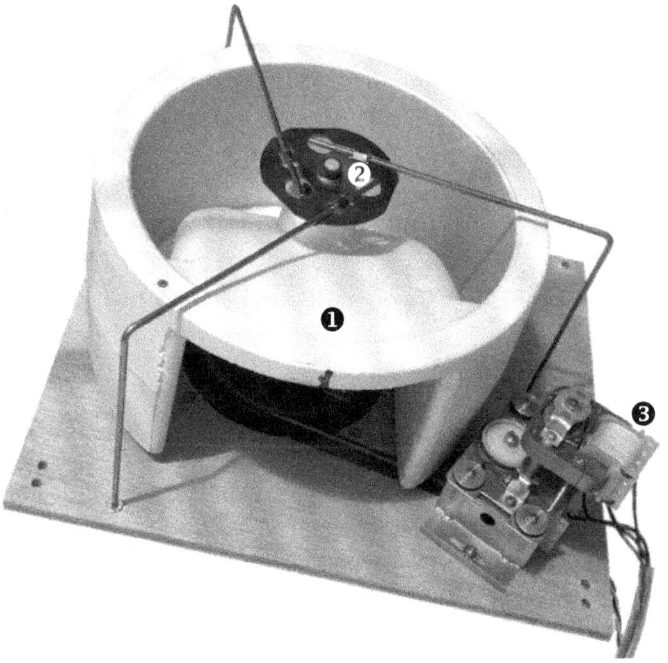

.

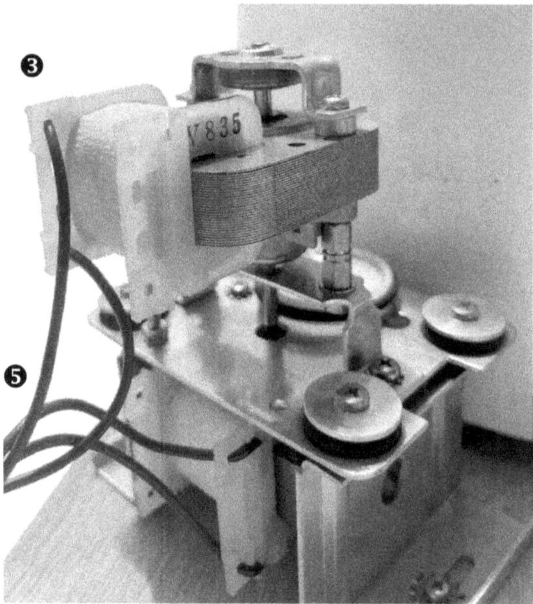

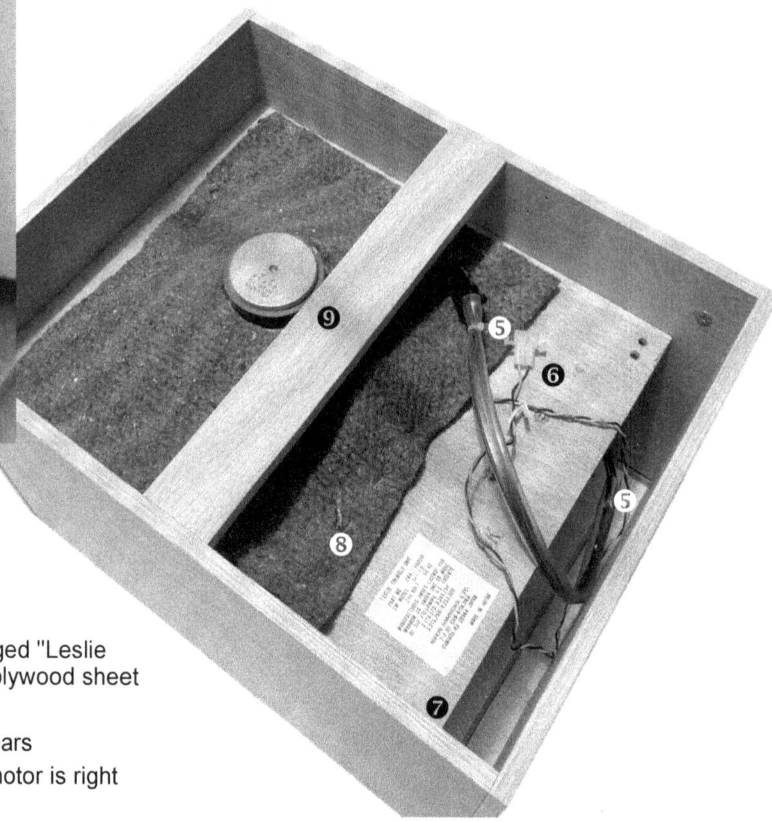

RIGHT, ABOVE, & PREVIOUS PAGE: The salvaged "Leslie Tremolo Unit" was self-contained, mounted on a plywood sheet

1) High-density polystyrene drum
2) The shaft with bearings and three mounting L-bars
3) The upper (high speed) motor; the low speed motor is right under
4) The upper motor's bearing
5) The four wires, two for each motor, terminated in a plug
6) The audio connector with two wires feeding the signal to the speaker's voice coil
7) One side of the original mounting board had to be cut a few millimeters to be able to fit snugly into the new cab
8) Acoustic insulation blanket covering the back of the speaker
9) As luck would have it, the speaker's magnet just cleared the cross bar of the new cabinet

RIGHT: Beware when hunting for possible donors among vintage organs. Despite controls that indicate a presence of a Leslie rotating assembly of some sort (given silly names such as "Sound-a-Rama"), some lower-end Hammond organs (such as model 123J3, whose three switches look promising) have only a fake, "electronic" tremolo effect that isn't even a frequency-modulated vibrato, let alone a Leslie effect. No rotating drums in those organs!

The woodwork: Bedside tables as cabinet frames

Browsing through K-mart Australia's website these Scallop bedside tables (reduced from AU$69 to AU$39) seemed worth taking the risk on, so we got a pair to see if they'd be suitable for this project; they were.

The two drawer fronts (1) were thick and sturdy, but the rest of the MDF frame was relatively thin. However, the rear crossbar and the snug-fitting Leslie mounting plywood base (photo on the previous page) would keep it all firmly together, as would the solid plywood back panel, permanently bolted to the frame, thus ensuring rigidity. On one of the enclosures, the one housing the Leslie rotating assembly module, we had to cut rectangular openings on the frame's sides (3) for the sound to be able to exit the enclosure remember, its front baffle has no openings at all (photo on the next page).

The base frames (4) and the two drawer fronts were discarded and front baffles and back covers were made out of 12mm plywood sheets. The front crossbar (2) was retained for support and rigidity.

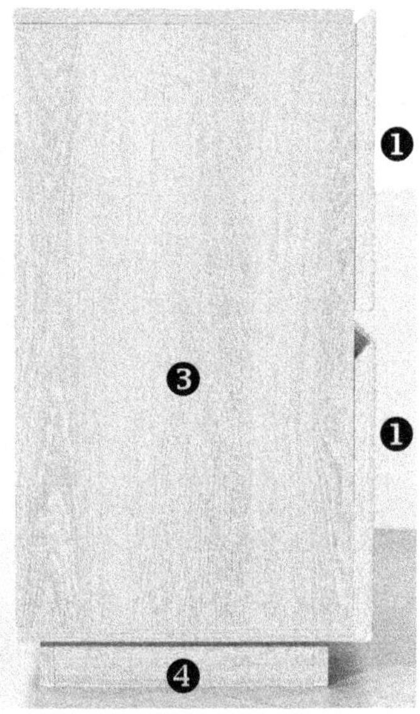

ABOVE: The funky-looking base plate of this wall clock was cut into four quarters and used to decorate the fronts of the two cabinets

LEFT: The bedside tables used as enclosure frames. The front baffles and the back covers had to be made to fit.

The alternative naming idea: The Atomium

These wall clocks were on clearance sale so we bought two. CNC-cut out of a solid aluminium plate, it depicts an electron's orbit around a nucleus. My first idea was to mount a speaker where the clock's dial was, but, after removing and discarding the clock mechanism, the hole in the middle was too small for even a 6" driver.

Never to admit defeat, I thought "Why don't we cut the pattern into quarters and thus obtain four identical logos that could be placed in the very corners of the rotating Leslie-style speaker enclosures?" For obvious reasons, the name was changed from "The Rotator" to "The Atomium".

BELOW: The finished "Rotator", a.k.a. "Atomium" cab, with permanently-attached speed-change footswitch (5). The IEC power inlet (6) with an integral switch and fuse is right below the speaker jack plate (7).

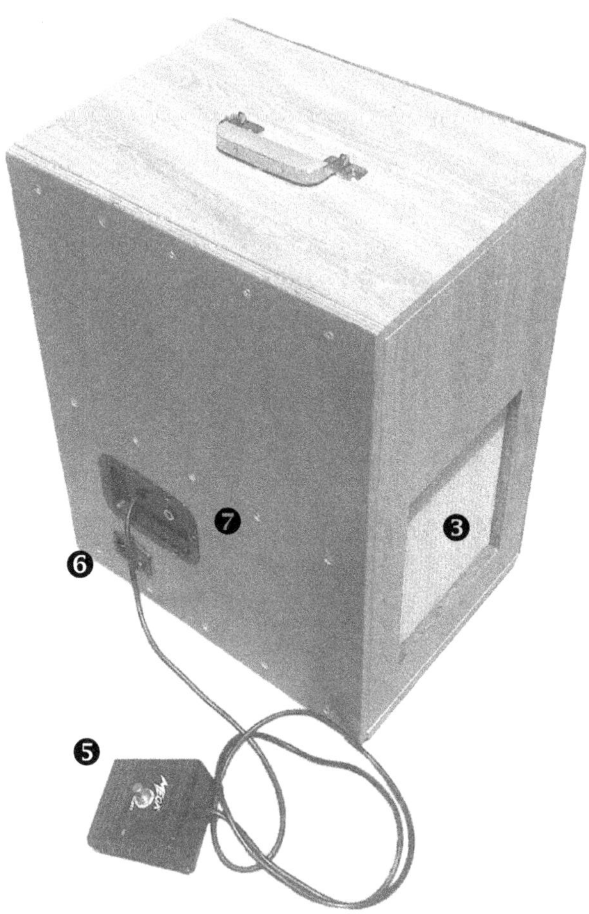

The fixed speaker cab and impedance-matching considerations

The bedside table frame was large enough to accommodate a 12" speaker but we had two 8" Japanese-made drivers (salvaged from the same Hammond organ), so we decided to reunite the family, so to speak. The DC resistance of each driver was 7.6 ohms, which is around 1.25 x 7.6 = 9.5 Ω nominal AC impedance.

Option A): Two in parallel in the fixed cab would give us approx. 4.75 Ω nominal AC impedance. Since this fixed cab was supposed to be used with the rotating cab, connecting all three in parallel would give us 9.5/3 = 3.17 Ω AC impedance, which is fine, tube amps with a 4 Ω tap would be able to drive it.

Option B): Connecting the two "fixed" drivers in series would give us 19 Ω AC impedance in this cabinet, which, if paralleled with the 9.5 Ω AC rotating cab, would result in the total impedance of 19||9.5 = 6.33 Ω total impedance. This can be driven by both 4Ω or 8Ω amplifier outputs.

As for the power distribution, the rotating cab would take twice as much power from the amp as the fixed cab (since its impedance is half of the fixed cab), meaning roughly one third of the amplifier's power output would go to the fixed cab and two thirds would be drawn by the rotating cab.

When the driver in the rotating cab is at full power (20 watts), the current through the series-connected fixed drivers is half of that current, meaning each will produce only half as much power as the rotating cab's driver, or 10 watts, so they will underutilized by 50%. The total power of the two cabs would be only 20W + 20W = 40 watts.

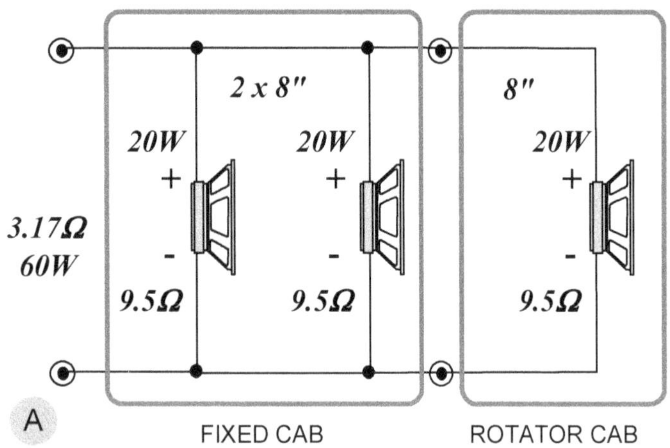

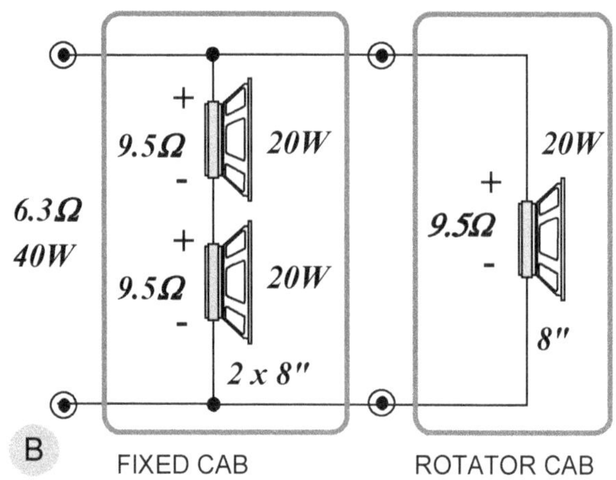

The front view of the fixed speaker cab (ABOVE LEFT) and the rear view (ABOVE RIGHT) with the back plate removed

THE END MATTER

14

Congratulations! You have persisted until the very end of the book (unless you've skipped right to it, naughty, naughty!).

Should you wish to expand your knowledge of the related topics, in this section you will find the descriptions of my other books and the further reading list.

I send you my best wishes, enjoy your music journey, and see you on the road ...

AUDIOPHILE TUBE AMPLIFIER BOOKS BY IGOR S. POPOVICH

Audiophile Vacuum Tube Amplifiers, Vol. 1:

- BASIC ELECTRONIC CIRCUIT THEORY
- ELECTRONIC COMPONENTS
- AUDIO FREQUENCY AMPLIFIERS
- PHYSICAL FUNDAMENTALS OF VACUUM TUBE OPERATION
- VOLTAGE AMPLIFICATION WITH TRIODES - THE COMMON CATHODE STAGE
- OTHER VOLTAGE AMPLIFICATION STAGES WITH TRIODES
- TETRODES AND PENTODES AS VOLTAGE AMPLIFIERS
- FREQUENCY RESPONSE OF VACUUM TUBE AMPLIFIERS
- IMPEDANCE-COUPLED STAGES AND INTERSTAGE TRANSFORMERS
- NEGATIVE FEEDBACK
- TONE CONTROLS, ACTIVE CROSSOVERS AND OTHER CIRCUITS
- PRACTICAL LINE-LEVEL PREAMPLIFIER DESIGNS
- PHONO PREAMPLIFIERS
- SINGLE-ENDED TRIODE OUTPUT STAGE
- PRACTICAL SINGLE-ENDED TRIODE AMPLIFIER DESIGNS
- PRACTICAL SINGLE-ENDED PSEUDO-TRIODE DESIGNS
- SINGLE-ENDED PENTODE AND ULTRALINEAR OUTPUT STAGES

Audiophile Vacuum Tube Amplifiers, Vol. 2:

- PRACTICAL SINGLE-ENDED PENTODE AND ULTRALINEAR DESIGNS
- PUSH-PULL OUTPUT STAGES
- PRACTICAL PUSH-PULL AMPLIFIER DESIGNS
- BALANCED, BRIDGE AND OTL (OUTPUT TRANSFORMERLESS) AMPLIFIERS
- THE DESIGN PROCESS
- FUNDAMENTALS OF MAGNETIC CIRCUITS AND TRANSFORMERS
- MAINS TRANSFORMERS AND FILTERING CHOKES
- POWER SUPPLIES FOR TUBE AMPLIFIERS
- AUDIO TRANSFORMERS
- TROUBLESHOOTING AND REPAIRING TUBE AMPLIFIERS
- UPGRADING & IMPROVING TUBE AMPLIFIERS
- SOUND CONSTRUCTION PRACTICES
- AUDIO TESTS & MEASUREMENTS
- TESTING & MATCHING VACUUM TUBES

Audiophile Vacuum Tube Amplifiers, Vol. 3:

- THE FRONT-END: SUPERIOR INPUT & DRIVER STAGES
- FROM SHOCKING TO SUBLIME: LESSONS FROM COMMERCIAL LINE STAGES
- DIY LINE-LEVEL PREAMPLIFIERS: $10,000 SOUND ON $500-$1,000 BUDGET
- THE STARS OF THE AUDION ERA: ANCIENT TUBES IN MODERN AMPS
- CHEAP & CHEERFUL: PREAMP & DRIVER TUBES FOR AUDIO EXPLORERS
- SLEEPING GIANTS: OUTPUT TUBES FOR THOSE WHO WANT TO BE DIFFERENT
- THE QUEEN OF HEARTS: SINGLE-ENDED AMPLIFIERS WITH 300B TRIODES
- TRIODES, PENTODES AND BEAM TUBES: MORE SINGLE-ENDED DESIGNS
- BIG BOTTLES: SET AMPLIFIERS WITH HIGH VOLTAGE TRANSMITTING TUBES
- THE WAY IT USED TO BE: VINTAGE PUSH-PULL AMPLIFIERS
- NEW? IMPROVED? MODERN PUSH-PULL AMPLIFIER DESIGNS
- CUTE, CLEVER OR CONTROVERSIAL? INTERESTING IDEAS FROM TUBE AUDIO'S PAST AND PRESENT
- THRIFTY TIPS & TRICKS: TIME & MONEY SAVING IDEAS
- OUTPUT AND INTERSTAGE TRANSFORMERS: FROM COMMERCIAL BENCHMARKS TO YOUR OWN DESIGNS
- MEASUREMENTS VERSUS LISTENING AND OTHER AUDIO DESIGN DILEMMAS

GUITAR AMPS AND RELATED BOOKS BY IGOR S. POPOVICH

Available from Amazon, Barnes & Noble, Book Depository and all other major online bookstores

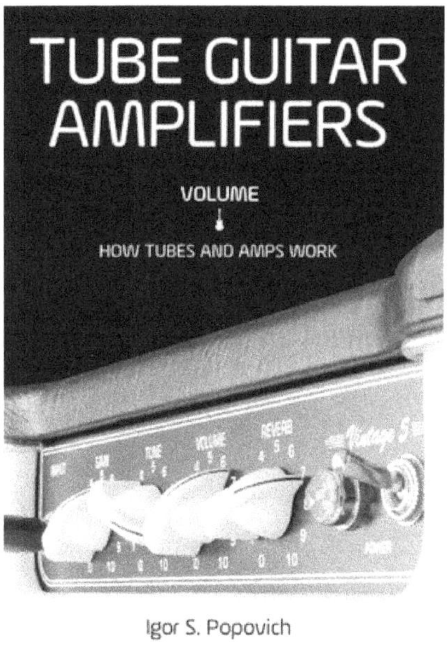

Tube Guitar Amplifiers Volume 1: How Tubes and Amps Work
ISBN: 978-0-9806223-5-5

- BASIC ELECTRONIC CIRCUIT THEORY
- AUDIO AMPLIFIERS
- ELECTRONIC COMPONENTS
- PHYSICAL FUNDAMENTALS OF VACUUM TUBE OPERATION
- TRIODES AS VOLTAGE AMPLIFIERS
- TETRODES, PENTODES AND BEAM-POWER TUBES
- INPUT CIRCUITS AND STAGES
- TONE CONTROLS
- ANALOG EFFECTS (TREMOLO, VIBRATO, REVERB) AND EFFECTS LOOPS
- POWER SUPPLIES FOR TUBE AMPLIFIERS
- SINGLE-ENDED TRIODE, PENTODE AND ULTRALINEAR OUTPUT STAGES
- PHASE SPLITTERS OR INVERTERS
- PUSH-PULL OUTPUT STAGES
- NEGATIVE FEEDBACK
- TRANSISTOR AND HYBRID GUITAR AMPLIFIERS

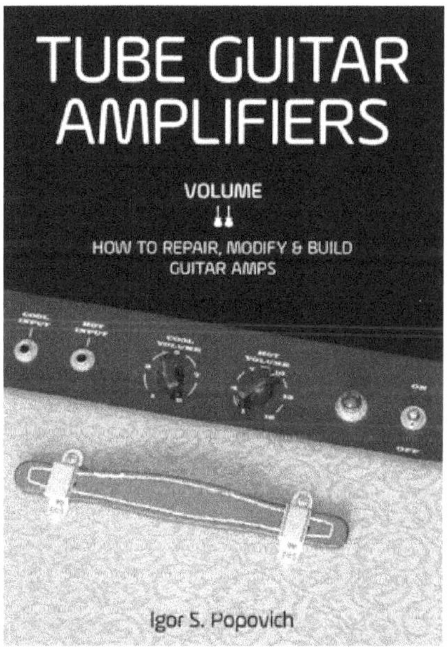

Tube Guitar Amplifiers Volume 2: How to Repair, Modify & Build Guitar Amps
ISBN: 978-0-9806223-6-2

- OUTPUT AND INTERSTAGE TRANSFORMERS FOR TUBE GUITAR AMPS
- LOUDSPEAKERS, OUTPUT ATTENUATORS & HEADPHONE CIRCUITS
- TROUBLESHOOTING AND REPAIRING TUBE GUITAR AMPLIFIERS
- WIRING, SOLDERING & MODIFICATION PRACTICES
- POWER SUPPLY MODIFICATIONS AND IMPROVEMENTS
- TONE TWEAKS
- MODERN PUSH-PULL AMPS
- DIY PROJECTS: CONVERTING SOLID STATE GUITAR AMPS TO TUBES
- DIY PROJECTS: ULTRA-SMALL AMPS
- REBUILDING COMMERCIAL AMPS IN A HANDWIRED (POINT-TO-POINT) FASHION
- DIY PROJECTS: QUIRKY & UNUSUAL DESIGNS
- CONVERTING VINTAGE TUBE GEAR INTO GUITAR AMPS

Audio Tests & Measurements: How to Test Electronic Components, Audiophile & Guitar Amplifiers and Loudspeakers Using Modern and Vintage Test Instruments
ISBN: 978-0-9806223-9-3

- TEST INSTRUMENTS, ERRORS, LIMITATIONS & SAFETY ISSUES
- SIGNAL SOURCES, TRACERS, POWER SUPPLIES AND FILTERS
- MULTIMETERS - TYPES, OPERATING PRINCIPLES AND FUNCTIONS
- OSCILLOSCOPES - HOW THEY WORK & HOW TO USE THEM
- TESTING PASSIVE ELECTRONIC COMPONENTS (RESISTORS, CAPACITORS & INDUCTORS)
- TESTING AUDIO AMPLIFIERS AND PREAMPLIFIERS
- DISTORTION MEASUREMENTS
- TRANSFORMER TESTS & MEASUREMENTS
- LOUDSPEAKER TESTS & MEASUREMENTS
- TRANSISTOR TESTERS AND CURVE TRACERS
- TESTING VACUUM TUBES (VALVES)

OTHER AUDIO-RELATED BOOKS BY IGOR S. POPOVICH

Available from Amazon, Barnes & Noble, Book Depository and all other major online bookstores

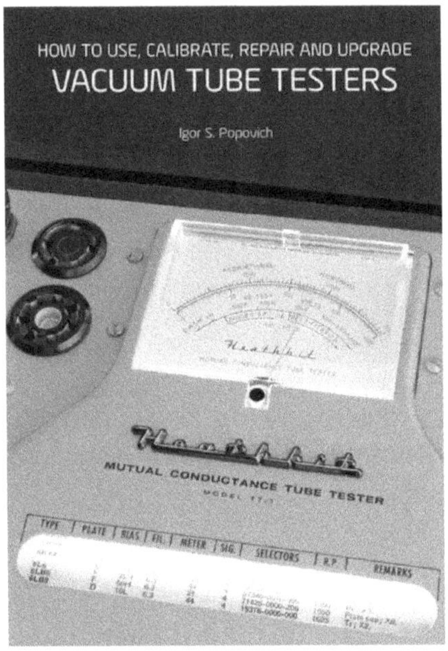

How to Use, Calibrate, Repair and Upgrade Vacuum Tube Testers
ISBN: 978-0-9806223-7-9

- HOW VACUUM TUBES WORK
- TESTING & MATCHING VACUUM TUBES
- EMISSION TESTERS
- GRID CIRCUIT TESTERS
- DYNAMIC CONDUCTANCE TESTERS
- PROPORTIONAL MUTUAL CONDUCTANCE TESTERS
- HICKOK-TYPE TESTERS
- TRUE MUTUAL CONDUCTANCE TESTERS
- REPAIRING & UPGRADING VINTAGE TUBE TESTERS
- TESTING & MATCHING TUBES WITHOUT A TUBE TESTER

Transformers For Tube Amplifiers: How to Design, Construct & Use Power, Output & Interstage Transformers and Chokes in Audiophile and Guitar Tube Amplifiers
ISBN: 978-0-9806223-8-6

- PHYSICAL FUNDAMENTALS OF MAGNETIC CIRCUITS AND TRANSFORMERS
- FILTERING CHOKES (INDUCTORS WITH DC CURRENT)
- TRANSFORMER MATERIALS, CONSTRUCTION METHODS AND ISSUES
- MAINS (POWER) TRANSFORMERS
- PHYSICAL FUNDAMENTALS OF AUDIO TRANSFORMERS
- SINGLE-ENDED OUTPUT TRANSFORMERS
- PUSH-PULL OUTPUT TRANSFORMERS
- SPECIAL MAGNETIC COMPONENTS: LOW POWER INPUT, PREAMP OUTPUT & DAC OUTPUT TRANSFORMERS, TRANSFORMER VOLUME CONTROL
- INTERSTAGE TRANSFORMERS, GRID & ANODE CHOKES
- OUTPUT AND INTERSTAGE TRANSFORMERS FOR TUBE GUITAR AMPS
- TRANSFORMER TESTS & MEASUREMENTS

Sound Improvement Secrets For Audiophiles
ISBN: 978-0-9806223-6-2

- BEFORE YOU BUY AN AUDIO SYSTEM OR COMPONENT
- WHAT DO WE LISTEN FOR AND WHAT DO WE ACTUALLY HEAR?
- CLEANING UP THE POWER SUPPLY NOISE, HUM, AND INTERFERENCE
- CABLES, FUSES, CONTACTS, AND CONNECTIONS
- UPGRADING & FINE-TUNING THE SOURCES: OPEN REEL RECORDERS, TURNTABLES, PHONO STAGES AND CD PLAYERS
- AUDIO AMPLIFIERS - HOW THEY WORK AND HOW TO IMPROVE THEM
- HEADPHONES AND HEADPHONE AMPLIFIERS
- LOUDSPEAKER TYPES, TESTS, AND IMPROVEMENTS
- COMPONENT MATCHING AND AUDIO SYSTEM INTEGRATION ISSUES
- LOUDSPEAKER POSITIONING
- OPTIMIZING YOUR LISTENING ROOM
- ACOUSTIC TREATMENTS
- MINIMIZING UNWANTED VIBRATIONS & OSCILLATIONS
- TROUBLESHOOTING YOUR AUDIO SYSTEM

LITERATURE

A few important points about the rating system

Rating other authors' work is not a pleasant task, and all such ratings are subjective and thus questionable. How we see a book's usefulness depends on many factors, among others, our prior level of knowledge and our expectations. I may have rated a particular title low, but you may find it useful and give it a much higher rating. Such a situation is possible but unlikely, if a reviewer has been realistic and consistent, and I have tried to be both.

Keep in mind that the ratings given here are for a few specific aspects, such as clear & concise explanations (as simple as possible), or "reader-friendliness" if you will, and, above all, practicality. Many otherwise valuable books are too theoretical for an average musician or amateur speaker builder or designer.

For instance, the Radiotron Designer's Handbook, a1482 page (4th edition only) classic on tube amps and related topics, an A-grade material on tube amps, only gets a D for its speakers section.

Apart from some titles being too theoretical (too much maths), almost all of these works focus on hi-fi (high fidelity) speakers and enclosures, and don't even mention issues pertinent to guitar and instrument speakers and cabs. As of this writing, I could only find two titles specifically covering guitar and instrument speaker cabinets.

Finally, being a relatively harsh reviewer, I didn't think any of these books actually deserved an "A", but that would had been unrealistic, so I had to scale up some ratings so that at least two or three end up with an "A-".

The rating system

Grade F - for "fail", totally useless, avoid at all costs

Grade D - read if you have nothing better to do with your life

Grade C - OK material but far from comprehensive or optimal, not very practical

Grade B - useful, some practical info, but with flaws and omissions

Grade A - recommended, a decent book

The titles and their authors

1st Book of Hi-Fi Loudspeaker Enclosures, B. B. Babani, 1974, 94 pages, C

2nd Book of Hi-Fi Loudspeaker Enclosures, B. B. Babani, 1974, 96 pages, C

Acoustical Engineering, H. Olson, 718 pages, 1957, B

Advanced Speaker Systems, Ray Alden, 124 pages, 1995, A

An Introduction to Loudspeakers and Enclosure Design, V. Capel, 1991, 148 pages, F

Audio Cyclopedia, 2nd edition, Howard M. Tremaine, 1969, 1757 pages, C

Building Pro Audio Loudspeaker Enclosures, Jeffrey A. Forsburg, 2002, 198 pages, C

Building Speaker Enclosures, Robert G. Middleton, 95 pages, 1975, C

Bullock on Boxes, Robert M. Bullock, 2017, 82 pages, C

Cabinet Handbook, Gilbert Arthur Briggs, 1962, 112 pages, C

Design and Construction of Guitar and Bass Speaker Systems, Robert C. Megantz, 2017, 92 pages, C

Designing, Building, and Testing Your Own Speaker System - With Projects, D. Weems, 1996, 224 pages, C

DIY speaker Cabinets for Musical Instrument Applications, Kevin O'Connor, 2001, 76 pages, C

Electroacoustics: Microphones, earphones and loudspeakers, M. L Gayford, 289 pages, 1970, D

Great Sound Stereo Speaker Manual, David B. Weems, 2000, 320 pages, B

High Performance Loudspeakers, Martin Colloms, 2018, 704 pages, C

Hi-Fi Loudspeakers and Enclosures, 2nd edition, Abraham B. Cohen, 1968, 438 pages, A-

High-Quality Loudspeaker Engineering, Martin James Lawrence, 2013, 792 pages, C

How To Build Speaker Enclosures, Alexis Badmaieff & Don Davis, 1966, 144 pages, B

How To Design, Build, & Test Complete Speaker Systems, David B. Weems, 1978, 336 pages, A-

Introduction to Electroacoustics & Audio Amplifier Design (3rd ed.), W. Marshall Leach, 2003, 298 pages, D
Introduction to Loudspeaker Design, 2nd edition, John L Murphy, 2014, 224 pages, C
Les haut-parleurs, 3rd edition, Jean Hiraga, 2000, 400 pages (in French), B
Loudspeaker and Headphone Handbook, 3rd edition, John Borwick, 2001, 732 pages, C
Loudspeaker Design Cookbook, Volume I, 8th edition, Vance Dickason, 2023, 398 pages, C
Loudspeaker Design Cookbook, Volume II, Vance Dickason, 2023, 128 pages, C
Loudspeaker Handbook, 2nd edition, John Eargle, 2003, 432 pages, B
Loudspeaker Handbook and Lexicon, Winslow Burhoe, 1978 (rev. 1997), 57 pages, F
Loudspeakers for Music Recording and Reproduction, Newell and Holland, 2006, 400 pages, C
Loudspeakers for Musicians, V. Capel, 1991, 178 pages, C
Loudspeakers: The Why and How of Good Reproduction, G. A. Briggs , 1990, 92 pages, C
More about Loudspeakers, Gilbert Arthur Briggs , 1963, 136 pages, C
Music, Physics and Engineering, Harry F. Olson, 1967, 480 pages, C
Radiotron Designer's Handbook, F. Langford-Smith (editor), 4th edition, 1952, 1482 pages, D
Rock Hardware: The Instruments, Equipment And Technology of Rock, T. Bacon, 1981, 224 pages, F
Speaker Building 201, Ray Alden, 2017, 158 pages, C
Techniques des haut-parleurs et enceintes acoustiques, 2nd edition, Pierre Loyez, 2003, 432 pages (in French), C
Testing Loudspeakers, J D`Appolito, 2018, 176 pages, C
Theory & Design of Loudspeaker Enclosures, J. E. Benson, 1996, 244 pages, C
The Science of Sound, Thomas Rossing, 3rd edition, 2001, 808 pages, B

NOTES

INDEX

A

B

C

D

E

F

G

H

I

K

L

M

N

O

INDEX, continued

NOTES

www.ingramcontent.com/pod-product-compliance
Ingram Content Group UK Ltd.
Pitfield, Milton Keynes, MK11 3LW, UK
UKHW061842041026
15591UKWH00014B/871